METHODS IN MOLECULAR BIOLOGY™

Series Editor
John M. Walker
School of Life Sciences
University of Hertfordshire
Hatfield, Hertfordshire, AL10 9AB, UK

For other titles published in this series, go to
www.springer.com/series/7651

Mass Spectrometry Imaging

Principles and Protocols

Edited by

Stanislav S. Rubakhin

*Beckman Institute, University of Illinois at Urbana-Champaign,
Urbana, IL, USA*

Jonathan V. Sweedler

*Department of Chemistry, University of Illinois at Urbana-Champaign,
Urbana, IL, USA*

Humana Press

Editors
Stanislav S. Rubakhin
Beckman Institute
University of Illinois
at Urbana-Champaign
N. Mathews Avenue 405
61801 Urbana Illinois
USA
roubakhi@illinois.edu

Jonathan V. Sweedler
Department of Chemistry
University of Illinois
at Urbana-Champaign
South Mathews Avenue 63-5 600
61801 Urbana Illinois
USA
jsweedle@illinois.edu

ISSN 1064-3745 e-ISSN 1940-6029
ISBN 978-1-60761-745-7 e-ISBN 978-1-60761-746-4
DOI 10.1007/978-1-60761-746-4
Springer New York Dordrecht Heidelberg London

Library of Congress Control Number: 2010930696

Printed on acid-free paper

Humana Press is part of Springer Science+Business Media (www.springer.com)

Preface

Why should you use the protocols described in this book? In order to characterize many biological systems, having knowledge of their chemical constituents, locations, and dynamics is important. Mass spectrometry (MS) provides unmatched capabilities for detection, characterization, and identification of analytes ranging from individual elements to complex multimolecular structures. A powerful enhancement to MS detection is the addition of spatial information; mass spectrometry imaging (MSI) combines the capabilities of modern MS with imaging. The distribution of hundreds of different analytes in a tissue can be determined in a single experiment. Unlike other imaging approaches, analyte preselection is not needed. Metabolites, peptides, proteins, and polynucleotides can be characterized. Distinct protocols have been developed for analysis of specimens in vivo, in vitro, and in situ.

This book is divided into three parts. The first section contains introductory chapters on MS and MSI. Chapter 1 provides an overview of MSI and focuses on current and future trends in the field. The success of a particular MSI experiment depends on the specific MS approach used. Therefore, the second chapter describes the basic principles of mass spectrometry relevant to MSI and includes cross-references to other chapters of this volume for easier navigation. The third chapter reviews the application of MSI to the study of elemental distributions. Following these introductory chapters, there are multiple protocols that describe qualitative and quantitative measurements of endogenous metabolites and xenobiotics as well as their identification and localization. The last section includes protocols for a variety of MSI approaches developed to study peptide and protein distributions. The experimental protocols presented herein encompass most MSI approaches and technologies for samples from a wide range of biological models including plants, invertebrates, and vertebrates.

The contributors to this volume include practitioners in academia, industry, and the clinic. MSI has great unmet potential to further the investigations in many disciplines, including molecular biology and neuroscience. This book is written for scientists who want to apply MSI methods to their research and would benefit from the detailed, step-by-step experimental protocols provided. Both novice and established MSI practitioners should find the volume a source of valuable methodological information. Included with the protocols are additional troubleshooting notes that highlight important nuances and potential pitfalls of the procedures outlined within each chapter.

We hope you enjoy reading this volume as much as we enjoyed putting it together.

Stanislav S. Rubakhin and Jonathan V. Sweedler

Contents

Contributors

JEFFREY N. AGAR • *Chemistry Department and Volen Center, Brandeis University, Waltham, MA, USA*
NATHALIE Y.R. AGAR • *Department of Neurosurgery, Brigham and Women's Hospital, Harvard Medical School, Boston, MA, USA*
JEREMY ALLEGOOD • *School of Chemistry and Biochemistry, Georgia Institute of Technology, Atlanta, GA, USA*
PEGGI M. ANGEL • *Mass Spectrometry Research Center, Vanderbilt University, Nashville, TN, USA*
MICHAEL BECKER • *Applications TOF-MS, Bruker Daltonik GmbH, Bremen, Germany*
ELENA S.F. BERMAN • *Los Gatos Research, Mountain View, CA, USA*
BEGOÑA CACHÓN-GONZÁLEZ • *Department of Medicine, University of Cambridge, Cambridge, UK*
M. CAMERON SULLARDS • *School of Chemistry and Biochemistry, School of Biology, Georgia Institute of Technology, Atlanta, GA, USA*
STEPHANIE S. CAPE • *School of Pharmacy and Department of Chemistry, University of Wisconsin-Madison, Madison, WI, USA*
RICHARD M. CAPRIOLI • *Departments of Chemistry and Biochemistry, Mass Spectrometry Research Center, Vanderbilt University, Nashville, TN, USA*
SUBHASH CHANDRA • *Cornell SIMS Laboratory, Department of Biomedical Engineering, Cornell University, Ithaca, NY, USA*
PIERRE CHAURAND • *Mass Spectrometry Research Center, Vanderbilt University, Nashville, TN, USA*
RUIBING CHEN • *School of Pharmacy and Department of Chemistry, University of Wisconsin-Madison, Madison, WI, USA*
YANFENG CHEN • *School of Chemistry and Biochemistry, Georgia Institute of Technology, Atlanta, GA, USA*
TIMOTHY M. COX • *Department of Medicine, University of Cambridge, Cambridge, UK*
SÖREN-OLIVER DEININGER • *Applications TOF-MS, Bruker Daltonik GmbH, Bremen, Germany*
MOHAMED EL AYED • *Laboratoire de Neuroimmunologie et Neurochimie Evolutives, FRE CNRS 3249, MALDI Imaging Team, Université Lille Nord de France, Université Lille 1, Villeneuve d'Ascq, France*
JEFFREY R. ENDERS • *Department of Chemistry, Vanderbilt Institute of Chemical Biology, and Vanderbilt Institute for Integrative Biosystems Research and Education, Vanderbilt University, Nashville, TN, USA*
ANDREW G. EWING • *Department of Chemistry, The Pennsylvania State University, University Park, PA, USA; Department of Chemistry, Göteborg University, Göteborg, Sweden*
JAMES S. FELTON • *Chemistry, Materials, Earth, and Life Sciences (CMELS) Directorate, Lawrence Livermore National Laboratory, Livermore, CA, USA*

LARISSA S. FENN • *Department of Chemistry, Vanderbilt Institute of Chemical Biology, and Vanderbilt Institute for Integrative Biosystems Research and Education, Vanderbilt University, Nashville, TN, USA*

SUSAN L. FORTSON • *Los Gatos Research, Mountain View, CA, USA*

ISABELLE FOURNIER • *Laboratoire de Neuroimmunologie et Neurochimie Evolutives, FRE CNRS 3249, MALDI Imaging Team, Université Lille Nord de France, Université Lille 1, Villeneuve d'Ascq, France*

JULIEN FRANCK • *Laboratoire de Neuroimmunologie et Neurochimie Evolutives, FRE CNRS 3249, MALDI Imaging Team, Université Lille Nord de France, Université Lille 1, Villeneuve d'Ascq, France*

TIMOTHY J. GARRETT • *GCRC Core Laboratory, Department of Medicine, University of Florida, Gainesville, FL, USA*

LIN HE • *Department of Chemistry, North Carolina State University, Raleigh, NC, USA*

MICHAEL L. HEIEN • *Assistant Professor, Department of Chemistry and Biochemistry, The University of Arizona, Tucson, AZ, USA*

YUNSHENG HSIEH • *Merck Research Laboratories, Department of Drug Metabolism and Pharmacokinetics, Kenilworth, NJ, USA*

SHELLEY N. JACKSON • *Structural Biology Unit, Cellular Neurobiology Branch, National Institute on Drug Abuse Intramural Research Program, National Institutes of Health, Baltimore, MD, USA*

VILMOS KERTESZ • *Organic and Biological Mass Spectrometry Group, Chemical Sciences Division, Oak Ridge National Laboratory, Oak Ridge, TN, USA*

WALTER A. KORFMACHER • *Department of Drug Metabolism and Pharmacokinetics, Merck Research Laboratories, Kenilworth, NJ, USA*

JANE-MARIE KOWALSKI • *Bruker Daltonics, Inc., Billerica, MA, USA*

PAUL J. KOWALSKI • *Bruker Daltonics, Inc., Billerica, MA, USA*

KRISTEN S. KULP • *Biosciences and Biotechnology Division, Lawrence Livermore National Laboratory, Livermore, CA, USA*

FANGBIAO LI • *Department of Drug Metabolism and Pharmacokinetics, Merck Research Laboratories, Kenilworth, NJ, USA*

LINGJUN LI • *School of Pharmacy and Department of Chemistry, University of Wisconsin-Madison, Madison, WI, USA*

QIANG LIU • *Department of Pathology & Lab Medicine, University of North Carolina at Chapel Hill, Chapel Hill, NC, USA*

YING LIU • *School of Biology, Georgia Institute of Technology, Atlanta, GA, USA*

A.F. MAARTEN ALTELAAR • *Biomolecular Mass Spectrometry and Proteomics Group, Utrecht University, Utrecht, The Netherlands*

JOHN A. MCLEAN • *Department of Chemistry, Vanderbilt Institute of Chemical Biology, and Vanderbilt Institute for Integrative Biosystems Research and Education, Vanderbilt University, Nashville, TN, USA*

ALFRED H. MERRILL, JR. • *School of Biology, School of Chemistry and Biochemistry, Georgia Institute of Technology, Atlanta, GA, USA*

PETER NEMES • *W.M. Keck Institute for Proteomics Technology and Applications, Department of Chemistry, George Washington University, Washington, DC, USA*

PAUL D. PIEHOWSKI • *Department of Chemistry, The Pennsylvania State University, University Park, PA, USA*

SANDER R. PIERSMA • *OncoProteomics Laboratory, Department of Medical Oncology, VUmc-Cancer Center Amsterdam, VU University Medical Center Amsterdam, Amsterdam, The Netherlands*
MARTIJN W.H. PINKSE • *Kluyver Laboratory, Department of Biotechnology, Netherlands Proteomics Center, Delft University of Technology, Delft, The Netherlands*
BRENDAN PRIDEAUX • *Analytical Sciences, Novartis Institutes for BioMedical Research, Basel, Switzerland*
MARIA C. PRIETO CONAWAY • *Thermo Fisher Scientific, San Jose, CA, USA*
MICHELLE L. REYZER • *Mass Spectrometry Research Center, Vanderbilt University, Nashville, TN, USA*
STANISLAV S. RUBAKHIN • *Department of Chemistry and Beckman Institute, University of Illinois at Urbana-Champaign, Urbana, IL, USA*
J. SABINE BECKER • *Central Division of Analytical Chemistry, Forschungszentrum Jülich, Jülich, Germany*
MICHEL SALZET • *Laboratoire de Neuroimmunologie et Neurochimie Evolutives, FRE CNRS 3249, MALDI Imaging Team, Université Lille Nord de France, Université Lille 1, Villeneuve d'Ascq, France*
SARAH A. SCHWARTZ • *David H. Murdock Research Institute, North Carolina Research Campus, Kannapolis, NC, USA*
MITSUTOSHI SETOU • *Mitsubishi Kagaku Institute of Life Sciences, Tokyo, Japan; Department of Molecular Anatomy, Hamamatsu University School of Medicine, Shizuoka, Japan*
DIETER STAAB • *Novartis Institutes for BioMedical Research, Basel, Switzerland*
JONATHAN STAUBER • *Laboratoire de Neuroimmunologie et Neurochimie Evolutives, FRE CNRS 3249, MALDI Imaging Team, Université Lille Nord de France, Université Lille 1, Villeneuve d'Ascq, France*
MARKUS STOECKLI • *Novartis Institutes for BioMedical Research, Basel, Switzerland*
KERSTIN STRUPAT • *Thermo Fisher Scientific GmbH, Bremen, Germany*
ROBERT M. STURM • *School of Pharmacy and Department of Chemistry, University of Wisconsin-Madison, Madison, WI, USA*
DETLEV SUCKAU • *Applications TOF-MS, Bruker Daltonik GmbH, Bremen, Germany*
YUKI SUGIURA • *Department of Bioscience and Biotechnology, Tokyo Institute of Technology, Kanagawa, Japan; Mitsubishi Kagaku Institute of Life Sciences, Tokyo, Japan*
J. SUSANNE BECKER • *Aeropharm GmbH, Rudolstadt, Germany*
JONATHAN V. SWEEDLER • *Department of Chemistry and Beckman Institute, University of Illinois at Urbana-Champaign, Urbana, IL, USA*
GARY J. VAN BERKEL • *Organic and Biological Mass Spectrometry Group, Chemical Sciences Division, Oak Ridge National Laboratory, Oak Ridge, TN, USA*
PETER D.E.M. VERHAERT • *Kluyver Laboratory, Department of Biotechnology, Netherlands Proteomics Center, Delft University of Technology, Delft, The Netherlands; Laboratory of Molecular Cell Biology, VIB, Flemish Institute of Biotechnology, Leuven, Belgium; BioMedical Research Center, University of Hasselt, Diepenbeek, Belgium*
AKOS VERTES • *W.M. Keck Institute for Proteomics Technology and Applications, Department of Chemistry, George Washington University, Washington, DC, USA*
ELAINE WANG • *School of Biology, Georgia Institute of Technology, Atlanta, GA, USA*
NICHOLAS WINOGRAD • *Department of Chemistry, The Pennsylvania State University, University Park, PA, USA*

MAXENCE WISZTORSKI • *Laboratoire de Neuroimmunologie et Neurochimie Evolutives, FRE CNRS 3249, MALDI Imaging Team, Université Lille Nord de France, Université Lille 1, Villeneuve d'Ascq, France*
JOHN H. WONG • *Chemistry Department and Volen Center, Brandeis University, Waltham, MA, USA*
AMINA S. WOODS • *Structural Biology Unit, Cellular Neurobiology Branch, National Institute on Drug Abuse Intramural Research Program, National Institutes of Health, Baltimore, MD, USA*
KUANG JEN J. WU • *Bioscience and Biotechnology Division, Physical and Life Sciences Directorate, Lawrence Livermore National Laboratory, Livermore, CA, USA*
LIGANG WU • *Seagate Technology US, Fremont, CA, USA*
YONGSHENG XIAO • *Department of Chemistry, University of California, Riverside, CA, USA*
RICHARD A. YOST • *Department of Chemistry, University of Florida, Gainesville, FL, USA*
TYLER A. ZIMMERMAN • *Department of Chemistry and Beckman Institute, University of Illinois at Urbana-Champaign, Urbana, IL, USA*

Part I

Introductory Reviews and Tutorials

Chapter 1

Imaging Mass Spectrometry: Viewing the Future

Sarah A. Schwartz and Richard M. Caprioli

Abstract

Imaging mass spectrometry (IMS) technology is an effective tool that is able to assess complex molecular mixtures in cells, tissues, or other sample types with high chemical specificity, allowing concurrent analysis of a variety of molecular species in a wide mass range, from small metabolites to large macromolecules such as proteins. Simultaneous localization of molecules, detection of post-translational modifications, and relative quantitative information can be obtained in a single experiment. Images generated by MS are unique because they are derived from direct molecular measurements and do not rely on target-specific reagents such as antibodies. Thus, the ability to map spatial distributions coupled with the mass accuracy and chemical specificity for MS-based detection makes IMS an effective discovery tool. Further structural assessment of compounds, including MS/MS fragmentation analysis, can be utilized in an imaging experiment to achieve accurate molecular identifications.

Key words: Imaging mass spectrometry, secondary ion mass spectrometry, matrix-assisted laser desorption ionization, desorption electrospray ionization, brain.

1. Introduction

The complexity of cells, tissues, organs, and whole systems is defined not only by the myriad of molecular events that occur but also by their distribution in space and time. Imaging mass spectrometry (IMS) technology provides a tool to assess these events with high chemical specificity, allowing concurrent analysis of a variety of molecular species in a wide mass range, from small metabolites to large macromolecules such as proteins. Simultaneous localization of molecules, detection of post-translational modifications, and relative quantitative information can be obtained in a single experiment. Although images

S.S. Rubakhin, J.V. Sweedler (eds.), *Mass Spectrometry Imaging*, Methods in Molecular Biology 656,
DOI 10.1007/978-1-60761-746-4_1, © Springer Science+Business Media, LLC 2010

generated by MS are similar in appearance to more traditional methodologies such as chemical staining, immunohistochemistry, and radiochemistry, they are unique because they are derived from direct molecular measurements and do not rely on target-specific reagents such as antibodies. Thus, the ability to map spatial distributions coupled with the mass accuracy and chemical specificity for MS-based detection makes IMS an effective discovery tool. Further structural assessment, including MS/MS fragmentation analysis, can be utilized for molecular identification (1).

IMS experiments may be performed using one of several MS ionization techniques: secondary ion mass spectrometry (SIMS), laser desorption/ionization (LDI), desorption electrospray ionization (DESI), and matrix-assisted laser desorption/ionization (MALDI) (2). In general, these techniques offer complementary capabilities (3). SIMS imaging is favored for higher spatial resolution imaging over a low mass range (<1,000 Da), MALDI IMS covers a much wider mass range (up to and over 100,000 Da) but at somewhat lower spatial resolution, and DESI uses ambient pressure for analysis of small molecules but at lower spatial resolution.

2. SIMS IMS

The imaging resolution and typically low mass range detected highlights SIMS IMS for small molecule imaging at sub-cellular resolution. SIMS utilizes a highly focused primary ion beam to impact the sample surface resulting in the sputtering of atoms and molecules. Ionization may occur in a dynamic mode, typically seen in elemental and isotopic analysis, in which high primary ion fluence is applied that erodes the sample surface and fragments molecular species. A low primary ion fluence (static mode) generates larger mass fragments and is used for the analysis of low molecular weight species. In some cases, sub-micron spatial resolution may be achieved, thereby allowing for sub-cellular molecular analysis (4). Ion maps are acquired by scanning the primary ion beam across the sample surface and collecting spectra as a function of beam position. Images are constructed for individual molecules or fragments of molecules of interest representing the relative abundance of the compound over the imaged surface. SIMS has been applied to imaging the cellular and subcellular distribution of a variety of compounds in biological samples including elements and lipids such as cholesterol (5–6).

For example, **Fig. 1.1** shows elemental images of brain sections from a rat bearing an intracranial 9L gliosarcoma tumor after treatment with *p*-boronophenylalanine (BPA) (7). Boron neutron

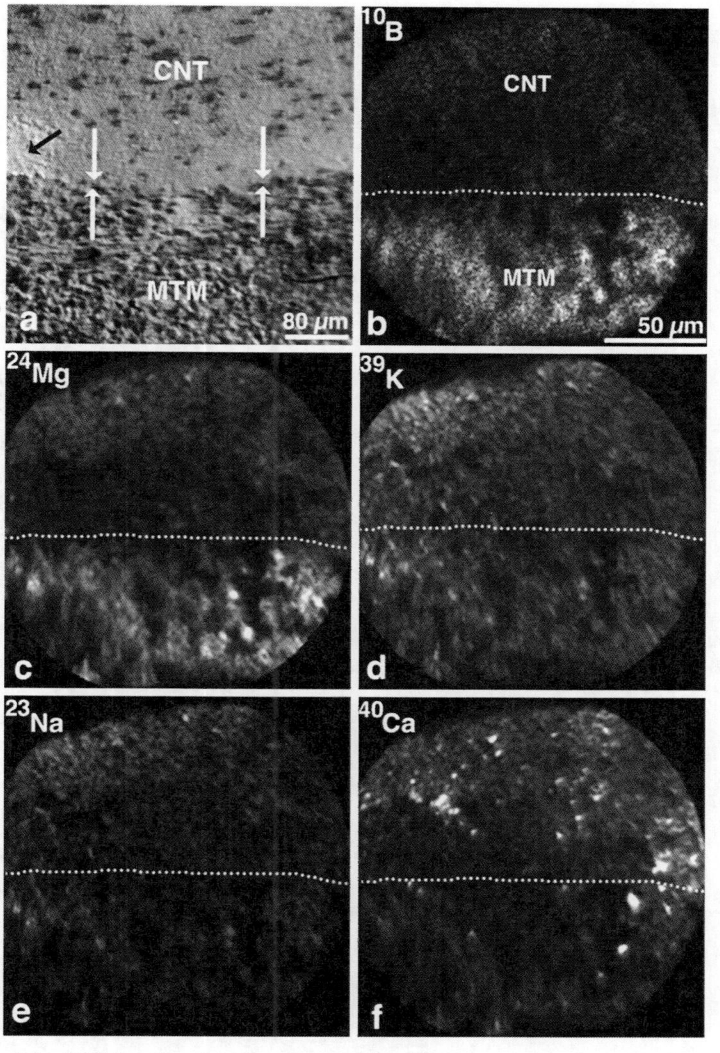

Fig. 1.1. SIMS imaging. Interface between the MTM and the normal brain of a male Fischer 334 rat bearing a 9L gliosarcoma. (**a**) H&E-stained adjacent 4-μm thick cryosection used for optical imaging. *White arrows* indicate the interface between the MTM (main tumor mass) and CNT (continuous normal tissue). The corresponding SIMS images for (**b**) 10B, (**c**) 24 Mg, (**d**) 39 K, (**e**) 23Na, and (**f**) 40Ca are presented from the same tissue region of analysis. *Dotted lines* indicate the interface between the MTM and CNT portions. (Reprinted with permission from ref. (7).)

capture therapy is dependent on the accumulation of boron-10 in tumor cells relative to the normal cells. SIMS imaging was used to monitor the distribution of boron-10 and other physiologically relevant species following BPA treatment.

Initial SIMS IMS techniques focused on low mass molecules, including elements, and were limited by low sensitivity. To enhance the ionization and detection of larger intact molecular ions, methods involving surface modifications have been used. These include adding energy absorbing compounds to the sample (8–15), the introduction of salts and acids (16), and sample metallization (16–18) to improve sensitivity and expand the mass range to include pharmaceuticals, lipids, and peptides. Some applications have utilized gallium or indium liquid metal ion guns (LMIGs) to obtain spatial resolution on the order of 1–2 μm. However, these pulsed ion beam devices have a low duty cycle that increases analysis time and have poor secondary ion pro-

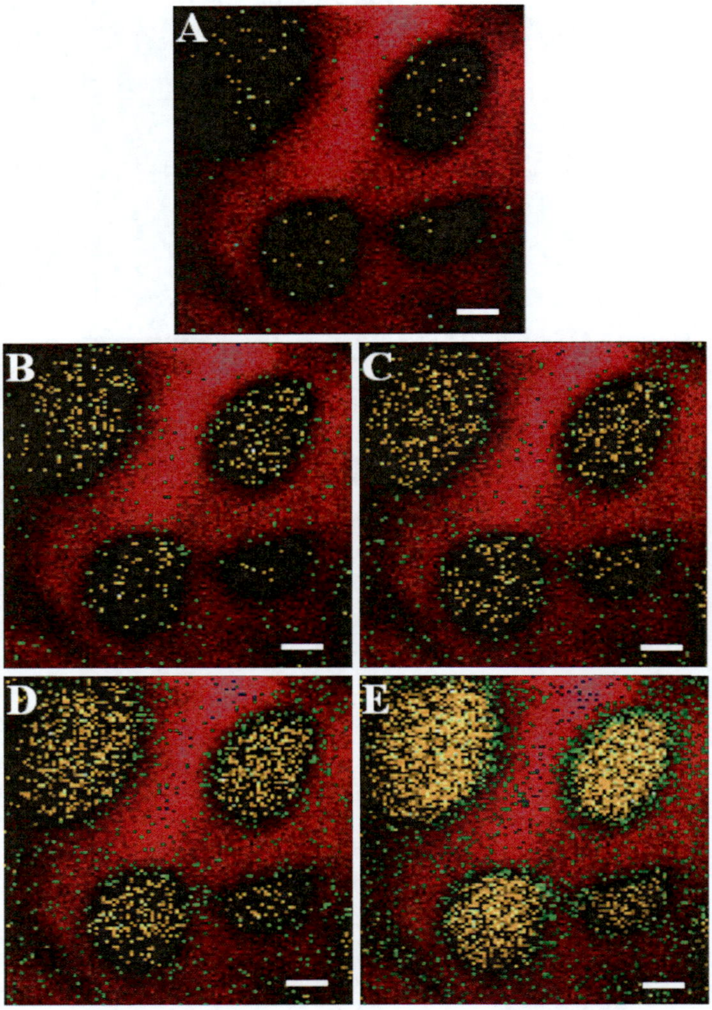

Fig. 1.2. SIMS imaging of macrophage cells. SIMS overlay images of the cholesterol fragment ions (*yellow*) and Na (*m/z* 23, *red*) are presented. *Green pixels* denote the presence of both ions on the macrophages. (**a**) Cholesterol [m–18] pseudomolecular ion at *m/z* 369, (**b**) cholesterol fragmentation ion at *m/z* 161, (**c**) cholesterol fragmentation ion at *m/z* 147, (**d**) cholesterol fragmentation ion at *m/z* 109, (**e**) cholesterol fragmentation ion at *m/z* 95. (Reprinted with permission from ref. (26).)

duction efficiency. A significant advance came from the use of primary ion beams of gold (Au_3^+) clusters. This technique was utilized to image mouse brain samples at a few micron resolution with an extended mass range. Images of ions from cholesterol, fatty acids, sulfatides, phosphatidylinositols, and triglycerides were obtained (19–20). Bismuth cluster beams have been implemented with improved spatial resolution (21) and primary Cs+ ions have been analyzed to enhance spatial resolution to less than 50 nm, although sensitivity is frequently an issue for lower resolution images (22).

New instrument configurations have been introduced (23) that eliminate the need to pulse the ion beam (24). More efficient polyatomic ion sources have also been implemented including C_{60}^+. These methods have created a new opportunity for molecular depth profiling and three-dimensional imaging and result in less bombardment-induced damage, a factor of 100 increase in ionization probability, and imaging at sub-micron resolution. Using these tools, intact molecules up to 1,000 Da, including lipids and small peptides, have been detected (21–25).

Figure 1.2 presents SIMS imaging results from an instrument equipped with an In^+ liquid metal ion source (26). This system provided increased spatial resolution and MS/MS capabilities for single cell imaging. Cholesterol localization was monitored from macrophage cells based on the primary In^+ ion and fragment ions. By evaluating the pseudomolecular ion for cholesterol imaging (m/z 369) and fragmentation ions at m/z 147, 161, 109, and 95, a more detailed image of the cell surface was obtained to determine cholesterol concentration at discrete locations on the cell surface.

3. MALDI IMS

The sensitivity and wide mass range achieved using MALDI MS makes this technique particularly appealing for tissue and whole-body section imaging. Sample preparation involves uniformly coating the sample with an energy absorbing matrix, either as discrete spots in an ordered array or a homogeneous coating, to aid in molecular desorption and ionization. MALDI IMS has predominantly been applied to imaging of proteins, peptides, lipids, and drugs in human and animal tissues (27). Applications of this technology have potential benefit for clinical diagnosis and treatment and have been applied to evaluation of diseased tissues including lung tumors (28), gliomas (29), breast cancer (30), ovarian tumors (31), and infections (32).

During drug discovery and development, it is important to track drugs and measure amounts of both drug and metabolites in organs and whole-body sections. Commonly, whole-body autoradiography of animal models is used for this purpose, although it does not readily provide information on the distribution of drug metabolites. IMS has been employed in such studies to assess both pharmaceutical and metabolite distribution in individual tissues or across whole-body animal cross sections.

Tandem mass spectrometry is utilized to monitor the transition from the precursor to selected product ions to provide sensitivity and specificity for small molecule detection (33) (for a review, see (34)). Images for larger molecule distribution, including peptides and proteins, can be measured to correlate drug distribution and therapeutic response within the same tissue.

MALDI IMS images from a whole rat sagittal tissue section have been reported for drug levels and disposition among organs (35). Animals were treated with olanzapine, a drug used to treat mood disorders in bipolar patients, and sections were collected for MALDI analysis at 2 h post–dose. Peptide and protein distributions were measured across the tissue sections as well as drug and metabolite localization over time. Results from the data from the 2-h time point indicate that the drug was detected in almost all tissues, with significant localization in the lung, spleen, bladder, kidney, liver, thymus, brain, and spinal cord. Metabolites including *N*-desmethylolanzapine and 2-hydroxymethyl olanzapine were detected in the bladder, liver, and kidney. Drug clearance in the brain and spinal cord was noted in images from the 6-h post-dose sections.

Whole-body MALDI MS imaging of proteins has also been performed and an example is presented in **Fig. 1.3**. Images from an embryonic day 14.5 mouse are shown: one stained and optically imaged and the other imaged by MALDI MS. In the latter, each of the five colors shown represents a different protein (unpublished data courtesy Peggi Angel, Vanderbilt University).

Careful sample preparation for MALDI IMS is essential to enhance signal quality and maintain accurate and reproducible molecular spatial resolution. It is important to apply the matrix solution in such a manner as to form homogeneous matrix crystals across the tissue surface without lateral displacement of the analyte molecules. Matrix deposition systems have been developed to support accurate and reproducible matrix deposition including matrix spotting (36), pneumatic (airbrush) (17, 37), electrospray (8), and capillary nebulization systems (38). Matrix seeding techniques have also been utilized to enhance crystal formation (39). Several automated matrix spotting devices are commercially available (e.g., the Labcyte Portrait, Shimadzu ChIP) that deposit picoliter-sized droplets in a grid across the sample surface. Here, the final spatial resolution of the image is defined by the droplet-to-droplet distance, typically 80–150 μm. Higher resolution MALDI images can be obtained using spraying devices that

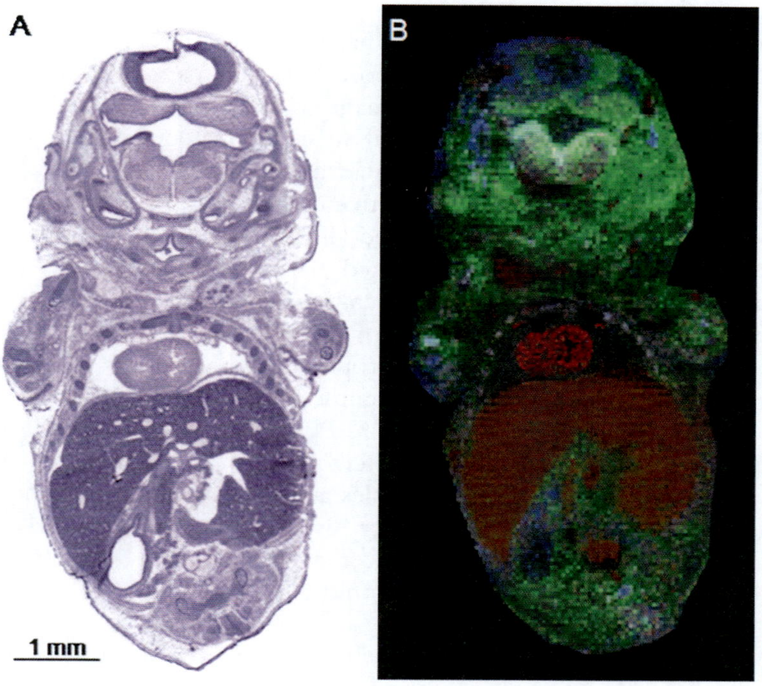

Fig. 1.3. MALDI imaging mass spectrometry of a coronal section from an embryonic day 14.5 mouse. (**a**) Visualization of anatomical features by hematoxylin and eosin staining of a neighboring section. (**b**) Color composite of merged ion images *m/z* 3,517 (*orange*), *m/z* 4,978 (*blue*), *m/z* 9,945 (*white*), *m/z* 10,011 (*red*), and *m/z* 11,403 (*green*). The MALDI IMS image was collected at an 80 μm lateral resolution. All masses are reported + 0.15%. (Courtesy Peggi Angel, Mass Spectrometry Research Center, Vanderbilt University School of Medicine.)

provide a homogeneous coating of the matrix (e.g., Bruker Daltonics ImagePrep and Leaptec TM-sprayer). Parameters including drying rate and tissue wetness need to be optimized to enhance sample coating and minimize molecule delocalization.

New approaches for sample preparation have also been reported that are aimed at enhancing signal quality. Combined MALDI matrices and a variety of tissue washing procedures (40–41) have been evaluated to improve molecule coverage. The addition of amphiphilic detergents or the use of ionic matrices has been reported to improve the protein signals obtained from imaging experiments (42–43). Spotting robots have been utilized to add chemical reagents on the tissue surface for chemical modifications or protein digestion (44), and the use of nanoparticles (45) and porous surfaces (46) has also been reported.

MALDI MS images can be correlated with histological information. This has been performed using a variety of approaches including comparing the image to a consecutive section that has been stained for histology, or staining the tissue section prior to imaging (47) or after the MALDI analysis (48). Advanced software has also been developed that allows an investigator to superimpose MALDI images over a macroscopic or microscopic optical image of the same section (47).

Initial work in the development of MALDI IMS methods used fresh frozen tissues without tissue fixation or using only simple ethanol fixation. At this time, formalin-fixed paraffin-embedded (FFPE) tissues could not be analyzed because of the cross-linked nature of the proteins in the fixed tissue. Sample preparation methods have since been developed for the analysis of FFPE tissue that utilize reactive matrices, including 2,4-dinitrophenylhydrazine, or enzymatic tissue digestion (31, 49). Recently, imaging MALDI MS has been employed for the analysis of FFPE tissue microarrays whereby hundreds of biopsies can be interrogated in a very short time span allowing banked tissue samples and corresponding patient outcomes to be used for biomarker discovery (50). This work describes the analysis of a lung cancer tissue microarray where different sub-classes of the disease could be differentiated based on their molecular phenotype.

Three-dimensional MALDI IMS has also been reported in recent papers (1, 51–52). Serial sections from a single specimen were taken and analyzed by IMS. The sections were collected at equal distances, enabling virtual z-stacks and three-dimensional volume rendering. The resulting images were then reconstructed to provide a depth dimension to the data set. Three-dimensional representations of analyte distribution throughout a sample were obtained to provide comprehensive spatial information for molecules of interest and support investigation of molecular localization patterns in individual tissue structures. This approach has been used to study molecular patterns within brain samples including peptide and protein localization within the substantia nigra and interpeduncular nucleus of a rat brain (51). Results from these techniques have been integrated with other imaging methodologies, including magnetic resonance imaging, to enhance the information obtained from these studies (53).

Recent technology research efforts have focused on optimizing spatial resolution, including utilizing optical lenses to better focus the laser beam, in some cases to sub-micron levels (54–55). Acquisition techniques have also been modified such as coupling complete sample ablation with smaller image array steps to improve spatial resolution (54). Several studies have reported improved spatial resolution through optimization of acquisition techniques and matrix deposition protocols. A stretched sample method has been described to improve imaging spatial resolution (56–57). In this method, a tissue section is adhered to a glass bead array embedded onto a parafilm M membrane and as the tissue is manually stretched, the tissue is separated into many small fragments. Computational tools were used to optimize analysis and reconstruct the whole tissue images.

IMS is commonly coupled with a linear time-of-flight (TOF) analyzer for the molecule separation and detection of proteins and reflector TOF analyzers for peptides. Ion trap quadrupole, TOF/TOF, and quadrupole TOF analyzers are typically used for drugs, metabolite, and other low MW analytes because of their MS/MS capabilities, providing high sensitivity in the presence of a complex matrix background (58).

Ion mobility-equipped instruments have recently been utilized for imaging and have the added capability of providing two-dimensional molecular separations based on molecular structure (59–60). **Figure 1.4** presents ion mobility imaging of endogenous lipids and peptides from rat brain tissue sections (59). Lipid signals, predominantly phosphatidylcholine species, were evaluated by selecting the ion mobility drift time and corresponding mass range for these species. Results demonstrate that these lipids are most prevalent in the tissue gray matter and are nearly absent in the corresponding white matter. In this case, a major advantage

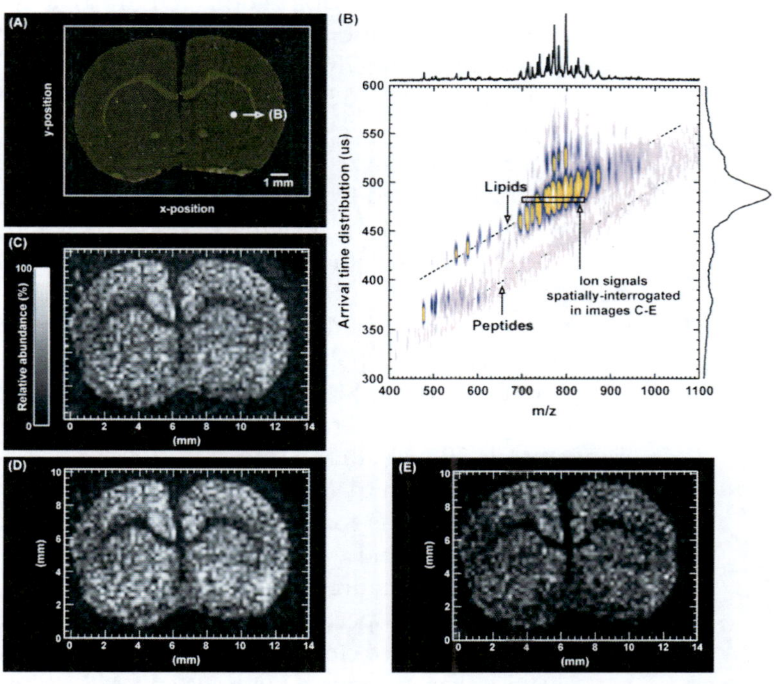

Fig. 1.4. Imaging IM-MS for mapping the spatial coordinates of analytes based on structure and molecular weight. (**a**) Optical image of a thin coronal rat brain tissue section adjacent the section analyzed by IMS. (**b**) Lipid imaging of the analytes indicated at an ion mobility arrival time of 480–484 μs and *m/z* 700–840 is illustrated in (**c**)–(**e**). Extracted ion density maps for two phospholipids at *m/z* 771–776 and 819–823, respectively. (Reprinted with permission from ref. (59).)

of ion mobility for imaging is the ability to generate separate images of two nominally isobaric molecular species, e.g., a peptide and a lipid having the same nominal *m/z* value.

A position-sensitive detector has been used for an instrument that is termed a mass microscope (61). Desorbed ions from discrete regions of the sample maintain their two-dimensional spatial distribution in the desorption/detection process. However, the entire spectrum is not obtained in the sample desorption event, with a few ion images produced in this process. Tissues are analyzed by irradiating relatively large areas, i.e., on the order of $150 \times 200\ \mu m^2$. The sample stage is moved to expose adjacent areas and the collected ion images are then stitched together to re-create the whole tissue image. Images with a spatial resolution on the order of 4–10 μm were reported using this approach.

4. Ambient Ionization Techniques

Desorption electrospray ionization (DESI) is a relatively new ambient surface ionization technique that provides advantages for analyte ionization directly from a sample surface, in the absence of a crystalline matrix (62). A pneumatically assisted electrospray ionization source is used to impact charged solvent droplets on the surface to be analyzed, resulting in analytes being picked up and desorbed/ionized from the sample surface. DESI MS imaging has been utilized for imaging of lipids (63) and drugs from whole-body tissue sections (64). While spatial resolution for this method is estimated at 200–400 μm, the advantages of performing under ambient conditions without matrix will make this a useful method for many applications (65–66).

An example of DESI MS imaging of pharmaceutical components is presented in **Fig. 1.5** (67). Lung tissues were collected from animals 30 min after dosing with clozapine. A tissue section was imaged using DESI MS across the *m/z* range of 200–1,100 to evaluate the sample for clozapine distribution and its metabolites. A variety of lipids were detected as well as clozapine and its *N*-desmethyl metabolite. This analysis and the results from LC-MS/MS analysis performed in a complementary study confirmed the presence of clozapine in the lung.

Several other ionization methods and MS technologies have been investigated for possible use for imaging. Atmospheric pressure infrared MALDI (AP IR-MALDI) and mid-infrared laser ablation with electrospray ionization (LAESI) have been reported and use focused laser radiation for ambient sampling with no sample preparation or chemical modification required. In these

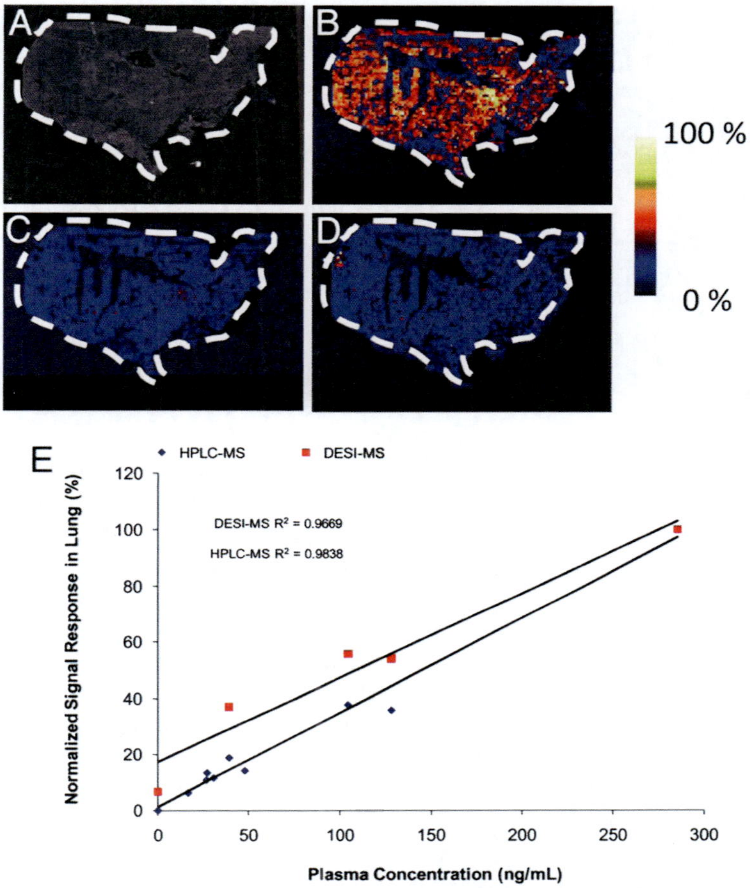

Fig. 1.5. DESI imaging of a drug in lung tissue. (**a**) Optical image. (**b**) Image of clozapine at *m/z* 327.1. (**c**) Image of desmethylclozapine at *m/z* 313.1. (**d**) Image of sodiated PC 16:0/16:0 at *m/z* 756.4. (**e**) DESI-MS imaging and LC/MS/MS results. The signal response in each method was normalized to the maximum response in each experiment and plotted against the clozapine plasma concentration as determined by LC/MS. (Reprinted with permission from ref. (67).)

methods, water serves as a matrix and therefore, water-rich samples can be evaluated directly. These techniques have been predominantly used for plant metabolite profiling (68–69).

5. Data Analysis and Image Evaluation

Due to the large amount of data produced, data processing and analysis are complex and time consuming, since many hundreds of images can be obtained for a single imaging experiment. Automated data processing and display tools as well as advanced data mining techniques have been developed to support this technique. Most notable is the data processing and visualization tool initially developed as The Image Tool (70) and subsequently further developed with improved automation, functionality, and visualization tools and termed BioMap (71) (http://www.maldi-msi.org/).

Image acquisition and processing software is also commercially available. Data processing and visualization mentioned include Tissueview by Applied Biosystems, ImageQuest by Thermo Scientific, FlexImaging by Bruker, and MassLynx by Waters. These tools typically include capabilities for data acquisition as well as pre-processing algorithms, data reduction tools such as principle component analysis (PCA), and image visualization tools.

Advanced statistical tools, including multivariate statistical methods, have been developed to efficiently extract signal pattern information from these complex data sets. These tools have been applied both to discover new molecular trends and to mine the molecular data for patterns that correlate to tissue subregions of interest. In general, data analysis methods utilize several key steps. Proteins that are differentially expressed between histomorphological defined groups are selected, using a variety of statistical tools. Reduction of the number of proteins found may be performed using strict cutoffs and statistical tests (72) and by utilizing tools such as principle component analysis (PCA) to remove molecular signals that do not provide significant information to define the data variance. Supervised and unsupervised methods can be utilized to classify regions of interest based on spectral patterns (73). When combined with data processing and visualization tools, these methods result in a powerful approach for analyzing the complex data sets obtained in IMS experiments.

6. Summary

The development of IMS has provided an invaluable tool to the field of molecular analysis. Direct analysis of tissues or individual cells offers a mechanism to determine molecular spatial orientation and to evaluate molecular signatures indicative of disease. The technology is commercially available and is widely used as a discovery tool to monitor the spatial distribution of elements, lipids, pharmaceuticals, metabolites, peptides, and proteins and to evaluate molecular activities in biological processes. Assessment of a variety of compounds can be performed in parallel, without the use of isotopic labeling. The potential of IMS has driven methodology and instrumentation developments such as improved imaging speed and spatial resolution. Sample preparation developments have improved reproducibility and data reliability, and data analysis techniques have been leveraged to evaluate large data sets obtained from complex samples such as tissues and even whole-body sections. It is clear that the understanding of health and disease and the evaluation of the molecular complexity that surrounds us and interacts with us daily will require advanced technologies of many kinds. Imaging mass spectrometry is one of these technologies that will bring this type of insight and understanding.

Acknowledgments

The authors thank Peggi Angel (Mass Spectrometry Research Center, Vanderbilt University) for the data presented in **Fig. 1.3** and to Erin Seeley for help in preparing this manuscript. We also gratefully acknowledge grants NIH R01 GM058008, DOD W81XWH-05-1-0179, and NIH/NCI CA68485-13.

References

1. Chen, R., Hui, L., Sturm, R. M., Li, L. (2009) Three dimensional mapping of neuropeptides and lipids in crustacean brain by mass spectral imaging. *J Am Soc Mass Spectrom*, **20**, 1068–1077.
2. Caprioli, R. M., Farmer, T. B., Gile, J. (1997) Molecular imaging of biological samples: localization of peptides and proteins using MALDI-TOF MS. *Anal Chem*, **69**, 4751–4760.
3. Monroe, E. B., Annangudi, S. P., Hatcher, N. G., Gutstein, H. B., Rubakhin, S. S., Sweedler, J. V. (2008) SIMS and MALDI MS imaging of the spinal cord. *Proteomics*, **8**, 3746–3754.
4. Chandra, S., Smith, D. R., Morrison, G. H. (2000) Subcellular imaging by dynamic SIMS ion microscopy. *Anal Chem*, **72**, 104A–114A.
5. Kurczy, M. E., Kozole, J., Parry, S. A., Piehowski, P. D., Winograd, N., Ewing, A. G. (2008) Relative quantification of cellular sections with molecular depth profiling ToF-SIMS imaging. *Appl Surf Sci*, **255**, 1158–1161.
6. Ostrowski, S. G., Kurczy, M. E., Roddy, T. P., Winograd, N., Ewing, A. G. (2007) Secondary ion MS imaging to relatively quantify cholesterol in the membranes of individual cells from differentially treated populations. *Anal Chem*, **79**, 3554–3560.
7. Smith, D. R., Chandra, S., Barth, R. F., Yang, W., Joel, D. D., Coderre, J. A. (2001) Quantitative imaging and microlocalization of boron-10 in brain tumors and infiltrating tumor cells by SIMS ion microscopy: relevance to neutron capture therapy. *Cancer Res*, **61**, 8179–8187.
8. Altelaar, A. F., van Minnen, J., Jimenez, C. R., Heeren, R. M., Piersma, S. R. (2005) Direct molecular imaging of *Lymnaea stagnalis* nervous tissue at subcellular spatial resolution by mass spectrometry. *Anal Chem*, **77**, 735–741.
9. Hanton, S. D., Cornelio Clark, P. A., Owens, K. G. (1999) Investigations of matrix-assisted laser desorption/ionization sample preparation by time-of-flight secondary ion mass spectrometry. *J Am Soc Mass Spectrom*, **10**, 104–111.
10. Luxembourg, S. L., McDonnell, L. A., Duursma, M. C., Guo, X., Heeren, R. M. (2003) Effect of local matrix crystal variations in matrix-assisted ionization techniques

for mass spectrometry. *Anal Chem*, **75**, 2333–2341.
11. McDonnell, L. A., Mize, T. H., Luxembourg, S. L., Koster, S., Eijkel, G. B., Verpoorte, E., de Rooij, N. F., Heeren, R. M. (2003) Using matrix peaks to map topography: increased mass resolution and enhanced sensitivity in chemical imaging. *Anal Chem*, **75**, 4373–4381.
12. McDonnell, L. A., Piersma, S. R., MaartenAltelaar, A. F., Mize, T. H., Luxembourg, S. L., Verhaert, P. D., van Minnen, J., Heeren, R. M. (2005) Subcellular imaging mass spectrometry of brain tissue. *J Mass Spectrom*, **40**, 160–168.
13. Nicola, A. J., Muddiman, D. C., Hercules, D. M. (1996) Enhancement of ion intensities in time-of-flight secondary ionization mass spectrometry. *J Am Soc Mass Spectrom*, **7**, 467–472.
14. Wittmaack, K., Szymczak, W., Hoheisel, G., Tuszynski, W. (2000) Time-of-flight secondary ion mass spectrometry of matrix-diluted oligo- and polypeptides bombarded with slow and fast projectiles: positive and negative matrix and analyte ion yields, background signals, and sample aging. *J Am Soc Mass Spectrom*, **11**, 553–563.
15. Wu, K. J., Odom, R. W. (1996) Matrix-enhanced secondary ion mass spectrometry: a method for molecular analysis of solid surfaces. *Anal Chem*, **68**, 873–882.
16. Nygren, H., Malmberg, P., Kriegeskotte, C., Arlinghaus, H. F. (2004) Bioimaging TOF-SIMS: localization of cholesterol in rat kidney sections. *FEBS* Lett, **566**, 291–293.
17. Altelaar, A. F., Klinkert, I., Jalink, K., de Lange, R. P., Adan, R. A., Heeren, R. M., Piersma, S. R. (2006) Gold-enhanced biomolecular surface imaging of cells and tissue by SIMS and MALDI mass spectrometry. *Anal Chem*, **78**, 734–742.
18. Delcorte, A., Bour, J., Aubriet, F., Muller, J. F., Bertrand, P. (2003) Sample metallization for performance improvement in desorption/ionization of kilodalton molecules: quantitative evaluation, imaging secondary ion MS, and laser ablation. *Anal Chem*, **75**, 6875–6885.
19. Sjovall, P., Lausmaa, J., Johansson, B. (2004) Mass spectrometric imaging of lipids in brain tissue. *Anal Chem*, **76**, 4271–4278.
20. Touboul, D., Halgand, F., Brunelle, A., Kersting, R., Tallarek, E., Hagenhoff, B., Laprevote, O. (2004) Tissue molecular ion imaging by gold cluster ion bombardment. *Anal Chem*, **76**, 1550–1559.
21. Brunelle, A., Touboul, D., Laprevote, O. (2005) Biological tissue imaging with time-of-flight secondary ion mass spectrometry and cluster ion sources. *J Mass Spectrom*, **40**, 985–999.
22. Guerquin-Kern, J. L., Wu, T. D., Quintana, C., Croisy, A. (2005) Progress in analytical imaging of the cell by dynamic secondary ion mass spectrometry (SIMS microscopy). *Biochim Biophys Acta*, **1724**, 228–238.
23. Fletcher, J. S., Vickerman, J. C. (2009) A new SIMS paradigm for 2D and 3D molecular imaging of bio-systems. *Anal Bioanal Chem*.
24. Fletcher, J. S., Rabbani, S., Henderson, A., Blenkinsopp, P., Thompson, S. P., Lockyer, N. P., Vickerman, J. C. (2008) A new dynamic in mass spectral imaging of single biological cells. *Anal Chem*, **80**, 9058–9064.
25. Touboul, D., Kollmer, F., Niehuis, E., Brunelle, A., Laprevote, O. (2005) Improvement of biological time-of-flight-secondary ion mass spectrometry imaging with a bismuth cluster ion source. *J Am Soc Mass Spectrom*, **16**, 1608–1618.
26. Piehowski, P. D., Carado, A. J., Kurczy, M. E., Ostrowski, S. G., Heien, M. L., Winograd, N., Ewing, A. G. (2008) MS/MS methodology to improve subcellular mapping of cholesterol using TOF-SIMS. *Anal Chem*, **80**, 8662–8667.
27. Cornett, D. S., Reyzer, M. L., Chaurand, P., Caprioli, R. M. (2007) MALDI imaging mass spectrometry: molecular snapshots of biochemical systems. *Nat Methods*, **4**, 828–833.
28. Yanagisawa, K., Shyr, Y., Xu, B. J., Massion, P. P., Larsen, P. H., White, B. C., Roberts, J. R., Edgerton, M., Gonzalez, A., Nadaf, S., Moore, J. H., Caprioli, R. M., Carbone, D. P. (2003) Proteomic patterns of tumour subsets in non-small-cell lung cancer. *Lancet*, **362**, 433–439.
29. Schwartz, S. A., Weil, R. J., Thompson, R. C., Shyr, Y., Moore, J. H., Toms, S. A., Johnson, M. D., Caprioli, R. M. (2005) Proteomic-based prognosis of brain tumor patients using direct-tissue matrix-assisted laser desorption ionization mass spectrometry. *Cancer Res*, **65**, 7674–7681.
30. Cornett, D. S., Mobley, J. A., Dias, E. C., Andersson, M., Arteaga, C. L., Sanders, M. E., Caprioli, R. M. (2006) A novel histology-directed strategy for MALDI-MS tissue profiling that improves throughput and cellular specificity in human breast cancer. *Mol Cell Proteomics*, **5**, 1975–1983.
31. Lemaire, R., Desmons, A., Tabet, J. C., Day, R., Salzet, M., Fournier, I. (2007) Direct analysis and MALDI imaging of

formalin-fixed, paraffin-embedded tissue sections. *J Proteome Res*, **6**, 1295–1305.
32. Corbin, B. D., Seeley, E. H., Raab, A., Feldmann, J., Miller, M. R., Torres, V. J., Anderson, K. L., Dattilo, B. M., Dunman, P. M., Gerads, R., Caprioli, R. M., Nacken, W., Chazin, W. J., Skaar, E. P. (2008) Metal chelation and inhibition of bacterial growth in tissue abscesses. *Science*, **319**, 962–965.
33. Stoeckli, M., Staab, D., Schweitzer, A., Gardiner, J., Seebach, D. (2007) Imaging of a beta-peptide distribution in whole-body mice sections by MALDI mass spectrometry. *J Am Soc Mass Spectrom*, **18**, 1921–1924.
34. Hsieh, Y., Chen, J., Korfmacher, W. A. (2007) Mapping pharmaceuticals in tissues using MALDI imaging mass spectrometry. *J Pharmacol Toxicol Methods*, **55**, 193–200.
35. Khatib-Shahidi, S., Andersson, M., Herman, J. L., Gillespie, T. A., Caprioli, R. M. (2006) Direct molecular analysis of whole-body animal tissue sections by imaging MALDI mass spectrometry. *Anal Chem*, **78**, 6448–6456.
36. Aerni, H. R., Cornett, D. S., Caprioli, R. M. (2006) Automated acoustic matrix deposition for MALDI sample preparation. *Anal Chem*, **78**, 827–834.
37. Schwartz, S. A., Reyzer, M. L., Caprioli, R. M. (2003) Direct tissue analysis using matrix-assisted laser desorption/ionization mass spectrometry: practical aspects of sample preparation. *J Mass Spectrom*, **38**, 699–708.
38. Chen, Y., Allegood, J., Liu, Y., Wang, E., Cachon-Gonzalez, B., Cox, T. M., Merrill, A. H., Jr., Sullards, M. C. (2008) Imaging MALDI mass spectrometry using an oscillating capillary nebulizer matrix coating system and its application to analysis of lipids in brain from a mouse model of Tay-Sachs/Sandhoff disease. *Anal Chem*, **80**, 2780–2788.
39. Puolitaival, S. M., Burnum, K. E., Cornett, D. S., Caprioli, R. M. (2008) Solvent-free matrix dry-coating for MALDI imaging of phospholipids. *J Am Soc Mass Spectrom*, **19**, 882–886.
40. Lemaire, R., Wisztorski, M., Desmons, A., Tabet, J. C., Day, R., Salzet, M., Fournier, I. (2006) MALDI-MS direct tissue analysis of proteins: improving signal sensitivity using organic treatments. *Anal Chem*, **78**, 7145–7153.
41. Seeley, E. H., Oppenheimer, S. R., Mi, D., Chaurand, P., Caprioli, R. M. (2008) Enhancement of protein sensitivity for MALDI imaging mass spectrometry after chemical treatment of tissue sections. *J Am Soc Mass Spectrom*, **19**, 1069–1077.
42. Lemaire, R., Tabet, J. C., Ducoroy, P., Hendra, J. B., Salzet, M., Fournier, I. (2006) Solid ionic matrixes for direct tissue analysis and MALDI imaging. *Anal Chem*, **78**, 809–819.
43. Norris, J. L., Porter, N. A., Caprioli, R. M. (2005) Combination detergent/MALDI matrix: functional cleavable detergents for mass spectrometry. *Anal Chem*, **77**, 5036–5040.
44. Groseclose, M. R., Andersson, M., Hardesty, W. M., Caprioli, R. M. (2007) Identification of proteins directly from tissue: *in situ* tryptic digestions coupled with imaging mass spectrometry. *J Mass Spectrom*, **42**, 254–262.
45. Ageta, H., Asai, S., Sugiura, Y., Goto-Inoue, N., Zaima, N., Setou, M. (2009) Layer-specific sulfatide localization in rat hippocampus middle molecular layer is revealed by nanoparticle-assisted laser desorption/ionization imaging mass spectrometry. *Med Mol Morphol*, **42**, 16–23.
46. Liu, Q., Xiao, Y., Pagan-Miranda, C., Chiu, Y. M., He, L. (2009) Metabolite imaging using matrix-enhanced surface-assisted laser desorption/ionization mass spectrometry (ME-SALDI-MS). *J Am Soc Mass Spectrom*, **20**, 80–88.
47. Chaurand, P., Schwartz, S. A., Billheimer, D., Xu, B. J., Crecelius, A., Caprioli, R. M. (2004) Integrating histology and imaging mass spectrometry. *Anal Chem*, **76**, 1145–1155.
48. Schwamborn, K., Krieg, R. C., Reska, M., Jakse, G., Knuechel, R., Wellmann, A. (2007) Identifying prostate carcinoma by MALDI-imaging. *Int J Mol Med*, **20**, 155–159.
49. Djidja, M. C., Francese, S., Loadman, P. M., Sutton, C. W., Scriven, P., Claude, E., Snel, M. F., Franck, J., Salzet, M., Clench, M. R. (2009) Detergent addition to tryptic digests and ion mobility separation prior to MS/MS improves peptide yield and protein identification for *in situ* proteomic investigation of frozen and formalin-fixed paraffin-embedded adenocarcinoma tissue sections. *Proteomics*, **9**, 2750–2763.
50. Groseclose, M. R., Massion, P. P., Chaurand, P., Caprioli, R. M. (2008) High-throughput proteomic analysis of formalin-fixed paraffin-embedded tissue microarrays using MALDI imaging mass spectrometry. *Proteomics*, **8**, 3715–3724.

51. Andersson, M., Groseclose, M. R., Deutch, A. Y., Caprioli, R. M. (2008) Imaging mass spectrometry of proteins and peptides: 3D volume reconstruction. *Nat Methods*, **5**, 101–108.
52. Crecelius, A. C., Cornett, D. S., Caprioli, R. M., Williams, B., Dawant, B. M., Bodenheimer, B. (2005) Three-dimensional visualization of protein expression in mouse brain structures using imaging mass spectrometry. *J Am Soc Mass Spectrom*, **16**, 1093–1099.
53. Sinha, T. K., Khatib-Shahidi, S., Yankeelov, T. E., Mapara, K., Ehtesham, M., Cornett, D. S., Dawant, B. M., Caprioli, R. M., Gore, J. C. (2008) Integrating spatially resolved three-dimensional MALDI IMS with *in vivo* magnetic resonance imaging. *Nat Methods*, **5**, 57–59.
54. Jurchen, J. C., Rubakhin, S. S., Sweedler, J. V. (2005) MALDI-MS imaging of features smaller than the size of the laser beam. *J Am Soc Mass Spectrom*, **16**, 1654–1659.
55. Spengler, B., Hubert, M. (2002) Scanning microprobe matrix-assisted laser desorption ionization (SMALDI) mass spectrometry: instrumentation for sub-micrometer resolved LDI and MALDI surface analysis. *J Am Soc Mass Spectrom*, **13**, 735–748.
56. Monroe, E. B., Jurchen, J. C., Koszczuk, B. A., Losh, J. L., Rubakhin, S. S., Sweedler, J. V. (2006) Massively parallel sample preparation for the MALDI MS analyses of tissues. *Anal Chem*, **78**, 6826–6832.
57. Zimmerman, T. A., Monroe, E. B., Sweedler, J. V. (2008) Adapting the stretched sample method from tissue profiling to imaging. *Proteomics*, **8**, 3809–3815.
58. Reyzer, M. L., Hsieh, Y., Ng, K., Korfmacher, W. A., Caprioli, R. M. (2003) Direct analysis of drug candidates in tissue by matrix-assisted laser desorption/ionization mass spectrometry. *J Mass Spectrom*, **38**, 1081–1092.
59. McLean, J. A., Ridenour, W. B., Caprioli, R. M. (2007) Profiling and imaging of tissues by imaging ion mobility-mass spectrometry. *J Mass Spectrom*, **42**, 1099–1105.
60. Trim, P. J., Henson, C. M., Avery, J. L., McEwen, A., Snel, M. F., Claude, E., Marshall, P. S., West, A., Princivalle, A. P., Clench, M. R. (2008) Matrix-assisted laser desorption/ionization-ion mobility separation-mass spectrometry imaging of vinblastine in whole body tissue sections. *Anal Chem*, **80**, 8628–8634.
61. Luxembourg, S. L., Mize, T. H., McDonnell, L. A., Heeren, R. M. (2004) High-spatial resolution mass spectrometric imaging of peptide and protein distributions on a surface. *Anal Chem*, **76**, 5339–5344.
62. Wiseman, J. M., Ifa, D. R., Venter, A., Cooks, R. G. (2008) Ambient molecular imaging by desorption electrospray ionization mass spectrometry. *Nat Protoc*, **3**, 517–524.
63. Wiseman, J. M., Puolitaival, S. M., Takats, Z., Cooks, R. G., Caprioli, R. M. (2005) Mass spectrometric profiling of intact biological tissue by using desorption electrospray ionization. *Angew Chem Int Ed Engl*, **44**, 7094–7097.
64. Kertesz, V., Van Berkel, G. J., Vavrek, M., Koeplinger, K. A., Schneider, B. B., Covey, T. R. (2008) Comparison of drug distribution images from whole-body thin tissue sections obtained using desorption electrospray ionization tandem mass spectrometry and autoradiography. *Anal Chem*, **80**, 5168–5177.
65. Kertesz, V., Van Berkel, G. J. (2008) Improved desorption electrospray ionization mass spectrometry performance using edge sampling and a rotational sample stage. *Rapid Commun Mass Spectrom*, **22**, 3846–3850.
66. Kertesz, V., Van Berkel, G. J. (2008) Improved imaging resolution in desorption electrospray ionization mass spectrometry. *Rapid Commun Mass Spectrom*, **22**, 2639–2644.
67. Wiseman, J. M., Ifa, D. R., Zhu, Y., Kissinger, C. B., Manicke, N. E., Kissinger, P. T., Cooks, R. G. (2008) Desorption electrospray ionization mass spectrometry: Imaging drugs and metabolites in tissues. *Proc Natl Acad Sci U S A*, **105**, 18120–18125.
68. Li, Y., Shrestha, B., Vertes, A. (2007) Atmospheric pressure molecular imaging by infrared MALDI mass spectrometry. *Anal Chem*, **79**, 523–532.
69. Nemes, P., Barton, A. A., Li, Y., Vertes, A. (2008) Ambient molecular imaging and depth profiling of live tissue by infrared laser ablation electrospray ionization mass spectrometry. *Anal Chem*, **80**, 4575–4582.
70. Stoeckli, M., Farmer, T. B., Caprioli, R. M. (1999) Automated mass spectrometry imaging with a matrix-assisted laser desorption ionization time-of-flight instrument. *J Am Soc Mass Spectrom*, **10**, 67–71.
71. Stoeckli, M., Staab, D., Staufenbiel, M., Wiederhold, K. H., Signor, L. (2002) Molecular imaging of amyloid beta peptides in

mouse brain sections using mass spectrometry. *Anal Biochem*, **311**, 33–39.

72. Caldwell, R. L., Caprioli, R. M. (2005) Tissue profiling by mass spectrometry: a review of methodology and applications. *Mol Cell Proteomics*, **4**, 394–401.

73. Deininger, S. O., Ebert, M. P., Futterer, A., Gerhard, M., Rocken, C. (2008) MALDI imaging combined with hierarchical clustering as a new tool for the interpretation of complex human cancers. *J Proteome Res*, 7, 5230–5236.

Chapter 2

A Mass Spectrometry Primer for Mass Spectrometry Imaging

Stanislav S. Rubakhin and Jonathan V. Sweedler

Abstract

Mass spectrometry imaging (MSI), a rapidly growing subfield of chemical imaging, employs mass spectrometry (MS) technologies to create single- and multi-dimensional localization maps for a variety of atoms and molecules. Complimentary to other imaging approaches, MSI provides high chemical specificity and broad analyte coverage. This powerful analytical toolset is capable of measuring the distribution of many classes of inorganics, metabolites, proteins, and pharmaceuticals in chemically and structurally complex biological specimens in vivo, in vitro, and in situ. The MSI approaches highlighted in this *Methods in Molecular Biology* volume provide flexibility of detection, characterization, and identification of multiple known and unknown analytes. The goal of this chapter is to introduce investigators who may be unfamiliar with MS to the basic principles of the mass spectrometric approaches as used in MSI. In addition to guidelines for choosing the most suitable MSI method for specific investigations, cross-references are provided to the chapters in this volume that describe the appropriate experimental protocols.

Key words: Mass spectrometry imaging, mass spectrometer, mass spectrum, ion source, mass analyzer, detector.

1. Introduction

Understanding the functioning of a cell, organ, or organism, whether normal or pathological, often relies on the ability to follow the spatial localization of molecules and atoms with sufficient spatial and temporal resolution. Success can be facilitated by mass spectrometry imaging (MSI), a robust analytical approach that combines the chemical information content of MS detection with chemical imaging technology. MSI allows the simultaneous

S.S. Rubakhin, J.V. Sweedler (eds.), *Mass Spectrometry Imaging*, Methods in Molecular Biology 656,
DOI 10.1007/978-1-60761-746-4_2, © Springer Science+Business Media, LLC 2010

localization of hundreds of analytes in a variety of biological samples, from defined morphological structures to entire organisms, without analyte preselection using antibody labeling. The metabolome, peptidome, and proteome can be investigated with MSI, providing spatial and chemical information for literally hundreds to thousands of analytes, ranging from small molecules to large proteins. Although implementation requires specialized hardware, software, and experimental protocols, including sample preparation methods, a range of features, along with recent advances in instrumentation, have contributed to an increased interest in using MSI for chemical imaging.

In MSI, a specimen is typically interrogated by a beam of light or ions called a microprobe. As a result, multiple analytes are moved from solid or liquid to gas phase, ionized, and characterized according to their mass-to-charge ratio. The microprobe is then moved to another location and the process repeated. If the microprobe is scanned across the surface at a sufficient number of regularly spaced points, chemical images for each analyte are created (**Fig. 2.1**). A chemical image not only presents informa-

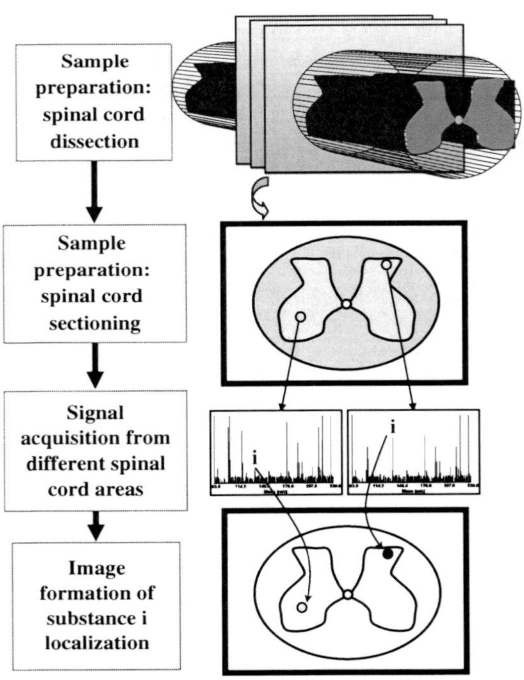

Fig. 2.1. Schematic of the generalized mass spectrometry imaging process utilizing the microprobe mode of operation. The major steps involved include sample preparation and interrogation with a probe, signal acquisition, processing, and final image formation. Distribution of a single or multiple analytes is determined in an array of spots covering the sample surface. Only two mass spectra containing signals from different analytes are shown in the signal acquisition step. Detected signal intensity is often encoded by color or grayscale. Here, the more intense signal is presented in the color *black*, which provides the most contrast from the background color.

tion about analyte localization but also provides signal intensity information, often encoded by color or grayscale. Depending on the analytical method used and sample properties, signal intensity often correlates with the absolute or relative concentration of each analyte. Although MSI approaches can differ in the software and sample preparation protocols used, they share many features of common MS measurements and thus are often employed in MS or proteomics facilities. However, MSI can easily be implemented in many biological and neuroscientific laboratories.

In addition to the discussion of the basics of mass spectrometric technologies presented herein, the reader will be directed, as appropriate, to cross-referenced chapters in this volume that provide more specifics on the MSI approaches. Even greater detail on both MS and MSI can be found in several excellent books and reviews (1–10).

2. Mass Spectrometry (MS)

Mass, considered a fundamental property of matter (11), is defined as a measure of a body's inertia or resistance to a change in body motion upon application of an external force. Having knowledge of a particle's exact mass and some details about its structure is sufficient to identify an element or polyatomic substance. MS is one of most precise, fast, and reliable methods available to determine the molecular and atomic masses of analytes with high accuracy in a single measurement.

2.1. Principles of Mass Spectrometry

A key element of MS is based on the ability of magnetic and/or electric fields to influence the motion of charged atoms and molecules, typically in a vacuum, in relation to their mass and charge. Newton's second law is a fundamental basis for MS, where the resistance of a particle to a change in its motion (inertia) exerted by a force, and the proportionality of this change to the intensity of the applied force, is used to determine a particle's mass. Indeed, the trajectories of two identically moving particles of the same charge will change differently under the influence of identical electric and/or magnetic field(s) if the particles have different masses. The motion of the heavier particle will be affected less than the lighter one. As a result, the final relative spatial positions of the moving particles will depend on their inertial mass (m) and, for differently charged particles, on their charge (z). A particle's charge is an important factor as the amount of force developed by a field relates directly to the charge. The motion of the higher-charged particle will be affected more by the fields than the lesser-charged particle's motion, when all parameters, except charge, are the same. Therefore, by controlling the forces, predicting particle trajectories, and selectively detecting particles, one can deconvolve the motions into each distinct particle's mass-to-charge ratio or m/z. Results of mass spectrometric investigation are typically presented in the form of mass spectra, where m/z values are plotted along the x-axis and the relative intensity of the charged particle signals against the y-axis.

In what follows, the major instrument components used in MS measurements are described. Why is knowledge about the mass spectrometric process important to a biologist interested in MSI? Although it may not seem obvious, understanding of the technology ensures selection of the optimum measurement approach and the most appropriate protocols for a specific project. As explained below, many of the measurement limitations experienced in an MSI investigation are directly related to the characteristics of the mass measurement process selected.

2.2. Mass Spectrometers

Mass spectrometers consist of several functionally distinct components (**Fig. 2.2**) – the ion source, mass analyzer, and detector – with resulting signal processing/output. As mentioned above, the characteristics of these components are what combine to give the overall figures of merit of the measurement and determine the most appropriate system for a given analytical challenge. More than one ion source, mass analyzer, and detector are often integrated into a single mass spectrometer, thereby providing greater flexibility in instrumental operating modes and enabling different types of analytes to be characterized. Importantly, most mass spectrometers can be used for MSI if the system allows ion signal acquisition from a number of discretely selected points and if the spatial positions of the specimen can be recorded.

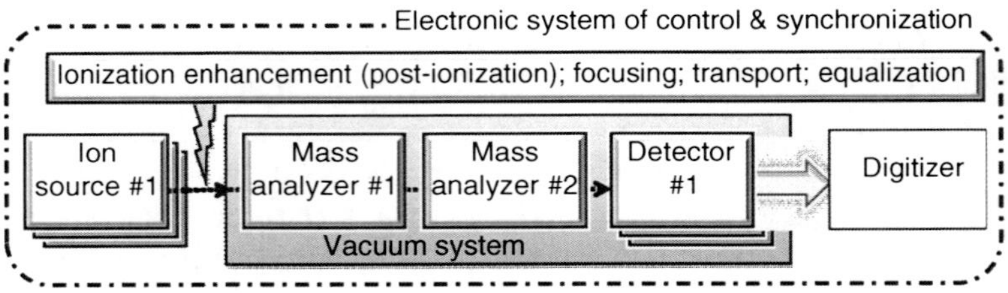

Fig. 2.2. Principal schematic of a mass spectrometer with atmospheric pressure ionization (API). In many mass spectrometers, the ion source operates under a vacuum, while in API, the ionization occurs under ambient/atmospheric pressure.

2.2.1. Ion Sources

Perhaps no other part of the measurement process affects the resulting data more significantly than the ionization step. Unless originally in the gas phase, analytes need to be vaporized from a solid or liquid phase, ionized, and transferred into the vacuum system of the mass analyzer. This vaporization process, often called desorption (sputtering), can be fairly energetic and can degrade/transform the analytes before they are characterized.

Vaporization can be achieved by a variety of techniques, including heating the samples, exposing them to a high electric field or laser irradiation, and/or via bombardment with fast atoms or atomic or molecular ions. Ionization of analytes may occur before, during, and after desorption. Therefore, the term desorption/ionization is commonly used to describe the entire event. Obviously, only desorption is required to produce gaseous neutrals, which can be ionized later using several post-ionization processes. Depending on the characteristics of the sample and desorption/ionization processes, different sets of neutral elements, smaller molecular fragments, or intact molecules, as well as their ions, are generated.

The resulting ions have either a positive or negative charge and so the polarity of the electric field applied at the ion source can determine which ions are injected into the mass analyzer. Importantly, some analytes are more likely to form positive ions and so are easier to characterize in the positive ion mode; others that form negative ions are more easily characterized in the negative ion mode. Moreover, the same sample analysis using different desorption/ionization approaches, even when coupled to the same mass analyzer, may yield different results.

Selecting the appropriate desorption/ionization approach is critical. That decision can result in analyte classes of a specific *m/z* range being either efficiently or poorly ionized, and it can often govern the spatial resolution of the MSI method, needed sample preparation steps, throughput and sensitivity of the analysis. Thus, many of the protocol chapters in this volume describe specific ionization approaches and their corresponding sample preparation steps in great detail.

Four of the most commonly used desorption/ionization methods for MSI are secondary ion mass spectrometry (SIMS), desorption electrospray ionization (DESI), matrix-assisted laser desorption/ionization (MALDI), and laser ablation (LA) with post-ionization. Other desorption/ionization approaches such as laser desorption/ionization (LDI) (*see* **Chapter 9**, (12)), desorption/ionization on silicon (DIOS) (13), electrospray ionization (ESI) (14), and nanostructure-initiator mass spectrometry (NIMS) (15, 16) also have great potential in MSI. Importantly, many mass spectrometers equipped with a MALDI ion source can be used with related ionization processes such as LDI, DIOS, LA, and laser-NIMS. Frequently, a specific ion source arrangement is optimized for a specific mass analyzer; for example, MALDI is often interfaced to a time-of-flight (TOF) mass analyzer (described below) although it can also be used with ICR-based instruments.

Although many instruments are designed for a specific ion source, several can integrate a range of sources. For instance, greater flexibility can be achieved with instruments that allow analysis of ions generated at atmospheric (ambient) pressure (AP), where the ion source does not require the vacuum conditions typically needed for the operation of most mass analyzers. AP sources enable investigation of samples containing volatile compounds that are incompatible with ion sources operating at vacuum. Some AP desorption/ionization systems provide interchangeable ion sources, including MALDI, ESI, and DESI, making these systems more flexible; however, this flexibility may be associated with trade-offs in performance specifications such as sensitivity. These modular ion sources are available from a variety of vendors, as well as manufactured according to researcher specifications (*see* **Chapter 9** for an example of a laboratory-built ion source). In the following, the principles of ion generation using different ion sources are briefly described. Only the major ionization modes used for MSI are highlighted here; other sources frequently applied to mass spectrometry are not common in imaging at this time.

2.2.1.1. SIMS

In SIMS, a beam of fast-moving ions is aimed at the sample (**Fig. 2.3**). On impact, these ions eject (or sputter) a variety of particles from the sample surface, including secondary ions. The bombardment process results in formation of gas phase neutrals, ions of both polarities, electrons, and photons. A number of primary ions are used in SIMS, including lightweight monoatomic ions such as Au^+, Ar^+, Xe^+, O^-, O^{2+}, Ga^+, and Cs^+. Larger polyatomic cluster ion beams, such as ${Au_n}^+$, ${Bi_n}^+$, ${C_{60}}^+$, and ${SF_5}^+$, are also employed; several of these result in an improved mass range of detection and sensitivity for heavier analytes. This increase in detection mass range is related to the characteristics of larger primary ions, which create higher ion currents and higher energy ion beams. Laser post-ionization has become an important tool for increased ion production with semisoft desorption by the

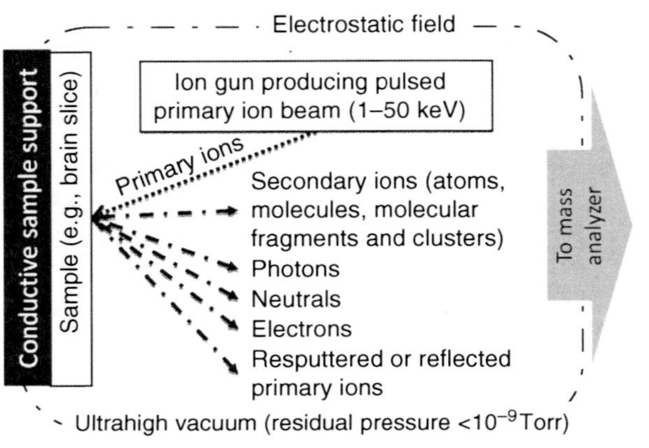

Fig. 2.3. Schematic of SIMS ion source operation.

cluster C_{60}^+ ion beam (17). Depending on the type of ion gun, the energy of the projectiles is in the keV–MeV range, which is sufficient to form atomic and molecular ions with *m/z* up to 2,000 in experiments with biological specimens. When working with biological tissues, molecular fragments, such as histidine-related *m/z* 110, phospholipid-related *m/z* 184, are often detected. Primary atomic ion beams can be focused to the nanometer diameter. As an example, 40-nm lateral resolution has been achieved with a Ga ion gun (18). Often, the larger cluster ion beams cannot be focused as well as their smaller ion counterparts and therefore, the probed sample spot can exceed 1 μm^2. A defocused ion beam is useful in MSI, with SIMS instruments capable of operating in the microscope mode. In this mode, the entire investigated area is probed with primary ions; secondary ions are transferred to the detector with their relative position preserved, resulting in the creation of a stigmatic 2D image. Much more common is the microprobe mode in which ion signals are collected from individual spots on an area of interest and the corresponding image recreated pixel by pixel.

Which primary ion flux should be considered for a particular MSI experiment? Investigation of biological specimens by SIMS is typically carried out in the static regime, which is when less than 1% of the sample surface, and about a nanometer of sample depth, is interrogated/modified by ion bombardment. It corresponds to a primary ion dose that does not exceed 10^{13} ions/cm^2. Higher-dose bombardment, or the dynamic mode, leads to greater sample alteration due to analyte layer mixing and more molecule fragmentation. The dynamic mode is useful for investigating elemental composition and depth profiling. Good results in sample depth profiling were recently achieved with large cluster primary ions (19). SIMS sputtering/ionization performance can be improved by coating the sample surface with metals and organic substances (*see* **Chapter 11**).

Mass spectrometers equipped with SIMS ion sources are broadly used in MSI of element, metabolite, and pharmaceutical localization (*see* **Chapters 4, 6, 11, 15**, and **16**), as well as quantitative measurements of drug localization in normal and pathology-affected tissues (*see* **Chapter 6**). SIMS imaging is very effective in imaging metabolite distribution in biological systems. Such investigation requires significant attention to preserving analyte localization. Therefore, sample preparation of hydrated specimens requires special consideration and is described in the above-mentioned chapters.

2.2.1.2. Laser Desorption Approaches

Several desorption/ionization approaches use light to vaporize and ionize analytes from liquid and solid samples. Ultraviolet (UV) and infrared (IR) lasers (e.g., from the 193-nm excimer to the 10.6-μm TEA-CO_2 laser) have been employed in laser-based ion sources. The appropriate approach is selected based on sample and/or analyte characteristics, such as mass and ionization properties; options include LA, LDI, laser-NIMS, DIOS, or MALDI.

The LA approach is popular in elemental analyses of specimens where the laser is set to produce light with a flux sufficient enough to completely vaporize a sample region (*see* **Chapter 3** and (10)). However, output of analyte ions can be low and post-ionization methods such as inductively coupled plasma (ICP) are needed.

LDI utilizes lower intensity (I) laser pulses, $I \leq 10$ MW/cm^2; metabolites and relatively large intact molecules, like peptides and some proteins, can be desorbed and ionized. An increase in energy leads to more molecules and atoms being desorbed into the gas phase, but results in a higher level of molecule fragmentation and loss of chemical information. IR LDI with electrospray post-ionization has proven to be a powerful approach to determine 2D and 3D localization of metabolites in live plant leaves (*see* **Chapter 9**). Switching to a shorter wavelength UV laser in the LDI ion source helps to improve imaging lateral resolution to 10 μm and has been used to investigate distribution of UV-absorbing metabolites in plant tissues (12). Interestingly, covering the sample with colloid silver and interrogating with LDI result in production of intact ions of metabolites that have bound silver atoms.

Since its introduction in the early 20th century by Sir Joseph John Thomson, mass spectrometry has been mostly applied to analyses of smaller analytes, with fewer successes in peptide and protein research. The limiting factor had been that it is difficult to vaporize and ionize intact proteins. However, this completely changed about 20 years ago with the invention of soft laser desorption by Koichi Tanaka (20), matrix-assisted laser desorption/ionization by Franz Hillenkamp and Michael Karas (21), as well as a critical demonstration of the applicability of electrospray ionization to analyses of large molecules by Bennett Fenn (22) (as described below); these ionization approaches opened the door to widespread use of MS to measure molecules ranging from low molecular weight metabolites to megadalton proteins and even to small structures such as viruses. Recognizing the impact of these achievements, Tanaka and Fenn were awarded the Nobel Prize in chemistry in 2002.

MALDI is widely employed in MSI. LDI and LA experiments have demonstrated that the surrounding analyte environment is critical for the desorption process. Therefore, mixing of analytes with a light-absorbing matrix was shown to be beneficial for the process (**Fig. 2.4**) (21). Upon laser irradiation, the matrix "explodes," carrying intact analytes, some of which are ionized. Importantly, for successful desorption/ionization, analyte particles should be in close proximity to the matrix/buffer molecules

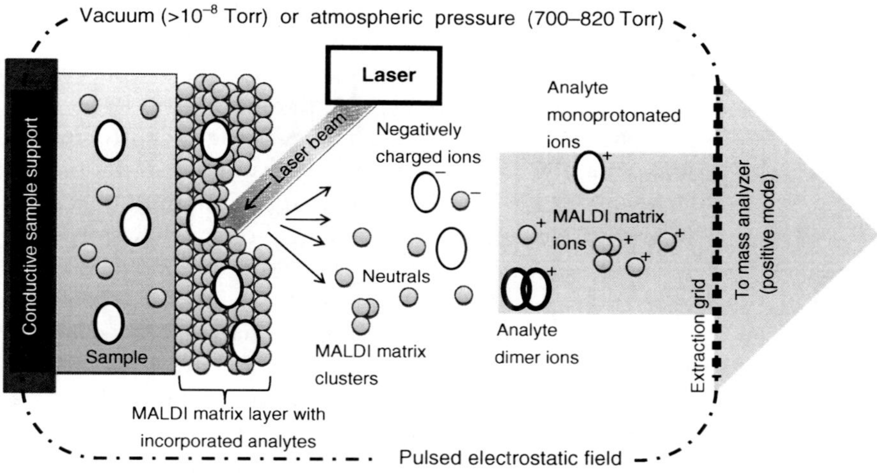

Fig. 2.4. Schematic of ion generation in the MALDI ion source.

and be spatially separated from each other, thus making the relative concentration of analyte-to-matrix ratio a critical factor to consider. More matrix is needed for analysis of the same number of larger molecules as compared to smaller ones. Application of matrix to the sample surface is a critical step in MSI because it may contribute to unwanted analyte spatial redistribution.

Which matrices should be used? For MSI, one often uses solid or liquid low molecular mass organic matrices such as 2,5-dihydroxybenzoic acid (DHB) and alpha-cyano cinnamic acid (CHCA) or their mixtures (23, 24). A variety of MALDI matrices have been introduced over the years. It is interesting that different matrices have some specificity in promoting desorption/ionization of particular types of analytes. For example, 9-aminoacridine (9-AA) is preferable for working with metabolites (25–27). DHB is often used for desorption/ionization of peptides and proteins, and 3-hydroxypicolinic acid (3-HPA) makes possible oligonucleotide analysis (28). For each MALDI matrix, analyte type and sample type, the polarity (positive or negative) of the MALDI ion source must be determined experimentally in order to achieve the best results. Consider for instance that metabolite analysis with 9-AA achieves better results in negative ion mode, while protein analysis with DHB performs better in positive ion mode. Mixing of the matrix solution with a solution containing analytes, or applying matrix solution onto a sample and allowing the analytes to mix with the matrix, precedes matrix analyte co-crystallization and interrogation with the laser. Solid matrix crystals can be also applied to dry or semidry samples with successful outcomes. However, relative spatial localization of analyte and matrix molecules is not always clear. Interestingly, DHB matrix solution not only works well for analyte extraction but can also serve as a preservation media. The antioxidant properties of DHB and the acidic environment of the media likely combine to achieve biologically active peptide preservation during several years of storage in the solution (29).

The MALDI process produces predominantly singly charged ions, thereby facilitating the processing and understanding of resulting data. Neutrals, singly charged and multiply charged ions, ions with elemental adducts, and those with water loss are also formed and may represent the dominant signal for a particular analyte. The ramification of producing singly charged ions is important in the discussion of mass analyzers that follows. One of the exemplary facets of MALDI is that it produces some of the best detection limits of any ionization method. Moreover, it performs well across a broad mass range in a wide number of analyte and sample types, including dried droplets on a metal target and tissue samples on a glass slide.

The diameter of the laser beam used to probe the sample surface typically determines the effective spatial resolution of a measurement performed in microprobe mode. Obviously, the laser beam diameter can be reduced by focusing the beam to smaller dimensions. However, as the laser beam diameter is reduced, it illuminates a smaller area, fewer molecules of each analyte are present within the probe beam, and so fewer molecules are ionized at each location. Therefore, smallest diameter beams are rarely practical because the amount of analyte that can be desorbed and ionized from a smaller sample area is not sufficient for detection and high-accuracy mass measurement. Consequently, the laser probe diameter for the analyses of proteins and peptides usually is larger than 10 μm.

2.2.1.3. Desorption Electrospray Ionization Mass Spectrometry

DESI, a relatively new ionization method that allows surfaces to be characterized under ambient conditions (30), offers a number of interesting characteristics for imaging. A dynamic liquid buffer interface is created on top of the sample by bombarding the sample surface with charged microdroplets, gas phase molecules, and charged clusters of solvents (e.g., a methanol/water/acetic acid mixture) (**Fig. 2.5**). This desorbs and ionizes analytes via several means, including electric field-related mechanisms. The analytes enter the electrostatic field formed between the spray emitter and the mass spectrometer inlet. More gaseous ions are generated at this stage by desorption/ionization, similar to what takes place during ESI; however, the precise mechanisms are still being determined (31, 32). DESI ion sources operate in either negative or positive ion modes. DESI produces singly charged ions as well as series of multiply charged ions for a single analyte. The larger the analyte, the more complex the multiply charged ion series will be. Even though the presence of an analyte in multiple charge states complicates qualitative and quantitative data

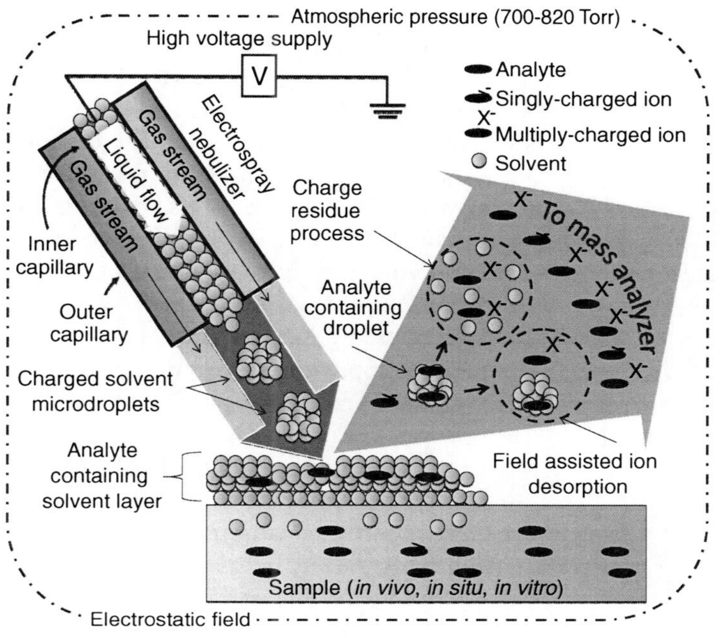

Fig. 2.5. Schematic of the desorption electrospray ionization (DESI) process (see text for details).

analysis, it often assists with fragmentation of analyte molecules, as well as permits the detection of analytes over an extended mass range. Perhaps most importantly, production of gaseous ions with DESI often does not require sample preparation and can be performed on almost any surface.

2.2.2. Mass Analyzers

2.2.2.1. General Principles

After the ions are created, the next step is to transfer them to the mass analyzer. Depending on ionization mechanisms, sampling requirements, and detection sensitivity and selectivity requisites, the injection of analyte ions into the mass analyzer can be done immediately after their desorption/ionization/post-ionization or after a delay. A delay is implemented for different purposes, including delivery of ions from the ion source to a distantly located mass analyzer (DESI), focusing of ions, or equalization of the energies of identical ions (MALDI). The reasons for some of these arrangements are discussed under the specific mass analyzers described below. After selecting the appropriate ionization source for an experiment, the second most important decision is choosing the right mass analyzer. Playing a key role in the separation of generated ion source ions according to their mass, charge, and, in some cases, shape, the mass analyzer also guides ions toward the detector or away from it. To be individually recognized, packets of ions of the same mass and charge are separated in space/time from ions of different mass and charge.

Two ions with the same mass-to-charge ratio cannot be distinguished by the mass analyzer, as would be the case with stereoisomers. However, these molecules can often be distinguished by breaking the ions apart and investigating the ion fragments with another mass analyzer. To accomplish this, mass spectrometers can be configured to store, filter, and fragment ions. Because each of these functions may require additional hardware, contemporary multiplexed mass spectrometer systems are often hybrids and include more than one mass analyzer, capable of multistage operation. This multiplex approach is generally referred to as tandem mass spectrometry, or MS/MS, to describe the two stages of analysis. Fragmentation of ions is important for analyte identification. Due to the existence of a variety of isobaric molecules, having knowledge of the exact mass is not enough to identify an analyte. Tandem MS improves identification via fragmentation because the tandem mass spectrum may provide a unique profile of the product ions, representing a molecular fingerprint. In fact, tandem MS of resulting fragments is also possible; in this case, the term $MS^3$is used. This concept can be carried out even further to create "MS to the *n*" (MS^n), applied when *n* stages of fragmentation/separation are involved in the analysis of a single substance. Although MS^{12} has been performed (33), MS^3 is sufficient for identification of some structurally similar lipid species (34). Historically, the names of many mass analyzers reflect the mechanisms of ion manipulation, details of hardware engineering, or principles of analyte mass determination. Multiplexed mass analyzers have led to the need to abbreviate the lengthy names of these MS systems. As a result, the lexicon of mass spectrometrists has become filled with QTOF, QqQ, oTOF, BEBE, QQQHQCQ, and others – acronyms which will be explained later in the text.

As mentioned above, MS uses electric and/or magnetic fields to determine the accurate mass of ionized atoms and molecules. The force developed by the fields on charged particles is described by Lorentz force law: $F = q\,(E + v \times B)$, where F is the force applied to the ion, q is the charge of the ion, E is the electric field, and $v \times B$ is the vector cross-product of the ion velocity and applied magnetic field. Change in motion of the charged particle under influence of Lorentz force follows Newton's second law: $F = ma$ where m is the mass of the charged particle and a is the acceleration. Combining these two equations gives an expression for the motion of charged particles under the influence of the fields: $(m/q)a = (E + v \times B)$.

It is evident from this equation that either mass-to-charge is measured or knowledge of the charge is required, in order to determine the mass using magnetic and/or electric fields. The mass spectra represent the distribution of signals of different mass-to-charge ratios, or m/z. The m/z is a dimensionless parameter where m is the mass of the particle and z represents the number of charges; z is proportional to the charge of a single proton (e; $z = q/e$). Therefore, separation and detection of ions in the mass spectrometer, as well as their representation in the resulting mass spectra, are done in order of their mass/charge ratio. Signals of heavier molecules with a larger charge may occur on the x-coordinate (m/z coordinate) of a mass spectrum earlier than the signal of a lighter but less-charged one. Obviously, if all analyzed ions were singly charged, this would not be important. For example, MALDI produces predominantly singly charged ions. However, multiply charged ions are formed by DESI (and at times with MALDI). Additional steps are needed to take into account the charge on a particle; this deconvolution process is typically done automatically for most instruments. However, for lower quality measurements, manual manipulation may be necessary.

One way that mass analyzers can be categorized is by the way they manipulate ions – continuous or pulsed. Continuous mass analyzers maintain a constant flow of ions into the analyzer and are well suited for interface to continuous ion sources such as DESI. MALDI uses a pulsed laser for desorption/ionization, making it compatible with pulsed mass analyzers. Several pulsed mass analyzers, such as ion cyclotron resonance, time-of-flight and ion traps, can accommodate continuous sources and accumulate the ions and introduce them in pulses. The accumulation/storage time depends on system capabilities and can range from microseconds to seconds. Magnetic sector (B), electric sector (E), and quadrupole (Q) mass filters are continuous mass analyzers, while time-of-flight (TOF), quadrupole ion trap (QIT), linear ion trap, Orbitrap, and ion cyclotron resonance (ICR) are discontinuous or pulsed analyzers. While several of the standard configurations have been mentioned, other combinations are possible; with effort, most ion sources can be linked to most mass analyzers.

Obviously, an ion needs to last as long as its analysis takes in the mass analyzer. Ion lifetime is related to the quality of the vacuum system, the requirements for which vary with each mass analyzer. The length of ion flight path and strength of the ion-guiding fields, as well as the measurement accuracy and resolution needed, combine to govern the quality of vacuum required. A longer flight path may need a higher vacuum so as to reduce the number of unwanted collisions between analyte ions and gaseous environmental particles. These collisions result in a change in ion motion trajectory and/or velocity and consequently, distortion of signal or even ion recombination and loss of signal. Although such effects can be negative, they can also be beneficial, depending on whether the ions need to be kinetically cooled or to achieve controlled fragmentation. Typically, fragmentation is performed at selected locations in the instrument by adding a collision gas into the ion flight path collision cell or directly into the mass analyzer.

Resolution is one of the primary figures of merit provided by a mass analyzer and is the parameter that often governs the purchase price of a system (**Fig. 2.6**). The higher the resolution, the better the instrument will perform when assigning chemical structures to specific signals. Another important characteristic of MS measurement is mass accuracy, which represents the closeness of the measured mass to the true mass. Mass resolution and mass accuracy for the same ions are predominantly determined by characteristics of the mass analyzers and ion detectors.

Resolution $R = \frac{M}{\Delta M_{FWHM}}$

Resolving power $R_P = \frac{\Delta M_{5\%}}{M}$

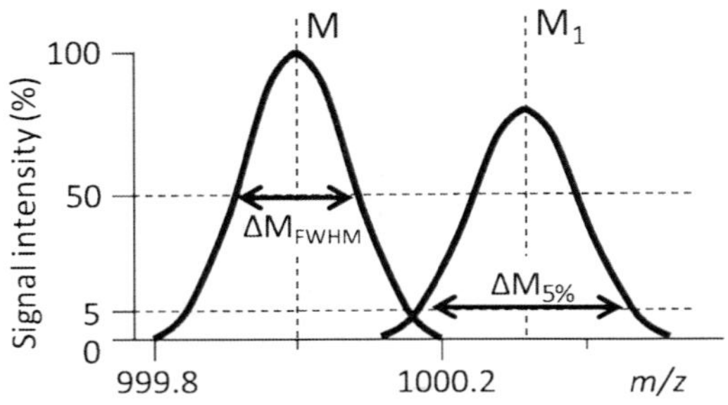

True mass of singly-charged ion = 999.97
Measured mass of singly-charged ion (M) = 1000.00
Difference (accuracy) = 0.03
Error = (0.03/999.97) x 10^6 = ~30 ppm

Peak width at 50% intensity (ΔM_{FWHM}) = 0.2
Resolution (R_{FWHM}) = 1000.00/0.2 = 5000
Resolving power ($R_{P\,5\%}$) = 0.38/1000 = 0.00038

Fig. 2.6. Figures of merit for defining mass analyzers: accuracy, resolution (R), and resolving power (R_P). These parameters are used to compare the performance of different MS approaches. An idealized mass spectrum containing two peaks with *m/z* values M and M_1 shown in the center of the figure. The resolution represents how close the *m/z* of two peaks can be before they overlap and cannot be distinguished. Resolving power is the inverse and is a measure of the ability of a mass spectrometer to separate different ions. Signal processing, such as filtering and baseline correction, influences final data sets and hence, the resolution value calculated from raw data will be different from that from processed data. ΔM can be measured at other signal intensity levels, e.g., 10%. ΔM_{FWHM} represents the full width of the peak at half its maximum (FWHM) height; ppm = parts per million. Importantly, there are a variety of resolution and resolving power definitions which may even contradict each other (http://goldbook.iupac.org, (7, 9)).

High mass resolution is important for detection of isotopic distributions (**Fig. 2.7a**). A majority of polyatomic compounds in biological specimens are composed of different isotopes of the same combination of atoms. For a singly charged molecule, a set of signals (peaks) differing by 1 *m/z* unit will appear in the mass spectra. The first peak in the distribution corresponds to a

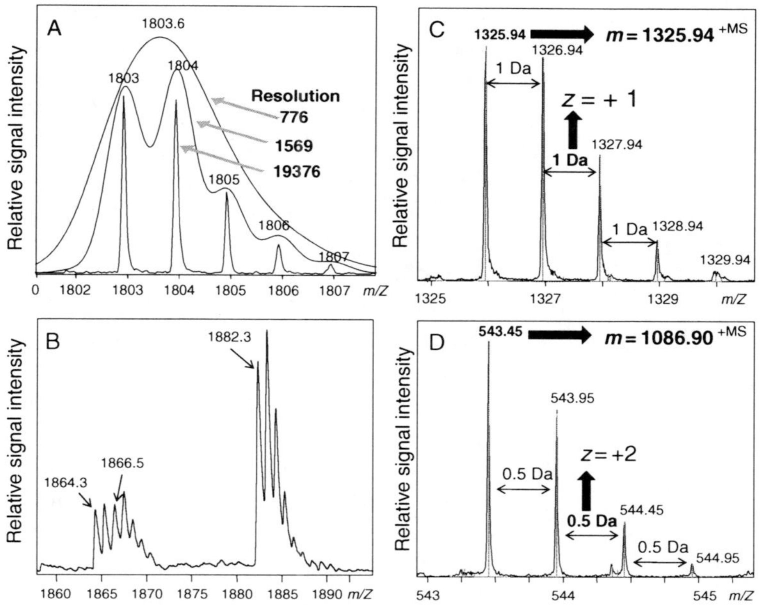

Fig. 2.7. High-resolution mass spectrometry enables determination of isotopic distributions. (**a**) Overlaid mass spectra of a compound with a monoisotopic mass of 1,802 Da (monoprotonated [M+H] = *m/z* 1,803) and an average mass of 1,802.6. Differing resolutions of detection are indicated. (**b**) Mass spectrum containing two different *m/z* ions and one putative ion at *m/z* 1,866.5. (**c**) Mass spectrum of the isotopic distribution of a singly charged molecular ion. (**d**) Mass spectrum of the isotopic distribution of a doubly charged molecular ion.

compound consisting of the lightest atomic isotopes, such as ^{12}C, ^{14}N, ^{16}O, and ^{1}H. For example, the isotope ^{12}C has a 98.9% abundance while the abundance of ^{13}C is only 1.1%. The mass of the molecule having the lightest atomic isotopes is called the monoisotopic mass. Of course, for multiply charged analytes, the spacing of these isotopic signals depends on the charge (e.g., for a doubly charged analyte, the spacing will be 0.5 Da instead of 1 Da). Knowledge of the isotopic distribution aids in analyte characterization. **Figure 2.7b** shows a mass spectrum with two unambiguously detected ions of *m/z* 1,864.3 and *m/z* 1,882.3, where the first ion is 18 units lighter, in good accordance with a loss of water molecule. Analyses of the peak pattern distribution in the signal series starting with *m/z* 1,864.3 indicate that a compound with putative *m/z* 1,866.5 exists in the sample. Unusual patterns of isotopic distribution also may occur due to poor instrument performance and/or incorrect instrument settings.

Determination of mass for multiply charged ions is possible by analysis of the isotopic pattern distribution, where spacing between peaks is related to ion charge. **Figure 2.7c** shows a mass spectrum with 1 *m/z* spacing, indicative of a singly charged ion, while **Fig. 2.7d** presents a pattern with 0.5 *m/z* spacing, corresponding to a doubly charged ion. The number of charges on the ion depends on many parameters, including the ionization approach used and the size of the particle. Larger proteins ionized with ESI may routinely have a charge of +30. Obviously, the mass analyzer requires exceptional resolving power to determine the isotopic distribution in this case, as the peaks are separated by 0.033 Da. Large ions with more charges are easier to manipulate by electric and/or magnetic fields, as well as fragment for sequence determination. However, because different molecules of the same compound may have different charge states, complex charge-state distributions create more complex mass spectra. This problem can be solved by forcing production of maximally charged ions with addition of glycerol or *m*-nitrobenzyl alcohol to the electrospray solution (35). Currently, the phenomenon of multiply charged ions exists mostly for DESI-MSI. MALDI, SIMS, and LA desorption/ionization approaches produce predominantly singly charged ions. Different principles are used to achieve high resolution, high throughput, and measurement accuracy among available mass analyzers. The following sections contain brief descriptions of the mass analyzers used in MSI.

2.2.2.2. Time-of-Flight (TOF) Instruments

The mechanism of operation of a TOF mass analyzer is fairly straightforward (**Fig. 2.8**). In what is known as the linear mode,

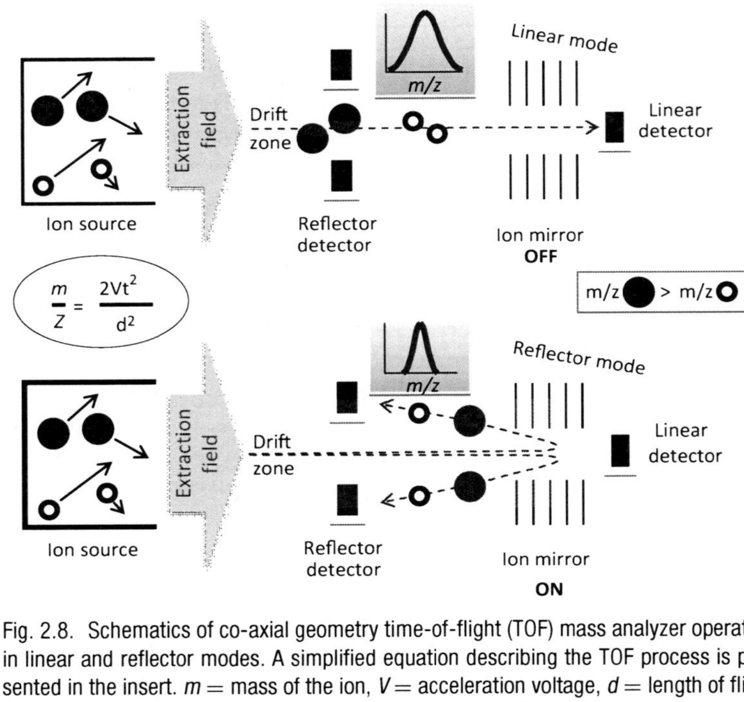

Fig. 2.8. Schematics of co-axial geometry time-of-flight (TOF) mass analyzer operation in linear and reflector modes. A simplified equation describing the TOF process is presented in the insert. m = mass of the ion, V = acceleration voltage, d = length of flight path, t = time from the moment of ion acceleration to the detection event.

ions are extracted from the ion source and unidirectionally accelerated by short pulses of electrostatic field, entering and moving in a drift space containing no field; all ions are accelerated with the same kinetic energy. Thus, lighter ions of the same charge will move faster than heavier ones and, therefore, come to the detector earlier; obviously, separation takes place in space and detection in time. One of the drawbacks of the linear mode is that all the ions initially do not have the exact same position and velocity, so that there is a spread in ion arrival times at the detector. This spatial distribution leads to formation of broad, lower amplitude signals at the detector, resulting in reduced resolution and at times lower sensitivity of detection.

There are two approaches used in TOF mass analyzers to improve resolution. One is delayed extraction (delayed injection into the mass analyzer, also called pulsed ion extraction) of the ions formed in the MALDI ion source. This delay reduces the energy spread of the same *m/z* ions. Delayed extraction parameters can be adjusted by the operator, with a longer delay time needed for improved detection of larger molecules. Another approach utilizes reflecting ion optics (reflector or reflectron mode) such as ion mirrors – a set of evenly spaced electrodes encompassing space on the ion path (**Fig. 2.8**). A single linear electric field with higher potential energy than the source potential is formed around each electrode. Ions are flying in these fields and are reflected (repulsed). As a result, resolution improves due to two factors. First, there is an increase in ion path length and thus, a greater distance between packets of ions; second, there is a reduction in the spread of kinetic energies of different particles of the same *m/z*. More energetic ions will travel longer paths in the field space than lower kinetic energy ions; therefore, they will be focused as they leave the ion mirror area, arriving at the detector in a more temporally compact packet. Both approaches are typically used when metabolites, peptides, and small- and medium-sized proteins are analyzed. However, the linear mode of operation with delayed extraction remains the preferred option for analysis of large molecules and molecular aggregates.

TOF analyzers are common in MSI applications because of their speed of operation and wide *m/z* range. They allow analysis of large singly charged molecular ions produced by MALDI (*see* **Chapters 7–11, 14–20, 22–24, 26**, and **27**).

2.2.2.3. Sector Instruments

Magnetic sector and electrostatic sector mass analyzers are well suited for operation with continuous ion sources; the trajectories of moving ions are curved by forces developed by the electric or magnetic fields (**Fig. 2.9**). The extent of this curvature depends on an ion's *m/z*. Sector analyzers can be used to monitor a single ion with high resolution. A narrow slit is installed between the detector and the ion analyzer; the position of the slit determines

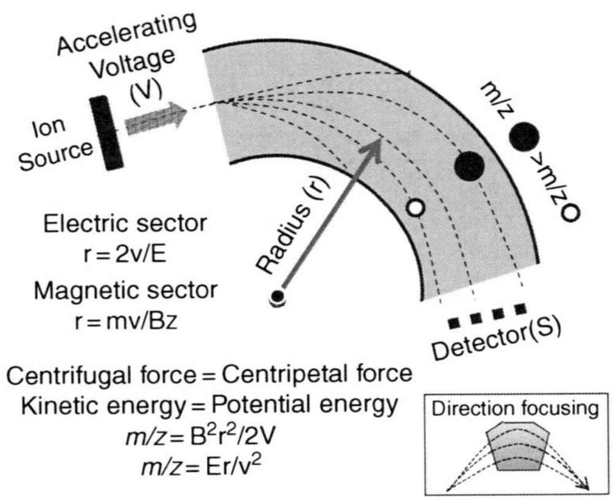

Fig. 2.9. General schematic of a sector mass analyzer. Ions extracted from the ion source are accelerated by an electrostatic field (accelerating potential, *V*) and enter the sector analyzer with velocity, *v*. Electric (electric flux density, *E*) or magnetic (magnetic flux density, *B*) fields bend the trajectory of the ions into curved paths with radius, *r*. Trajectories of ions with larger *m/z* are affected more than smaller ones. An illustration of the direction-focusing ion beam approach in a magnetic sector mass analyzer is shown in the insert. Due to the dependence of the radius of an ion's trajectory on its kinetic energy (*E*) in the electrostatic sector mass analyzer and on its momentum (mv) in the magnetic sector mass analyzer, the systems are also referred to as ion energy and ion momentum filters.

which ion is detected. A narrower slit improves mass resolution but decreases sensitivity. Mass resolution is also dependent on the cross section of the incoming ion beam, the *m/z* ion kinetic energy spread, and the radius of ion trajectory. Different *m/z* ions can be recorded simultaneously by using multiple detectors (or a detector array). Due to fast ion transmission and low level of interaction between ions in the beam (e.g., minimal space-charge effects leading to ion–ion repulsions), sector analyzers are capable of quantitative measurements. Ions approach the sector analyzer as focused or defocused beams. The latter can be refocused with a direction-focusing approach using a magnetic sector mass analyzer (**Fig. 2.9**, inset). Electrostatic mass analyzers are efficient ion kinetic energy filters whereas magnetic sector analyzers are capable of filtering ions with differing momentum (**Fig. 2.9**). Therefore, hybrid instruments combining these two mass analyzers enable double focusing and may achieve 100,000 resolution. Double-focusing instruments have been employed to image the distribution of elements by LA-ICP MSI (*see* **Chapter 3**). Magnetic sector SIMS instruments (*see* **Chapter 6**), like the CAMECA IMS 7f, are capable of distinguishing such ions as ^{56}Fe and $^{28}Si_2$ and allow direct ion microscopy (stigmatic mode imaging) and scanning microprobe mode imaging.

2.2.2.4. Quadrupole (Q) Instruments

The quadrupole analyzer is compatible with continuous ion sources such as DESI. Although the upper *m/z* range of the quadrupole is not high, neither is DESI well suited for desorbing high molecular weight analytes and so the two approaches work well together. The quadrupole, as its name suggests, consists of four precisely aligned metal hyperbolic or cylindrical rods (**Fig. 2.10**). Superimposed direct current (DC) and oscillating radiofrequency (RF) electric fields are used to create conditions where ions of only a certain *m/z* (typically a 1 *m/z* mass transmission window) will have a stable trajectory inside the device and therefore pass through it. Other ions of lower and higher *m/z* will leave the analyzer prematurely or collide with the rods and skeleton. Therefore, the quadrupole mass analyzer is a form of mass filter. The ion path in the quadrupole starts as circular transforms to complex spiral-like propagation inside the field. Depending on the analyzer's design, stable trajectories for ions of a particular *m/z* can be achieved in an oscillating electric field by setting the appropriate RF frequency and RF and DC voltage amplitudes. Importantly, simultaneous change of DC and RF voltage ampli-

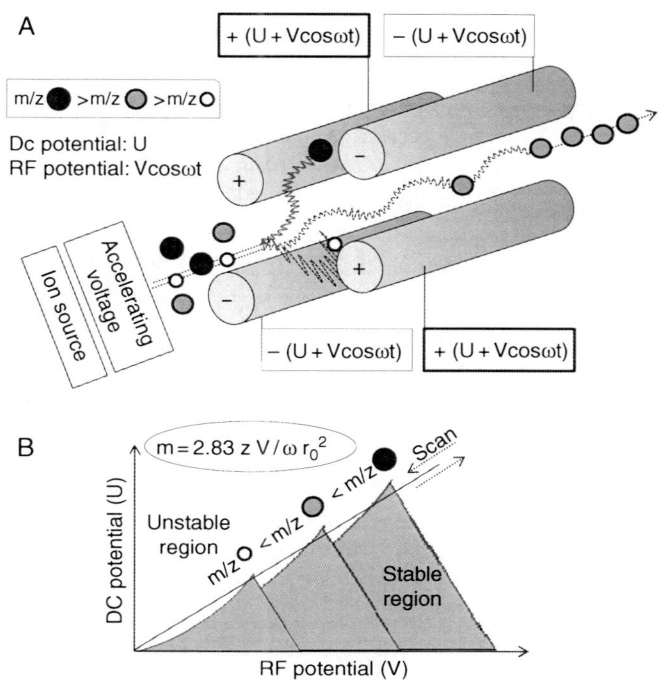

Fig. 2.10. The quadrupole mass analyzer is an ion filtering device that creates an oscillating electric field between four rods. (**a**) Schematic of a quadrupole mass analyzer. $-(U+V\cos\omega t)$ and $+(U+V\cos\omega t)$ are cumulative potential created by superimposed direct current potential (U) and radiofrequency current potential ($V\cos\omega t$). (**b**) The encircled equation describes an area of stability for a particular *m/z* ion trajectory (ω = circular frequency; r_0 = field radius) depicted on the graph.

tudes allows transmission of ions of differing *m/z*. Keeping the RF/DC amplitude ratio constant, while gradually and simultaneously changing the amplitudes, helps to scan/transmit ions in a broad *m/z* range with a high level of selectivity. Simultaneous increase in RF and DC amplitudes is necessary to transmit ions with larger *m/z*. Altering the optimal ratio will increase or decrease the *m/z* window, thus impacting the selectivity of detection and mass resolution.

The quadrupole mass analyzer is relatively small, efficient, and affordable. They are widely used in gas chromatography MS (GC-MS) and liquid chromatography MS (LC-MS). A variety of hybrid instruments exist with multiple quadrupole mass filters installed in the ion path, including systems as complex as QQQHQCQ (Q – quadrupole lens; H – homogeneous magnetic sector; C – cylindrical electric sector). However, triple quads are most common. In this case, the first Q selects a narrow *m/z* range, the second Q fragments it with RF field energy and/or collision gas, and the third Q scans and passes the resulting fragments toward the detector. The application of quadrupole analyzers to MSI typically has been in the role of ion filter or collision cell in hybrid instruments (*see* **Chapters 9** and **23**).

2.2.2.5. Ion Traps

It was recognized that the approach of creating a stable ion path using DC and RF electric fields could also be used for ion storage. Ion traps can accumulate and spatially contain ions, as well as release them for detection. Extended storage times allow opportunities for ion fragmentation, detection, and sorting. How does one prevent the ion cloud from dissipating and colliding with mass analyzer surfaces? Additional fields guide the ions along complex trajectories in a relatively small space. Different approaches for forming these forces distinguish the types of ion traps. Quadrupole ion traps (linear and 3D) utilize electrostatic fields and RF potentials (a Paul ion trap), and the Orbitrap (discussed in detail below) operates with electrostatic fields (a Kingdon trap).

The 3D quadrupole ion trap (QIT) consists of three hyperbolic-shaped electrodes encompassing the region where the electric field is formed by RF potential applied to the central ring electrode, and DC, supplementary RF, or ground potential is established on two end-cap electrodes (**Fig. 2.11a**). Ions of an *m/z* range, determined by amplitude and frequency of the RF fields, are trapped in this region. The trapped ions have stable oscillating trajectories in the QIT until a controlled destabilizing change in the potentials is introduced. Such destabilization can be done for cleanup of QIT space from multiple unwanted ions using broadband waveforms or for sending predetermined *m/z* ions toward the detector by an RF scan. Similar to the Q analyzer, superimposing RF and DC potentials on the ring electrode

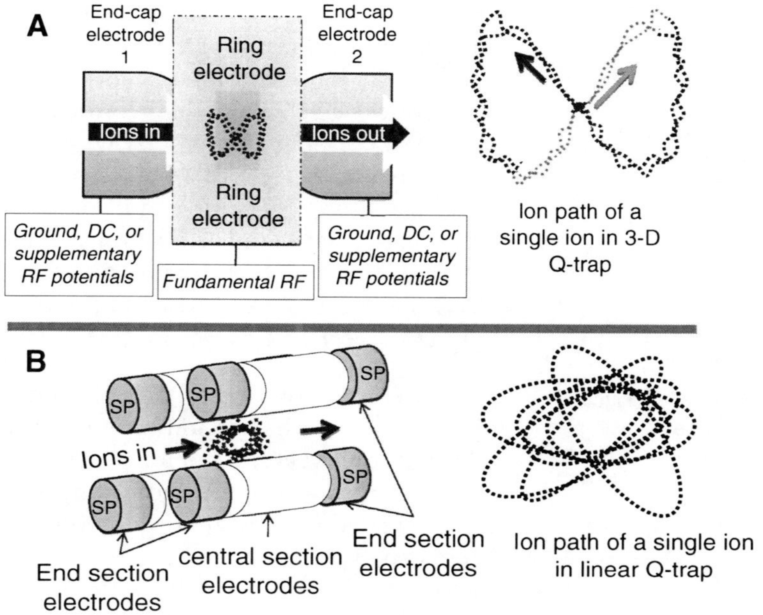

Fig. 2.11. Quadrupole ion traps provide a versatile approach for storage, fragmentation, and selection of ions. (**a**) Schematic of 3D quadrupole ion trap consisting of a ring electrode and two end-cap electrodes. Only one ion ejection pathway is shown. A combination of DC and RF potentials is applied to the electrodes. 1 mtorr of helium is typically added to the mass analyzer. (**b**) Schematic of a linear quadrupole ion trap. During ion trapping, DC and RF ion-guiding potentials are maintained on the central electrodes while the end section electrodes maintain stop potential, SP.

allows storage of single *m/z* ions. This feature is important in MS^n experiments where individual analytes can be selected and fragmented, the fragments selected for detection, and/or a specific fragment selected for another round of fragmentation. Typically, MS^5 can be achieved. Ion capacity is important for MS^n experiments because each subsequent MS stage has less material for analysis than the previous stage. Increased ion storage volume also allows for reduction of space-charge effects that may occur in situations where the concentration of ions is high enough that they start to repel each other. The space-charge effect often determines low mass resolution due to broadening of analyte signals and also for formation of artifacts in mass spectra related to interactions between ions. QIT use for MSI is described in **Chapter 13**.

A variant of the ion trap is the linear version that creates a cylindrical space for the ion cloud, linear quadruple ion traps (**Fig. 2.11b**) provide greater ion storage volume compared to QITs. As a result, more ions can be stored with greater efficiency, allowing detection with high resolution and increased sensitivity for a particular sample. Linear ion traps are called 2D because they utilize a 2D RF field to confine ions radially. This field is developed between four centrally located hyperbolic or cylindrical rods surrounded on both sides by two sets of quadrupole end electrodes, which maintain DC stop potential, thereby preventing ions from leaving the trap axially. Both 2D and 3D quadrupole ion traps can be part of a hybrid set up or serve as the sole mass analyzer. Pulsed ion sources are common for ion traps in MS analyses that require longer operating sequences. However, different instrumental configurations, as well as high speed operation, allow ions traps to be linked with continuous ion sources without loss of sample (for an example in MSI, *see* **Chapter 25**).

2.2.2.6. Ion Cyclotron Resonance (ICR)

Ion cyclotron resonance uses the principle of ions orbiting in an ICR cell, with each ion having a characteristic frequency that depends on its m/z. A combination of magnetic and electric fields allows separation of ions according to their m/z. The magnetic field, created by a magnet which surrounds the ICR cell, is static and may have different geometries (**Fig. 2.12a**). The cell has six electrodes, four of which are used for ion containment and ion path manipulation and two as trapping electrodes. The ions, whether injected or generated inside the cell, are constrained by the magnetic field as well as the electric fields established on the trapping electrodes. Ions continuously orbit while static electric potential restricts their axial movement. At this stage, the ions form a diffuse ion cloud, comprised of a mix of ions of differing m/z. By applying a sequence of RF pulses to the excitation plates, specific ions are resonantly activated and start to move in a well-defined packet while maintaining the cyclotron trajectory. Using sequential RF pulses, the ion packets can be synchronized. These packets of ions are measured as they repeatedly pass near the detection plates. The detection plates are used to measure image current from the ions, where each m/z has a unique cyclotron frequency that depends on the mass and charge of the ions. Higher m/z ions will appear less frequently. The timing of the image currents represents a frequency recording of the ions in the detection cell; the more transits of each ion, the more accurately the frequencies can be determined. The frequencies are converted into a more conventional mass spectrum using a Fourier transform (FT).

ICR is known for achieving the highest mass resolution ($m/\Delta m_x$, quoted as resolving power in some publications), as one example, reaching 8 million for 8.6 kDa bovine ubiquitin (36). ICR resolution increases linearly with the increasing strength of the magnetic field. Therefore, large magnets, up to 25 T, are employed in FT-ICR MS (37). A resolution ($m/\Delta m_{50\%}$) of 200,000 for a 14.5 T instrument at m/z 400 at a 1 Hz measurement rate has been reported (38). ICR mass analyzers became

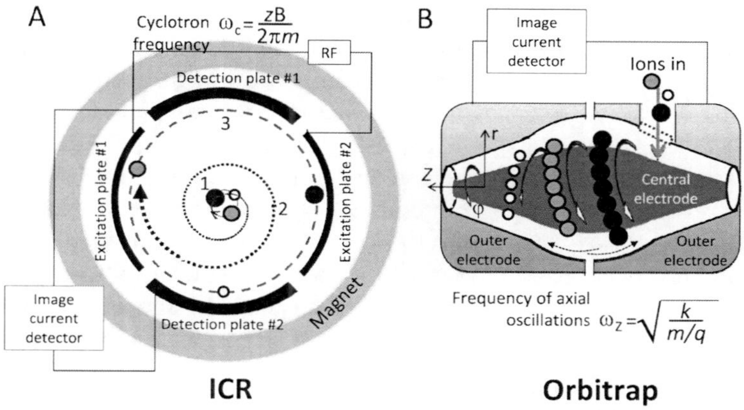

Fig. 2.12. High-resolution mass spectrometric analyses are achieved with ion cyclotron resonance (ICR) and Orbitrap mass analyzers. (**a**) Schematic of an ICR mass analyzer consisting of a magnet surrounding a cylindrical ICR cell formed by three pairs of electrodes (two trapping electrodes are not shown), electronic circuitries for detection of the ion image current, and generation of RF and DC potentials. Three stages of ion separation and detection are marked with numbers: 1 – initial electrostatic trapping (no RF is applied), 2 – cyclotron motion when RF is used to resonantly excite ions of a particular *m/z* and move them to higher orbit, and 3 – detection phase when RF is turned off and detection plates are engaged to sense ion packets. The equation shows the dependence of cyclotron frequency on mass (*m*) and charge (*z*) of an analyte ion moving in the magnetic field of strength, *B*. (**b**) Schematic representation of an Orbitrap mass analyzer. A central electrode is surrounded by one outer electrode, which is divided into two halves by nonconductive space. Although different approaches are implemented for ion detection with the Orbitrap, only one is shown here. A simplified equation describing the frequency of axial oscillations in the mass analyzer demonstrates that this parameter – ω_Z – can be used for determination of mass (*m*) to charge (*q*) ratio. k = field curvature; ϕ = rotation; *Z* and *r* = coordinate axes. Both analyzers are operating at vacuum conditions.

powerful tools in proteomics investigations with the advent of top-down (intact proteins are analyzed) and bottom-up (peptides of enzymatically digested proteins are analyzed) approaches. A large variety of ion fragmentation approaches are available for ICR, although these are less common in MSI. This is due in part to its limited *m/z* range (typically not over ~10 kDa); however, ICR works well with ESI when multiply charged ions are generated so that a much higher mass range can be interrogated. This limitation is much more severe in MSI when using MALDI because it generates mostly singly charged ions.

2.2.2.7. The Orbitrap

Recently, a compact but powerful mass analyzer – the Orbitrap – was introduced (**Fig. 2.12b**) (39). Using a balance between the centripetal influence of an electrostatic field developed on a central electrode and the opposite centrifugal force of ions rapidly injected and moving in the mass analyzer, the Orbitrap produces packets of ions according to their *m/z*. The trajectory of ion motion inside the Orbitrap resembles a complex spiral orbiting alongside the central electrode. The axial component of this motion is dependent on the mass and charge of a particle. Therefore, recording the image current generated by the motion of different *m/z* ion packets produces a complex record of change of current amplitude over time. Again, a Fourier transform allows the conversion into relative signal intensity versus *m/z*. The Orbitrap employs relatively new technology and at its current stage of development is capable of producing a resolution of >100,000 at *m/z* 400 for a 1.5 s acquisition time or 60,000 at a 1 Hz acquisition rate. This capability allows determination of the localization of analytes with similar *m/z* values, in particular using the protocol described in **Chapter 25**.

2.2.3. Detectors

After the ions are separated or processed, they need to be detected. For several mass analyzers, ion detection is integral to mass analyzer operation as is true for the ICR instrument. Here we briefly describe one of the most common ion detectors.

Moving ions produce signals in detectors on impact or by creating electric currents. A number of types of electron multipliers are used in mass spectrometers. As an example, a microchannel electron multiplier plate has a large array of 5–10 μm diameter channels which may occupy an area >10 cm^2. Ions strike the detector's surface inside an individual channel, thereby inducing secondary electron and photon emission. The efficiency of this process depends on the kinetic energy of incoming ions, incident angle of impact, and detector surface properties. Secondary electrons continue to move toward the detector and strike tunnel walls, again inducing more electrons. This effect can multiply the number of ions by more than a million fold. The process continues along the channel until electrons reach a conductor, which transmits this electric current to amplifiers and further signal processing. Detectors operate at vacuum conditions usually below 10^{-5} torr. However, new technologies for generation of curved channels in channel electron multipliers have resulted in an ion detector that is operational at 10^{-2} torr (http://www.detechinc.com/em/quad.htm). Importantly, the ion detectors of some mass spectrometers can be replaced/updated, translating into significant improvements in detection capabilities.

Why is it important to understand how the detector works? Many mass spectrometers have the option of adjusting detector sensitivity. Increasing the sensitivity may help to observe more compounds and aid in the detection of low-abundance analytes. Unfortunately, detector operation at a higher sensitivity setting can increase the noise level and reduce the lifetime of the detector. In addition, if the detector is set at a higher sensitivity and intense signals are present, this can degrade detector performance. High spatial resolution MS imaging requires significant dynamic range and often the detector is optimized for low analyte levels at each probed spot. Thus, detector performance can degrade during the hundreds of thousands of acquisitions that take place when creating a number of larger ion images.

What is next? The signals produced by detectors are digitized, processed, and converted to mass spectra. MS imaging experiments may generate thousands of mass spectra containing information on hundreds of signals acquired from specimens. Obviously, it is not possible to check individual mass spectrum quality. Therefore, data analysis in MSI experiments is mostly done using final ion images. Such analysis may generate false-positive signal detection due to a variety of factors, including issues as shown in **Fig. 2.13**. Different baseline correction, denoising, and intel-

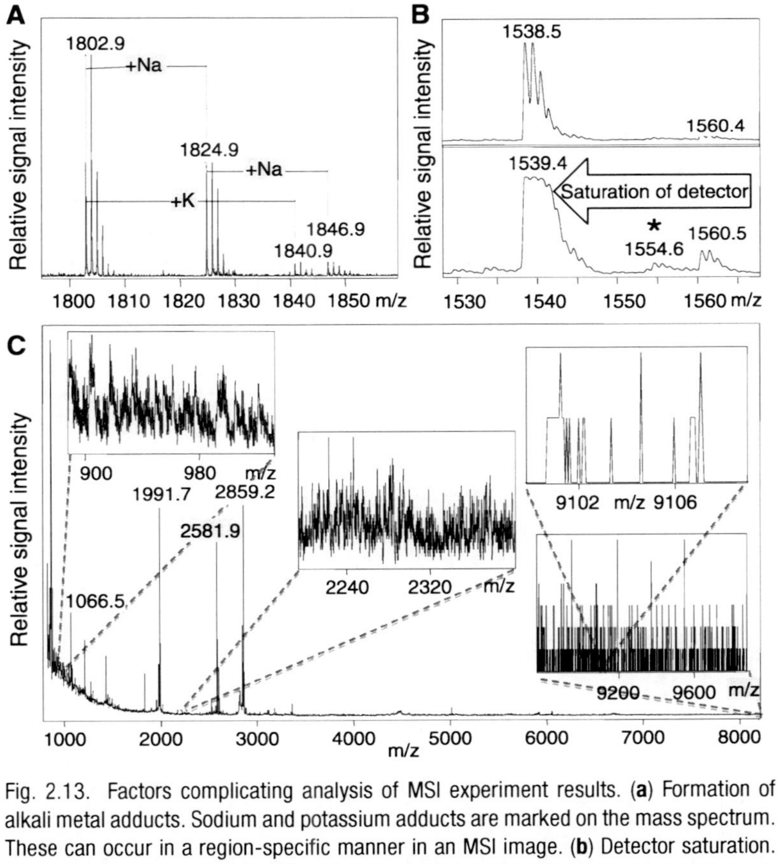

Fig. 2.13. Factors complicating analysis of MSI experiment results. (**a**) Formation of alkali metal adducts. Sodium and potassium adducts are marked on the mass spectrum. These can occur in a region-specific manner in an MSI image. (**b**) Detector saturation. Mass spectra acquired with different laser fluencies (the total energy per unit area). Note lost resolution on the major peak and improved detection of lower intensity peaks as a result of fluency increase (*lower trace*). (**c**) *Curved baseline* shape, chemical, and/or digitization noise may produce false-positive peak detection during the automatic peak picking process. Mass spectrum shows an elevated baseline and high level of chemical noise (*left inset*) in the lower *m/z* range, mixed chemical and digitization noises in the middle range (*central inset*), and only digitization noise in the higher *m/z* region.

ligent peak picking algorithms are implemented for batch mass spectra processing in MSI (40, 41) (*see* **Chapters 1, 22**, and **27**). Nevertheless, a random check of several of the original mass spectra is an important step during data analysis.

2.2.4. Calibration

Mass spectrometers are complex systems and many parameters can be optimized according to experimental requirements. These adjustments, as well as possible drift in the calibration of different components, may lead to measurement of analyte *m/z* with a systematic error. Therefore, calibration of the mass scale is often performed using standards, often more than three; this compensates for these deleterious effects and results in a more accurate *m/z* scale. During automatic or manual calibration, standards with known masses and predictable charge are measured, differences between determined mass and true mass are found, and a set of calibration constants are created. These constants become part of the *m/z* calibration during subsequent measurements performed with the same instrumental settings with unknown analytes.

Internal, external, or mass defect-based calibration can be performed (42). External calibration is done with mixtures of standards measured separately from analyzed unknown samples. In contrast, internal calibration requires the presence of calibrants in the sample. Importantly, if the *m/z* of endogenous analytes are known, they can be used to calibrate the mass spectrometer. This type of calibration is preferable because no increase in complexity of sample occurs and influences from different parameters, such as sample thickness and analyte environment, are accounted for. As an example, trypsin autocleavage molecular ions are used for calibration in proteomics research (43), as is mass defect-based calibration, where peptides of enzymatically cleaved proteins are used for calibration and protein identifications (42, 44). The latter approach is also useful for determination and subsequent elimination of nonpeptide species from peak lists.

Calibrant selection is governed by several requirements, including the nature of targeted analytes, investigated *m/z* range, type of mass spectrometer, and calibration approach. Unique, exogenous calibrants are preferable for internal calibration when endogenous calibrants are not available. Using analytes that can be endogenously present in an investigated sample for calibration is discouraged due to the possibility of ambiguous experimental outcomes, as well as accidental contamination of other samples. High-quality calibration is achieved with calibrants that bracket the mass of the analyte of interest. After their acquisition, individual mass spectra can be recalibrated employing different calibration profiles. However, recalibration of complex MS^n spectra, or an entire set of mass spectra obtained during an MS imaging experiment, can be difficult. Therefore, it is important to verify calibration before performing an imaging experiment using the appropriate standards.

3. Conclusions

Mass spectrometry provides a set of tools with versatile and powerful figures of merit. When used under the control of appropriate operating protocols, many MS-platforms can be used for MSI. The large number of data-points in an image, and the complex matrix of working directly with tissue samples, can place severe performance constraints on an MSI experiment. Nonetheless, the rapid increase in the performance of ion sources, mass analyzers, and detectors has led to a new generation of sensitive, high-resolution, and relatively easy-to-use mass spectrometers. These are now providing unmatched capabilities for chemical imaging. The mature mass spectrometry field provides biologists with a variety of novel imaging modalities that are appropriate for a number of applications. While there is no single MS-based tool that is ideal for all analytes, the assortment of instruments now available allows a broad range of samples to be probed. In the following chapters, specific protocols are described in much greater detail.

Acknowledgments

The authors would like to thank Kevin Tucker for helpful discussion of the chapter and Stephanie Baker for her help with the manuscript preparation. This material is based on work supported by Award No. P30DA018310 from the National Institute on Drug Abuse (NIDA). The content is solely the responsibility of the authors and does not necessarily represent the official views of NIDA or the National Institutes of Health.

References

1. Burlingame, A. L. (2005) Mass spectrometry: modified proteins and glycoconjugates, *Methods in Enzymolology*. Elsevier Academic Press, Amsterdam, Boston, MA.
2. Chance, M. (ed.) (2008) *Mass Spectrometry Analysis for Protein–Protein Interactions and Dynamics*. John Wiley & Sons, Hoboken, NJ.
3. Downard, K. (ed.) (2007) *Mass Spectrometry of Protein Interactions*. Wiley Interscience, Hoboken, NJ.
4. Lipton, M. S., Páya-Tolic, L. (2009) *Mass Spectrometry of Proteins and Peptides: Methods and Protocols*. Humana Press, Springer, distributor, New York, NY, London.
5. Matthiesen, R. (ed.) (2007) *Mass Spectrometry Data Analysis in Proteomics*. Humana Press, Totowa, NJ.
6. Murphy, R. C. (1993) *Mass Spectrometry of Lipids*. Plenum Press, New York, NY.
7. Sparkman, O. D. (2006) *Mass Spec Desk Reference*. 2nd ed. Global View Publishing, Pittsburgh, PA.
8. Wanner, K. T., Höfner, G. (eds.) (2007) *Mass Spectrometry in Medicinal Chemistry:*

Applications in Drug Discovery. Wiley-VCH, John Wiley (distributor), Weinheim, Chichester.
9. McLafferty, F. W., Turecek, F. (1993) *Interpretation of Mass Spectra*. University Science Books, Mill Valley, CA.
10. Becker, J. S. (2007) *Inorganic Mass Spectrometry: Principles and Applications*. John Wiley & Sons, Chichester, Englandm Hoboken, NJ.
11. De Podesta, M. (2001) *Understanding the Properties of Matter*. Taylor & Francis, London, New York, NY.
12. Holscher, D., Shroff, R., Knop, K., Gottschaldt, M., Crecelius, A., Schneider, B., Heckel, D. G., Schubert, U. S., Svatos, A. (2009) Matrix-free UV-laser desorption/ionization (LDI) mass spectrometric imaging on the single-cell level: distribution of secondary metabolites of *Arabidopsis thaliana* and *Hypericum* species. *Plant J*, **60,** 907–918.
13. Liu, Q., Guo, Z., He, L. (2007) Mass spectrometry imaging of small molecules using desorption/ionization on silicon. *Anal Chem*, **79**, 3535–3541.
14. Van Berkel, G. J., Kertesz, V., Koeplinger, K. A., Vavrek, M., Kong, A. N. (2008) Liquid microjunction surface sampling probe electrospray mass spectrometry for detection of drugs and metabolites in thin tissue sections. *J Mass Spectrom*, **43**, 500–508.
15. Northen, T. R., Yanes, O., Northen, M. T., Marrinucci, D., Uritboonthai, W., Apon, J., Golledge, S. L., Nordstrom, A., Siuzdak, G. (2007) Clathrate nanostructures for mass spectrometry. *Nature*, **449**, 1033–1036.
16. Yanes, O., Woo, H. K., Northen, T. R., Oppenheimer, S. R., Shriver, L., Apon, J., Estrada, M. N., Potchoiba, M. J., Steenwyk, R., Manchester, M., Siuzdak, G. (2009) Nanostructure initiator mass spectrometry: tissue imaging and direct biofluid analysis. *Anal Chem*, **81**, 2969–2975.
17. Willingham, D., Kucher, A., Winograd, N. (2008) Molecular depth profiling and imaging using cluster ion beams with femtosecond laser postionization. *Appl Surf Sci*, **255**, 831–833.
18. Sakamoto, T., Koizumi, M., Kawasaki, J., Yamaguchi, J. (2008) Development of a high lateral resolution TOF-SIMS apparatus for single particle analysis. *Appl Surf Sci*, **255**, 1617–1620.
19. Wucher, A., Cheng, J., Zheng, L., Winograd, N. (2009) Three-dimensional depth profiling of molecular structures. *Anal Bioanal Chem*, **393**, 1835–1842.
20. Tanaka, K. (2003) The origin of macromolecule ionization by laser irradiation (Nobel lecture). *Angew Chem Int Ed Engl*, **42**, 3860–3870.
21. Hillenkamp, F., Karas, M. (1990) Mass spectrometry of peptides and proteins by matrix-assisted ultraviolet laser desorption/ionization. *Methods Enzymol*, **193**, 280–295.
22. Fenn, J. (2002) Electrospray ionization mass spectrometry: how it all began. *J Biomol Tech*, **13**, 101–118.
23. Cornett, D. S., Reyzer, M. L., Chaurand, P., Caprioli, R. M. (2007) MALDI imaging mass spectrometry: molecular snapshots of biochemical systems. *Nat Methods*, **4**, 828–833.
24. Kaletas, B. K., van der Wiel, I. M., Stauber, J., Guzel, C., Kros, J. M., Luider, T. M., Heeren, R. M. (2009) Sample preparation issues for tissue imaging by imaging MS. *Proteomics*, **9**, 2622–2633.
25. Seetharaman, V., Royston, G. (2007) Quantitative detection of metabolites using matrix-assisted laser desorption/ionization mass spectrometry with 9-aminoacridine as the matrix. *Rapid Commun Mass Spectrom*, **21**, 2072–2078.
26. Rachal, L. V.-S., David, M. H. (2002) 9-Aminoacridine as a matrix for negative mode matrix-assisted laser desorption/ionization. *Rapid Commun Mass Spectrom*, **16**, 1575–1581.
27. Edwards, J. L., Kennedy, R. T. (2005) Metabolomic analysis of eukaryotic tissue and prokaryotes using negative mode MALDI time-of-flight mass spectrometry. *Anal Chem*, **77**, 2201–2209.
28. Liu, Y., Sun, X., Guo, B. (2003) Matrix-assisted laser desorption/ionization time-of-flight analysis of low-concentration oligonucleotides and mini-sequencing products. *Rapid Commun Mass Spectrom*, **17**, 2354–2360.
29. Romanova, E. V., Rubakhin, S. S., Sweedler, J. V. (2008) One-step sampling, extraction, and storage protocol for peptidomics using dihydroxybenzoic acid. *Anal Chem*, **80**, 3379–3386.
30. Takats, Z., Wiseman, J. M., Gologan, B., Cooks, R. G. (2004) Mass spectrometry sampling under ambient conditions with desorption electrospray ionization. *Science*, **306**, 471–473.
31. Manicke, N. E., Wiseman, J. M., Ifa, D. R., Cooks, R. G. (2008) Desorption electrospray ionization (DESI) mass

spectrometry and tandem mass spectrometry (MS/MS) of phospholipids and sphingolipids: ionization, adduct formation, and fragmentation. *J Am Soc Mass Spectrom*, **19**, 531–543.
32. Wiseman, J. M., Ifa, D. R., Venter, A., Cooks, R. G. (2008) Ambient molecular imaging by desorption electrospray ionization mass spectrometry. *Nat Protoc*, **3**, 517–524.
33. Lundstrom, S. L., D'Alexandri, F. L., Nithipatikom, K., Haeggstrom, J. Z., Wheelock, A. M., Wheelock, C. E. (2009) HPLC/MS/MS-based approaches for detection and quantification of eicosanoids. *Methods Mol Biol*, **579**, 161–187.
34. Houjou, T., Yamatani, K., Nakanishi, H., Imagawa, M., Shimizu, T., Taguchi, R. (2004) Rapid and selective identification of molecular species in phosphatidylcholine and sphingomyelin by conditional neutral loss scanning and MS^3. *Rapid Commun Mass Spectrom*, **18**, 3123–3130.
35. Iavarone, A. T., Jurchen, J. C., Williams, E. R. (2001) Supercharged protein and peptide ions formed by electrospray ionization. *Anal Chem*, **73**, 1455–1460.
36. Shi, S. D., Hendrickson, C. L., Marshall, A. G. (1998) Counting individual sulfur atoms in a protein by ultrahigh-resolution Fourier transform ion cyclotron resonance mass spectrometry: experimental resolution of isotopic fine structure in proteins. *Proc Natl Acad Sci U S A*, **95**, 11532–11537.
37. Shi, S. D. H., Drader, J. J., Hendrickson, C. L., Marshall, A. G. (1999) Fourier transform ion cyclotron resonance mass spectrometry in a high homogeneity 25 tesla resistive magnet. *J Am Soc Mass Spectrom*, **10**, 265–268.
38. Schaub, T. M., Hendrickson, C. L., Horning, S., Quinn, J. P., Senko, M. W., Marshall, A. G. (2008) High-performance mass spectrometry: Fourier transform ion cyclotron resonance at 14.5 tesla. *Anal Chem*, **80**, 3985–3990.
39. Makarov, A., Denisov, E., Kholomeev, A., Balschun, W., Lange, O., Strupat, K., Horning, S. (2006) Performance evaluation of a hybrid linear ion trap/orbitrap mass spectrometer. *Anal Chem*, **78**, 2113–2120.
40. Jardin-Mathe, O., Bonnel, D., Franck, J., Wisztorski, M., Macagno, E., Fournier, I., Salzet, M. (2008) MITICS (MALDI Imaging Team Imaging Computing System): a new open source mass spectrometry imaging software. *J Proteomics*, **71**, 332–345.
41. Broersen, A., van Liere, R., Altelaar, A. F., Heeren, R. M., McDonnell, L. A. (2008) Automated, feature-based image alignment for high-resolution imaging mass spectrometry of large biological samples. *J Am Soc Mass Spectrom*, **19**, 823–832.
42. Hjerno, K., Hojrup, P. (2007) Calibration of matrix-assisted laser desorption/ionization time-of-flight peptide mass fingerprinting spectra. *Methods Mol Biol*, **367**, 49–60.
43. Luo, Q., Nieves, E., Kzhyshkowska, J., Angeletti, R. H. (2006) Endogenous transforming growth factor-beta receptor-mediated Smad signaling complexes analyzed by mass spectrometry. *Mol Cell Proteomics*, **5**, 1245–1260.
44. Wolski, W. E., Farrow, M., Emde, A. K., Lehrach, H., Lalowski, M., Reinert, K. (2006) Analytical model of peptide mass cluster centres with applications. *Proteome Sci*, **4**, 18.

Chapter 3

Imaging of Metals, Metalloids, and Non-metals by Laser Ablation Inductively Coupled Plasma Mass Spectrometry (LA-ICP-MS) in Biological Tissues

J. Sabine Becker and J. Susanne Becker

Abstract

The determination of the localization and distribution of essential and beneficial metals (e.g., Cu, Fe, Zn, Mn, Co, Ti, Al, Ca, K, Na, Cr and others), toxic metals (like Cd, Pb, Hg, U), metalloids (e.g., As, Se, Sb), and non-metals (such as C, S, P, Cl, I) in biological tissues is a challenging task for life science studies. Over the past few years, the development and application of mass spectrometric imaging (MSI) techniques for elements has been rapidly growing in the life sciences in order to investigate the uptake and the transport of both essential and toxic metals in plant and animal sections. Laser ablation inductively coupled plasma mass spectrometry (LA-ICP-MS) is a very sensitive and efficient trace, surface, and isotopic analytical technique for biological samples. LA-ICP-MS is increasingly utilized as an elemental mass spectrometric technique using double-focusing sector field (LA-ICP-SFMS) or quadrupole mass spectrometers (LA-ICP-QMS) to produce images of detailed regionally specific element distributions in thin biological tissue sections. Nowadays, MSI studies focus on brain research for studying neurodegenerative diseases such as Alzheimer's or Parkinson's, stroke, or tumor growth, or for the imaging of cancer biomarkers in tissue sections.

The combination of the mass spectrometry imaging of metals by LA-ICP-MS with proteomics using biomolecular mass spectrometry (such as MALDI-MS or ESI-MS) to identify metal-containing proteins has become an important strategy in the life sciences. Besides the quantitative imaging of metals, non-metals and metalloids in biological tissues, LA-ICP-MS has been utilized for imaging metal-containing proteins in a 2D gel after electrophoretic separation of proteins. Recent progress in applying LA-ICP-MS in life science studies will be reviewed including the imaging of thin slices of biological tissue and applications in proteome analysis in combination with MALDI/ESI-MS to analyze metal-containing proteins.

Key words: Biological tissues, mass spectrometric imaging, laser ablation inductively coupled plasma mass spectrometry, metal distribution, non-metal images, quantification, Se, gel electrophoresis.

S.S. Rubakhin, J.V. Sweedler (eds.), *Mass Spectrometry Imaging*, Methods in Molecular Biology 656,
DOI 10.1007/978-1-60761-746-4_3, © Springer Science+Business Media, LLC 2010

1. Introduction

Inorganic mass spectrometry has had a significant impact on chemical analysis, especially since the adoption of inductively coupled plasma mass spectrometry (ICP-MS) for the determination of element concentrations down to the trace and ultratrace level and for isotope ratio measurements on biological samples due to its very high sensitivity and low detection limits (1–3). The direct analysis of biological samples without time-consuming sample preparation became possible by coupling a laser ablation (LA) device to ICP-MS, thus enabling information on spatial distributions of elements to be obtained (4, 5). LA-ICP-MS is today one of the most important surface mass spectrometric techniques in the life sciences for quantitative studies of metal distribution in biological tissues (human brain, animal, or plant). Compared to alternative methods in biological and clinical research for the visualization of metal distribution in tissues such as specific chemical staining, immunohistochemical staining (tags) techniques, or radiolabels for visualizing and identifying metal and molecular tags, LA-ICP-MS provides multi-element capability; high sensitivity and low detection limits for most metals, metalloids and several non-metals. X-ray spectroscopic techniques for biological tissues (6), scanning electron microscopy with energy-dispersive X-ray analysis (SEM-EDX), proton-induced X-ray emission (PIXE) and autoradiography are often not sensitive enough for trace metal imaging. The application of an X-ray fluorescence nanoprobe using the Synchrotron Radiation Facility allows highly spatially resolved metal images of tissues, but the equipment and measurements are very expensive (7).

Other surface inorganic mass spectrometric techniques, such as secondary ion mass spectrometry (SIMS), have been established not only for the imaging of metals but also for studying the distribution of small molecules in biological systems (8–16). Different applications of SIMS for mass spectrometric imaging are described in detail in **Chapters 4, 6, 11, 15** and **16**. In spite of its advantages, issues occur in SIMS if suitable matrix-matched reference materials are not available. Consequently, quantification of the analytical data is difficult due to inherent matrix effects that can be many orders of magnitude. Compared to SIMS, matrix effects are significantly lower (factor: 2–3) in LA-ICP-MS and the detection limits of imaging LA-ICP-MS on biological tissues observed at the sub-microgram per gram level are in general better (1). In addition, LA-ICP-MS possesses superior features for the quantification of analytical data in quite different systems and can be employed as a powerful and sensitive imaging technique to produce images of detailed regionally specific element distributions in thin tissue sections of different sizes (17–21).

As the most frequently used mass spectrometric imaging technique for biological tissues, matrix-assisted laser desorption/ionization mass spectrometry (MALDI-MS) (22–25) is well established for the imaging of small molecules and large biomolecules up to the m/z range over 100,000 Da within biological systems (8, 22). The application of MALDI-IMS as an MS-based imaging technique has grown rapidly, enhanced by the production of commercial instrumentation and devices for sample preparation, data acquisition, and analysis (24) described in several chapters of this book.

The determination of element distributions (imaging or mapping) from biological sample surfaces, such as on thin sections of biological tissues and also from nanobioelectronic devices, is of increasing relevance for a multitude of life science studies and also the development of biosensors. The analysis of essential metals (such as Cu, Zn, Fe, Ni, Mn, Mo, Mg, Ca, Na, K, Cr, B and others), metalloids (e.g., Se, As, Sb), or non-metals (like C, S, O, N, P, I, Cl and S), which are of vital importance in biological systems, is an issue in modern life science studies. Whereas a deficit of essential elements can result in various deficiency diseases, too high concentrations of these elements can be toxic. On the other hand, heavy toxic metals (like Cd, As, Cr (VI), Pb, Tl, Hg, U, Th and others) are important from the public health viewpoint and, in the low concentration range, can influence biological processes in living organisms and thus cause various illnesses. For example, the heavy metal lead (Pb^{2+}) may be substituted for Ca^{2+} in Ca-dependent biological processes involved in synaptic transmission that are essential for learning, memory, growth and differentiation of nerve cells and motor function (26). Since all nutrient elements and toxic metals are in general inhomogeneously distributed in biological tissues, imaging studies require powerful analytical techniques with both good spatial resolution and high signal/noise ratio (low detection limits).

As a mass spectrometric imaging technique, LA-ICP-MS enables the distribution of metals, metalloids and non-metals and isotopes to be measured (5, 17, 27). In the past, LA-ICP-MS was developed as the method of choice for the imaging of essential metals (such as Cu, Zn, Pb, Th and U – often at trace concentration levels), metalloids (Se) and selected non-metals (P, S, C, I) from microtome cut thin tissue sections (optimum at 20 μm thickness) of biological tissues (e.g., rat or human brains) (5, 17, 27–30). Advantages of LA-ICP-MS include high sample throughput, high sensitivity and accuracy and precision of the analytical data. In addition, no charging-up effects on surfaces occurs and fewer matrix effects allow straightforward quantification of analytical data. In general quantification can be performed using standard reference samples of the same matrix composition. If no suitable certified standard reference materials for quantification procedures are available, alternative calibration strategies have been developed in the author's laboratory. If no reference material is available for quantitative LA-ICP-MS analysis on biological tissues, matrix-matched laboratory standards can be prepared and employed (19, 31, 32). However, solution-based calibration also has been created as an alternative quantification procedure by inserting a micronebulizer into the laser ablation chamber (31). By means of mass spectrometric imaging analysis using LA-ICP-MS, inhomogeneous (often layered), site-specific metal distributions can be obtained in tissue sections (as demonstrated, for example, in human brain tissues for the hippocampus) or in tumor-infected regions or control brain. Furthermore, LA-ICP-MS has been utilized for a fast screening of 1D and 2D gels for the detection of metals, metalloids and non-metals in protein bands or protein spots separated by 1D and 2D gel electrophoresis (33–36).

This chapter reviews the progress and applications of LA-ICP-MS with and without a collision cell in comparison to the more sensitive double-focusing sector field LA-ICP-MS (21, 37) when used for imaging analysis from biological tissues for life science studies.

2. LA-ICP-MS Instrumentation

For mass spectrometric imaging studies by LA-ICP-MS, various commercial laser ablation systems are available on the analytical market mostly using a Nd:YAG laser (e.g., from NewWave, Fremont, CA, USA, or from CETAC Technologies, Omaha, NE, USA working at wavelengths of 266 and 213 nm). The development of different laser ablation cells for LA-ICP-MS from quite different materials including the imaging of biological tissues and also gels is described in the literature (1). In commercial laser ablation cells, tissues with a maximum sample size of about 20×20 mm can be analyzed. Larger laser ablation chambers (so-called SuperCell^{TM}) have been designed by NewWave Research and also CETAC to analyze large thin sections of tissues as illustrated on a human brain hemisphere in **Section 4** of this chapter, as well as 2D gels from the gel electrophoretic separation of proteins (and also for thin petrographic sections of geological samples and layered materials).

For LA-ICP-MS measurements of dried biological tissue the laser ablation is normally performed at room temperature. For fresh tissues (e.g., plant tissues), the application of a cooled laser ablation chamber – commercially available from selected laser ablation companies or as developed in the author's lab using two Peltier elements behind the target holder made of aluminum (38)– is recommended.

The laser ablation chamber is coupled directly via a connection tube to the ICP torch of a quadrupole-based inductively coupled plasma mass spectrometer (ICP-QMS) with or without a collision cell, to a double-focusing sector field ICP-MS with single ion collection (ICP-SFMS), or to an multiple ion collector ICP-MS (MC-ICP-MS) (1). In LA-ICP-MS, different types of mass spectrometers with single ion collection are employed. For imaging ICP-MS measurements of tissues, quadrupole mass spectrometers are common. **Figure 3.1** shows a schematic of an imaging mass spectrometer (LA-ICP-MS) for elemental mapping using a laser ablation system for the ablation of biological tissue in an inert atmosphere, an inductively plasma ion source and a quadrupole ion separation system. Similarly, the laser ablation system was coupled to double-focusing sector field ICP-MS or time-of-flight analyzers. LA-ICP-MS with sector fields offers the highest sensitivity achievable for the imaging of selected trace elements (for difficult to analyze elements such as Se) (27) or for spatially resolved measurements in the nanometer range (39–41). In our experiments with LA-ICP-MS, we used laser ablation systems (Nd:YAG at a wavelength of 266 or 213 nm from CETAC and from NewWave) coupled to ICP-QMS Agilent (with an octopole reaction cell), Elan 6100, or to double-focusing sector field ICP-SFMS (with reverse Nier–Johnson geometry) Element from Thermo Fisher Scientific. Time-of-flight (ToF) mass analyzers in LA-ICP-MS were applied for fast measurements of transient signals. Due to its relatively low sensitivity, LA-ICP-ToF-MS is restricted in its imaging of trace elements in biological tissues.

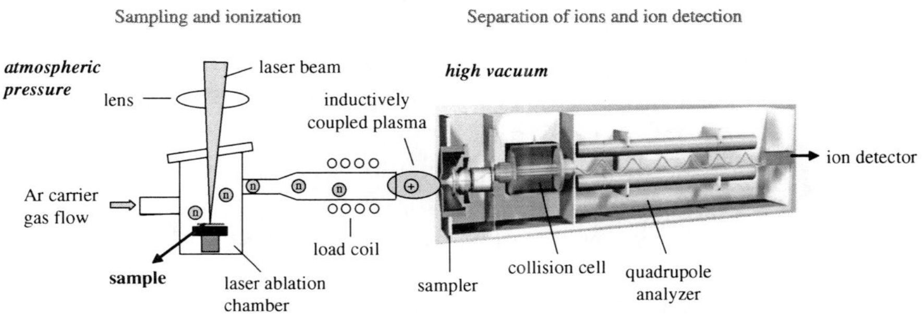

Fig. 3.1. Instrumental outline of LA-ICP-MS (laser ablation inductively coupled plasma mass spectrometer), the laser ablation system is coupled to an ICP quadrupole mass spectrometer with hexapole collision cell.

2.1. Measurement Procedure of Imaging LA-ICP-MS

Because soft materials' biological tissues are easy to ablate from a glass substrate, an Nd:YAG laser with a wavelength of 266 nm is sufficient for complete ablation of sample. The optimum operating laser power density in all our experiments was found to be 1×10^9 W cm^{-2}.

Mass spectrometric measurements by LA-ICP-MS for 2D imaging of biological tissues are performed by line scan (line per line) ablation of thin tissue sections with a focused laser (Nd:YAG) beam. The diameter of the laser crater can be varied during imaging mass spectrometric measurements of the thin tissue section between 50 and about 200 μm. The measurement time for imaging LA-ICP-MS of biological tissues (up to several hours) depends on the size of the tissue area analyzed or gel section selected and the laser scan speed applied (varies between 20 and 60 μm s^{-1}). The spot size and laser scan speed were optimized to obtain images with high spatial resolution. In our routine measurements on rat/mouse brain tissues by LA-ICP-QMS, we applied a spatial resolution of 120 μm and a scan speed of 30 μm s^{-1}. In our experiments, the distance between the lines was only 10 μm, so that consequently most biological tissue was ablated. Due to the significantly higher sensitivity of LA-ICP-SFMS compared to LA-ICP-QMS the spatial resolution was reduced to 50 μm. A defined sample area (in general, several cm^2) of a thin section of brain tissue (thickness: 20 μm) was scanned with a focused laser beam. Ion intensities of the analytes, e.g., of $^{64}Zn^+$, were measured by LA-ICP-MS within the area of interest in different regions of the human brain tissue (e.g., human hippocampus). In order to compare measurements it is important that the experimental conditions are constant and the data obtained have high reproducibility.

For the imaging of elemental distribution in protein spots blotted onto membranes of a conventional size, Jakubowski and coworkers (35) investigated different laser ablation cell geometries for the continuous ablation of nitrocellulose (NC) membranes during linear translation of the cell. The final cell has a volume of about 11.3 cm^3 in which single-shot signals are washed out in less than a second so that the translation velocities of up to 1.5 mm s^{-1} can be applied to baseline separate structures with a distance of less than 2 mm, as is demonstrated using the example of a phosphoprotein mixture separated by SDS-PAGE and blotted onto a NC membrane.

The whole procedure from the sample preparation of thin sections of biological tissue by cryo-cutting via the imaging procedure in LA-ICP-MS, including the scanning (line by line) of tissue in the selected brain area of mouse brain by a focused laser beam to measure ion intensities for analyte ions as a function of time and for a final evaluation of data in order to produce

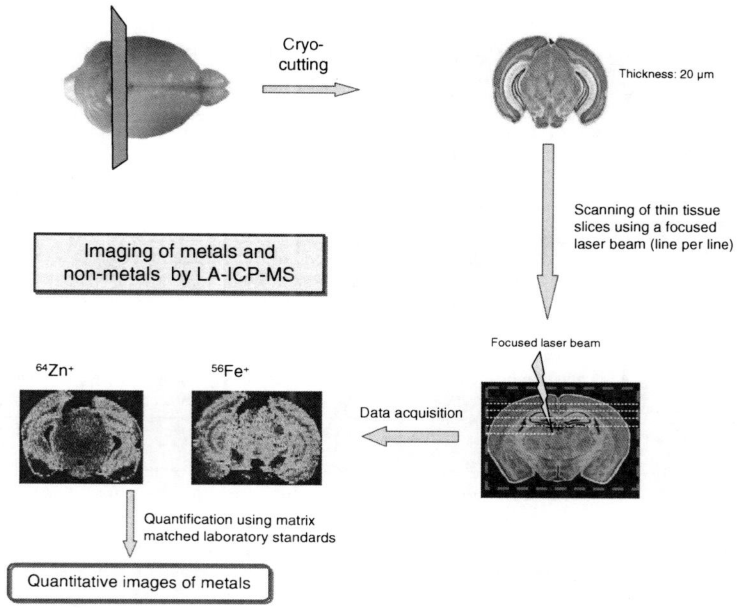

Fig. 3.2. Schematic of mass spectrometric imaging procedure by LA-ICP-MS on thin section of mouse brain tissue.

quantitative images of metals, is illustrated in **Fig. 3.2**. The metal images obtained, e.g., the Zn and Fe distributions, in mouse brain tissue (Parkinson's disease) shown in this figure correlates with immunostained and autoradiographic images and photographs of the investigated slice.

2.2. Evaluation and Quantification of Analytical Data

No commercial software is yet available for the evaluation of analytical LA-ICP-MS data in order to obtain quantitative images of biological tissues. We developed different software packages with the aim of obtaining 3D metal or non-metal distributions, e.g., in rat brain, in order to correlate the data with MRI (magnet resonance images), immunostained, autoradiographic, and histochemical images of the same rat brain obtained at Institute of Medicine of Forschungszentrum Jülich, Germany. Furthermore, new quantification strategies were developed and adapted to the biological matrix investigated (21).

2.2.1. Quantification Using Synthetic Matrix-Matched Laboratory Standards

Quantitative images of elements can be obtained using prepared matrix-matched laboratory standards for calibration as demonstrated in several previous papers (1, 21, 42). The preparation protocol for synthetic matrix-matched laboratory standards for the calibration of imaging LA-ICP-MS for biological tissues is summarized in **Fig. 3.3**. In general, different synthetic laboratory standards with elements of interest in well-defined concentrations were prepared. For example, five slices of the same biological tissue were spiked with selected standard solutions at

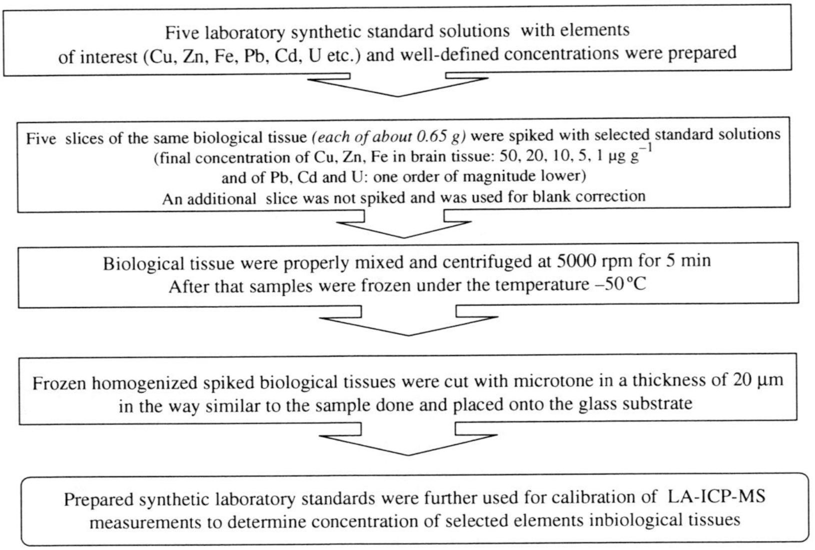

Fig. 3.3. Preparation protocol of synthetic matrix-matched laboratory standards for calibration of LA-ICP-MS images.

defined concentration, mixed and frozen at a temperature of –50°C, cut in thin slices, and placed onto the glass substrate. An additional unspiked slice was used for blank correction. The laboratory standards for the calibration of LA-ICP-MS images were measured together with biological tissues in the same measurement cycles, and consequently an ideal matrix matching was obtained.

Matrix-matched homogenized laboratory standards with well-defined element concentrations of analytes were prepared and utilized for the quantification of LA-ICP-MS data in imaging brain tissues in routine mode in our laboratory. Matrix-matching standards were used to constitute calibration curves, where the regression coefficient of the calibration curves obtained was typically >0.9 for all analytes investigated.

2.2.2. Quantification Using Standard Reference Materials

Jackson et al. (43) proposed using pressed pellets of biologically certified reference materials for calibrations of rat brain images. The authors determined the elemental distributions and concentrations of Cu, Zn and Fe by LA-ICP-QMS (laser spot size: 60 μm and scan speed: 120 μm s^{-1}) in whole rat brain sections (100 μm thickness) and the physiologically important elements P and S were also analyzed. The distributions and concentration ranges for these elements demonstrate the utility of this technique for rapid mapping of brain thin sections (43).

Kidness et al. (29) described an analytical procedure for 2D mapping of copper and zinc in a section of sheep liver. For quantification purposes, the authors applied a homogeneous Certified Reference Material (CRM; LGC 7112, pig liver).

To quantify images of parts of heavy-metal-tolerant plants (leaves of *Elsholtzia splendens*) in order to study the hyperaccummulation of Cu and Zn, the standard reference material, NIST SRM 1515 apple leaves, was doped, mixed and homogenized with analytes of defined concentration. The quantification was performed via an external calibration procedure by measuring the calibration curves. Another standard reference material, NIST SRM 1547 Peach Leaves, was used to validate the analytical procedure (44, 45).

2.2.3. Solution-Based Calibration in LA-ICP-MS

Several solution-based calibration techniques (external calibration, standard addition and isotope dilution technique) developed in the author's laboratory represent new calibration strategies for the quantification of analytical data in LA-ICP-MS (1). During the laser ablation of a thin section of biological tissue, defined standard solutions with increasing concentration were nebulized, whereby the calibration of the analytical method was performed by a standard addition mode (21).

Online isotope dilution technique LA-ICP-MS is proposed as the method of choice for quantitative element mapping and imaging even if no reference materials are available (42). For online solution-based calibration with LA-ICP-MS, a microflow total consumption nebulizer DS-5 (CETAC Technologies, Omaha, NE, USA) was inserted into the laser ablation chamber (21). It is more convenient to use solution-based calibration because such calibration can be performed easily, quickly and in any concentration range for many elements. The application of the online isotope dilution technique in LA-ICP-MS using a microflow nebulizer inserted into the laser ablation chamber to biological standard reference materials (e.g., apple leaves) yielded accurate analyte concentrations (42).

3. Figures of Merit of LA-ICP-MS Imaging on Biological Tissues

As a powerful trace element analytical technique with multi-element capability, LA-ICP-MS allows the production of a multitude of metals, metalloids and non-metal images quasi-simultaneously. Ion images were evaluated on the basis of the imaging procedure by the newly developed LA-ICP-MS technique. In order to validate the ion images, at least two isotopes were analyzed if available (e.g., $^{63}Cu^+$ and $^{65}Cu^+$ or $^{64}Zn^+$ and $^{66}Zn^+$), and it is proposed that the elemental distribution in neighboring tissue slices should be performed under the same experimental parameters. Possible isobaric interferences of single-charged analyte ions with polyatomic ions or double-charged ions were studied carefully and can be generally minimized using double-focusing sector field ICP-MS at the required mass resolution or by applying collision-induced reactions in quadrupole-based ICP-MS with a collision/reaction cell. In general, fewer interference problems occur in the LA-ICP-MS analysis of thin tissues in dry plasma compared to ICP-MS measurements of aqueous solutions.

3.1. Study of Reproducibility of Imaging LA-ICP-MS

The reproducibility of the imaging LA-ICP-MS method for the analysis of thin cross sections of biological tissues was studied by Zoriy and Becker (20). A roughly 3% reproducibility of the newly developed LA-ICP-MS procedure for homogeneous tissues (e.g., thin sections of matrix-matched laboratory (44) standards) was observed (see top right of **Fig. 3.4**). Dependences of LA-ICP-MS signal intensities of $^{63}Cu^{+}$ on the laser beam diameter studied using synthetically prepared homogeneous brain laboratory standard doped with 10 μg g^{-1} of Cu is shown on the left in **Fig. 3.4**. In further studies, several adjacent inhomogeneous human brain sections of a thickness of 20 μm were cut from the same tissue and screened to compare the distribution of the elements of interest (e.g., Zn, Cu and C) and analyzed by LA-ICP-MS to discover how reproducible these profiles are within the neighboring sections. **Figure 3.4** shows the quantitative Zn image of inhomogeneous human brain tissue and the results of reproducibility studies

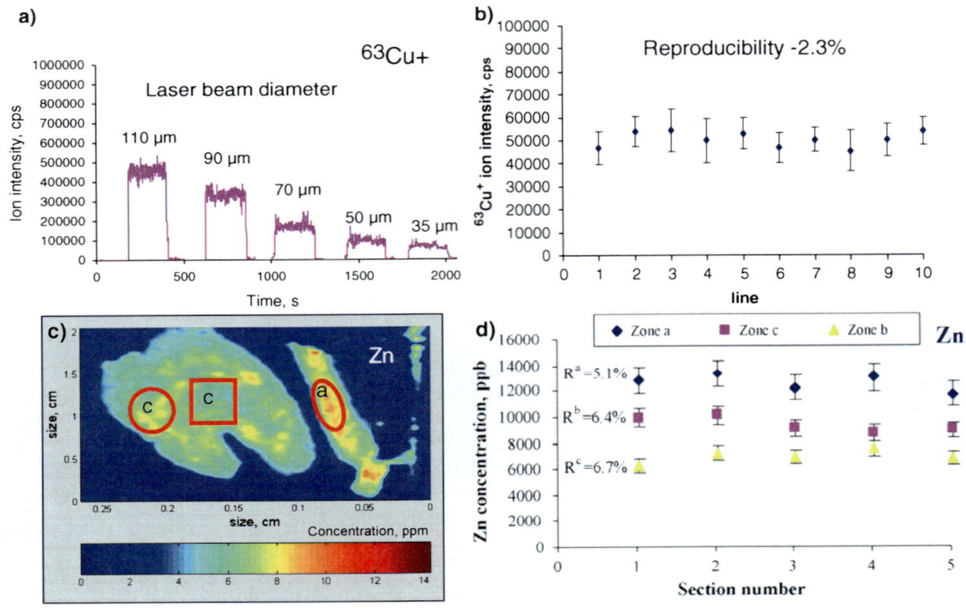

Fig. 3.4. (**a**) Measured ion intensity of $^{63}Cu^{+}$ for different laser beam diameters and (**b**) reproducibility of imaging LA-ICP-MS on homogenized synthetic laboratory standard. A (**c**) Zn image of an inhomogeneous human brain tissue and (**d**) reproducibility of five adjacent slices of these brain tissue. (**a**) and (**b**) reprinted with permission from Elsevier.

for Zn imaging by LA-ICP-MS of five neighboring sections. The values obtained for the reproducibility in the three selected zones (marked in the left-hand figure) were in the range of 5.1–6.7%.

Kidness et al. (29) characterized laser ablation inductively coupled plasma time-of-flight mass spectrometry (LA-ICP-MS) for imaging of biological tissues in terms of precision. Precisions for thin sections of sheep liver reached 27–59% (raster scan) and 9–47% (line scan) RSD for copper, whereas the precision for zinc was significantly better, 8–18% (raster scan) and 4–21% (line scan) RSD.

3.2. Detection Limits

The detection limits obtained for images measured by LA-ICP-MS depend on the experimental equipment and operating parameters employed and thus differ for different elements. The detection limits varied between microgram per gram and nanogram per gram. For example, the radionuclides Th and U were detected in biological tissues with highest sensitivity. Both radioactive metals were found to be homogeneously distributed at the ultratrace level in human hippocampus. The detection limits were determined for both elements in the low-nanogram per gram range.

Jackson et al. (43) achieved sub-microgram per gram detection limits by imaging ICP-QMS on rat brain tissues (100 μm thickness) for the essential trace elements Cu and Zn.

In the following section, selected examples of different applications of the LA-ICP-MS imaging technique will be discussed.

4. Application Fields of Mass Spectrometric Imaging by LA-ICP-MS on Biological Tissues and Gels

LA-ICP-MS enables imaging of thin biological tissue sections of plants and animals to study the distribution of essential and beneficial elements such as Zn, Cu, Fe, S, P, Se and Mn as well as of toxic and also radioactive metals (e.g., Hg, Pb, Cd, Th and U) with a spatial resolution in the micron range. In contrast to imaging by MALDI-MS, no application of matrix is required in LA-ICP-MS (nor in SIMS). A complete laser ablation of biological tissue is performed at higher laser energy in an atmospheric laser ablation chamber. The advantage of imaging LA-ICP-MS compared to SIMS, the latter operating in a high-vacuum ion source, is that no charging of the sample surface by the interaction of a focused laser beam with non-conducting material occurs and significantly fewer matrix effects are observed. These properties of LA-ICP-MS allow an easy quantification of the measured ion intensity.

In the following section, the application of LA-ICP-MS for the imaging of metals, Se and non-metals will be discussed with selected examples concerning brain samples, animal, or plant tissues. In addition, the new strategy of combining LA-ICP-MS imaging to detect metal-containing proteins and phosphoproteins and their elucidation by biomolecular mass spectrometry will be described.

In **Fig. 3.5** different application fields of mass spectrometry imaging by LA-ICP-MS are summarized. In the life sciences, in the author's laboratory these applications focus mainly on the quantitative imaging of essential, beneficial, and toxic metals; metalloids and non-metals in thin tissue sections, e.g., in order to study diseased neurodegenerative tissue (Alzheimer's and Parkinson's disease) or tumor growth and stroke in the brain compared to normal tissue. In a second step, metalloproteins and phosphoproteins are observed after the gel electrophoretic separation of proteins (in 1D or 2D gels) by LA-ICP-MS, whereby MALDI-MS and/or ESI-MS is employed for the identification of metalloproteins (phosphoproteins) in gels after spotting and tryptic digest. Nowadays, LA-ICP-MS imaging is employed to characterize the distribution of metal-containing proteins in gels (5).

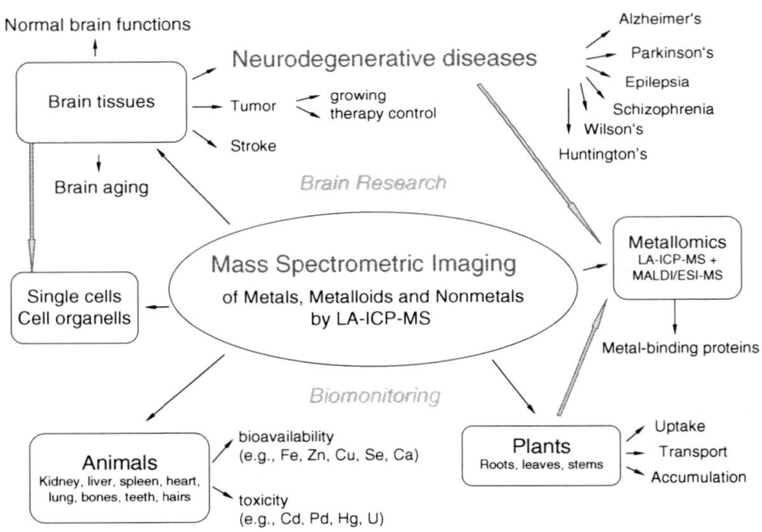

Fig. 3.5. Application fields of mass spectrometric imaging by LA-ICP-MS in brain research and for biomonitoring of essential and toxic metals.

4.1. Multielemental Imaging by LA-ICP-MS on Brain Tissues

In the first example, the multielement capability of LA-ICP-MS imaging will be demonstrated. The element distribution of selected essential metals in rat brain tissues was determined in the routine mode using a quadrupole ICP-MS (Agilent 7500 ce) with the laser UP 266 ablation system from NewWave with a lateral resolution in the 120 μm range. As illustrated in **Fig. 3.6**, the distribution of analyzed metals (ion images are shown here) such as

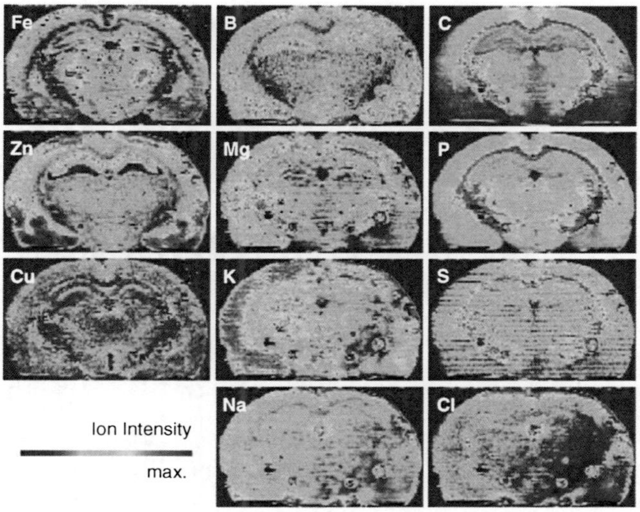

Fig. 3.6. Images of $^{56}Fe^+$, $^{64}Zn^+$, $^{63}Cu^+$, $^{11}B^+$, $^{24}Mg^+$, $^{39}K^+$, ^{23}Na, $^{13}C^+$, $^{31}P^+$, $^{34}S^+$, and $^{35}Cl^+$ ions of control rat brain measured by LA-ICP-QMS (Agilent 7500 ce and NewWave UP 266).

$^{56}Fe^+$, $^{64}Zn^+$, $^{63}Cu^+$, $^{11}B^+$, $^{24}Mg^+$, $^{39}K^+$ and ^{23}Na and selected non-metals like $^{13}C^+$, $^{31}P^+$, $^{34}S^+$ and $^{35}Cl^+$ is mostly different (Becker et al. (2009), unpublished data). A layered structure was found for Fe and Zn in the rat brain cortex. However, the distribution of metal and non-metals also differs in the hippocampus. Whereas the highest Zn content was detected in the hippocampus (CA3 region), correlating with a slight enrichment of Mg, for other elements depletion in the hippocampus tended to be found. The phosphorus image measured by LA-ICP-MS using a quadrupole analyzer shows a similar distribution to that of carbon (*see* **Fig. 3.6**). The presented ion images of selected elements are sufficient to demonstrate the quite different element distribution in the brain section investigated and may yield insights on the role of these elements in selected brain regions. The quantitative Zn and Cu distribution in several sub-regions of the human brain hemisphere was determined by LA-ICP-SFMS. Synthetic matrix-matched laboratory standards were applied for quantification procedures of metal images (19). The distribution of Zn and Cu in a human brain hemisphere measured by LA-ICP-SFMS using a large laser ablation chamber from Cetac is illustrated in **Fig. 3.7**. As a result of LA-ICP-MS studies, we found a layered structure with higher Zn and Cu content in gray matter compared to white matter.

The concentration profiles of Zn and Cu in the hippocampus area and in tissues from other regions of the human brain were investigated using LA-ICP-MS by Becker et al. (19, 21) It was found that copper and zinc are localized differently in the brain

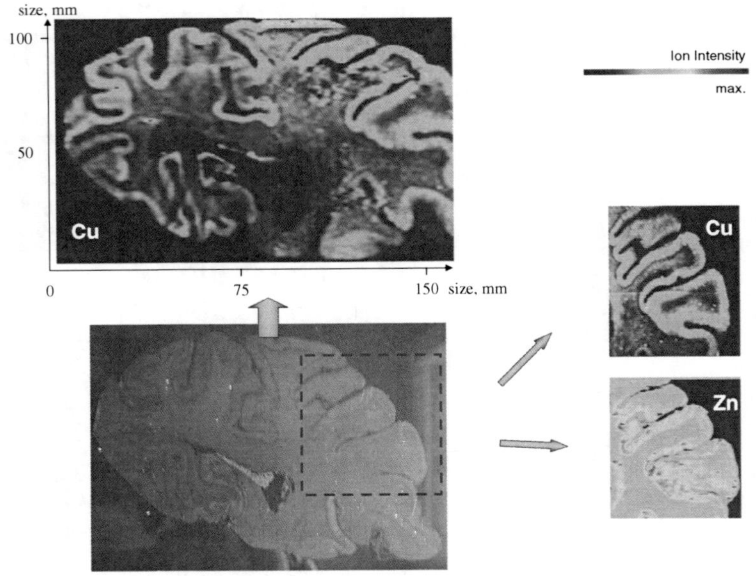

Fig. 3.7. Images of Cu and Zn in part of human brain and Cu distribution in whole hemisphere tissue measured by LA-ICP-SFMS using a large laser ablation chamber (Element and Cetac LSX 200).

regions analyzed with the highest concentration in the hippocampus. The distribution of toxic Pb was first measured in the human hippocampus in the 0.2 $\mu g\ g^{-1}$ concentration range (19).

4.2. Study of Metal Distribution in Diseased Brain Region by LA-ICP-MS

The study of the distribution of metals in brain tissue is of special relevance in the tumor region compared to control tissue. It is well known that metals can also catalyze cytotoxic reactions leading to DNA modification (such as the Fenton reaction) or toxic processes at high concentrations that might be involved during tumor progression. Metal distribution in human brain tumor (human glioblastoma multiforme – GBM) and also tumor growth after injecting cultured tumor cells into rat brains were studied in tissue sections by imaging LA-ICP-MS in the author's lab (31, 46, 47).

A double-focusing sector field LA-ICP-SFMS (Element, Cetac LSX 200) was employed to measure the ion intensities of two essential elemental (Cu and Zn) and toxic metals (Pb and U) within the tumor area and the surrounding region invaded by GBM as well as in control tissue. The size of the brain samples was approximately 1 cm^2. The quantitative determination of copper, zinc, lead and uranium distribution in brain tissues by LA-ICP-MS was performed again using prepared matrix-matched laboratory standards doped with these elements of interest. An inhomogeneous distribution was observed for all the metals measured. The limits of detection (LODs) obtained for Cu and Zn were 0.34 and 0.14 $\mu g\ g^{-1}$, respectively, while LODs of 12.5 and 6.9 $ng\ g^{-1}$ were determined for Pb and U, respectively. A correlation was found between LA-ICP-MS and receptor-autoradiography results. The receptor-autoradiography technique using A1AR and pBR for labeling was employed. Regarding A1AR, we used the specific A1 adenosine receptor (A1AR)–ligand, 3H-CPFPX [^{3}H-cyclopentyl-3-(3-fluoropropyl)-1-propylxanthine], which has been shown to specifically label the invasive zone around GBMs. The peripheral benzodiazepine receptor was labeled with ^{3}H-Pk11195 [^{3}H-1-(2-chlorophenyl)-*N*-methyl-*N*-(1-methylpropyl)-3-isoquinoline-carboxamide]. **Figure 3.8** illustrates the lateral distributions of Cu, Zn, Pb and U in thin sections of the glioblastoma human brain sample measured by LA-ICP-SFMS together with the autoradiographic images: (a) peripheral benzodiazepine receptor (pBR) and (b) A1 adenosine receptor ligand (A1A).

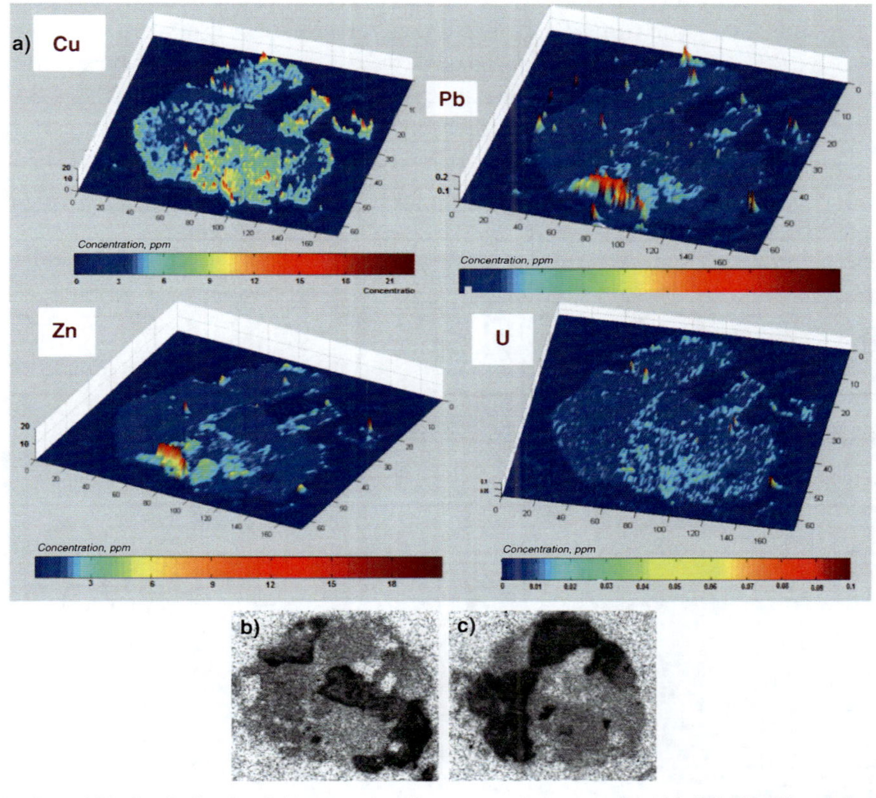

Fig. 3.8. (**a**) Cu and Pb distribution in glioblastoma multiforme sample measured by LA-ICP-MS, (**b**) peripheral benzodiazepine receptor (pBR) and (**c**) A_1 adenosine receptor ligand (A_1A).

Metal concentrations were normalized for the cellular density, which was 1.3–4.8 times higher in tumors in comparison to control tissue. After normalization, the concentrations of copper and zinc were 3.9–9.1 and 2.4–9.6 times higher in the peritumoral zone compared to the solid tumor, respectively, whereas concentrations of lead and uranium were not elevated. Dehnhardt et al. (47) demonstrated that the increased copper and zinc levels were found in a zone which is characterized by a relatively high A1AR (adenosine receptor) density.

A systematic study of several different rat brain samples infected with tumor cells (different incubation time) was performed by Zoriy et al. (46) The analyzed sections were quantitatively imaged for their P, S, Fe, Cu and Zn content. In addition, $^{13}C^+$ was monitored in all LA-ICP-MS measurements and its application as an internal standard was evaluated. The results of the measurements were compared with respect to the difference of the element content in control and tumor tissues. As an example, in **Fig. 3.9** the lateral distribution images of Cu (b) and Zn (c) measured by LA-ICP-MS are presented. The small tumor at the bottom is made visible by a slight enrichment of Zn and a depletion of Cu (marked). After tumor growth (upper figures), the tumor region is characterized by a significantly changed metal distribution compared to the normal hemisphere (right). The locations of tumor region in the photographs and LA-ICP-MS images are in agreement, with the average concentrations of Cu and Zn in the tumor areas determined as 15 and 17 μg g^{-1}, respectively. The Cu and Zn contents were partly enriched or

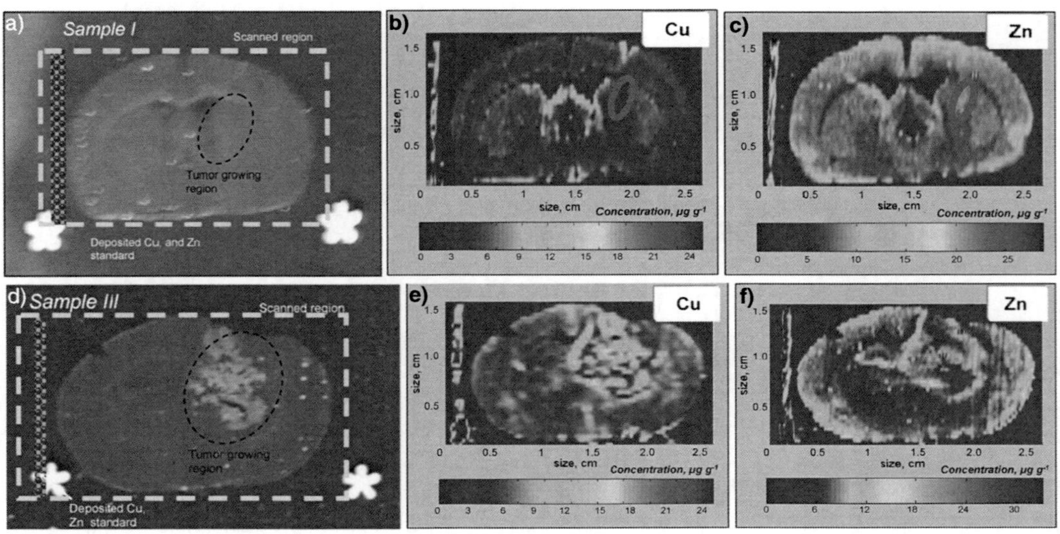

Fig. 3.9. Images of Cu [(**b**) and (**e**)] and Zn [(**c**) and (**f**)] distribution in cross section of rat brain sample with two different tumor sizes compared to photograph of these slices [(**a**) and (**d**)] (reprinted with permission from Elsevier).

depleted in the tumor tissue (right side of each image) in comparison to the control (left side) as shown in **Fig. 3.9**.

At present, we are studying the distribution of metals (via the measurement of $^{64}Zn^+$, $^{63}Cu^+$, $^{56}Fe^+$, $^{49}Ti^+$, $^{23}Na^+$, $^{24}Mg^+$ and $^{39}K^+$) and non-metals (via $^{13}C^+$, $^{31}P^+$, $^{34}S^+$ and $^{35}Cl^+$) in thin tissue sections of rat brain (20 μm tissue thickness, maximum analyzed area ~180 mm^2) after a photoinduced cortical infarct (medical rat brain studies on photothrombosis were carried at Forschungszentrum Jülich by Langen et al. (48, 49)) using LA-ICP-QMS in order to construct 3D metal and non-metal images of the rat brain (Becker et al. (2009), unpublished data). Examples of ion images of selected metals (Fe, Zn, Cu, and Na) and non-metals (C, S, P, and Cl) measured on a rat brain section with a photoinduced infarct by imaging LA-ICP-MS are summarized in **Fig. 3.10**. The position of the measured slice in the rat brain with photothrombosis is marked in **Fig. 3.10**. This analytical technique was also employed to measure the very abundant alkali metals K and Na in control rat brain, as well as difficult-to-ionize halogens like chlorine. A strong correlation of Cl with the Na distribution in rat brain tissue was found. In the infarct region,

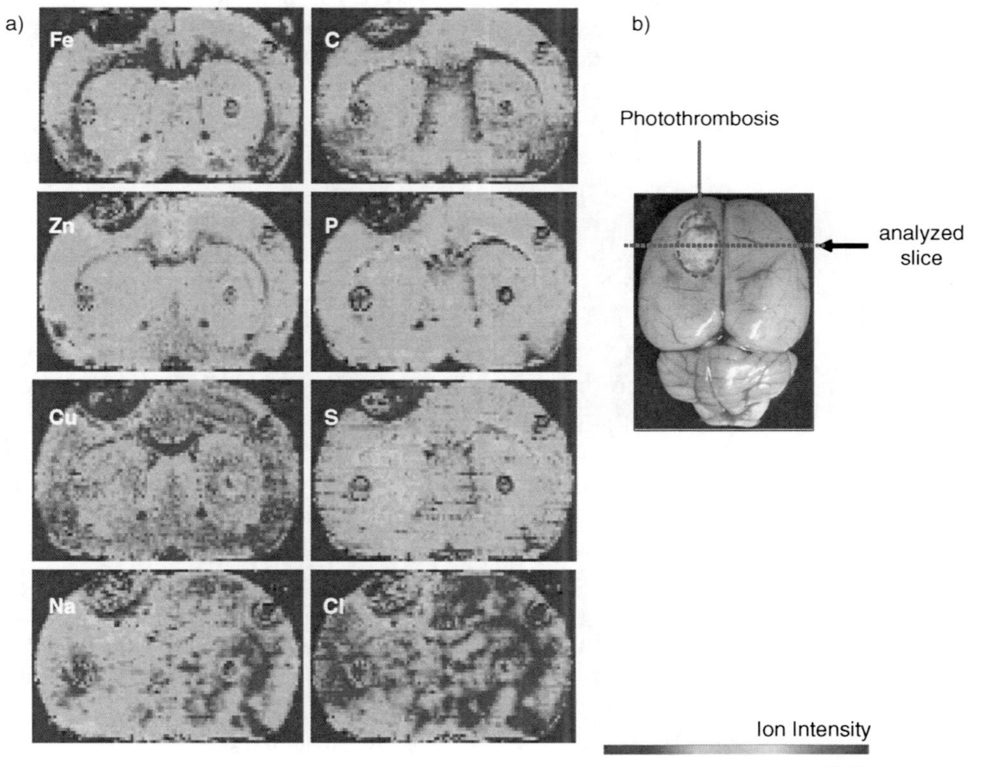

Fig. 3.10. (**a**) Distribution of Fe, Zn, Cu, Na, C, P, S and Cl in rat brain tissue with photoinduced stroke measured by LA-ICP-QMS (Agilent 7500 ce and NewWave UP 266) and (**b**) photograph of rat brain with photothrombosis (46).

selected elements (Fe, Zn, P, Mg and Ti) are clearly enriched, whereas K is depleted. These imaging mass spectrometric results of thin sections of rat brain tissues provide novel information on the distribution of elements in rat brain tissues compared to control samples and may also be an important tool for analyzing brain tumor metabolism and pathogenesis.

Investigations of metal distribution in brain tissue from a mouse with a tumor treated with cisplatin were produced by scanning 14 μm thin sections of kidney tissue using sector field LA-ICP-SFMS as described by Zoriy et al. (50). Platinum was inhomogeneously distributed in the tissue with concentrations in the range of 10–25 $\mu g\ g^{-1}$ where an enrichment in the tumor region was observed (50).

As a microanalytical technique, LA-ICP-MS was applied in the author's laboratory for evaluating the absorption of Pt by individuals undergoing cancer therapy using cisplatin. The concentration of Pt along a single strand of hair from a patient who had been treated with cisplatin as a cytostatic drug was monitored by LA-ICP-MS (51). For quantification, Pt-spiked hair strands were prepared and analyzed. By scanning the whole strand width by LA-ICP-MS small variations of Pt concentration along the hair strand can be observed. The maximum concentrations of Pt found along the hair strands in the measured transient signals were 26.9 ± 5.3, 14.7 ± 3.3, 20.9 ± 3.9, and 26.1 ± 3.8 $\mu g\ g^{-1}$, which corresponded to four treatments of cisplatin administered to the patient at 3 week intervals. The platinum distribution found in the analyzed hair may contribute to the optimization of cisplatin therapy.

In many neurodegenerative diseases, abnormal metal deposition has been observed within the brain (e.g., in Alzheimer's, Parkinson's, or Wilson's disease) (4, 52). We studied the iron content and metal distribution in Parkinson's rat brain samples by LA-ICP-MS (53, 32). Numerous biochemical abnormalities have been discovered in the brain since MPTP (methylphenytetrahydropyridine) has been used as a neurotoxin-induced form of Parkinson's disease by destruction of specific neurons in the substantia nigra (SN). The oxidative stress-induced neuronal damage of tissue is combined with a significant accumulation of iron in SN (54). In **Fig. 3.11**, the iron distribution compared to zinc and copper is illustrated in Parkinson's mouse brain treated with MPTP. An increased Fe enrichment in SN is clearly seen, whereby we found that the Fe content is significantly higher compared to normal brain tissue (32). Zinc measured in the same slice is enriched in the hippocampus.

In addition, poisoning by toxic heavy metals like Pb, Cd, or Hg, even in very low amounts, can cause permanent damage to the brain and nervous system (55). More details on the role of metals in the pathophysiology and pathogenesis of

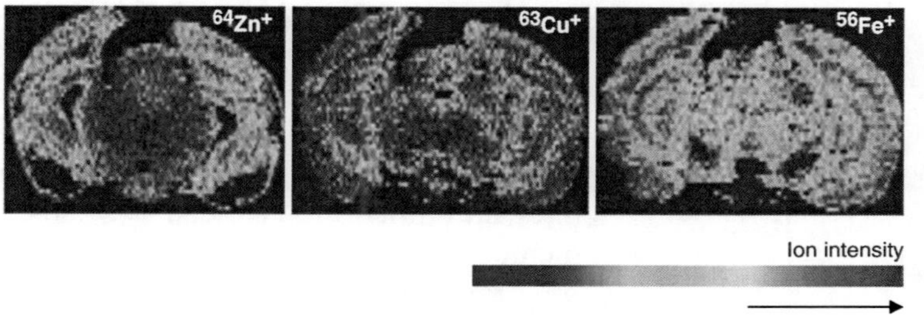

Fig. 3.11. Distribution of Zn, Cu and Fe in Parkinson's mouse brain measured by LA-ICP-QMS (Agilent 7500 ce and NewWave UP 266).

neurodegenerative disorders are described in the book *Neurodegenerative Diseases and Metal Ions* edited by Sigel et al. (56).

The specific toxicity of trace metals and compounds largely depends on their bioavailability in different organs or compartments of the organism considered. Imaging LA-ICP-MS with a spatial resolution in the 100 μm range was applied to study heavy metal distribution (neodymium and uranium) in brain tissues for toxicological screening (57). Rat brain postmortem tissues were stained in an aqueous solution of neodymium (metal concentration 100 μg g^{-1}) for 3 h. The incubation of heavy metal in thin slices of brain tissue was observed by an imaging mass spectrometric LA-ICP-MS technique. The LA-ICP-SFMS images of neodymium (57) on stained rat brain tissue (thickness 30 μm) are shown in **Fig. 3.12**. Imaging LA-ICP-MS allows structures of interest to be identified and the relevant dose range to be estimated.

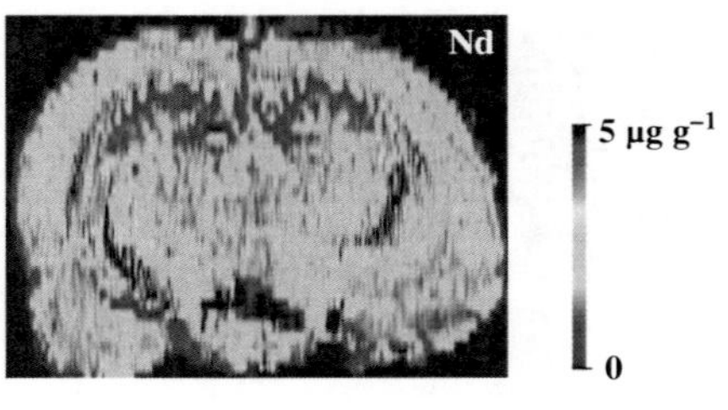

Fig. 3.12. New staining technique in medicine: neodymium distribution in rat brain tissue after treatment in neodymium solution.

The determination of the metal concentration (excess or deficiency) compared to normal tissue, the binding to proteins and the quantitative distribution of metals in brain tissues is of the highest significance for the study and treatment of neurodegenerative diseases and is linked to the development of mass spectrometric techniques on biological complex systems.

4.3. Imaging of Metals Bound to Proteins in 2D Gels

Metallomics and metalloproteomics are emerging fields addressing the role, uptake, transport, and storage of trace metals essential for protein functions. The methodologies utilized in metallomics and metalloproteomics to provide information on the identity, quantity, and function of metalloproteins are discussed here. As an elemental mass spectrometric technique, LA-ICP-MS has mostly been employed to identify the metal bound to a protein and MALDI/ESI-MS to elucidate the structure, dynamics, and function of a metal–protein complex. Other approaches include X-ray absorption and X-ray fluorescence spectroscopy and bioinformatics sequence analysis. X-ray absorption spectroscopy utilizing a synchrotron radiation source is a powerful tool for providing a direct analysis of metal bound to proteins and proteomic metal distribution in biological matrices. With the advent of genome sequencing, a large database of protein primary structures has been established, and specific tools have been developed to identify metalloproteins in the genome sequences (58).

A challenging analytical strategy in life science studies consists of the combined application of element imaging mass spectrometric techniques like LA-ICP-IMS and biomolecular mass spectrometry such as ESI- or MALDI-MS to elucidate the structure and sequence of metal- or phosphorus-containing proteins. In **Fig. 3.13**, the combination of imaging mass spectrometry by LA-ICP-MS and proteome analysis is shown schematically with the example of one part of the human brain (hippocampus). This new analytical strategy starts by imaging thin sections of tissues to first detect metals, metalloids and non-metals in order to study neurodegenerative diseases or tumor growth via the quantitative (mostly abnormal) metal distribution of essential elements (e.g., Cu, Zn, Fe, Se) and also toxic metals (like Pb, Cd). In our

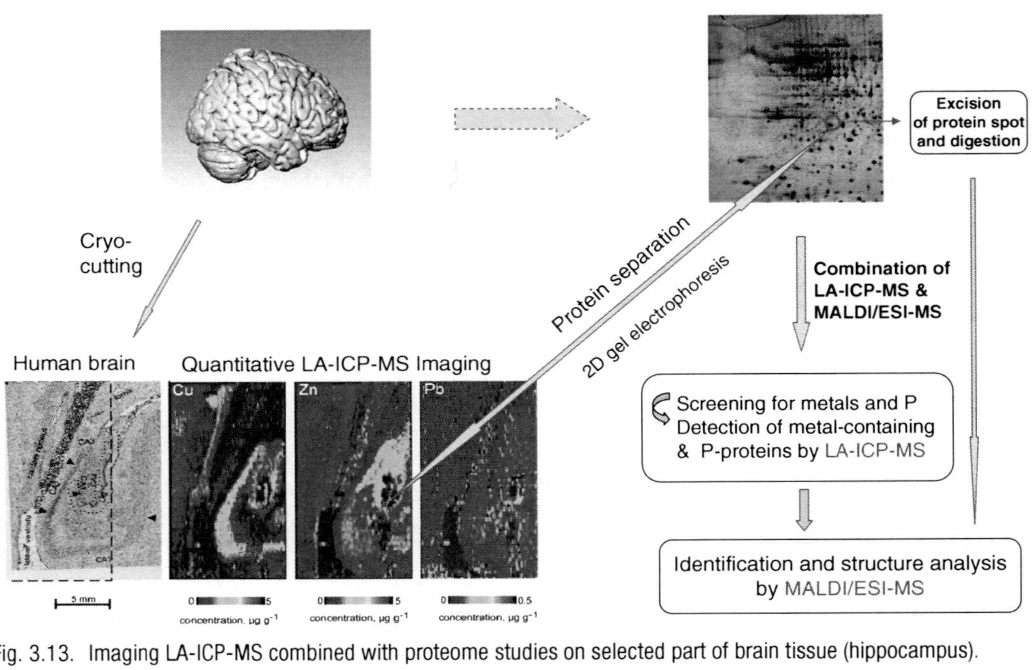

Fig. 3.13. Imaging LA-ICP-MS combined with proteome studies on selected part of brain tissue (hippocampus).

example (*see* **Fig. 3.13**), the hippocampus was quantitatively analyzed with respect to copper. The zinc distribution is shown and lead was detected at a low concentration level. In the second step, the proteins were separated from the selected region of interest by 1D or 2D gel electrophoresis. Metalloproteins and phosphoproteins were detected by LA-ICP-MS using a new powerful and sensitive screening technique. The protein spots containing phosphorus or metals were cut out from a second gel produced under exactly the same conditions, and after tryptic digestion the proteins were identified and the sequence determined by MALDI- or ESI-MS.

For example, the binding of Cu and Zn on bovine serum proteins was studied using a tracer experiment by LA-ICP-MS measurements (33). 1D BN-PAGE gels with separated bovine serum proteins after gel electrophoresis were doped with an enriched isotope copper tracer (^{65}Cu) as a function of time (from 30 s up to 24 h). In several protein bands in the 1D gel before the tracer experiment metal ions (mostly Zn) were found by LA-ICP-MS, but after experiments using enriched ^{65}Cu a fast exchange of Zn bonded on bovine serum albumin by copper was observed. This experimental finding demonstrates the formation of copper-binding proteins during tracer experiments in the 1D gel. Furthermore, proteins of bovine serum were separated by 2D blue native gel electrophoresis with native conditions in the first and the second dimension. The 2D BN-Page in the pI range of 4–7 after colloidal Coomassie staining is illustrated on the right-hand side of **Fig. 3.14**. The wide band in the middle of the gel was identified as bovine serum albumin by MALDI-TOF-MS. $^{34}S^+$, $^{63}Cu^+$ and $^{64}Zn^+$ bonded to the protein were detected by

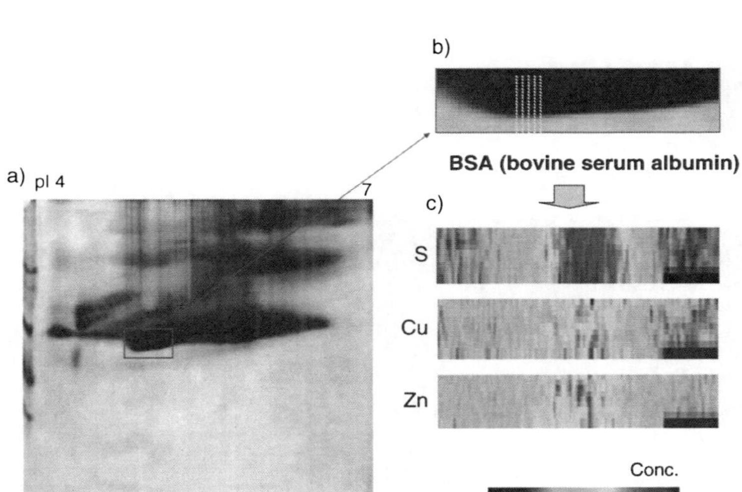

Fig. 3.14. (**a**) 2D BN-PAGE of bovine serum stained with colloidal Coomassie. (**c**) S, Cu and Zn bonded to the protein were detected by (**b**) LA-ICP-MS in five line scans from the (BSA) protein band.

LA-ICP-MS of bovine serum albumin in five line scans of 2D BN gel as is illustrated on the left in **Fig. 3.14**. A correlation of the maximum intensity of sulfur from protein with the maximum intensity of $^{63}Cu^{+}$ and $^{64}Zn^{+}$ was observed.

2D gel electrophoresis (2D GE) is well known as a powerful separation technique, which allows the separation of thousands of proteins from complex protein mixtures. The high-resolution capability of 2D GE can be obtained by independent protein separation steps in the first and second dimensions. The first dimension of gel electrophoresis is isoelectric focusing (IEF), whereby the individual proteins of a protein mixture move to their isoelectric point in a pH gradient. The proteins therefore lose their net charge and their electrophoretic mobility. In the second dimension, these proteins are separated orthogonally by SDS-PAGE according to their molecular weight (MW). The proteins separated in 2D gels are visualized, for example, by the silver staining technique. LA-ICP-MS was employed as a useful screening technique to detect metals and/or phosphorus in separated protein spots. In **Fig. 3.15** one cut of a 2D SDS-PAGE gel from a human brain sample from the somatomotor cortex is shown. The

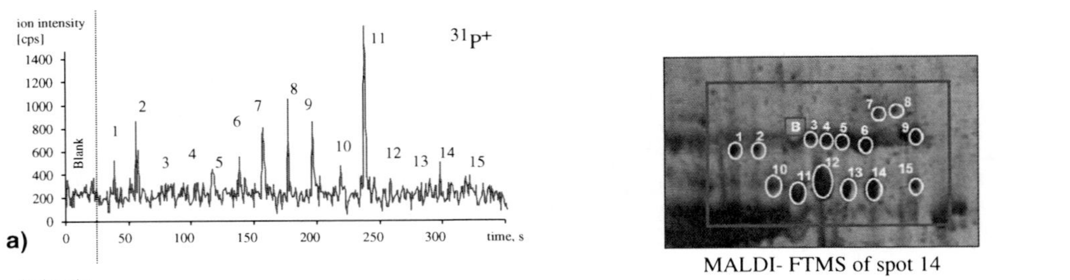

Fig. 3.15. Detection of P-, Zn- and Cu-containing proteins in human brain somatomotor cortex on a 2D gel (SDS-PAGE) LA-ICP-SFMS and identification of selected protein spot 14 by MALDI-FTICR-MS (34).

cut was scanned with LA-ICP-MS for phosphorus, copper and zinc. Phosphorus or metals was found in several of the analyzed gel spots. Also a MALDI-FTICR-MS spectrum with the identified peptides of spot 14 is shown in **Fig. 3.15**. A phosphopeptide was also identified in this spot.

A BN-PAGE gel for protein separation can be used instead of IEF as the first dimension of the gel electrophoresis. This technique was used on mitochondria samples from baker's yeast as shown on the left in **Fig. 3.16**. On this gel, different cuts were scanned for elements of interest like phosphorus and metals. The results from the LA-ICP-MS measurement of one cut can be seen in **Fig. 3.16** on the right-hand side. For example, the protein in spot 59 was identified by a database search after MALDI-FTICR-MS (*see* **Fig. 3.16**) as Aac2p (major mitochondrial ATP/ADP translocator).

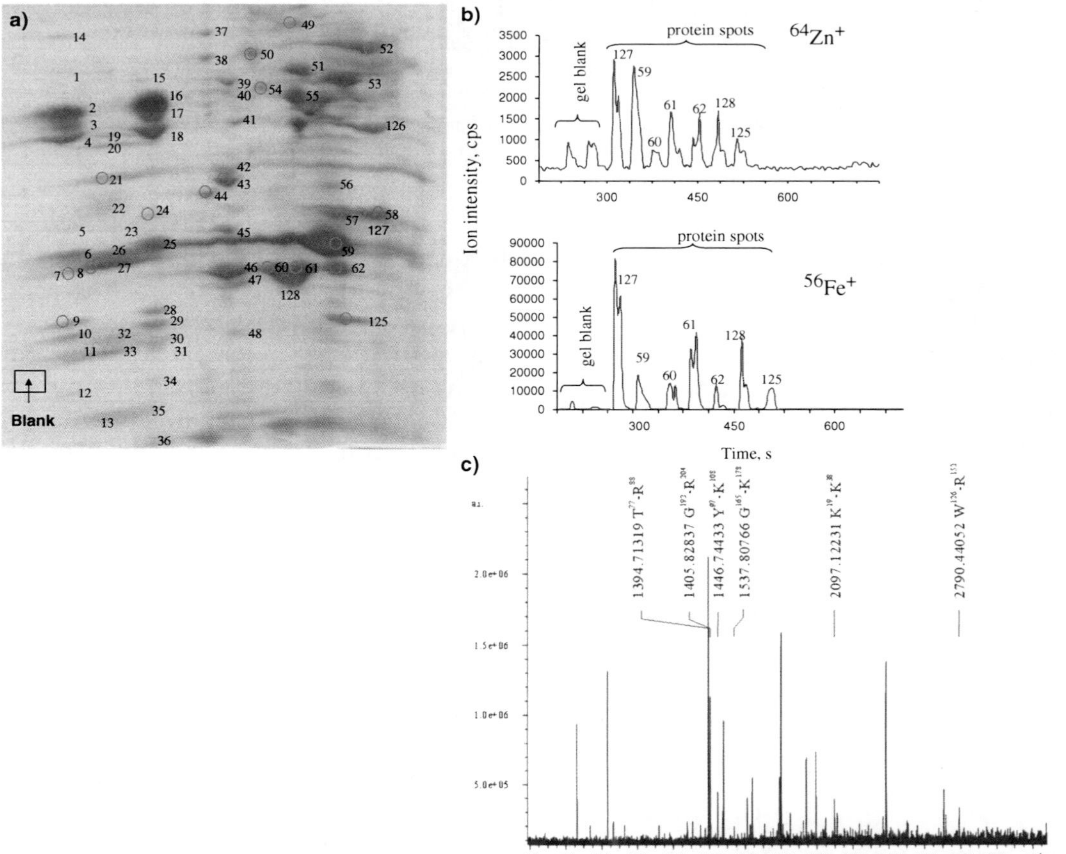

Fig. 3.16. Transient signals of $^{64}Zn^{+}$- and $^{56}Fe^{+}$-containing proteins in yeast mitochondria on a 2D gel (blue native/SDS-PAGE) LA-ICP-SFMS and identification of selected protein spot 59 by MALDI-FTICR-MS (63).

Using this new analytical strategy, it would be possible to image the metal or phosphorus-containing protein spots in 2D gel in order to demonstrate their lateral distribution. The application of the imaging technique developed for tissues to the metal distribution in protein spots on gels allows additional valuable information to be obtained, for example, on the quality of the separation of metal- and phosphorus-containing proteins. This is demonstrated for Zn-binding proteins in a 2D BN-PAGE gel of rat kidney water extract. In **Fig. 3.17**, the lateral distribution of Zn on 2D BN-Page gel in several protein spots was measured by LA-ICP-QMS (using ICP-QMS Agilent 7500 ce and laser ablation system UP 266 from NewWave).

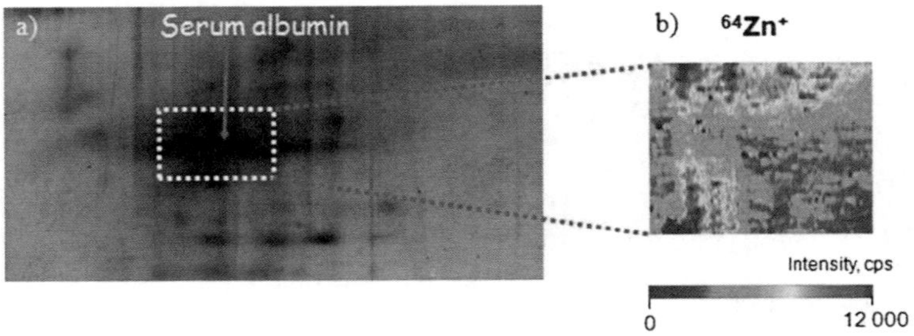

Fig. 3.17. (**a**) A 2D BN-PAGE gel (rat kidney water extract) (**b**) showing the detection of (**b**) Zn-containing proteins by imaging of the gel section using LA-ICP-SFMS (5).

4.4. Further Studies by Imaging LA-ICP-MS on Biological Tissues

Of special interest is the quantitative determination of selenium distribution in biological tissues. Over the past three decades, selenium has been intensively investigated as an antioxidant trace element (59). It is widely distributed throughout the body, but is particularly retained in the brain, even in the case of prolonged dietary selenium deficiency. Changes in selenium concentration in blood and brain have been reported in Alzheimer's disease and brain tumors. The functions of selenium are believed to be carried out by selenoproteins, in which selenium is specifically incorporated as the amino acid, selenocysteine. Several selenoproteins are expressed in the brain and possess antioxidant activities and the ability to promote neuronal cell survival (59). An analytical imaging technique was developed to analyze thin brain sections by LA-ICP-SFMS. Due to the low selenium concentration in brain tissue, the technique was verified on slug tissues. For the quantification of selenium, again matrix-matched standard reference materials doped with selenium were prepared and analyzed. The selenium distribution is illustrated, together with the photograph of cross section of a slug, in **Fig. 3.18**. By LA-ICP-MS measurements of selenium in snail tissue a layered structure with a higher selenium concentration, e.g., in the skin and different organs was found. The selenium concentration in a 100 μm thin section of

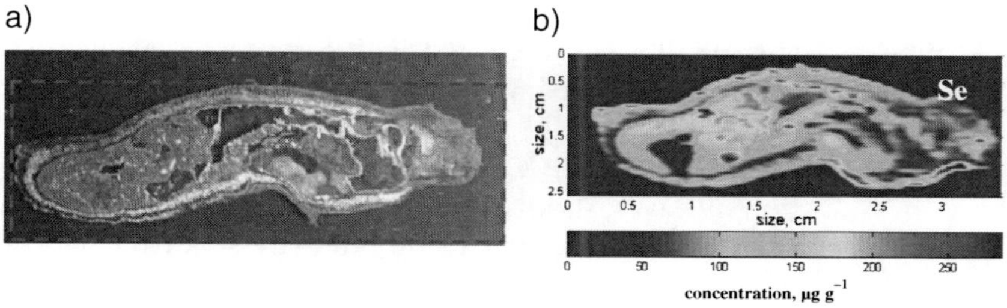

Fig. 3.18. (**a**) Selenium distribution in slug tissue measured by LA-ICP-SFMS (Element and Cetac LSX 200) and (**b**) a photograph of the tissue slice.

snail tissue was observed to be up to 25 μg g^{-1}. The detection limit for selenium was found to be 150 ng g^{-1} (27).

LA-ICP-MS was employed to study biomonitoring of metal contamination in longitudinal tissue sections of the marine snail (60). The results of imaging of essential metals (Cu, Fe and, Zn), toxic metals (Pb, Cd, Hg, and As), radioactive metals (U and Th) and two halogens (I and Cl) in snail tissue measured by LA-ICP-MS compared to a photograph of the slice are shown in **Fig. 3.19**. The mass spectrometric analysis yielded an inhomogeneous distribution for all elements investigated. The detection limits for the distribution analysis of Cu, Zn, Cd, Hg and Pb measured by LA-ICP-MS were in the microgram per gram range.

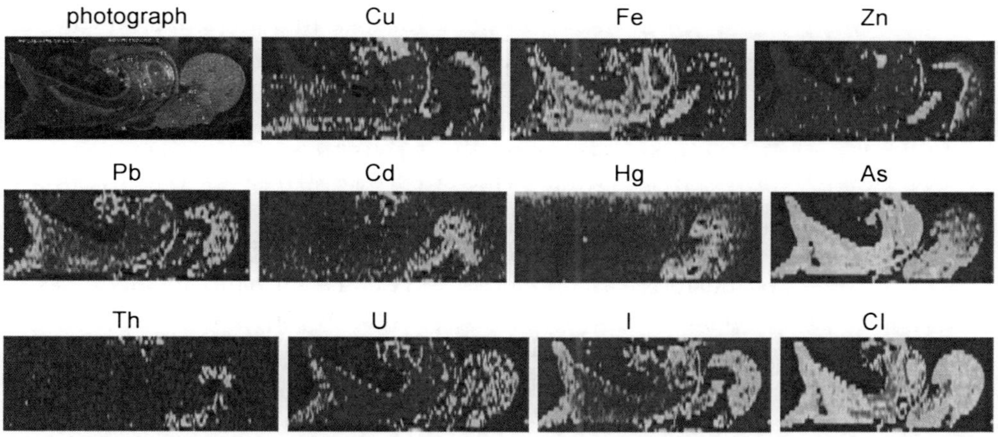

Fig. 3.19. Imaging of essential and toxic metals in snail tissue measured by LA-ICP-MS (Agilent 7500 ce and NewWave) compared to photograph of slice (*left top*).

Metal images in the leaves, shoots and roots of plant tissues (fixed onto glass slides) were produced by LA-ICP-MS (ICP-QMS Elan 6100 from Perkin Elmer/Sciex and Cetac LSX 200) in the author's laboratory (61). Thin sections (30 and 40 μm) of tobacco (*Nicotiana tobaccum*) plant tissues were analyzed by LA-ICP-QMS with respect to Mg, Mn, Fe, Cu and Zn together with the laboratory standards under the same experimental conditions. Together with the nutrient metals, toxic metals like Cd and Pb were detected in the tobacco leaf investigated. For example, the maximum Cd concentration found in the tobacco leaves was 5 $\mu g\ g^{-1}$. Quantitative imaging of the selected elements by LA-ICP-QMS revealed their inhomogeneous distribution in leaves, shoots, and roots.

Certain terrestrial plants known as "metal hyperaccumulators" can accumulate high concentrations of potentially toxic metallic elements (such as zinc, manganese, nickel, cobalt, copper, selenium, cadmium or arsenic) in their leaves and stems without suffering from any impairment of growth. High-resolution secondary ion mass spectrometry (NanoSIMS) was employed for studies of metal distribution on longitudinal sections of *Alyssum lesbiacum* leaves (62). Smart et al. observed high concentrations of nickel in the peripheral regions of the large unicellular stellate leaf hairs (trichomes) and in the epidermal cell layer. Electron probe microanalysis (EPMA) was used to provide independent confirmation of elemental distribution in the specimens, but the superior spatial resolution and high chemical sensitivity of the nanoSIMS technique provided a more detailed image of elemental distribution in these biological specimens at the cellular level. Whereas EMPA allowed a quantification of the metal concentration in the percentage and sub-percentage range, SIMS only yielded qualitative ion images of a small analysis area but at excellent lateral resolution in the 50 nm range.

Quantitative imaging of essential nutrients in the plant leaves was conducted using LA-ICP-MS to study the accumulation and distribution of metals and non-metals by metal-tolerant/hyperaccumulator plants, that means the uptake and transport of nutrients in plants which are able to hyperaccumulate toxic metals. After ^{65}Cu treatment, the leaves of the Cu-tolerant plant *E. splendens* were scanned directly with a focused laser beam in a laser ablation chamber to investigate the accumulation of Cu and other essential nutrients in the leaves. The ablated material was transported with argon as the carrier gas to a quadrupole-based ICP-MS (ICP-QMS), and the ion intensities of $^{65}Cu^+$,$^{39}K^+$, $^{55}Mn^+$, $^{31}P^+$ and $^{11}B^+$ were measured by ICP-QMS. For quantification purposes, synthetic laboratory standards were prepared from standard reference material (NIST SRM 1515 Apple Leaves) spiked with defined concentrations of analytes and used for calibration. The standards were measured together with the leaf samples. The quantification procedure was validated by standard reference material SRM NIST 1547 Peach Leaves using one-point calibration. The resulting calibration curves from the prepared standards showed good linearity with the correlation coefficients (R^2>0.996). The distribution profiles of Cu, K, Mn, P and B in the leaves of *E. splendens* were quantified using these calibration curves. As shown in **Fig. 3.20**, the shape and structure

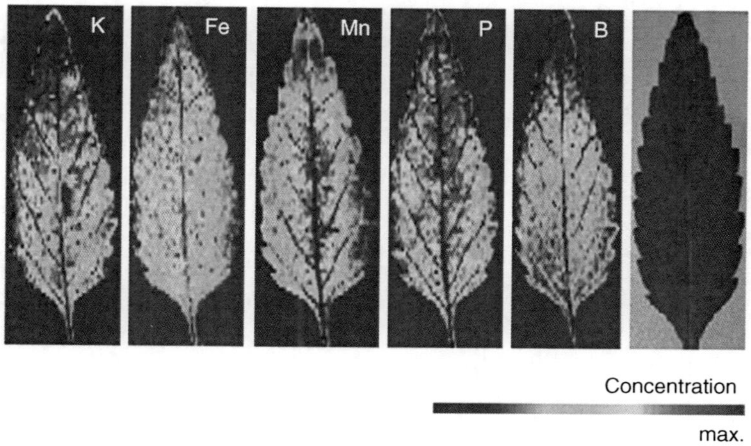

Fig. 3.20. Quantitative images of nutrient elements K, Fe, Mn, P and B measured by LA-ICP-MS in the leaves of *E. splendens* after Cu treatment.

of the leaves was clearly given by LA-ICP-MS imaging of these elements. After treatment, the maximum Cu content was found in the veins near the petiole and at the bottom edge around the petiole of the newly formed leaves. In a further study, the ligand binding with Cu in the leaves will be studied by biomolecular mass spectrometry.

Newly developed imaging techniques using LA-ICP-MS as an elemental analytical technique in combination with MALDI-MS as a biomolecular mass spectrometric technique offer the capability to provide new information on analyzed tissue samples in order to understand and explain basic processes in the life sciences.

5. Future Developments of Imaging by LA-ICP-MS in the Nanometer Range

In order to improve the lateral resolution of LA-ICPMS down to the nanometer-scale range, near-field LA-ICP-MS NF-LA-ICP-MS was created by Becker et al. (39–41). This technique uses the near-field enhancement effect at the tip of a thin silver needle in a laser beam (Nd:YAG laser, wavelength – 532 nm) on the sample surface. The thin silver needle was etched electrolytically in an electrochemical cell using a droplet of citric acid as the electrolyte. A robust needle etching procedure was established to produce the thin needles with a tip diameter in the hundreds of nanometer range. The “sample-to-tip” distance was controlled via the measurement of a tunnel current between the needle and the sample surface (39). For nanolocal analysis by NF-LA-ICP-MS on soft matter (for example, on 2D gels and biological samples), a small-volume transparent laser ablation chamber was constructed and coupled to a double-focusing sector field ICP mass spectrometer. A small amount of soft sample material was ablated at atmospheric pressure by a single laser shot in the near field of the silver tip in the defocused Nd:YAG laser beam. By single-shot analysis on 2D gels and biological surfaces doped with uranium, an enhancement of the ion intensities of the transient signals of up to factor 60 was observed compared to a background signal. Using the near-field effect in LA-ICP-MS, a nanolocal analysis will be possible on biological samples with nanometer-scale spatial resolution. The present experiments on near-field LA-ICP-MS open up a new and challenging path for future applications in imaging elements in the nanometer range in the life sciences, biology and medicine, e.g., for analyses of single cells, cell organelles, or biological structures in the nanometer range in order to detect disease, but also in materials science, nanotechnologies, and nanoelectronics.

6. Conclusions

Imaging mass spectrometric techniques using LA-ICP-MS was developed for imaging the distribution of metals and non-metals in thin sections of biological tissues. The quantitative imaging analysis of essential and toxic elements in biological tissues allows studies to be performed of element distribution, transport processes, bioavailability, and possible contamination.

The results of imaging mass spectrometry using LA-ICP-MS in combination with biomolecular mass spectrometry provide novel information on the distribution of elements and element species in biological tissues and enable the identification of protein structures.

Acknowledgments

The authors would like to thank A. Matusch (Philipps University of Marburg), M. Wagner (Goethe University Frankfurt, Germany) and D. Salber (Forschungszentrum Jülich, Germany) for providing biological tissues and M. Zoriy and A. Zimmermann (Forschungszentrum Jülich, Germany) and B. Wu (Zhejiang University, Hangzhou, China) for technical support with LA-ICP-MS measurements.

References

1. Becker, J. S. (2007)*Inorganic Mass Spectrometry: Principles and Applications.* John Wiley & Sons, New York, NY.
2. Becker, J. S. (2002) Determination of trace elements in small amounts of environmental samples by ICP-MS: a review. *Can J Anal Sci Spectr*, **47**, 98–108
3. Pozebon, D., Dressler, V., Becker, J. S., Matusch, A., Zoriy, M., Becker, J. S. (2008) Biomonitoring of essential and toxic elements in small biological tissues by ICP-MS. *J Anal At Spectrom*, **23**, 1281–1284.
4. Hutchinson, R. W., Cox, A. G., McLeod, C. W., Marshall, P. S., Harper, A., Dawson, E. L., Howlett, D. R. (2005) Imaging and spatial distribution of ß-amyloid peptide and metal ions in Alzheimer's plaques by LA-ICP-MS. *Anal Biochem*, **346**, 225.
5. Becker, J. S., Zoriy, M., Wu, B., Matusch, A., Becker, J. Su. (2008) Imaging of essential and toxic elements in biological tissues by LA-ICP-MS. *J Anal At Spectrom*, **23**, 1275–1280.
6. Punshon, T., Jackson, B. P., Lanzirotti, A., Hopkins, W. A., Bertsch, P. M., Burger, J. (2005) Application of synchrotron X-ray microbeam spectroscopy to the determination of metal distribution and speciation in biological tissues. *Spectrosc Lett*, **38**, 343–363.
7. Carmona, A., Cloetens, P., Devès, G., Bohic, S., Ortega, R. (2008) Nano-imaging of trace metals by synchrotron X-ray fluorescence into dopaminergic single cells and neurite-like processes. *J Anal At Spectrom*, **23**, 1083–1088.
8. Todd, P. J., Schaaf, T. G., Chaurand, P., Caprioli, R. M. (2001) Organic ion imaging of biological tissue with secondary ion mass spectrometry and matrix-assisted laser desorption/ionization. *J Mass Spectrom*, **36**, 355.
9. Chandra, S. (2003) SIMS ion microscopy as a novel, practical tool for subcellular chemical imaging in cancer research. *Appl Surf Sci*, **203–204**, 679–683.
10. Touboul, D., Halgand, F., Brunelle, A., Kersting, R., Tallarek, E., Hagenhoff, B., Laprevote, O. (2004) Tissue molecular ion imaging by gold cluster ion bombardment. *Anal Chem*, **76**, 1550.
11. Broersen, A., van Liere, R., Altelaar, A. F. M., Heeren, R. M. A., McDonnell, L. A. (2008) Automated, feature-based image alignment for high-resolution imaging mass spectrometry of large biological samples. *J Am Soc Mass Spectrom*, **19**, 823–832.
12. McDonnell, L. A., Piersma, S. R., Altelaar, A. F. M., Mize, T. H., Luxembourg, S. L., Verhaert, P. D. E. M., van Minnen, J., Heeren, R. M. A. (2005) Subcellular imaging mass spectrometry of brain tissue. *J Mass Spectrom*, **40**, 160–168.
13. Brunelle, A., Laprevote, O. (2007) Recent advances in biological tissue imaging with time-of-flight secondary ion mass spectrometry: polyatomic ion sources, sample preparation, and applications. *Curr Pharm Des*, **13**, 3335–3343.
14. Brunelle, A., Touboul, D., Laprévote, O. (2005) Biological tissue imaging with time-of-flight secondary ion mass spectrometry and cluster ion sources. *J Mass Spectrom*, **40**, 985–999.
15. Heeren, M. A., Mc. Donnell, L. A., Amstalden, E., Luxaebourg, S. L., Altelaar, A. F. M., Piersma, S. R. (2006) Why don't biologists use SIMS? A critical evaluation of imaging MS. *Appl Surf Sci*, **252**, 6827–6835.
16. Chandra, S., Tjarks, W., Lorey, D. R., Barth, R. F. (2007) Quantitative subcellular imaging of boron compounds in individual mitotic and interphase human glioblastoma cells with imaging secondary ion mass spectrometry (SIMS). *J Microsc*, **229**, 92–103.
17. Becker, J. S., Zoriy, M., Becker, J. Su., Dobrowolska, J., Matusch, A. (2007) Imaging mass spectrometry by laser ablation inductively coupled plasma mass spectrometry in biological tissues and proteomics. *J Anal At Spectrom*, **22**, 736–744.
18. Becker, J. S., Becker, J. Su., Zoriy, M., Dobrowolska, J., Matusch, A. (2007) Imaging mass spectrometry in biological tissues by laser ablation inductively coupled plasma mass spectrometry. *Eur J Mass Spectrom*, **13**, 1–6.
19. Dobrowolska, J., Dehnhardt, M., Matusch, A., Zoriy, M., Koscielniak, P., Zilles, K., Becker, J. S. (2008) Quantitative imaging of zinc, copper and lead in three distinct regions of the human brain by laser ablation inductively coupled plasma mass spectrometry. *Talanta*, **74**, 717–723.
20. Becker, J. S., Zoriy, M., Matusch, A., Salber, D., Palm, C., Becker, J. Su. (2010) Bioimaging of metals by laser ablation inductively coupled plasma mass spectrometry (LA-ICP-MS). *Mass Spectrom Rev*, **29**, 156–175.
21. Becker, J. S., Zoriy, M., Pickhardt, C., Palomero-Gallagher, N., Zilles, K. (2005) Imaging of copper, zinc, and other elements

in thin section of human brain samples (hippocampus) by laser ablation inductively coupled plasma mass spectrometry. *Anal Chem*, 77, 3208–3216.

22. Caprioli, R. M., Farmer, T. B., Gile, J. (1997) Molecular imaging of biological samples: localization of peptides and proteins using MALDI-TOF MS. *Anal Chem*, **69**, 4751–4760.
23. Stoeckli, M., Chaurand, P., Hallahan, D. E., Caprioli, R. M. (2001) Imaging mass spectrometry: a new technology for the analysis of protein expression in mammalian tissues. *Nat Med*, 7, 493–496.
24. Norris, J. L., Cornett, D. S., Mobley, J. A., Andersson, M., Seeley, E. H., Chaurand, P., Caprioli, R. M. (2007) Processing MALDI mass spectra to improve mass spectral direct tissue analysis. *Int J Mass Spectrom*, **260**, 212–221.
25. Chaurand, P., Cornett, D. S., Caprioli, R. M. (2006) Molecular imaging of thin mammalian tissue sections by mass spectrometry. *Curr Opin Biotech*, **17**, 431–436.
26. Pohl, H. R., Abadin, H. G., Risher, J. F. (2006) Neurotoxicity of cadmium, lead and mercury. *Metal Ions in Life Sciences* (Sigel, A., Sigel, H., Sigel, R. K. O. eds.), 395–425, Wiley & Sons, New York, NY.
27. Becker, J. S., Matusch, A., Depboylu, C., Dobrowolska, J., Zoriy, M. (2007) Quantitative imaging of selenium, copper, and zinc in thin sections of biological tissues (slugs-genus arion) measured by laser ablation inductively coupled of plasma mass spectrometry. *Anal Chem*, **79**, 6074–6080.
28. Feldmann, J., Kindness, A., Ek, P. (2002) Laser ablation of soft tissue using a cryogenically cooled ablation cell. *J Anal At Spectrom*, **17**, 813–818.
29. Kindness, A., Sekaran, N., Feldmann, J. (2003) Two-dimensional mapping of copper and zinc in liver sections by laser ablation-inductively coupled plasma mass spectrometry. *Clin Chem*, **49**, 1916–1923.
30. Ghazi, A. M., Wataha, J. C., O'Dell, N. L., Singh, B. B., Simmons, R., Shuttleworth, S. (2002) Quantitative concentration profiling of nickel in tissues around metal implants: a new biomedical application of laser ablation sector field ICP-MS. *J Anal At Spectrom*, **17**, 1295.
31. Becker, J. S., Zoriy, M., Becker, J. Su., Pickhardt, C., Damoc, E., Juhacz, G., Palkovits, M., Przybylski, M. (2005) Determination of phosphorus copper and zinc containing human brain proteins by LA-ICP-MS and MALDI-FTICR-MS. *Anal Chem*, 77, 5851–5860.
32. Matusch, A., Depboylu, C., Palm, C., Wu, B., Höglinger, G. U., Schäfer, M. K.-H., Becker, J. S. (2010) Cerebral bio-imaging of Cu, Fe, Zn and Mn in the MPTP mouse model of Parkinson's disease using laser ablation inductively coupled plasma mass spectrometry (LA-ICP-MS). *J Am Soc Mass Spectrom*, **21**, 161–171.
33. Becker, J. Su., Pozebon, D., Dressler, V., Lobinski, R., Becker, J. S. (2008) LA-ICP-MS studies of zinc exchange by copper in bovine serum albumin using an isotopic enriched copper tracer. *J Anal At Spectrom*, **23**, 1076–1082.
34. Becker, J. Su., Mounicou, S., Zoriy, M. V., Becker, J. S., Lobinski, R. (2008) Analysis of metal-binding proteins separated by non-denaturating gel electrophoresis using matrix-assisted laser desorption/ionization mass spectrometry (MALDI-MS) and laser ablation inductively coupled plasma mass spectrometry (LA-ICP-MS). *Talanta*, **76**, 1183–1188.
35. Feldmann, I., Koehler, C. U., Roos, P. H., Jakubowski, N. (2006) Optimisation of a laser ablation cell for detection of heteroelements in proteins blotted onto membranes by use of inductively coupled plasma mass spectrometry. *J Anal At Spectrom*, **21**, 1006–1015.
36. Becker, J. Su., Zoriy, M., Przybylski, M., Becker, J. S. (2007) Study of formation of Cu- and Zn-containing tau protein using isotopic-enriched tracers by LA-ICP-MS and MALDI-FTICR-MS. *J Anal At Spectrom*, **22**, 63–68.
37. Becker, J. S., Zoriy, M., Dehnhardt, M., Pickhardt, C., Zilles, K. (2005) Copper, zinc, phosphorus and sulfur distribution in thin section of rat brain tissues measured by laser ablation inductively coupled plasma mass spectrometry: possibility for small-size tumor analysis. *J Anal At Spectrom*, **20**, 912–917.
38. Zoriy, M., Kayser, M., Izmer, A., Pickhardt, C., Becker J. S. (2005) Determination of uranium isotopic ratios in biological samples using laser ablation inductively coupled plasma double focusing sector field mass spectrometry with cooled ablation chamber. *Int J Mass Spectrom*, **242**, 297–302.
39. Zoriy, M., Kayser, M., Becker, J. S. (2008) Possibility of nano-local element analysis by near-field laser ablation inductively coupled plasma mass spectrometry (LA-ICP-MS): new experimental arrangement and first application. *Int J Mass Spectrom*, **273**, 151–155.

40. Zoriy, M., Becker, J. S. (2009) Near-field laser ablation inductively coupled plasma mass spectrometry (LA-ICP-MS): a novel elemental analytical technique at nanometer scale. *Rapid Commun Mass Spectrom*, **23**, 23–30.
41. Becker, J. S., Gorbunoff, A., Zoriy, M., Izmer, A., Kayser, M. (2006) Evidence of near-field laser ablation inductively coupled plasma mass spectrometry (NF-LA-ICP-MS) at nanometre scale for elemental and isotopic analysis on gels and biological samples. *J Anal At Spectrom*, **21**, 19–25.
42. Pickhardt, C., Izmer, A., Zoriy, M., Schaumloffel, D., Becker, J. S. (2006) On-line isotope dilution in laser ablation inductively coupled plasma mass spectrometry using a microflow nebulizer inserted in the laser ablation chamber. *Int J Mass Spectrom*, **248**, 136–141.
43. Jackson, B., Harper, S., Smith, L., Flinn, J. (2006) Elemental mapping and quantitative analysis of Cu, Zn, and Fe in rat brain sections by laser ablation ICP-MS. *Anal Bioanal Chem*, **384**, 1618.
44. Zoriy, M., Becker, J. S. (2007) Imaging of elements in thin cross sections of human brain samples by LA-ICP-MS: A study on reproducibility. *Int J Mass Spectrom*, **264**, 175–180.
45. Wu, B., Zoriy, M., Chen, Y., Becker, J. S. (2009) Imaging of nutrient elements in the leaves of Elsholtzia splendens by laser ablation inductively coupled plasma mass spectrometry (LA-ICP-MS). *Talanta*, **78**, 132–137.
46. Zoriy, M., Dehnhardt, M., Matusch, A., Becker, J. S. (2008) Comparative imaging of P, S, Fe, Cu, Zn and C in thin sections of rat brain tumor as well as control tissues by laser ablation inductively coupled plasma mass spectrometry. *Spectrochim Acta B*, **63**, 375–382.
47. Dehnhardt, M., Zoriy, M., Khan, Z., Reifenberger, G., Ekstrom, T. J., Becker, J. S., Zilles, K., Bauer, A. (2008) Element distribution is altered in a zone surrounding human glioblastoma multiforme. *J Trace Elem Med Biol*, **22**, 17–23.
48. Langen, K. J., Salber, D., Hamacher, H., Stoffels, G., Reifenberger, G., Pauleit, D., Coenen, H., Zilles, K. (2007) Detection of secondary thalamic degeneration after cortical infarction using cis-4-18F-fluoro-D-proline. *J Nucl Med*, **48**, 1482–1491.
49. Langen, K. J., Hamacher, K., Weckesser, M., Floeth, F., Stoffels, G., Bauer, D., Coenen, H. H., Pauleit, D. (2006) *O*-(2-[18F]fluoroethyl)-L-tyrosine: uptake mechanisms and clinical applications. *Nucl Med Biol*, **33**, 287–294.
50. Zoriy, M., Matusch, A., Spruss, T., Becker, J. S. (2007) Laser ablation inductively coupled plasma mass spectrometry for imaging of copper, zinc, and platinum in thin sections of a kidney from a mouse treated with cis-platin. *Int J Mass Spectrom*, **260**, 102–106.
51. Pozebon, D., Dressler, V., Matusch, A., Becker, J. S. (2008) Monitoring of platinum in a single hair by laser ablation inductively coupled plasma mass spectrometry (LA-ICP-MS) after cisplatin. *J Mass Spectrom*, **272**, 57–62.
52. Qureshi, G. A., Syed, S. A., Parvez, S. H. (2007) Role of selenium, iron, copper and zinc. *Oxidative Stress and Neurodegenerative Disorders*, **719**, Elsevier, Maryland Heights, MO.
53. Depboylu, C., Matusch, A., Tibl, F., Zoriy, M., Michel, P., Riederer, P., Gerlach, M., Becker, J. S., Örtel, W. H., Höglinger, G. U. (2007) Glia protects neurons against extracellular human neuromelanin. *Neurodegener Dis*, **4**, 218–226.
54. Gerlach, M., Double, K. L., Götz, M. F., Youdim, M. B. H., Riederer, P. (2006) The role of iron in the pathogenesis of Parkinson's disease. *Neurodegenerative Diseases and Metal Ions* (Sigel, A., Sigel, H., Sigel, R. K. O. eds.), John Wiley & Sons, New York, NY.
55. Faa, G., Lisci, M., Caria, M. P., Ambu, R., Sciot, R. N., Nurchi, V. M.,, Silvagni, R., Diaz, A, Crisponi, G. (2001) Brain copper, iron, magnesium, zinc, calcium, sulfur and phosphorus storage in Wilson's disease. *J Trace Elem Med Biol*, **15**, 155–160.
56. Sigel, A., Sigel, H., Sigel, R. K. O. (2006) *Neurodegenerative Diseases and Metal Ions.* John Wiley & Sons, New York, NY.
57. Becker, J. S., Dobrowolska, J., Zoriy, M., Matusch, A. (2008) Imaging of uranium on rat brain sections using LA-ICP-MS: a new tool for the study of critical substructures affined to heavy metals in tissues. *Rapid Commun Mass Spectrom*, **22**, 2768–2772.
58. Shi, W., Chance, M. R. (2008) Metallomics and metalloproteomics. *Cell Mol Life Sci (CMLS)*, **65**, 3040–3048.
59. Chen, J., Berry, M. J. (2003) Selenium and selenoproteins in the brain and brain diseases. *J Neurochem*, **86**, 1–12.
60. Santos, M. C., Wagner, M., Wu, B., Oehlmann, J., Cadore, S., Becker, J. S. (2009) Biomonitoring of metal contamination in a marine prosobranch

snail (Nassarius reticulatus) by imaging laser ablation inductively coupled plasma mass spectrometry (LA-ICP-MS). *Talanta*, **80**, 428–433.

61. Becker, J. S., Dietrich, R. C., Matusch, A., Pozebon, D., Dressler, V. L. (2008) Quantitative images of metals in plant tissues measured by laser ablation inductively coupled plasma mass spectrometry. *Spectrochim Acta*, **B63**, 1248–1252.
62. Smart, K. E., Kilburn, M. R., Salter, C. J., Smith, J. A. C., Grovenor, C. R. M. (2007) NanoSIMS and EPMA analysis of nickel localisation in leaves of the hyperaccumulator plant Alyssum lesbiacum. *Int J Mass Spectrom*, **260**, 107–114.
63. Becker, J. S., Zoriy, M., Krause-Buchholz, U., Becker, J. S., Pickhardt, C., Przybylski, M., Pompe, W., Roedel, G. (2004) In-gel screening of phosphorus and copper, zinc and iron in proteins of yeast mitochondria by LA-ICP-MS and identification of phosphorylated protein structures by MALDI-FT-ICR-MS after separation with two-dimensional gel electrophoresis. *J Anal At Spectrom*, **19**, 1236–1243.

Part II

Protocols for MS Imaging of Distribution of Small Molecules Including Metabolites and Pharmaceuticals

Chapter 4

Lipid Detection, Identification, and Imaging Single Cells with SIMS

Michael L. Heien, Paul D. Piehowski, Nicholas Winograd, and Andrew G. Ewing

Abstract

Time-of-flight secondary ion mass spectrometry (ToF-SIMS) can be utilized to map the distribution of various molecules on a surface with submicrometer resolution. Many of its biological applications have been in the study of membrane lipids, such as phospholipids and cholesterol. For these studies, the effectiveness of chemical mapping is limited by low signal intensity from various biomolecules. Because of the high-energy nature of the SIMS ionization process, many molecules are identified by detection of characteristic fragments. Cluster ion sources are able to increase ionization, leading to increased information collected from a surface. In this chapter, we highlight the utility of SIMS to image lipids at single cells. Particularly, we will describe sample preparation, data collection, and the analysis of lipids for two systems; rat oligodendrocytes and *Tetrahymena thermophila*. SIMS spectra yield information regarding lipid identity and concentration across cell surface.

Key words: Time-of-flight (ToF), secondary ion mass spectrometry (SIMS), freeze-fracture, freeze-etch, imaging, mass spectrometry, lipids, single cell.

1. Introduction

Mass spectrometry imaging with time-of-flight secondary ion mass spectrometry (ToF-SIMS) has revealed the spatial distribution of chemicals on a surface (1). When applied to biological samples, this method offers spatial information on biologically relevant small molecules (<1,000 Da). Indeed, ToF-SIMS imaging has been shown to be a powerful analytical tool for mapping the distribution of lipids on a cell membrane at the single-cell

S.S. Rubakhin, J.V. Sweedler (eds.), *Mass Spectrometry Imaging*, Methods in Molecular Biology 656,
DOI 10.1007/978-1-60761-746-4_4,

level (2–11). This method combines submicrometer spatial resolution, high chemical specificity, and surface sensitivity, making it a promising tool for the study of lipids in cellular membranes. Briefly, a pulsed beam of primary ions is directed at the analysis surface, causing the sputtering of ions and molecules. The ions are extracted into a ToF mass analyzer generating a mass spectrum. To obtain an image, the primary beam is raster-scanned across the surface, recording a mass spectrum for each pixel. A particular m/z is then selected from the spectrum, and its intensity is mapped across the imaging area.

When combined with cryogenic sample preparation techniques, imaging ToF-SIMS permits the detailed study of membrane lipids during dynamic processes such as membrane fusion. Mass spectrometric imaging with freeze-fracture sample preparation allows the study of lipid domain formation in native membranes without chemical labels. This chapter outlines the methods necessary to perform ToF-SIMS imaging. We outline sample preparation methods for two types of cells, rat oligodentrocytes and *T. thermophila*, an adhering and non-adhering cell type, respectively. The cells are cultured, mounted onto a silicon substrate, frozen, fractured, and analyzed. An alternate sample treatment, freeze-etching (8), is briefly described in the **Section 3**. These methods can be adapted for a given system. Images are then generated, and major lipid fragments are identified during data analysis.

2. Materials

2.1. Cell Culture

1. Tetrahymena cultures were obtained from Craig T. VanBell at Edinboro University, Edinboro, PA, USA.
2. PPYGFe medium: 2% proteose peptone (Becton, Dickinson and Company, Sparks, MD, USA), 0.1% yeast extract (Becton, Dickinson and Company), 0.2% yeast extract (Becton, Dickinson and Company), and 0.003% ferric EDTA. Prepare immediately prior to use.
3. Tris buffer solution: 10 mM Tris–HCl, pH adjusted to 7.4 using 1 M NaOH. Store at room temperature.
4. 15 ml centrifuge tubes.
5. 2 L baffled polycarbonate Erlenmeyer flasks with filtered cap (Sigma Aldrich, Allentown, PA, USA).
6. Primary Sprague–Dawley rat pup oligodendrocytes were obtained from Gong Chen, The Pennsylvania State University.

7. Minimum Essential Medium (Invitrogen).
8. N1 supplement (Sigma Aldrich).
9. 50 ml polystyrene tissue culture flasks.

2.2. Sample Preparation

1. HPLC grade hexane (EMD chemicals, Gibbstown, NJ, USA).
2. Absolute ethanol (Pharmco-AAPER, Brookfield, CT, USA).
3. Deionized water (18 MΩ).
4. 5×5 mm silicon shards (Ted Pella, Redding, CA, USA).
5. 20 ml scintillation vials.
6. Ethane 99%.
7. Liquid nitrogen (LN_2).
8. Storage dewar, 50 l minimum.
9. London finder grid (Ted Pella, Redding, CA, USA).
10. Bath Sonicator.
11. Fluoroware wafer carrier box (Fluoroware Inc., Chaska, MN, USA).
12. Colloidal silver liquid (Ted Pella).

2.3. Instrumentation

1. Kratos Prism ToF-SIMS spectrometer (Manchester, UK) or similarly designed static SIMS imaging instrument.
2. Indium liquid metal ion source (Beaverton, OR, USA).
3. Microchannel plate detector (Galileo Co., Sturbridge, MA, USA).
4. Channeltron detector (Burle, Lancaster, PA, USA).
5. Vertical illuminator microscope (Olympus, Melville, NY, USA).
6. Spot RT CCD Camera (Diagnostic Instruments, Sterling Heights, MI, USA).
7. Keithley 485 Auto Ranging Picoammeter (Keithley, Inc., Solon, OH, USA).
8. Custom-made freeze-fracture sample preparation chamber, **Fig. 4.2c** (12).

3. Methods

As stated in the introduction, we outline sample preparation methods for two types of cells, rat oligodendrocytes and *T. thermophila*. The culture protocols for *T. thermophila* are discussed;

however, this does not limit the analysis to this type of cell. The cells are cultured, mounted onto a silicon substrate, frozen, fractured, and analyzed. An alternate sample treatment, freeze-etching, is also described. These methods are transferable to imaging other single cells, and it is assumed that the reader has some familiarity with SIMS mass spectrometry.

3.1. Cell Culture

1. Autoclave PPYGFe medium.
2. The two complementary mating types of *T. thermophila*, B2086 and Cu428.1, are maintained separately in autoclaved PPYGFe medium. Cells are grown overnight in 2 L flasks at 30°C with shaking (3).
3. Prior to mating, the cells are washed by centrifugation, 400×*g* for 5 min, in 10 mM Tris buffer (pH 7.4). The cells are resuspended in the Tris buffer and shaken overnight at 30°C.
4. To initiate mating, the complementary strains are mixed in equal proportions and incubated, without shaking, at 30°C for 4 h.
5. Rat oligodendrocytes are maintained in MEM with N1 supplement and incubated at 37.4°C until maturation was observed under a microscope.

3.2. Preparation of Cells for Mass Spectrometry Imaging

1. Before use the silicon shards must be cleaned (*see* **Note 1**). To do this, the shards are placed in a glass scintillation vial and covered with approximately 10 ml of HPLC grade hexane (*see* **Note 2**). The vial is placed in a bath sonicator for 10 min. Drain the hexane and repeat. A third rinse using absolute ethanol should be conducted in the same manner.
2. After cleaning the shards are rinsed with DI water and blown dry with nitrogen (*see* **Note 3**). Shards are stored in airtight Fluoroware containers until use.
3. For adherent oligodendrocyte, the Si substrates must be added to the culture dish to allow adhesion. Rat oligodendrocytes are dislodged from the culture flask by agitation, most often by tapping gently against the side of the culture flask. The culture medium, containing dislodged cells, is drawn into a pipette. The solution is slowly added to a sterile culture dish containing the cleaned substrates. The dish is then incubated at 37.4°C for 30 min.
4. While the samples are in the incubator the liquid ethane for plunge-freezing is prepared. To collect liquid ethane, an ethane tank is fitted with a regulator and held in a ring stand. The regulator is attached to a length of flexible PVC tubing with a graduated pipette in the opposite end to create a small aperture for the hose. The pipette is placed in a LN_2-cooled

beaker. The ethane is leaked into the beaker. The flow rate is adjusted to maximize flow without letting the ethane escape. It takes a few minutes to collect a sufficient amount of liquid ethane. When the desired amount is acquired, the beaker is removed from the liquid nitrogen. The ethane will last about 10 min.

5. After incubation the substrates are removed from the medium and quickly (3 s or less), rinsed in DI water. A top shard is placed *directly* on the substrate to create a sandwich. The arrangement is shown in **Fig. 4.1**. The excess water is wicked away using a Kimwipe (*see* **Note 4**). The sandwiches are then plunged in liquid ethane for approximately 5 s, or until the ethane stops boiling. Once the shard is frozen it is placed under LN_2 for storage until use.
6. *T. thermophila* are non-adherent cells. 200 μl of suspended cell solution is dropped onto a clean substrate. The sample is covered with a top shard by placing it *directly* on the sample. The excess water is wicked away using a Kimwipe. The sample is then plunge-frozen in liquid ethane for 5 s, or until the ethane stops boiling. The sample sandwich is then placed under LN_2 for storage until use (3).

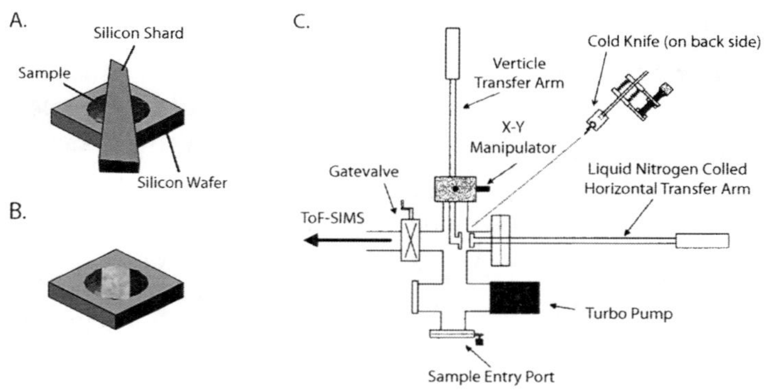

Fig. 4.1 Sample on silicon shards. (**a**) Silicon substrate with sample. (**b**) Silicon substrate with sample after freeze-fracture, showing the exposed area. (**c**) Schematic diagram of the freeze-fracture chamber attached to the TOF-SIMS instrument. Thermocouple attached to the bottom of the vertical transfer rod is not shown. ((**c**) Reproduced from (12) with permission from the American Chemical Society.)

3.3. Sample Transfer into the Vacuum Environment

1. Before entering the sample into the mass spectrometer, all of the components that contact the sample must be cooled (*see* **Note 5**). The instrument has five dewars connected to it. The instrument is cooled by flowing gaseous N_2, at a pressure of $\approx$ 40 psi, through copper tubing that is immersed in the dewars. All five dewars are filled with LN_2 and the temperature is monitored using thermocouple feedthroughs attached

to the cooled components. In addition, the instrument has a sorption pump that evacuates the fast entry port. This pump must be cooled to LN_2 temperature prior to introducing the sample.

2. The sample must be mounted onto the analysis block under LN_2. A London finder grid is first attached to the sample block using conductive silver paste. Then, a Styrofoam cooler is filled with LN_2 and the sample block, samples, and fast entry apparatus are placed in the cooler. When the LN_2 stops boiling, the sample can be mounted.
3. To mount the sample, a shard is fit in the recessed sample area with the top shard facing out. Two screws with washers are used to affix the sample to the block (*see* **Note 6**). The sample block is mounted on the entry port post and fit inside the metal sheath. The sheath is then screwed onto the fast entry port arm, while keeping the sample assembly under nitrogen.
4. Introducing the sample must be done as quickly as possible to minimize warming as well as condensation. The arm assembly is quickly inserted into the entry port, and the port is evacuated using the sorption pump for 30 s. After 30 s, the sample is placed into the instrument and quickly transferred to the pre-cooled sample transfer arm (**Fig. 4.2c**).

3.4. In Vacuo *Freeze-Fracture*

1. The sample is then transferred to the freeze-fracture stage. The temperature and pressure are crucial during the fracture process (13). It is usually necessary to wait several minutes for the vacuum to recover after introducing the sample.
2. To be able to fracture the sample, it is important that the block is mounted with the fracture shard extending above the substrate on the side of the fracture knife. The knife blade is positioned above the top shard. In the *z*-direction, the blade is positioned such that it will fall between the top shard and the sample block when brought down.
3. When the fracture stage reaches the desired temperature of –106.5°C, the sample is fractured by turning the blade counterclockwise with a quick flick of the wrist (*see* **Note 7**). This should result in the top shard snapping off. It is important to design a cold stage to catch the shard to prevent water desorption changing the vacuum and depositing ice on the sample.
4. After the fracture, the blade is withdrawn from the sample approximately 5 mm and positioned directly in front of the sample. The knife blade acts as a cold trap, because it is colder than the sample, catching excess water vapor caused from the fracture.

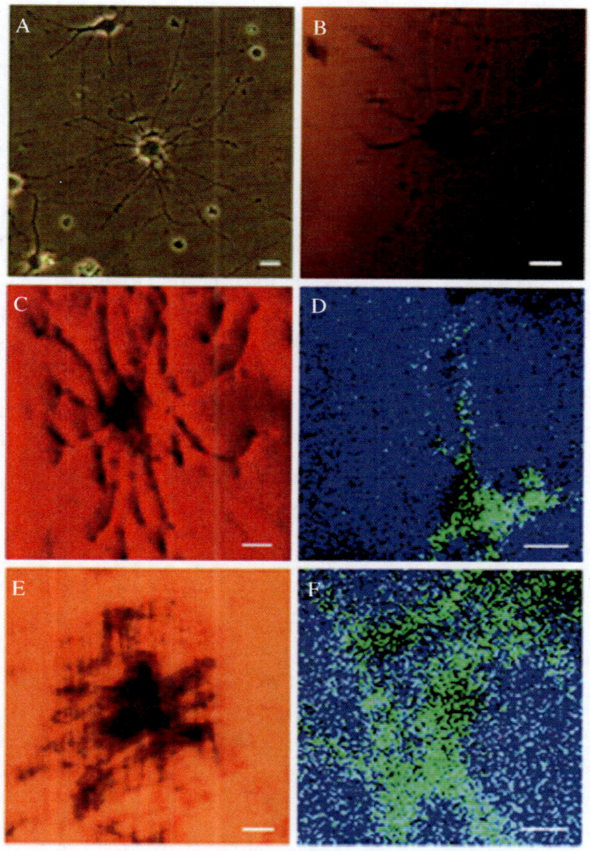

Fig. 4.2 (**a**) Brightfield image of rat oligodendrocyte in culture. (**b**) Brightfield image of an oligodendrocyte cultured on silicon and cryogenically preserved using the described sample preparation method. (**c**) SIM (secondary ion microscopy) image of rat oligodendrocyte. (**d**) SIMS image of a preserved oligodendrocyte, *m/z* 184 (PC) *green*, *m/z* 28 (Si) *blue*, intensity scale 0–4. (**e**) Brightfield image of oligodendrocyte that has been warmed to allow crystallization of water. (**f**) SIMS image of a corresponding area shown is in (**e**), *m/z* 184 (PC) *green*, *m/z* 28 (Si) *blue*, intensity scale 0–2. Images were obtained using an In^+ primary ion beam; all scale bars are 20 μm. (Reproduced from (8) with permission from the American Chemical Society.)

5. When the preparation chamber pressure has recovered from the fracture, the sample can then be transferred to the sample analysis chamber.

3.5. Preparation for Imaging

1. In the case of adherent cells it is useful to do a freeze-etch step prior to imaging (8). This is achieved by removing the dewar from the sample stage cooling line. This will result in the sample stage warming at a rate of approximately 5°C/min. The stage is warmed to –80°C to remove

excess surface water and then quickly returned to LN_2 temperature.

2. The sample beam is focused. A standard sample, a London finder grid, attached to the sample block is used to obtain the desired focus. The primary ion beam is focused to a point by two electrostatic lenses in series. The focal point of these lenses can be manipulated by increasing and decreasing the potential that is applied to them. The London finder grids have distinct features of various sizes and are made to high specifications. The focal point of the beam is manipulated until the user can visually distinguish features of the same size scale as they intend to measure. The interface width (the width measured at which the signal decreases from 85 to 15%) of a sharp interface is a good estimate of the primary ion beam spot diameter. For optimal imaging, this width should be less than or equal to the width of the pixels to be used for imaging.
3. Before cell imaging, the SIMS primary ion beam must be aligned with the microscope mounted on the instrument so that cells to be imaged can be identified and positioned in front of the primary ion beam. To do this, a digital micrograph of the grid is taken using the camera on the microscope. Then a SIMS image of the grid is obtained. The two images are overlaid to determine the imaging area in the eyepiece.
4. Mass calibration of the mass spectrum is important to obtain the proper peak assignments. Calibration can be achieved by using the spectrum obtained from the SIMS image of the grid. A good three-point calibration can be obtained, from Na^+, Cu^+, and Ag^+. The Ag^+ signal comes from the conductive silver paste that is used to mount the grid to sample block.

3.6. Collecting Images

1. When collecting SIMS images the static limit should be used to insure a reliable representation of the surface. The static limit is defined as 1×10^{13} ions/cm^2. To calculate the limit you must know the pulse width, primary ion current, field-of view, and the pixel size. A discussion of each of these is found below.
2. Primary ion current – to measure the primary ion current, a picoammeter is connected to the analysis stage through the stage voltage feedthrough while the beam is in the DC mode. Higher primary ion currents result in faster image collection times, as the static limit is reached faster. However, higher currents are obtained at the expense of increase in spot size, which leads to decreased spatial resolution.
3. Spot size – this is the diameter of the primary ion beam at the focal point; it is the ultimate limit to spatial resolution.

Smaller spot sizes can be obtained through rigorous beam alignment and generally result in lower primary ion currents.

4. Pixel size – the pixel size is the field-of-view (FOV) divided by the amount of pixels chosen for the image (*see* **Note 8**). Choosing fewer pixels results in quicker acquisition times at the expense of spatial resolution.
5. Mass range – the mass range that is collected for an experiment depends on what the analyst is looking for. Traditionally the SIMS analyst collects from 10 to 1,000 amu.

3.7. Data Analysis

1. The first step in data analysis is producing mass-specific images. Images are produced by selecting the mass of interest from the mass spectrum and plotting the intensity, the number of counts per pixel, versus the spatial position. This produces an *x–y* plot of pixels, which corresponds to the distribution of particular atoms or molecules in the specimen. Signal intensity is conveyed through the use of a false-color scale. A table of characteristic fragments for common lipids is given in **Section 3.8.** Because of the high-energy nature of the SIMS ionization process, these fragments are generally of low mass < 200 Da. In addition, lipid species, with the exception of cholesterol, are identified by the lipid headgroup fragments.
2. Localization – because of the complexity and difficulty of the sample preparation, images must be analyzed to verify that the cells are properly preserved. **Figure 4.2** contains both optical and SIMS images of cells which have been freeze-etched. These contain well-prepared cells (**Fig. 4.2a–d**) and cells that have been damaged during the process (**Fig. 4.2e–f**). The first check to perform on cells is to examine the localization of cellular fragments, i.e., phosphocholine (PC), phosphoethanolamine (PE), cholesterol, sphingomyelin SM, and m/z^+ 69. These fragments should be found in the cell region and only in the cell region. Cell size and shape can be gleaned from brightfield images taken before and after analysis. Anti-localization is also important in verifying cell preservation. Sodium and silicon should be anti-localized to the cell, meaning that these ions should only be found surrounding the cell and not in it. Anti-localization of potassium, however, is indicative of a ruptured cell.
3. Line scans are used for the relative quantification of the distribution of lipids. A line scan is a plot of the signal intensity for a given ion, as a function of its lateral position on a line, drawn by the analyzer. The width and length of the line are chosen based on the feature to be displayed. This yields a method that can be used to quantify data. **Figure 4.3** contains line scans taken on a pair of mating

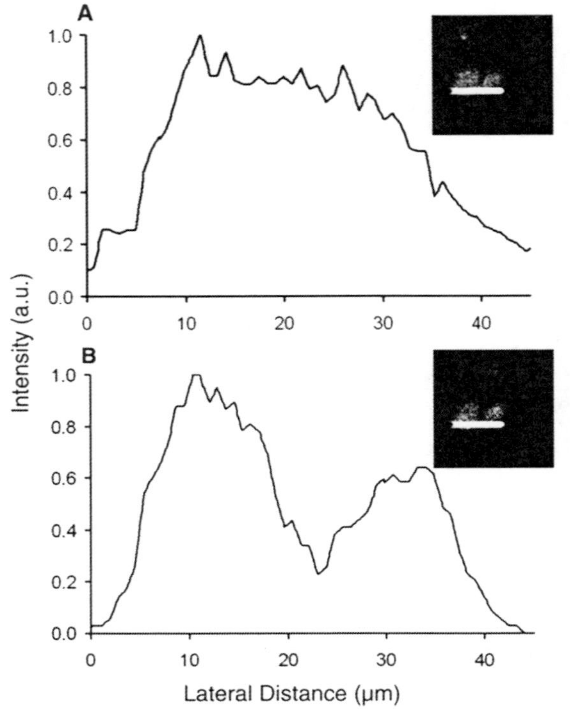

Fig. 4.3. Line scans of the molecule-specific images of *Tetrahymena* graphically support the idea that PC decreases at the conjugation junction between mating cells. Data points were collected every 120 nm. (**a**) Line scan for *m/z* 69 through the conjugation junction, illustrating that the total lipid content is relatively constant across the mating cells. The *inset* shows the SIMS image for *m/z* 69, highlighting the pixels used for the line scan. (**b**) Line scan for *m/z* 184 through the junction, demonstrating a sharp decrease in signal at the conjugation junction. The *inset* shows the SIMS image for *m/z* 184, highlighting the pixels used for the line scan. (Reproduced from (3) with permission from the Association for the Advancement of Science.)

cells (*T. thermophila*). They demonstrate the formation of a PC-depleted domain in the region where the two cells join (**Fig. 4.3b**). The absence of this domain in the line scan of the ubiquitous hydrocarbon fragment (*m/z* 69, **Fig. 4.3a**) indicates that the cell membranes are continuous in this region.

4. The signal intensity of a given ion is affected by many factors including surface concentration, topography, sample matrix, and primary ion beam current fluctuations. To account for these factors a pseudo-internal standard is used to standardize line scans, $C_5H_9^+$ or *m/z* 69. This ion is derived from the acyl chains of phospholipids and thus is constant across the cell. The intensity distribution can be used to demonstrate that intensity fluctuations are due to surface concentration as opposed to the other factors mentioned above.

Standardization is achieved by dividing the intensity of the ion being analyzed by the intensity of *m/z* 69, for each pixel.

3.8. Lipid Identification

1. Owing to the high-energy nature of SIMS, molecular ions for lipids have been less often observed to date. Lipids can often be identified, however, by identification of specific fragment ions. Fragment ions are determined by generating standard films of the lipid of interest, and then collecting the resultant mass spectra (5). Mass spectra collected from a biological specimen can then be compared to the mass spectra for the fragmentation of a lipid standard for identification.
2. **Table 4.1** (5) contains common fragment ions for some lipids. These can be used to identify lipids present on the cell surface (*see* **Note 9**).

Table 4.1
Fragment ions for common lipids

Lipid	Fragment	Mass (*m/z*)
Phosphatidylcholine	$C_5H_{12}N^+$	86.0970
Phosphatidylcholine	$C_5H_{15}NPO_4{}^+$	184.0739
Phosphatidylcholine	$C_8H_{19}NPO_4{}^+$	224.1052
Phosphatidylethanolamine	$C_2H_7NPO_3{}^+$	124.0164
Phosphatidylethanolamine	$C_2H_9NPO_4{}^+$	122.0007
Phosphatidylethanolamine	$C_2H_7NPO_4{}^-$	140.0113
Phosphatidylglycerol	$C_3H_9PO_6Na^+$	195.0035
Phosphatidylinositol	$C_6H_{10}PO_8{}^-$	241.0114
Phosphatidylinositol	$C_6H_{12}PO_9{}^-$	259.0219
Phosphatidylinositol	$C_9H_{16}PO_9{}^-$	299.0533
Phosphatidylinositol	$C_3H_8NPO_6Na^+$	207.9988
Cholesterol	$C_{27}H_{45}{}^+$	369.3521
Cholesterol	$C_{27}H_{46}O^+$	386.3549
Sulfatide	$C_6H_9SO_8Na^+$	263.9916

4. Notes

1. Cleaning the substrates is crucial, as SIMS is very sensitive to contaminants, which can interfere with signals of interest as well as internal standard calibration.

2. Glass Scintillation vials should not be capped as solvent vapor can extract polymers from the cap, contaminating the solvent.
3. Once the shards have been cleaned they should only be handled by the edges using solvent-rinsed (acetone) tweezers.
4. Wicking away the excess water after the top shard is applied, greatly increases the likelihood of a successful fracture.
5. It is good to cover the styrofoam cooler containing the sample block, samples, and fast entry apparatus whenever you are not working with the sample to prevent condensation, which can damage the vacuum chamber. In addition, it is important to insure that the LN_2 completely covers all cooled parts; otherwise any exposed metal will collect ice quickly.
6. When tightening the screws holding the silicon substrate, it is important to stop as soon as resistance is met. Over tightening will result in broken substrates. For screwing the sample to block, we acquired a small spring-loaded grabber, used by computer technicians, for holding the screw and washer together under LN_2.
7. The fracture temperature is critical to obtaining a good fracture. If the sample is too cold, condensation dominates, leading to an ice-covered sample; too warm and sublimation dominates. If a sample has been fractured too warm, there will be little or no visible ice on the surface and SIMS images display significant smearing of lipids. In the case of a cold fracture, there is extensive ice visible on the surface and the SIMS spectrum will be dominated by water clusters.
8. To maximize lateral resolution and obtain the most signal, it is good to match the spot size to the pixel size. The data can later be down-binned to obtain the necessary signal intensity. This is done off-line by combing adjacent pixels and summing their mass spectra.
9. Some contaminants may interfere with the signals listed in the table. You must ensure these are not present in the sample.

Acknowledgments

This work was supported by the National Institutes of Health (R01EB002016). A.G.E. is supported as a Marie Curie Chair from the European Union 6th Framework.

References

1. Johansson, B. (2006) ToF-SIMS imaging of lipids in cell membranes. *Surf Interface Anal*, **38**, 1401–1412.
2. Roddy, T. P., Donald M. Cannon, J., Meserole, C. A., Winograd, N., Ewing, A. G. (2002) Imaging of freeze-fractured cells with *in situ* fluorescence and time-of-flight secondary ion mass spectrometry. *Anal Chem*, **74**, 4011–4019.
3. Ostrowski, S. G., Bell, C. T. V., Winograd, N., Ewing, A. G. (2004) Mass spectrometric imaging of highly curved membranes during tetrahymena mating. *Science*, **305**, 71–73.
4. Sostarecz, A. G., McQuaw, C. M., Ewing, A. G., Winograd, N. (2004) Phosphatidylethanolamine-induced cholesterol domains chemically identified with mass spectrometric imaging. *J Am Chem Soc*, **126**, 13882–13883.
5. Ostrowski, S. G., Szakal, C., Kozole, J., Roddy, T. P., Xu, J. Y., Ewing, A. G., Winograd, N. (2005) Secondary ion MS imaging of lipids in picoliter vials with a buckminsterfullerene ion source. *Anal Chem*, 77, 6190–6196.
6. Ostrowski, S. G., Kurczy, M. E., Roddy, T. P., Winograd, N., Ewing, A. G. (2007) Secondary ion MS imaging to relatively quantify cholesterol in the membranes of individual cells from differentially treated populations. *Anal Chem*, **79**, 3554–3560.
7. McQuaw, C. M., Zheng, L. L., Ewing, A. G., Winograd, N. (2007) Localization of sphingomyelin in cholesterol domains by imaging mass spectrometry. *Langmuir*, **23**, 5645–5650.
8. Piehowski, P. D., Kurczy, M. E., Willingham, D., Parry, S., Heien, M. L., Winograd, N., Ewing, A. G. (2008) Freeze-etching and vapor matrix deposition for ToF-SIMS imaging of single cells. *Langmuir*, **24**, 7906–7911.
9. Piehowski, P. D., Carado, A. J., Kurczy, M. E., Ostrowski, S. G., Heien, M. L., Winograd, N., Ewing, A. G. (2008) MS/MS methodology to improve subcellular mapping of cholesterol using TOF-SIMS. *Anal Chem*, **80**, 8662–8667.
10. Kurczy, M. E., Piehowski, P. D., Parry, S., Jiang, M., Chen, G. A., Ewing, A. G., Winograd, N. (2008) Which is more important in bioimaging SIMS experiments- the sample preparation or the nature of the projectile? *Appl Surf Sci*, **255**, 1298–1304.
11. Fletcher, J. S., Rabbani, S., Henderson, A., Blenkinsopp, P., Thompson, S. P., Lockyer, N. P., Vickerman, J. C. (2008) A new dynamic in mass spectral imaging of single biological cells. *Anal Chem*, **80**, 9058–9064.
12. Colliver, T. L., Brummel, C. L., Pacholski, M. L., Swanek, F. D., Ewing, A. G., Winograd, N. (1997) Atomic and molecular imaging at the single cell level with ToF-SIMS. *Anal Chem*, **69**, 2225–2231.
13. Cannon, D. M. J., Pacholski, M. L., Winograd, N., Ewing, A. G. (2000) Molecule specific imaging of freeze-fractured, frozen hydrated model membrane systems using mass spectrometry. *J Am Chem Soc*, **122**, 603–610.

Chapter 5

The Application and Potential of Ion Mobility Mass Spectrometry in Imaging MS with a Focus on Lipids

Amina S. Woods and Shelley N. Jackson

Abstract

Tissue profiling and imaging by MALDI mass spectrometry has allowed the direct analysis and localization of biomolecules in tissue. However, due to the in situ nature of this technique, the complexity of tissue, and the need for a chemical matrix in MALDI, the signal recorded can be extremely complex and difficult to assign. Combining ion mobility with matrix-assisted laser desorption/ionization is a very powerful technique for fast separation and analysis of biomolecules in complex mixtures (such as tissue and cells), as isobaric lipid, peptide, and oligonucleotide molecular ions are pre-separated in the mobility cell before mass analysis. Differences in drift time of as much as 30% are obtained in a timescale of hundreds of microseconds. Molecular ions of the same biochemical family fall along trend lines when plotted in 2D plots of mobility drift time as a function of *m/z*. In this chapter ion mobility MALDI-MS ability to analyze various biomolecules in tissue, that is, lipids and proteins, as well as its ability to separate species from all of the major phospholipid classes from tissue and extracts, the effects that radyl chain length, degree of unsaturation, head group composition have upon their ion's cross section in the gas phase, and how it can be used not only to distinguish them from other biochemical groups in a mixture but also to differentiate them from other lipid species will be illustrated.

Key words: MALDI, ion mobility, lipids, imaging, peptides, drugs, profiling.

1. Introduction

The lipid membrane maintains a cell's shape and its connection to other cells. It also allows certain molecules into and out of the cell, so it is selectively permeable. The membrane is a sea of lipids in which proteins and glycoproteins that govern the cell's interactions are floating; thus it is the gate of the cell. The main lipid components of the cell's membrane are phospholipids,

S.S. Rubakhin, J.V. Sweedler (eds.), *Mass Spectrometry Imaging*, Methods in Molecular Biology 656,
DOI 10.1007/978-1-60761-746-4_5, © Springer Science+Business Media, LLC 2010

cholesterol, and sphingolipids. The development of easy to use, highly accurate, and affordable mass spectrometers has made the study of lipids more accessible to many researchers and has led to the rapid growth of the field of lipidomics (1–6). Matrix-assisted laser desorption/ionization mass spectrometry (MALDI-MS) has become a valuable technique for imaging of biomolecules directly from tissue (*see* **Chapters 1** and **2**); although initially used for peptides and proteins, lipids have begun to be imaged recently.

The combining of MALDI-MS with ion mobility (IM) mass spectrometry has allowed for a wide range of samples to be analyzed by MALDI-IM MS (7, 8). In MALDI-IM MS, ions formed by the (MALDI) process enter the electric field of a gas-filled mobility drift cell region where they acquire an average drift velocity based on their collision cross section (Ω) to charge ratio, thus allowing the rapid separation of molecules according to their conformation (7–12). We have previously demonstrated that IM MS separates biomolecules along trend lines according to their composition, thus allowing the assignments of the ions' chemical types based on their specific *m/z* and mobility drift time (12). Due to the rapid gas-phase separation of molecules in ion mobility, it offers unique advantages for mass spectrometry imaging of tissues and has led to some recent studies using MALDI-IM MS for imaging (13–15). Examples of the use of MALDI-IM MS will be discussed below.

2. Materials

2.1. Ion Mobility Mass Spectrometer Hardware and Software

1. MALDI-IM-TOFMS (Ionwerks, Inc., Houston, TX, USA).
2. X–Y sample stage (Model MM-3 M-F-2, National Aperture, Inc., Salem, NH, USA).
3. Nd:YLF UV laser (Crystalaser, Reno, NV, USA).
4. IDL software (Research Systems, Boulder, CO, USA).

2.2. Matrix and Standard Solutions

1. 2,5-Dihydroxy benzoic acid, DHB (Fluka, Switzerland), prepared at 30 mg/ml in 50% ethanol for tissue profiling or at 100 mg/ml in 90% ethanol.
2. 6-Aza-2-thiothymine, ATT (Sigma, St. Louis, MO, USA), prepared as a saturated solution in 50% ethanol.
3. Sinapic acid, SA (Fluka), prepared as a saturated solution in 50% ethanol.
4. 2,6-Dihydroxyacetophenone, DHA (Fluka), prepared at 30 mg/ml in 50% ethanol (*see* **Note 1**).
5. 2,4,6-Trihydroxyacetophenone, THA (Fluka), prepared as a saturated solution in 50% ethanol.

6. PSG gold colloid (2.0 nm, 2.6×10^{15} particles/ml) (nanoComposix, San Diego, CA, USA) diluted 1:4 v/v in ethanol.
7. Lipid standard tissue extracts (Avanti Polar Lipids, Alabaster, AL, USA): PCs from porcine brain, PIs from bovine liver, PSs from porcine brain, and SMs from porcine brain were prepared in chloroform:methanol (2:1 v/v) at a concentration of 10 nmol/μl. Further dilution was done directly in the matrix solutions.
8. Cocaine hydrochloride (Sigma) was prepared at 20 mg/kg free base in normal saline.

2.3. Tissue Harvesting, Sectioning, Preparation

1. Male Sprague–Dawley rats (Harlan Industries, Indianapolis, IN, USA).
2. Isoflurane (Baxter, Deerfield, IL, USA).
3. Isopentane (J T Baker, Phillipsburg, NJ, USA).
4. Cryostat (CM 3050 S; Leica Microsystems Nussloch GmbH, Nussloch, Germany).
5. Artistic airbrush14 (Aztek A470/80 Airbrush System; Testor Corporation, Rockford, IL, USA).

3. Methods

3.1. Ion Mobility Mass Spectrometry

The ion mobility mass spectrometer used in this study is a periodic focusing MALDI-IM-TOFMS instrument with positive ion mode (*see* **Note 2**) and is illustrated in **Fig. 5.1**. A mobility resolution of 30 (FWHM of drift time) and a mass resolution of 3,000 for *m/z* 1,000 using an orthogonal time-of-flight mass spectrometer (o-TOFMS) are routinely achieved on calibration standards. Ions are drawn into the mobility cell by a voltage between the sample and the exit of the helium-filled IM cell so that the ion's drift velocity is proportional to the density (surface area) of the ion. The length of the mobility cell is 15 cm. It is operated at 1,700 V with 3.5 torr helium pressure. An X–Y sample stage provides 1 μm accuracy in beam positioning and sample scanning. A Nd:YLF UV laser ($\lambda = 349$ nm at 200 Hz) is used to generate ions in the source at the operating pressure of the mobility cell.

Separation of ions with the same mass but different surface areas is thus possible. Small dense ions quickly traverse the IM cell and enter the mass spectrometer where a pulsed voltage applied to the orthogonal extraction region deflects the ion beam into a time-of-flight mass spectrometer equipped with a reflectron, which measures the ions' *m/z* directly and their ion mobility

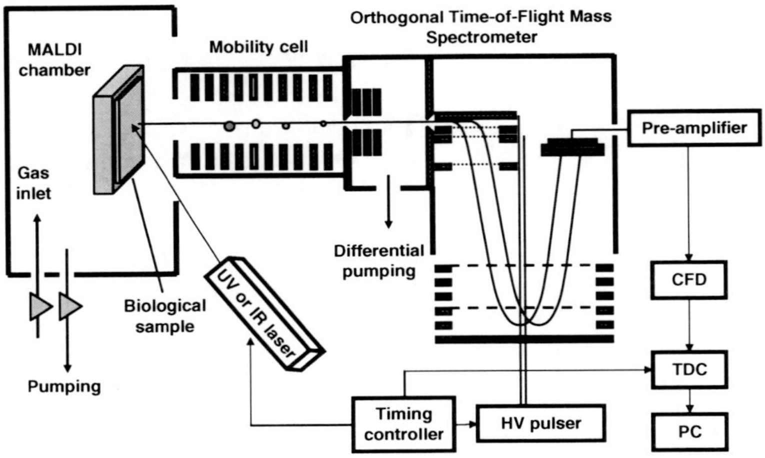

Fig. 5.1. Schematic diagram of MALDI-IM-TOFMS instrument.

elution time indirectly. This indirect determination of the IM time is accomplished by "tagging" each such o-TOF mass spectrum with the elapsed time since the firing of laser pulse onto the sample. Ion drift times are typically under 1 ms for m/z 5,000 yet the time of flight in the mass spectrometer is only a few tens of microseconds. Thus after each UV laser desorption pulse, several hundred mass spectra can be obtained from the eluted ions separated in the IM cell.

In this work, the data is presented as 2D contour plots of ion intensity as a function of ion mobility drift time (y-axis) and m/z (x-axis). In the 2D contour plots of ion mobility versus mass, compounds that have the same molecular weight but different structures are observed along ion groupings, which have different slopes. All contour plots were produced using IDL software. The images presented below were acquired with a 200 μm spatial resolution and a 3 s acquisition time per step. The laser spot size is less than 50 μm and the angle of incidence is 60° (tilt measured from the surface normal). Thus within the 3 s acquisition, a swath of 100 μm or less is cut across the pixel. A specific region of interest in both mass and mobility dimensions is preselected within the software. Offsets in both mass and mobility data acquisition parameters limit incoming data to a narrow mass range of approximately 100 Da and a mobility drift time range spanning 16 μs (at m/z 800). The region of interest mass bins and corresponding intensities are saved to a file along with the positional coordinates.

3.2. Tissue Sectioning and Preparation

Appropriate tissue preparation is important to maintaining the spatial resolution of the biomolecules of interest. In this study, male Sprague–Dawley rats between 300 and 420 g were euthanized with isoflurane (*see* **Note 3**). Brains were quickly removed from the skull and frozen in dry ice-chilled isopentane for 15 s at –30°C, prior to storage at –80°C. For cocaine experiments, rats received ip injection of cocaine hydrochloride (20 mg/kg free base in normal saline) and were euthanized with an overdose of isoflurane 15–25 min later. The brain collection procedure was identical as described above.

Next, frozen brain tissue was cut into thin sections, between 10 and 20 μm, in a cryostat. Typically, tissue samples are attached to the cryostat sample stages using optimal cutting temperature compound (OCT). However, care must be taken not to contaminate the tissue with OCT, because of previous studies (16) showing that OCT interference can reduce the quality of the mass spectra. This is especially important for lipids in which the m/z range of interest is under 1,000 (*see* **Note 4**). After cutting, tissue sections were collected directly onto MALDI sample targets. For profiling tissue experiments, 0.1–0.3 μl (according to the size of the anatomical region of interest) matrix solution was spotted onto the tissue section prior to insertion into the mass spectrometer.

The two matrices used to spray coat the tissue sections were DHB and 2.0 nm gold particles. The matrix solutions were sprayed on the tissue sections with an artistic airbrush14 (Aztek A470/80 Airbrush System, the Testors Corporation, Rockford, IL, USA). Each target insert was sprayed with 5–6 spray cycles. A 2 min wait between spray cycles was observed to ensure sufficient evaporation of organic solvents from the target insert surface. One spray cycle is defined as an organized rastering of spraying, which results in complete coverage of the target insert surface by the spraying mist. The distance between the nozzle of the airbrush and the target plate was kept between 15 and 20 cm. The matrix was perpendicularly sprayed onto the target plate.

3.3. Biomolecules Separation by Ion Mobility in Tissue Profiling Experiments

Coupling MALDI-MS with ion mobility results in the fast sorting of biomolecules along trend lines, which provides a powerful tool for the analysis of complex samples such as tissue. **Figure 5.2** illustrates an MALDI-IM spectrum of a mixture containing sphingomyelins, dynorphin peptides, and the oligonucleotide, CATG, with ATT matrix in positive ion mode. As can be observed, each type of biomolecule falls along its own trend line. Overall, the differences in mobility drift times between isobaric ions from the three types of biomolecules tested (lipid, peptide, oligonucleotide) are approximately 15–20%. These results have been repeated for other standards and in tissue with the same pattern of mobility separation, in which the hydrophilic oligonucleotides have the fastest drift times, followed by peptides and proteins, and finally the hydrophobic lipids that usually have the slowest drift times (12, 13, 17).

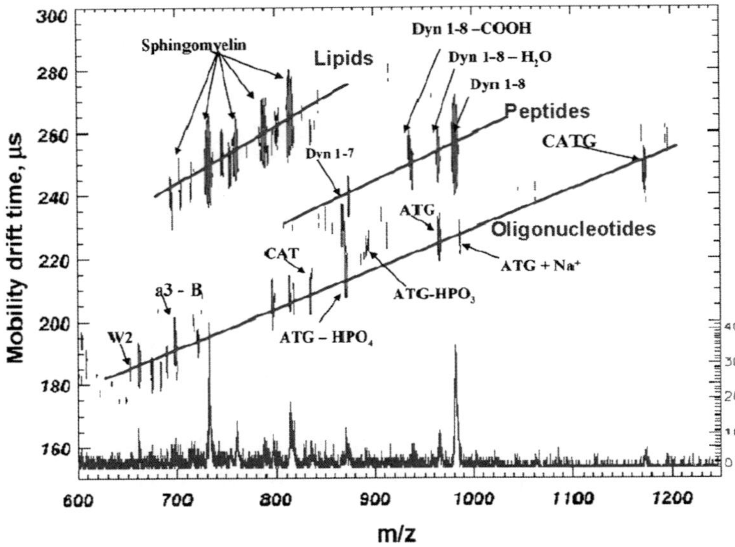

Fig. 5.2. MALDI-IM 2D plot of a mixture of lipids, peptides, and oligonucleotides with ATT matrix. Note the separation of the different types of biomolecules and their fragments along familial trend lines. Reproduced from (12) with permission from American Chemical Society.

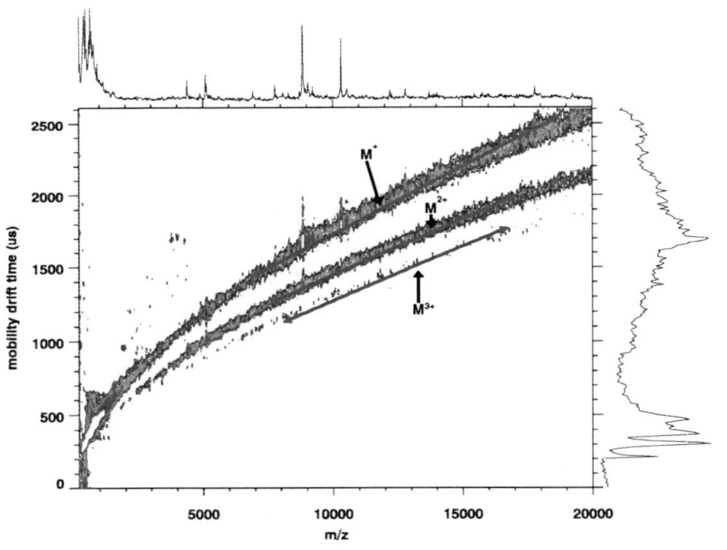

Fig. 5.3. MALDI-IM 2D plot of the cerebral caudate–putamen region in rat brain tissue with SA matrix. Note the separation of charge states for proteins and peptides along trend lines.

Figures 5.3 and **5.4** show two examples where mobility separation is valuable in direct tissue analysis. **Figure 5.3** contains an MALDI-IM 2D plot of the cerebral caudate–putamen region in rat brain tissue with SA matrix in positive ion mode. Several mass peaks-associated peptides and proteins in the brain

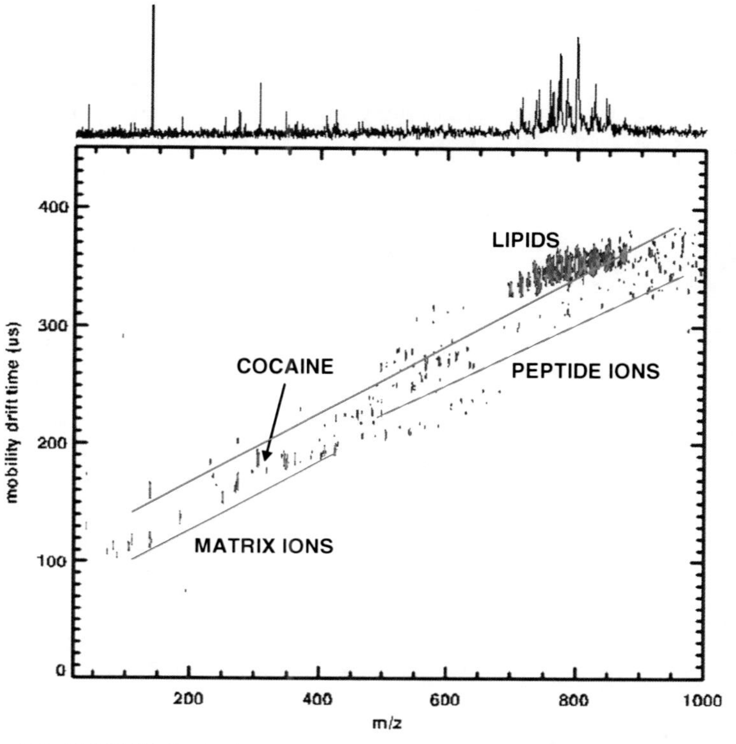

Fig. 5.4. MALDI-IM 2D plot from the cerebrum of a cocaine-injected rat (20 mg of free base/kg), 20 min after ip injection with DHB matrix. Note the mobility separation of cocaine from matrix interference ions.

are recorded in the *m/z* range between 1,000 and 20,000. In this *m/z* range, three clear trend lines are observed corresponding to singly, doubly, and triply charged ions. Thus, using ion mobility MS we are able to resolve monomers and multimers of peptides and proteins, which make identification easier and allows for the imaging of pure singly charged protein ions. **Figure 5.4** illustrates an MALDI-IM spectrum from the cerebrum of a cocaine-injected rat with DHB matrix. Clear mobility separation is observed between lipids and peptides but more importantly for this sample is the mobility separation of the cocaine molecular ion (M+H = 304.1) from the DHB matrix ions. Without this separation it would be quite difficult to differentiate the cocaine signal from matrix ions. Matrix interference is a major disadvantage for MALDI analysis of molecules below 500 Da.

MALDI-IM MS is useful for the separation of major groups of biomolecules such as lipids, peptides, nucleic acids, but it is also capable of separating different classes of molecules within a group, for example, phospholipids. Phospholipids are present in abundance in all biological membranes and represent over half the lipid content of the brain (18). The general structural requirement for phospholipids is the presence of an alcohol, phosphate, and a fatty acid chain. Major classes of phospholipids include phosphatidylcholine (PC), phosphatidylinositols (PI), phosphatidylserine (PS), and sphingomyelin (SM). **Figure 5.5a**

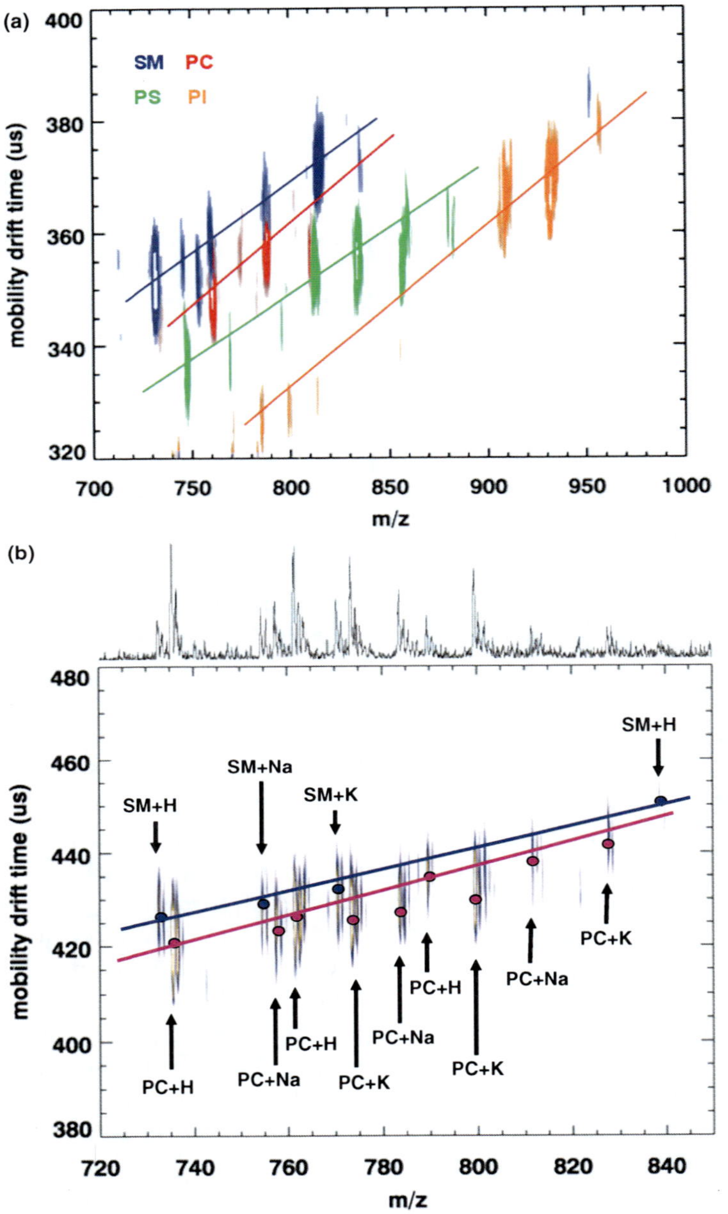

Fig. 5.5. (**a**) Overlay of MALDI-IM 2D plots of brain extracts of PCs (*red*), PSs (*green*), SMs (*blue*), and liver extracts of PIs (*orange*) with THA matrix. Note the mobility separation of the different phospholipid classes based upon head groups, acyl chains, and alkali salt adducts. (**b**) MALDI-IM 2D plot from rat cerebral tissue with DHB matrix. Note the mobility separation of PC and SM species and between protonated and alkali adducts.

shows an overlay of MALDI-IM 2D plots of brain extracts of PCs (red), PSs (green), SMs (blue), and liver extracts of PIs (orange) with THA matrix in positive ion mode. The sample spot contained a 100 pmol from each phospholipid extract. As seen in **Fig. 5.5a**, different phospholipid classes occupy different drift time *m/z* space, which simplifies their assignment in complex mixtures. This result is due to changes on an ion's cross section in the gas phase because of differences in head groups, acyl chains, and salt adducts among the different phospholipid classes. This phenomenon has been observed in detail previously (19). **Figure 5.5b** illustrates an MALDI-IM 2D plot from rat cerebral tissue with DHB matrix in positive ion mode. Several species of PC and SM were recorded as protonated, sodiated, and potassiated adducts. One clear observation from this figure is that SM species have a slightly higher drift time compared to PC species. PC and SM both contain a phosphocholine head group; however, the SM ceramide backbone differs slightly from the glycerol backbone of PC. This difference accounts for the difference in drift time and demonstrates the ability of ion mobility to separate very similar classes of molecules.

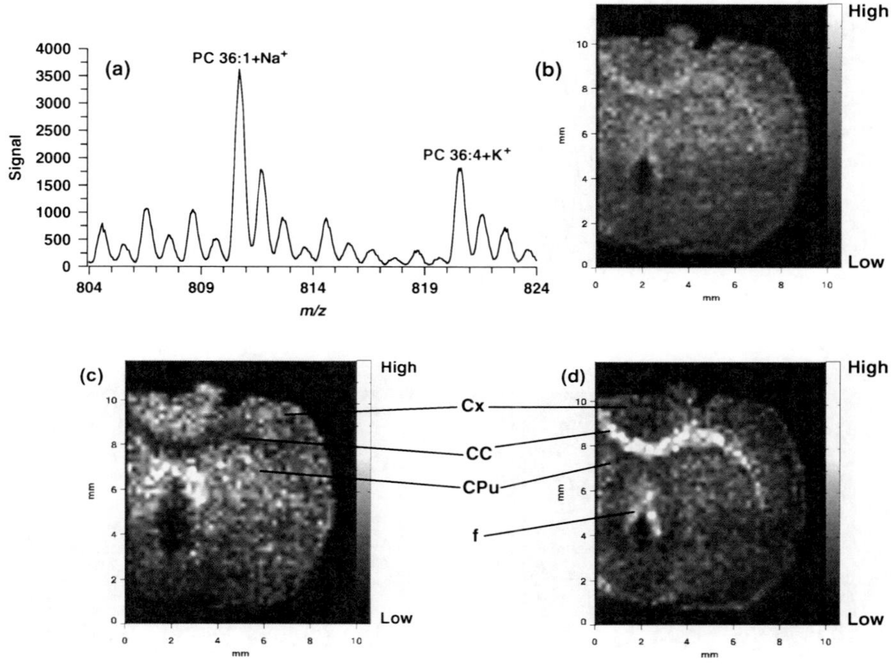

Fig. 5.6. MALDI-IM images using DHB matrix in positive ion mode. (**a**) 1D mass spectrum obtained from section. (**b**) Image of *m/z* range 804–824. (**c**) Image of mass peak at *m/z* 810.6 (PC 36:1+Na). (**d**) Image of mass peak at *m/z* 820.5 (PC 36:4+K). Images are 51 × 60 pixels. Abbreviations: Cx – cerebral cortex; CC – corpus callosum; CPu – caudate–putamen; f – fornix.

3.4. Imaging of Lipids by MALDI-IM-TOFMS

Since MALDI-IM MS separates different groups of biomolecules according to drift time and our imaging software has a drift time window of 16 μs, we are able to generate images from a single ion with no interference from other isobaric ions or corresponding dimers or multimers. **Figure 5.6** shows the results of an MALDI-IM imaging run recorded with DHB matrix from rat cerebral tissue in positive ion mode. **Figure 5.6a** represents the total 1D mass spectrum used to generate the images in **Fig. 5.6b–d**. The image in **Fig. 5.6b** is produced using the total ion count for the 1D mass spectrum in **Fig. 5.6a**. **Figure 5.6c** shows the distribution of m/z 810.6 (PC 36:1+Na), while **Fig. 5.6d** illustrates the distribution of m/z 820.5 (PC 36:4+K). Anatomically distinct regions are detected in **Fig. 5.6d** with more signal recorded in white matter regions including the corpus callosum and fornix, while in **Fig. 5.6c** stronger signal is observed in the cerebral cortex and caudate–putamen, both gray matter regions.

In order to demonstrate the importance of matrices in detecting different classes of lipids (*see* **Note 5**), **Fig. 5.7** illustrates the

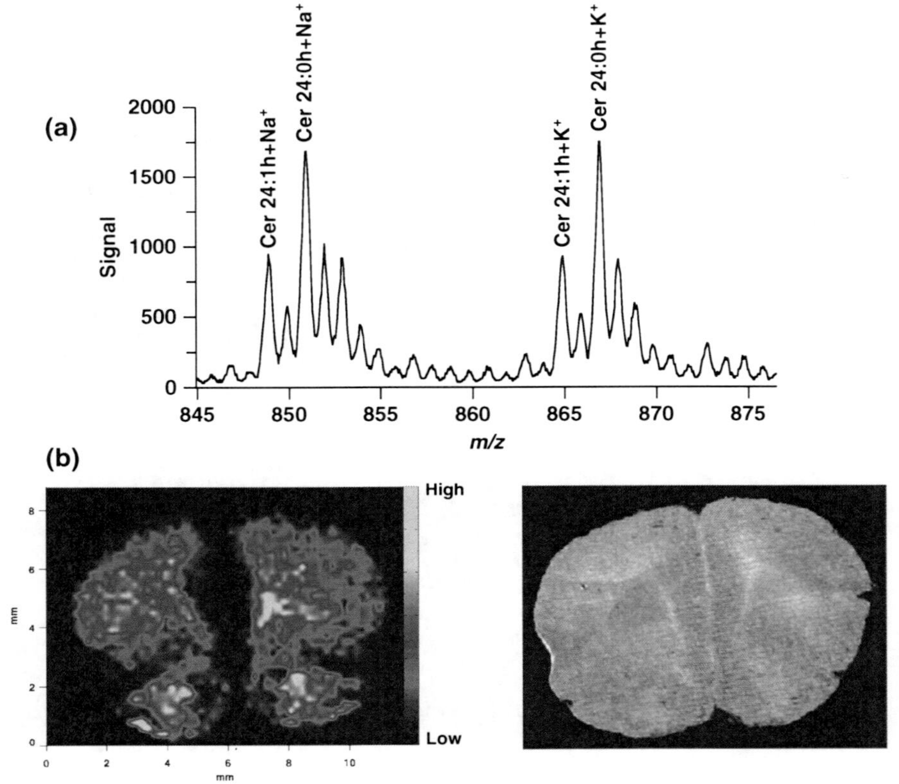

Fig. 5.7. MALDI-IM image using 2.0 nm Au particles in positive ion mode. (**a**) 1D mass spectrum recorded from the tissue section. (**b**) Image of potassiated cerebroside 24:0 OH at m/z 866.8. The image is a 45 × 60 pixel. The adjacent frame is a photograph of the rat brain section.

results of an MALDI-IM MS image obtained using 2.0 nm gold nanoparticles as a matrix for a rat brain tissue section in positive ion mode. Previous studies have used implanted gold clusters (20, 21) and gold nanoparticles (13) to analyze cerebrosides in tissue; compared to traditional organic acid matrices, such as DHB, in which PC species are the dominant signals in this lipid mass range in positive ion mode. **Figure 5.7a** shows a 1D mass spectrum for the whole tissue section, in which strong signals corresponding to cerebroside 24:0 OH at *m/z* 850.7 and 866.8 and the sodiated and potassiated cerebroside 24:1 OH at *m/z* 848.7 and 864.8 were observed. **Figure 5.7b** shows the distribution of potassiated cerebroside 24:0 OH at *m/z* 866.8, which is highly concentrated in white matter regions in the brain.

4. Notes

1. One note of caution is that DHA matrix sublimes under high vacuum pressure and may last only for 15–30 min according to the concentration and vacuum pressure of the ion source.
2. Although the current configuration of the MALDI-IM mass spectrometer used in this study has only positive ion mode, negative ion mode is the preferred mode for several lipid classes. In general for direct tissue analysis, PCs and SMs are easier to ionize in positive ion mode, while negative ion mode is preferred for the analysis of PEs, PSs, PIs, PGs, PAs, sulfatides, gangliosides, and cardiolipins.
3. All the animal work in this study abides by the Guide for the Care and Use of Laboratory Animals (NIH).
4. One alternative to OCT is to attach the tissue samples to the cryostat sample stages using ice slush made from distilled water (17). In this method, the ice slush only comes in contact with the tissue blocks at the surface opposing the sample stages and is frozen into a thin layer of ice within 5 s.
5. Proper matrix selection is one of the keys to a successful MALDI experiment and is especially important in direct tissue analysis due to the complex nature of the sample. For lipid analysis in tissue, we typically use DHA, DHB, and gold nanoparticles according to what class of lipids we are probing. In positive ion mode, both DHA and DHB work well for PCs and SMs, while gold nanoparticles are effective for targeting cerebrosides in tissue. In negative ion mode, all three work well for sulfatides, while DHA and DHB detect PIs, and DHA offers the ability to easily detect PSs, PEs, PGs, and gangliosides directly from tissue.

Acknowledgments

This research was supported by the Intramural Research Program of the National Institute on Drug Abuse, NIH. The authors thank the Office of National Drug Control Policy (ONDCP) for instrumentation funding, without which this and other projects could not have been accomplished and Dr. J Albert Schultz and Tom Egan (Ionwerks, Inc.) for assistance with the ion mobility mass spectrometer and the imaging computer software.

References

1. Holthuis, J. C. M., Levine, T. P. (2005) Lipid traffic: floppy drives and a superhighway. *Nat Rev Mol Cell Biol*, **6**, 209–220.
2. Waston, A. D. (2006) Lipidomics: a global approach to lipid analysis in biological systems. *J Lipid Res*, **47**, 2101–2111.
3. van Meer, G. (2005) Cellular lipidomics. *EMBO J*, **24**, 3159–3165.
4. Wenk, M. R. (2005) The emerging field of lipidomics. *Nat Rev Drug Discov*, 7, 594–610.
5. Piomelli, D., Astarita, G., Rapaka, R. (2007) A neuroscientist's guide to lipidomics. *Nat Rev Neurosci*, **8**, 743–753.
6. Han, X. (2007) Neurolipidomics: challenges and developments. *Front Biosci*, **12**, 2601–2615.
7. Gillig, K. J., Ruotolo, B., Stone, E. G., Russell, D. H., Fuhrer, K., Gonin, M., Schultz, J. A. (2000) Coupling high-pressure MALDI with ion mobility/orthogonal time-of-flight mass spectrometry. *Anal Chem*, **72**, 3965–3971.
8. von Helden, G., Wyttenbach T., Bowers, M. T. (1995) Conformation of macromolecules in the gas phase: use of matrix-assisted laser desorption methods in ion chromatography. *Science*, **267**, 1483–1485.
9. Baumbach, J. I., Eiceman, G. A. (1999) Ion mobility spectrometry: arriving on site and moving beyond a low profile. *Appl Spectrosc*, **53**, 338A–355A.
10. McDaniel, E. W., Mason, E. A. *The Mobility and Diffusion of Ions in Gases.* Wiley, New York, NY, 68–72, 1973.
11. Collins, D. C., Lee, M. L. (2002) Developments in ion mobility spectrometry-mass spectrometry. *Anal Bioanal Chem*, **372**, 66–73.
12. Woods, A. S., Ugarov, M., Egan, T., Koomen, J. Gillig, K. J., Fuhrer, K., Gonin, M., Schultz, J. A. (2004) Lipid/peptide/nucleotide separation with MALDI-ion mobility-TOF MS. *Anal Chem*, **76**, 2187–2195.
13. Jackson, S. N., Ugarov, M., Egan, T., Post, J. D., Langlais, D., Schultz, J. A., Woods, A. S. (2007) MALDI-ion mobility-TOFMS imaging of lipids in rat brain tissue. *J. Mass Spectrom*, **42**, 1093–1098.
14. McLean, J. A., Ridenour, W. B., Caprioli, R. M. (2007) Profiling and imaging of tissues by imaging ion mobility-mass spectrometry. *J Mass Spectrom*, **42**, 1099–1105.
15. Trim, P. J., Henson, C. M., Avery, J. L., McEwen, A., Snel, M. F., Claude, E., Marshall, P. S., West, A., Princivalle, A. P., Clench, M. R. (2008) Matrix-assisted laser desorption/ionization-ion mobility separation-mass spectrometry imaging of vinblastine in whole body tissue sections. *Anal Chem*, **80**, 8628–8634.
16. Schwartz, S. A., Reyzer, M. L., Caprioli, R. M. (2003) Direct tissue analysis using matrix-assisted laser desorption/ionization mass spectrometry: practical aspects of sample preparation. *J Mass Spectrom*, **38**, 699–708.
17. Jackson S. N., Wang H-Y. J., Woods, A. S., Ugarov M., Egan T., Schultz J. A. (2005) Direct tissue analysis of phospholipids in rat brain using MALDI-TOFMS and MALDI-ion mobility-TOFMS. *J Am Soc Mass Spectrom*, **16**, 133–138.
18. Agranoff, B. W., Benjamins, J. A., Hajra, A. K. (1999) *Basic Neurochemistry Molecular, Cellular and Medical Aspects.* 6th ed. Siegel, G. J. et al., Eds., Lippincott Williams & Wilkins, Philadelphia, PA, 47–67.
19. Jackson, S. N., Ugarov, M., Post, J. D., Egan, T., Langlais, D., Schultz, J. A., Woods, A. S. (2008) A study of phospholipids by ion

mobility TOFMS. *J Am Soc Mass Spectrom*, **19**, 1655–1662.

20. Novikov, A., Caroff, M., Della-Negra, S., Lebeyec, Y., Pautrat, M., Schultz, J. A., Tempez, A., Wang H-Y. J., Jackson, S. N., Woods, A. S. (2004) Matrix-implanted laser desorption/ionization mass spectrometry. *Anal Chem*, **76**, 7288–7293.
21. Tempez A., Ugarov, M., Egan, T., Schultz, J. A., Novikov, A., Della-Negra, S., Lebeyec, Y., Pautrat, M., Caroff, M., Smentkowski, V. S., Wang, H-Y. J., Jackson, S. N., Woods, A. S. (2005) Matrix implanted laser desorption ionization (MILDI) combined with ion mobility-mass spectrometry for bio-surface analysis. *J Proteome Res*, **4**, 540–545.

Chapter 6

Quantitative Imaging of Chemical Composition in Single Cells by Secondary Ion Mass Spectrometry: Cisplatin Affects Calcium Stores in Renal Epithelial Cells

Subhash Chandra

Abstract

A detailed protocol for quantitative single cell mass spectrometry imaging (MSI) analysis is described in this chapter with examples of the treatment of cells with anticancer drug, cisplatin. Cisplatin, *cis*-diamminedichloridoplatinum ii (CDDP), is widely used for the treatment of many malignancies, including testicular, ovarian, bladder, cervical, head and neck, and small cell and non-small cell lung cancers. The possibility of renal injury by cisplatin treatment is a major dose-limiting factor in this cancer therapy. At present, the mechanisms of cisplatin-induced renal cytotoxicity are poorly understood. In this work, secondary ion mass spectrometry (SIMS) was used for investigating cisplatin-induced alterations in intracellular chemical composition in a well-established model (LLC-PK_1 cell line) for studying renal injury. The cells were cryogenically prepared by the sandwich freeze-fracture method for subcellular imaging analysis of chemical composition (total concentrations of K^+, Na^+, and Ca^{2+}) in individual cells. The single cell analysis of these diffusible ions necessitates the use of reliable cryogenic sample preparations for SIMS. The sandwich freeze-fracture method offers a simple approach for cryogenically preserving diffusible ions and molecules inside the cells for SIMS analysis. A CAMECA IMS-3f SIMS ion microscope instrument capable of producing chemical images of single cells with 500-nm spatial resolution was used in the study. In cisplatin-treated cells, SIMS imaging showed the presence of detectable amount of platinum at mass 195, as $^{195}Pt^+$ secondary ions in individual cells. SIMS observations also revealed that individual cells differed in their response to cisplatin. While the chemical composition of some cells was unaffected by cisplatin, others showed a reduction in cytoplasmic calcium stores that was not associated with changes in their intracellular K or Na concentrations. Another population of cells displayed an increase in cytoplasmic calcium concentration that was associated with higher levels of intracellular Na and a reduction in K concentration of the same cells. Since the loss of intracellular K and the gain of Na and Ca are typical symptoms of cell injury, it is plausible that the initial response of the cell to cisplatin treatment is the reduction in cytoplasmic calcium pool in stores. If, somehow, the calcium stores are compromised with cisplatin, then maintenance of free Ca^{2+} homeostasis would become uncontrollable in the cell. These observations open new avenues of research for understanding of the mode of action of cisplatin in cell injury. This study also demonstrates the need and vast potential of single cell imaging mass spectrometry techniques in cell biology and medicine.

S.S. Rubakhin, J.V. Sweedler (eds.), *Mass Spectrometry Imaging*, Methods in Molecular Biology 656,
DOI 10.1007/978-1-60761-746-4_6,

Key words: Cisplatin, renal injury, calcium stores, freeze-fracture, cellular potassium and sodium, SIMS, imaging mass spectrometry, cryogenic sample preparation for SIMS.

1. Introduction

The chemotherapeutic cisplatin, *cis*-diamminedichloridoplatinum ii (CDDP), is widely used for the treatment of many malignancies, including testicular, ovarian, bladder, cervical, head and neck, and small cell and non-small cell lung cancers (1). There is a large amount of literature on the mechanisms of cisplatin-induced cell death, but its mode of action still remains unclear. Cisplatin may initiate the programmed cell death (apoptosis) via its binding to DNA, which may activate p53-dependent apoptotic pathways. Cisplatin, however, can also kill cancer cells with mutated p53 (2–4). Cisplatin may kill cells via multiple modes like apoptosis, necrosis, and perturbation of calcium homeostasis (3–6). Understanding the mechanisms of cisplatin resistance and toxicity is an area of active research in cancer therapy.

Due to the renal excretion of cisplatin, the kidney is exposed to a higher concentration of cisplatin than any other organ. Nephrotoxicity of cisplatin is one of the most important dose-limiting factors in cancer therapy. A single therapeutic dose of cisplatin may cause kidney damage within 48 h (7). Therefore, the chemotherapeutic dose for cancer therapy is limited by the risk of chronic or acute renal failure (8, 9). Renal proximal tubular cells are the main target for cisplatin toxicity in the kidney, since cisplatin nephrotoxicity manifests primarily as proximal tubule dysfunction (8–10). Some of these dysfunctions may include the altered synthesis of nucleic acids and proteins, disorganization of intracellular cytoskeleton, and defective transport of organic and inorganic solutes (9). The in vitro models have been useful in studying the mode of action of cisplatin based on single cell observations with optical/laser confocal fluorescence microscopy (8, 10, 11–13). These studies have implicated cisplatin-induced injury with DNA fragmentation and activation of apoptotic pathways, endoplasmic reticulum stress, mitochondrial dysfunction, and the production of reactive oxygen metabolites. The perturbation of intracellular calcium homeostasis has also been observed with cisplatin treatments (14). More work is needed for understanding cisplatin-induced alterations in intracellular chemical composition of physiologically relevant cations, such as K^+, Na^+, and Ca^{2+}, since they are intimately involved in the maintenance of membrane potential of the cell. This study shows that the SIMS-based single cell imaging technique of ion microscopy can provide a valuable tool for this line of research in chemotherapy.

Imaging mass spectrometry techniques are becoming valuable tools in biology and medicine due to their capabilities of in situ measurements of elements, isotopes, and molecules in single cells and at subcellular scale resolution (15–20). As discussed in various chapters of this book, imaging mass spectrometry techniques differ widely in their capabilities of detecting elements (isotopes) or molecules and have wide range of applications in biological research. In magnetic sector dynamic SIMS instruments, the recent technological innovations allow isotopic imaging capabilities of negative secondary ions at 50-nm spatial resolution with a liquid metal Cs^+ primary ion beam (18, 21, 22). However, in NanoSIMS the detection of positive secondary ions needed for the imaging of physiologically relevant cations (K^+, Na^+, etc.) has remained at submicron scale resolution with the O^- primary ion beam (22). A major challenge for imaging mass spectrometry originates from the sample preparation, which needs to immobilize "The analyte" faithfully in its native location to withstand the high vacuum conditions of SIMS instruments (15, 17, 22, 23). This study focuses on the utility of a CAMECA IMS-3f SIMS instrument, which is capable of producing secondary ion images with a 500-nm spatial resolution in the ion microscope mode for studying changes in the chemical composition of renal epithelial cells in response to cisplatin.

2. Materials

2.1. Cell Culture, Growth Medium, Cisplatin, Isopentane, Liquid Nitrogen, and Compressed Dry Nitrogen

1. The pig kidney LLC-PK_1 cell line has been used as an effective model for testing renal injury upon exposure to cisplatin (12, 13). This is a non-excitable cell type with characteristics of mainly the proximal tubules (24). LLC-PK_1 cells (American Type Culture Collection (ATCC) CRL1392, Manassas, VA, USA) were maintained in Medium 199 supplemented with 3% fetal bovine serum (ATCC, Manassas, VA, USA) in a humidified atmosphere of 95% air and 5% CO_2 at 37°C.
2. Cisplatin, *cis*-diamminedichloridoplatinum (ii) (Sigma, St. Louis, MO, USA).
3. Isopentane, 2-methylbutane (Sigma).
4. Liquid nitrogen.
5. A tank of compressed dry nitrogen.

2.2. Silicon Substrate for Cell Growth and Latex Beads as Spacers

1. An electrically conducting cell growth substrate is required for imaging analysis with a CAMECA IMS-3f SIMS ion microscope because the sample is held at ± 4,500 V in the high vacuum sample chamber of the instrument.

Polished high-purity N-type semiconductor-grade silicon wafers were used for cell growth (Silicon Quest International, Santa Clara, CA, USA). For cell growth, the silicon wafers were cut into small pieces of random shapes of approximately 1 cm^2 surface area using a diamond-tipped scribe.

2. The latex beads serve as spacers in the sandwich freeze-fracture method of sample preparation (see later). Latex beads with approximate diameter of 11 μm (Duke Scientific, Palo Alto, CA, USA) were used in the study.

2.3. Freeze-Drier, Reflected Light Microscope, and Air-Tight Sample Chamber

1. A TIS-U-Dry freeze-drier (FTS Systems Inc., Stone Ridge, NY, USA) is used for freeze-drying the cells.
2. An Olympus Microscope with reflected light Nomarski optics (Olympus America Inc., Center Valley, PA, USA) is used for photographing the freeze-dried cells on silicon substrates.
3. A silicon substrate containing the freeze-dried cells is transferred from a desiccator to a homemade air-tight Teflon chamber to avoid any rehydration during optical microscopy measurements. A photo of this simple chamber is shown in **Fig. 6.1**. The silicon chip containing the freeze-dried cells is placed in a rectangular flat indentation in the Teflon. A glass coverslip is placed over the indentation in Teflon and sealed with hot wax as shown in **Fig. 6.1**.

Fig. 6.1. Photograph of an air-tight sample chamber for making reflected light microscopy measurements on freeze-dried cells on a silicon substrate. A silicon chip containing the cells is isolated from the normal atmosphere by a glass coverslip which is glued to a Teflon support. Reproduced from ref. (30).

2.4. Scanning Electron Microscope (SEM)

For observing structural preservation of freeze-dried cells at greater details than provided by optical microscopy, a JOEL JSM 35 scanning electron microscope (JEOL Ltd., Tokyo, Japan) was used in this study. For improving the quality of SEM images, the freeze-dried cells were coated with a thin layer of Au/Pd in a sputter coater (Ted Pella Inc., Redding, CA, USA).

2.5. Secondary Ion Mass Spectrometry (SIMS) Instrumentation

A CAMECA IMS-3f magnetic sector dynamic SIMS instrument (Cameca Inc., Paris, France), which is capable of producing isotopic (elemental) images with 500-nm spatial resolution in the ion microscope mode, was used in this study. The original 3f SIMS instrument has been upgraded and equipped with a primary ion beam mass filter, a 5f Hall Probe control chassis, and a Charles Evans & Associates model PC-1CS computer interface system for control of instrument operation. The instrument is integrated to a CCD camera capable of 14 bits per pixel image digitization (Photometrics, Tucson, AZ, USA).

2.6. Computer Image Processing

Computer image processing was performed using DIP Station software (Haydon Image Processing Group, Boulder, CO, USA).

3. Methods

3.1. The Cell Growth and Cisplatin Treatment

1. The LLC-PK_1 cells were maintained in Medium 199 supplemented with 3% fetal bovine serum at 37°C and 5% CO_2 atmosphere. The cells were grown on the polished surface of high-purity N-type semiconductor-grade silicon pieces. The silicon substrate is not toxic to cells and has been used for SIMS studies of ion transport and localization of calcium stores and anticancer drugs at subcellular scale resolution (25–27). Silicon pieces were thoroughly washed in water and sterilized prior to cell seeding. The cells were seeded at a density of 2.5×10^5 cells per 50-mm plastic Petri dish (*see* **Note 1**). Each dish contained six to eight silicon substrate pieces. After the cells reached about 80% confluency on silicon substrates, approximately 25,000 non-reactive 11-μm latex beads were added to each Petri dish (*see* **Note 2**). The beads were allowed to settle for approximately 30 min prior to the experiments. The addition of beads is a necessity because they act as spacers in the sandwich freeze-fracture method of sample preparation and protect the cells from squashing during cryogenic sampling (28). For exposure of LLC-PK_1 cells to the drug, 6-μM cisplatin was added to each Petri dish for 4 h. The Petri dishes

without the drug treatment contained control cells. The 4-h short exposure with clinically relevant low concentration of 6-μM cisplatin was chosen for investigating the early cytotoxic response of cells to cisplatin (9, 14).

3.2. Cryogenic Sample Preparation

1. After respective treatments, the cells were cryogenically prepared with the sandwich freeze-fracture method (28). In brief, the method involved the following steps: (i) the silicon pieces containing the cells are removed from the nutrient medium; (ii) excessive nutrient medium is removed from the cells by touching the edges of the silicon piece with a filter paper (*see* **Note 3**); (iii) a new clean silicon piece is placed on top, sandwiching the cells (and beads) between the two silicon pieces; (iv) the sandwich is fast frozen in super-cooled isopentane (*see* **Note 4**); (v) the frozen sandwich is quickly transferred to liquid nitrogen (*see* **Note 5**); (vi) the fracture is exposed by prying apart the two halves of the silicon cell and bead-silicon sandwich under liquid nitrogen with simple tools that can be immersed in liquid nitrogen (23, 28). A photo of these simple tools is shown in **Fig. 6.2**.
2. For complimentary replica or membrane fracture studies, the two silicon pieces from both sides of the sandwich can be analyzed in the frozen hydrated state with scanning electron microscopy techniques (29). The substrate side of the sandwich contains groups of cells fractured at the plasma membrane. The apical half-membrane and the overlaying growth medium were removed to the non-substrate side of the sandwich. The cells fractured at the dorsal surface

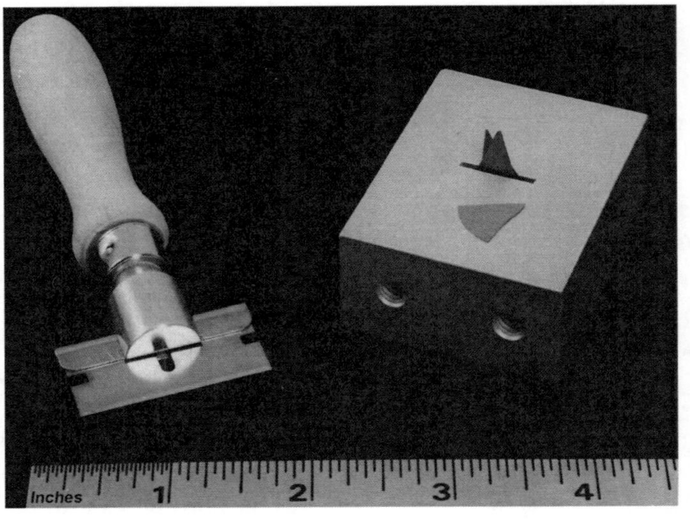

Fig. 6.2. Simple tools used for fracturing cell cultures grown on the silicon substrate under liquid nitrogen in sandwich freeze-fracture method. Reproduced from ref. (28).

(apical membrane) produced by this method are nearly the whole cells without the EF-leaflet of the plasma membrane (29). These fractured cells are used for SIMS imaging studies of chemical composition after freeze-drying. The purity and simplicity of this method is recognized by the fact that cells remained unperturbed in their growth medium throughout the sampling procedure that takes about 15 s. This sandwich fracture method is equally feasible to cells grown on other planer surfaces such as germanium wafer pieces, glass, and plastic substrates.

3.3. Freeze-Drying

1. The silicon substrates containing the fractured cells are transferred from liquid nitrogen and onto the pre-chilled sample stage of the freeze-drier. For this step, a thin aluminum plate of about three square inches may serve as a base for transferring the silicon pieces first under liquid nitrogen and onto the aluminum plate. The aluminum plate is then transferred to the pre-chilled sample stage of the freeze-drier. Depending on the type of freeze-drier, the sample stage can be pre-chilled between the –65 and the –90°C. In our laboratory a TIS-U-Dry freeze-drier is used for freeze-drying the cells. After the overnight freeze-drying, the temperature of the sample stage of the freeze-drier is gradually increased to 40°C to avoid any rehydration during venting of the freeze-drier. The freeze-drier is vented to dry nitrogen, and the aluminum plate containing the samples is quickly transferred to a desiccator. These samples can now be stored for a desired time for optical (and/or SEM) and SIMS analyses.

3.4. Light Microscopy Measurements

A silicon substrate containing the freeze-dried cells is transferred from a desiccator to an air-tight Teflon chamber to avoid any rehydration during optical microscopy measurements (**Fig. 6.1**). An Olympus microscope with reflected light Nomarski optics was used for photographing the fractured freeze-dried cells. **Figure 6.3** shows an example of how to find fractured cells on silicon substrates under reflected light Nomarski imaging. The low magnification image contains a fractured area on the silicon substrate, which is shown with a dotted line (**Fig. 6.3a**). This area appears clean since the fracture plane has passed through the cell surfaces and overlaying nutrient medium has been removed by this process. At least hundreds of cells are contained within this area. Such fractured areas scatter randomly on a silicon substrate and may occupy only a small fraction of the total surface area of the substrate (*see* **Note 6**). A higher magnification view of individual cells in fractured areas shows well-preserved nuclei, nucleoli, and cytoplasmic structures (**Fig. 6.3b**).

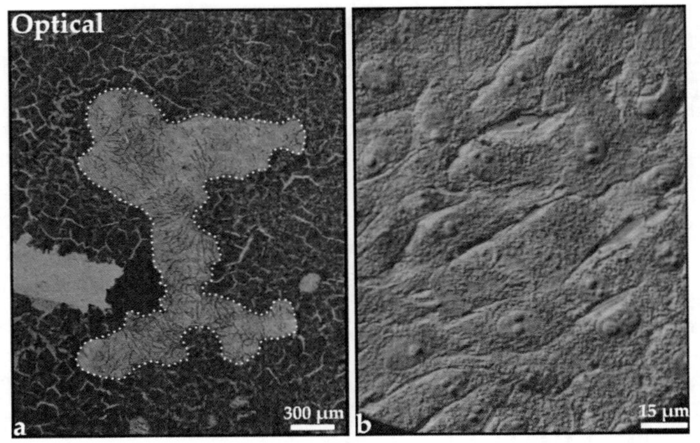

Fig. 6.3. Recognition of fractured renal epithelial LLC-PK_1 cells in freeze-dried samples with a reflected light microscope: Fractured cells can be recognized on silicon substrates from their clean appearance. (**a**) A low magnification reflected light Nomarski image illustrating a fractured area within the boundary highlighted by a *dotted line*, containing hundreds of fractured cells; (**b**) a high magnification reflected light microscopy image shows individual cells with discernible nucleus, nucleolus, and cytoplasmic structures.

3.5. Scanning Electron Microscopy (SEM) Measurements

The SEM provides an invaluable technique for observing structural preservation of fractured cells at greater details and to correlate SEM observations of single cell analysis with SIMS imaging of chemical composition (30, 31). An example of structural preservation of fractured cells is shown here. For improving the quality of SEM images, the cells were coated with a thin layer of Au/Pd in a sputter coater. **Figure 6.4** shows SEM images of fractured cells. An arrow in the low magnification SEM image points to a spacer latex bead in the fractured area containing many cells (**Fig. 6.4a**). A higher magnification view of fractured cells, recorded after tilting the sample stage of the SEM by 40°, shows well-preserved cytoplasmic structures of mitochondrial dimensions. A metaphase cell in the field of view shows the presence of sister chromatids (arrow) in some chromosomes which are laying perpendicular to the incident beam (**Fig. 6.4b**). The tilting of the SEM sample stage can significantly improve the recognition of microfeatures in surface topography of fractured cells.

3.6. SIMS Imaging of Chemical Composition in Single Cells

1. A CAMECA IMS-3f magnetic sector SIMS instrument, capable of producing isotopic (elemental) images with 500-nm spatial resolution in the ion microscope mode, was used in this study. A 5.5 keV primary ion beam of O_2^+ (approximately 100 nA beam current with a spot size of 60 µm in diameter) was used in this study. The primary ion beam was raster scanned over a 250 μm^2 region. A 60-µm contrast aperture was used for imaging positive secondary ion images.

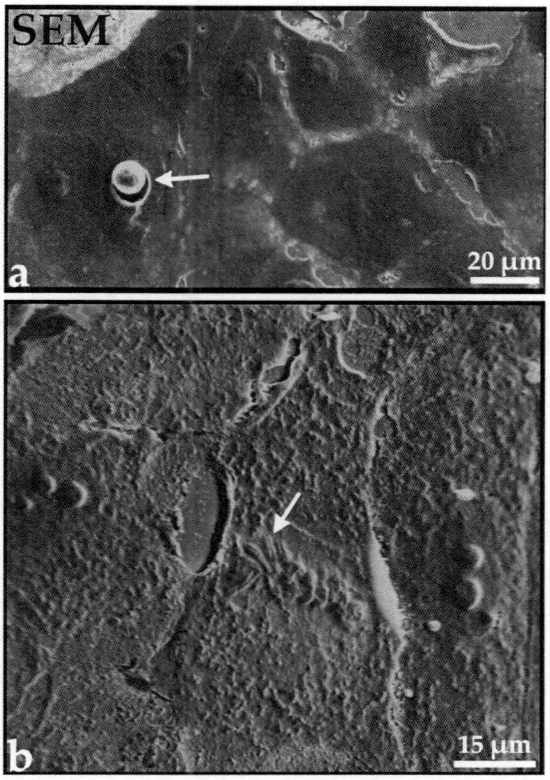

Fig. 6.4. SEM images of fractured freeze-dried renal epithelial LLC-PK_1 cells (**a**, **b**). An *arrow* in image (**a**) points to a spacer bead in a fractured area. In image (**b**) sister chromatids can be recognized in some metaphase chromosomes of the dividing cell. The thread-like cytoplasmic structures plausibly represent mitochondria in cells.

2. The cells were coated with a thin layer of Au/Pd to enhance their electrical conductivity for SIMS analysis. In the positive secondary ion detection mode, images of masses 12, 23, 39, 40, and 195 provided the subcellular distribution of $^{12}C^+$, $^{23}Na^+$, $^{39}K^+$, $^{40}Ca^+$, and $^{195}Pt^+$ secondary ions, respectively. A few minutes of pre-sputtering is required for secondary signal stabilization, fine image focusing, and the removal of Au/Pd coating from the cell surface prior to the recording of SIMS images on a CCD camera. These few minutes of pre-sputtering represent an initial step of the instrument tuning and signal stabilization under the bombardment of a reactive primary ion beam, like O_2^+ used here. In the CAMECA IMS-3f SIMS ion microscope instrument, SIMS images are recorded one image at a time with continuous sputtering of the cell surface in the *Z*-axis. In general, the sputtering times were 0.4 s for integration of $^{39}K^+$ and $^{23}Na^+$ images and 2 min for $^{12}C^+$, $^{40}Ca^+$, and $^{195}Pt^+$ images. It should be noted that topographic artifacts can be easily distinguished

in the analysis of cultured cells, as they cause stretching and blurring of SIMS images.

3. An example of the SIMS imaging of chemical composition in control renal epithelial LLC-PK_1 cells is shown in **Fig. 6.5**. SIMS images represent spatially resolved distribution of the total concentration (both free and bound forms) of designated elements via the detection of their respective major isotopes. The level of brightness within a SIMS image is directly proportional to the local concentration of the analyte. The ^{39}K SIMS image reveals a nearly homogeneous distribution for intracellular potassium. The individual cells in the field of view are easily recognized as they are separated by dark lines representing inter-cellular spaces (**Fig. 6.5a**). These cells also reveal high K-low Na signals (**Fig. 6.5b**) with a K/Na ratio of approximately 11, which is indicative of the analysis of well-preserved healthy cells. The ^{23}Na image appears dark since it was integrated for the same amount of time (0.4 s) as the ^{39}K image. This approach provides a direct visual comparison for evaluating the health status of the analyzed individual cells in the field of view. The ^{40}Ca image reveals lower concentration of total calcium in the nucleus as compared to the cytoplasm (**Fig. 6.5c**).

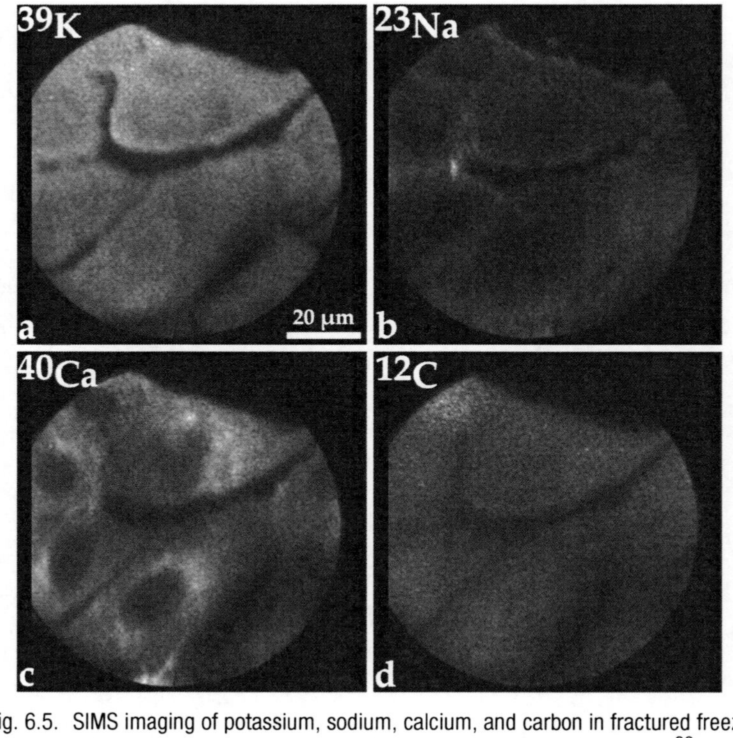

Fig. 6.5. SIMS imaging of potassium, sodium, calcium, and carbon in fractured freeze-dried control renal epithelial LLC-PK_1 cells (panels **a–d**). Individual images of $^{39}K^+$ and $^{23}Na^+$ were integrated on the CCD camera for 0.4 s and $^{12}C^+$ and $^{40}Ca^+$ for 2 min.

The calcium storing organelle, the endoplasmic reticulum, is a major contributor to the cytoplasmic pool of calcium in SIMS images (31). A bright-intensity perinuclear region is also present in the cytoplasm of each cell in the ^{40}Ca image (**Fig. 6.5c**). This region has been identified as the Golgi apparatus in this cell line (32). A carbon image is also shown to illustrate the distribution of carbon in these cells (**Fig. 6.5d**). The ^{12}C image is recorded since spatially resolved carbon intensities in relation to the signals of other analytes provide the foundation for quantitative imaging at subcellular scale resolution (see later). There was no detectable signal at mass 195, indicating that background for $^{195}Pt^{+}$ secondary ions was minimal in these control cells.

4. An example of the SIMS analysis of chemical composition in cisplatin-treated LLC-PK_1 cells is shown in **Fig. 6.6**. Individual cells in SIMS images are identified by numbers for the ease of their matching between the ^{39}K, ^{23}Na, and ^{40}Ca images (**Fig. 6.6a–c**). Individual cells display great variations in their response to cisplatin treatment, as it is clearly visu-

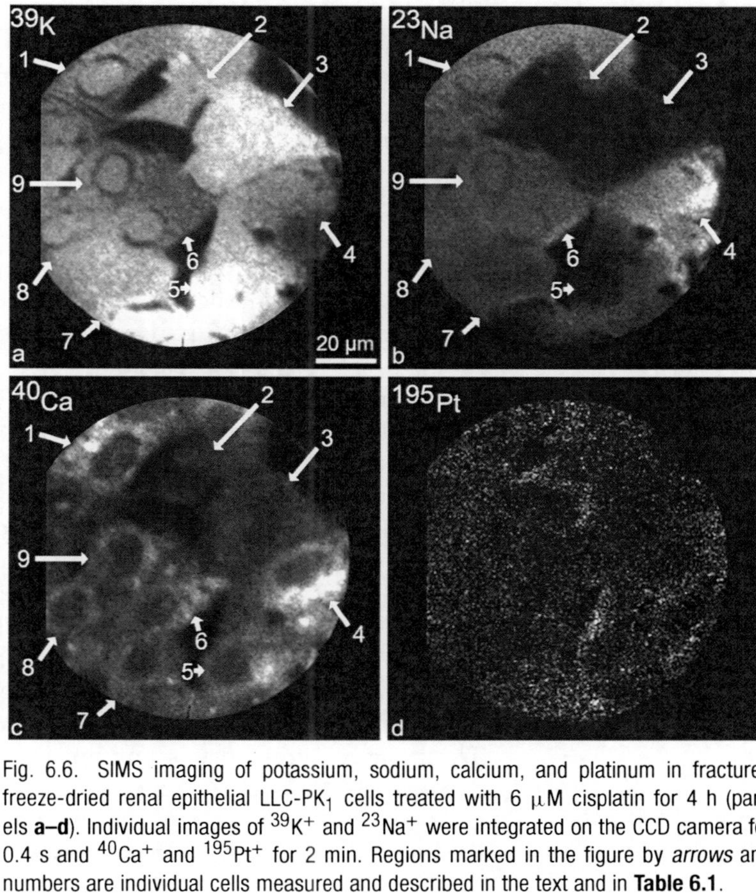

Fig. 6.6. SIMS imaging of potassium, sodium, calcium, and platinum in fractured freeze-dried renal epithelial LLC-PK_1 cells treated with 6 μM cisplatin for 4 h (panels **a–d**). Individual images of $^{39}K^{+}$ and $^{23}Na^{+}$ were integrated on the CCD camera for 0.4 s and $^{40}Ca^{+}$ and $^{195}Pt^{+}$ for 2 min. Regions marked in the figure by *arrows* and numbers are individual cells measured and described in the text and in **Table 6.1**.

alized in ^{39}K, ^{23}Na, and ^{40}Ca SIMS images (**Fig. 6.6a–c**). For example, while the majority of cells have gained significant quantities of sodium in response to cisplatin, but some (e.g., cells 2 and 3) still maintain the high K-low Na signature. The distribution of calcium stores has been altered completely in these cells. For example, the higher concentrations of calcium in Golgi apparatus can no longer be visualized after cisplatin treatment. Also, in general, the cells with higher Na signals also show a higher level of accumulation in their cytoplasm. It is plausible that the cisplatin response is cell cycle dependent, and, therefore, such a large variation is observed in these asynchronously growing cells. A much better understanding of these cisplatin-induced changes in chemical composition can be made by single cell analysis of each cell, quantitatively, by digital image processing (see below). It should also be noted that SIMS is capable of imaging Pt in cells. The presence of Pt was detected at mass 195, as $^{195}Pt^{+}$ secondary ions (**Fig. 6.6d**), even though SIMS' sensitivity for Pt detection is poor due to its high ionization potential.

3.7. Quantification of SIMS Ion Microscopy Images

Quantification of SIMS ion microscopy images in fractured freeze-dried cells has been reviewed previously (15). In brief, the illumination of a dynamic SIMS ion microscopy image, I, is related to the concentration of the imaged element M by the equation (33):

$$I = tC_{\mathrm{M}}Si_{\mathrm{p}}a_0$$

where t is the practical ion yield of element M (the ratio of the number of M^{+} ions collected to the number of M atoms removed from the sample), C_{M} is the atomic concentration of element M corrected for its isotopic abundance, S is the total sputtering yield (number of atoms of any kind removed from the sample per incoming primary ion), i_{p} is the primary beam current density, and a_0 is the analyzed area of the sample surface. While i_{p} and a_0 are controllable, t and S depend on many factors. The practical ion yield of an element is inversely proportional to the exponential of the element's first ionization potential and depends on the chemical state of the element, the chemical and physical properties of the sample matrix, instrumental transmission, and sampling conditions. Variations of t and S within the imaged field have been considered as matrix effects, along with mass interferences due to the matrix which can augment measured intensities. The basic requirements of any image quantification scheme for dynamic SIMS images are (i) the sample preparation method must preserve native elemental distributions of the sample, (ii) the evaluation of matrix effects must be made, (iii) the quantification standard should have the matching matrix composition as the sample, and (iv) calibration of the ion microscope's imaging system.

First, the reliability of sandwich freeze-fracture sample preparation method in preserving intracellular chemical composition has been confirmed in many studies (e.g. see reviews in ref. (15, 23)). Both structures and chemical composition are well preserved for subcellular scale SIMS studies in fractured freeze-dried cells. Second, the evaluation of matrix effects between the nucleus and the cytoplasm of freeze-fractured freeze-dried cells revealed no significant differences (34, 35). This observation was not surprising since the major components of the cell matrix in mammalian cell cultures are C, H, N, and O, and their cellular distributions are largely homogeneous at the submicron scale (34). Third, the quantification standards were generated from cell culture samples themselves and the relative sensitivity factors (RSF) for desired analytes were determined with respect to the cell matrix $^{12}C^{+}$ carbon signals (36). In the last step of image quantification, the imaging system was calibrated for pixel-by-pixel image quantification by ratioing the analyte signals to the cell matrix $^{12}C^{+}$ signals in the same spatial registration (36). The absolute dry weight concentrations obtained by this method can be converted into wet weight millimolar concentrations by assuming 85% cell water content in mammalian cells (37). Subcellular quantification of SIMS ion images of ^{39}K, ^{23}Na, ^{40}Ca, and ^{10}B from fractured freeze-dried cells has provided invaluable information in studies of organelle level calcium stores (27, 31, 32) and screening of boronated anticancer compounds in boron neutron capture therapy of cancer (20, 26, 38, 39).

3.8. Digital Image Processing of SIMS Images for Quantitative Imaging

SIMS isotope images were digitized directly from the microchannel plate/fluorescent screen assembly of the ion microscope with a slow scan CCD camera capable of 14 bits per pixel image digitization. SIMS images of positive secondary ions of ^{12}C, ^{23}Na, ^{39}K, and ^{40}Ca were recorded from the same cell. SIMS image integration times varied according to their intensities. In general, the image integration times were 0.4 s for ^{39}K and ^{23}Na and 2 min for ^{12}C and ^{40}Ca images. The variations in time of exposure for various SIMS isotope images were compensated for quantification in relation to the time of exposure of the ^{12}C image. Computer image processing was performed using DIP Station software. SIMS images were quantified using relative sensitivity factors to the cell matrix element $^{12}C^{+}$ signals in the same spatial registration (36). The registration of analyte signals to the $^{12}C^{+}$ signals in the same spatial location within a cell in SIMS images compensates for variations in microchannel plate response and primary ion beam heterogeneity. The concentrations of ^{39}K and ^{23}Na were calculated on the single cell basis since their distributions are nearly homogeneous. Total calcium was measured in the nucleus and the cytoplasm by selecting regions of interests (ROIs) in individual cells.

3.9. Cisplatin-Induced Alterations of K, Na, and Ca Concentrations in Single Cells

The corresponding concentration of K, Na, and Ca in control and cisplatin-treated renal epithelial LLC-PK_1 cells shown in SIMS images in **Figs. 6.5** and **6.6**, respectively, is listed in **Table 6.1**. Physiologically relevant concentrations of K and Na were observed in control cells (**Table 6.1**). The cell cytoplasm contains nearly twofold higher concentration of calcium stores than nucleus in control cells (**Table 6.1**). The endoplasmic reticulum is mainly responsible for storing the majority of cytoplasmic calcium in this cell line (31). Cisplatin-treated cells display large variations in their chemical composition. For example, cells 2, 3, and 5 show intracellular concentrations of K and Na very comparable to the healthy control cells, but their cytoplasmic calcium stores have been reduced substantially (**Table 6.1**). Cells 1, 4, 6, 8, and 9 have gained toxic levels of intracellular Na (over 70 mM) and accumulated substantially higher Ca concentrations in both cytoplasm and nucleus (**Table 6.1**). Since these observations were reproducible (not shown), a few patterns of cisplatin-induced cytotoxic response can be deduced from single cell observations. First, the cells with healthy K and Na levels but reduced

Table 6.1
Total K, Na, and Ca concentrations calculated from SIMS ion microscopy images of single cells shown in control (Fig. 6.5) and cisplatin-treated (Fig. 6.6) renal epithelial LLC-PK_1 cells. Concentrations in control cells are listed as the mean ± SD for all cells shown in Fig. 6.5. For cisplatin-treated cells, concentrations are listed in individual cells identified by the cell numbers in Fig. 6.6. Total Ca concentrations reflect the level of calcium stores in designated subcellular compartments

Treatment	Total Ca (Nucleus) (mM)	Total Ca (Cytoplasm) (mM)	Cellular K (mM)	Cellular Na (mM)
Control cells	0.51 ± 0.07	1.03 ± 0.16	168 ± 13	15 ± 3
Cisplatin				
Cell 1	0.80	1.74	139	91
Cell 2	0.44	0.69	179	12
Cell 3	0.48	0.71	183	14
Cell 4	0.77	1.63	97	96
Cell 5	0.45	0.76	146	16
Cell 6	0.81	1.68	133	73
Cell 7	0.53	0.87	157	33
Cell 8	0.75	1.47	133	72
Cell 9	0.82	1.45	128	71

cytoplasmic calcium stores (Cell 2, 3, and 5) probably reflect the early response to cisplatin as the reduction in ER calcium pool. This may be a defense mechanism of the cell against cisplatin, which causes a shift in fundamental calcium homeostasis. Second, a persistent perturbation of calcium homeostasis may raise intracellular ionized Ca^{2+} levels in the toxic range. At this stage, mitochondria will participate in Ca^{2+} buffering by unphysiological loading of calcium for protecting the cells from injury (or death). It is plausible that cells 1, 4, 6, 8, and 9 are at this stage of cisplatin-induced injury and demonstrate cytoplasmic loading of calcium in SIMS analysis. Further signs of toxicity are also evident in these cells by the gain of Na^+ and the loss of K^+ (**Table 6.1**). Such alterations in intracellular K^+ and Na^+ concentrations will result in destabilization of the membrane potential. SIMS observations shown here provide strong support to previous studies that have implicated ER stress, mitochondrial dysfunctions, and perturbation of calcium homeostasis in cisplatin-treated cells (12–14). This work also demonstrates that mass spectrometry-based single cell imaging techniques can provide valuable tools in cell biology and medicine for characterization of chemotherapeutic agents.

4. Notes

1. In general, two Petri dishes per treatment are sufficient to provide enough number of silicon substrate pieces for cryogenic sampling and SIMS analysis.
2. After the beads are added, a gentle shaking of the Petri dish by rotating it in your hand is recommended for dispersing the beads evenly in the nutrient medium.
3. It is not recommended to dry the cells in this step. A thin layer of overlaying nutrient medium on top of the cells is a requirement for sandwich freeze-fracture.
4. Super-cooled isopentane can be made easily by taking about 50–80 ml of isopentane in a 200-ml beaker. The beaker containing isopentane is placed in a bigger container filled with liquid nitrogen to less than half the height of the isopentane beaker. This assembly is placed on a magnetic stir. The isopentane is stirred by putting a small magnet in the beaker. In approximately 60–90 s of stirring, such a cooling results in clouding of isopentane prior to its freezing. This cloudy isopentane is known as "super-cooled isopentane." Individual users may modify this procedure according to their needs. As a caution, it is recommended that this step be done after wearing safety goggles.

5. Frozen silicon sandwiches can be stored under liquid N_2 for desired periods of time.
6. It is also possible that any given silicon substrate may not contain any fractured areas. In general, six to eight substrate pieces per treatment would provide sufficient number of fractured cells for SIMS analysis. Also, the cells grown to higher confluency (over 70%) on silicon substrates improve the chance of fracturing them as a monolayer in large areas. Although we have characterized this fracture by matching compliments of the same individual cells on the substrate and the other side of the sandwich (29), it cannot be ruled out that some fractured areas contain cells where the fracture plane has passed through the cell surface without splitting the apical membrane.

Acknowledgments

This work was supported in part by Biological and Environmental Research Program (BER), US Department of Energy, grant number DE-FG02-91ER 61138. Partial supports from a NIH grant R01CA129326 (National Cancer Institute), Cornell Core Facilities, and NYSTAR Program are also acknowledged. (The content discussed in this chapter is solely the responsibility of the author and does not necessarily represent the official views of the National Cancer Institute or National Institutes of Health.) The NIH/NSF Development Resource for Biophysical Imaging Opto-electronics (DRBIO) is acknowledged in culturing the cells used for SIMS experiments in this study.

References

1. Rosenberg, B. (1999) Platinum complexes for the treatment of cancer: why the search goes on Cisplatin. *Chemistry and Biochemistry of a Leading Anticancer Drug* (Lippert, B. ed.) Wiley-VCH, Basel, Switzerland, 3–27.
2. Zamble, D. B., Jacks, T., Lippard, S. J. (1998) p53-dependent and -independent response to cisplatin in mouse testicular teratocarcinoma cells. *Proc Natl Acad Sci U S A*, **95**, 6163–6168.
3. Pestell, K. E., Hobbs, S. M., Titley, J. C., Kelland, L. R., Walton, M. I. (2000) Effect of p53 status on sensitivity to platinum complexes in a human ovarian cancer cell line. *Mol Pharm*, **57**, 503–511.
4. Gonzalez, V. M., Fuertes, M. A. Alonso, C., Perez, J. M. (2001) Is cisplatin-induced cell death always produced by apoptosis? *Mol Pharmacol*, **59**, 657–663.
5. Lieberthal, W., Triaca, V., Levine, J. (1996) Mechanisms of death induced by cisplatin in proximal tubular epithelial cells: apoptosis vs. necrosis. *Am J Physiol*, **270**, F700–F708.
6. Liang, X., Shen, D., Chen, K. G., Wincovitch, S. M., Garfield, S. H., Gottesman, M. M. (2005) Trafficking and localization of platinum complexes in cisplatin-resistant cell lines monitored by fluorescence-labeled platinum. *J Cell Physiol*, **202**, 635–641.
7. Osman, A. M., El-Sayed, E. M., El-Demerdash, E., Al-Hyder, A., El-Didi, M.,

Attia, A. S., Hamada, F. M. A. (2000) Prevention of cisplatin-induced nephrotoxicity by methimazole. *Pharmacol Res*, **41**, 113–119.

8. Nowak, G. (2002) Protein kinase C-α and ERK1/2 mediate mitochondrial dysfunction, decrease in active Na^+ transport, and cisplatin-induced apoptosis in renal cells. *J Biol Chem*, **277**, 43377–43388.
9. Rabik, C. A., Dolan, M. E. (2007) Molecular mechanisms of resistance and toxicity associated with platinating agents. *Cancer Treat Rev*, **33**, 9–23.
10. Cumming, B. S., Schnellmann, R. G. (2002) Cisplatin-induced renal cell apoptosis: caspase 3-dependent and -independent pathways. *JPET*, **302**, 8–17.
11. Horky, M., Wurzer, G., Kotala, V., Anton, M., Vojtesek, B., Vacha, J., Wesierska-Gadek, J. (2000) Segregation of nucleolar components coincides with caspase-3 activation in cisplatin-treated HeLa cells. *J Cell Sci*, **114**, 663–670.
12. Park, M. S., Leon, M. D., Devarajan, P. (2002) Cisplatin induces apoptosis in LLC-PK_1 cells via activation of mitochondrial pathways. *J Am Soc Nephrol*, **13**, 858–865.
13. Liu, H. and Beliga, R. (2005) Endoplasmic reticulum stress-associated caspase 12 mediates cisplatin-induced LLC-PK_1 cell apoptosis. *J Am Soc Nephrol*, **16**, 1985–1992.
14. Splettstoesser, F., Florea, A-M., Busselberg, D. (2007) IP_3 receptor antagonist, 2-APB, attenuates cisplatin induced Ca^{2+}-influx in HeLa-S3 cells and prevents activation of calpain and induction of apoptosis. *Brit J Pharmacol*, **151**, 1176–1186.
15. Chandra, S., Smith, D. R., Morrison, G. H. (2000) Subcellular imaging by dynamic SIMS ion microscopy. *Anal Chem*, **72**, 104A–114A.
16. Rubakhin, S. S., Jurchem, J. C., Monroe, E. B., Sweedler, J. V. (2005) Imaging mass spectrometry: fundamentals and applications to drug discovery. *Drug Discov Today*, **10**, 823–837.
17. Winograd, N. (2005) The magic of cluster SIMS. *Anal Chem*, 77, 142A–149A.
18. Lechene, C., F. Hillion, F., McMahon, G., Benson, D., Kleinfeld, A. M., Kampf, J. P., Distel, D., Lyuten, Y., Bonventre, J., Hentschel, D. Park, K. M., Ito, S., Schwartz, M., Benichou, G., Slodzian, G. (2006) High resolution quantitative imaging of mammalian and bacterial cells using stable isotope mass spectrometry. *J Biol*, **5**, 20–30.
19. Gazi, E., Lockyer, N. P., Vickerman, J. C., Gardner, P, Dwyer, J., Hart, C.A., Brown, M. D., Clarke, N. W., Miyan, J. (2004) Imaging ToF and synchrotron-based FT-IR microspectroscopic studies of prostate cancer cell lines. *Appl Surf Sci*, **231–232**, 452–456.
20. Chandra, S., Tjarks, W., Lorey, II, D. R., Barth, R. F. (2008) Quantitative subcellular imaging of boron compounds in individual mitotic and interphase human glioblastoma cells with imaging secondary ion mass spectrometry (SIMS). *J Microsc*, **229**, 92–103.
21. Slodzian, G., Daigne, B., Girard, F., Boust, F., Hillion, F. (1992) Scanning secondary ion mass spectrometry with parallel detection. *Biol Cell*, **74**, 43–50.
22. Boxer, S. J., Kraft, M. L., Weber, P. K. (2009) Advances in imaging secondary ion mass spectrometry for biological samples. *Annu Rev Biophys*, **38**, 53–74.
23. Chandra, S. (2008) Challenges of biological sample preparation for SIMS imaging of elements and molecules at subcellular resolution. *Appl Surf Sci*, **255**, 1273–1284.
24. Hull, R. N., Cherry, W. R., Weaver, G. W. (1976) The origin and characteristics of a pig kidney strain, LLC-PK_1. *In Vitro*, **12**, 670–677.
25. Chandra, S., Morrison, G. H. (1985) Imaging elemental distribution and ion transport in cultured cells with ion microscopy. *Science*, **228**, 1543–1544.
26. Chandra, S., Kabalka, G. W., Lorey, II, D. R., Smith, D. R., Coderre, J. A. (2002) Imaging of fluorine and boron from fluorinated-boronophenylalanine in the same cell at organelle resolution by correlative SIMS ion microscopy and confocal laser scanning microscopy. *Clin Cancer Res*, **8**, 2675–2683.
27. Chandra, S. (2008) Subcellular imaging of RNA distribution and DNA replication in single mammalian cells with SIMS: the localization of heat shock induced RNA in relation to the distribution of intranuclear bound calcium. *J Microsc*, **232**, 27–35.
28. Chandra, S., Morrison, G. H., Wolcott, C. C. (1986) Imaging intracellular elemental distribution and ion fluxes in cultured cells with ion microscopy: a freeze-fracture methodology. *J Microsc*, **144**, 15–37.
29. Chandra, S., Morrison, G. H. (1997) Evaluation of fracture planes and cell morphology in complimentary fractures of cultured cells in the frozen-hydrated state by field-emission secondary electron microscopy: feasibility for ion localization and fluorescence imaging studies. *J Microsc*, **186**, 232–245.
30. Chandra, S., Morrison, G. H. (1992) Sample preparation of animal tissues and cell cultures for secondary ion mass spectrometry (SIMS) microscopy. *Biol Cell*, **74**, 31–42.

31. Chandra, S. (2005) Quantitative imaging of subcellular calcium stores in mammalian LLC-PK_1 epithelial cells undergoing mitosis by SIMS ion microscopy. *Eur J Cell Biol*, **84**, 783–797.
32. Chandra, S., Kable, E. P. W., Morrison, G. H., Webb, W. W. (1991) Calcium sequestration in the Golgi apparatus of cultured mammalian cells revealed by laser scanning confocal microscopy and ion microscopy. *J Cell Sci*, **100**, 747–752.
33. Morrison, G. H., Slodzian, G. (1975) Ion microscopy. *Anal Chem*, **47**, 932A–943A.
34. Chandra, S., Ausserer, W. A., Morrison, G. H. (1987) Evaluation of matrix effects in ion microscopic analysis of freeze-fractured, freeze-dried cultured cells. *J Microsc*, **148**, 223–239.
35. Ausserer, W. A., Chandra, S., Morrison, G. H. (1988) Morphological and elemental integrity of freeze-fractured, freeze-dried cultured cells during ion microscopic sampling. *J Microsc*, **154**, 39–57.
36. Ausserer, W. A., Ling, Y-C., Chandra, S., Morrison, G. H. (1989) Quantitative imaging of boron, calcium, magnesium, potassium, and sodium distributions in cultured cells with ion microscopy. *Anal Chem*, **61**, 2690–2695.
37. Buchanan, R. A., Leapman, R. D., O'Connell, M. F., Reese, T. S., Andrews, S. B. (1993) Quantitative scanning transmission electron microscopy of ultrathin cryosections: subcellular organelles in rapidly frozen liver and cerebellar cortex. *J Struct Biol*, **110**, 244–255.
38. Chandra, S., Lorey, II, D. R., Smith, D. (2002) Quantitative subcellular dynamic SIMS imaging of boron-10 and boron-11 isotopes in the same cell delivered by two combined BNCT drugs: *in vitro* studies on human glioblastoma T98G cells. *Radiat Res*, **157**, 700–710.
39. Chandra, S., Lorey, II, D. R. (2007) SIMS ion microscopy imaging of boronophenylalanine (BPA) and $^{13}C^{15}N$-labeled phenylalanine in human glioblastoma cells: relevance of subcellular scale observations to BPA-mediated boron neutron capture therapy of cancer. *Int J Mass Spectrom*, **260**, 90–101.

Chapter 7

Imaging MALDI Mass Spectrometry of Sphingolipids Using an Oscillating Capillary Nebulizer Matrix Application System

Yanfeng Chen, Ying Liu, Jeremy Allegood, Elaine Wang, Begoña Cachón-González, Timothy M. Cox, Alfred H. Merrill, Jr., and M. Cameron Sullards

Abstract

Matrix deposition is a critical step in tissue imaging by matrix-assisted laser desorption/ionization mass spectrometry (MALDI-MS). It greatly affects the quality of MALDI imaging, especially for the analytes (such as lipids) that may easily dissolve in the solvent used for the matrix application. This chapter describes the use of an oscillating capillary nebulizer (OCN) to spray small droplets of matrix aerosol onto the sample surface for improved matrix homogeneity, reduced crystal size, and controlled solvent effects. This protocol allows visualization of many different lipid species and, of particular interest, sphingolipids in tissue slices of Tay-Sachs/Sandhoff disease by imaging MALDI-MS. The structures of these lipids were identified by analysis of tissue extracts using electrospray ionization in conjunction with tandem mass spectrometry (MS/MS and MS^3). These results illustrate the usefulness of tissue imaging MALDI-MS with matrix deposition by OCN for the molecular analysis in normal physiology and pathology. In addition, the observation of numerous lipid subclasses with distinct localizations in the brain slices demonstrates that imaging MALDI-MS could be effectively used for "lipidomic" studies.

Key words: Imaging MALDI mass spectrometry, sphingolipids, oscillating capillary nebulizer, matrix application.

1. Introduction

Imaging matrix-assisted laser desorption/ionization mass spectrometry (MALDI-MS) is a powerful tool that may be used to determine the spatial distribution and relative abundance of specific molecules in biological samples such as histological slices of

S.S. Rubakhin, J.V. Sweedler (eds.), *Mass Spectrometry Imaging*, Methods in Molecular Biology 656,
DOI 10.1007/978-1-60761-746-4_7,

tissues (1–4). Matrix deposition is a critical step in tissue imaging by MALDI-MS. It greatly affects the quality of MALDI imaging in terms of mass resolution, detection sensitivity, spatial resolution, and reproducibility. The effectiveness of the matrix is determined by the size, density, and homogeneity of the clusters/crystals that form on the surface. The extent to which deposition of the matrix perturbs the localization of molecules in the sample, such as the lateral diffusion of the analytes (especially lipids that can dissolve in the solvent used for matrix deposition), is also a critical factor for the success of imaging MALDI-MS experiments.

The oscillating capillary nebulizer (OCN) (5–7) is a low-cost device and is capable of providing a uniform matrix coating for accurate mass analysis, good sensitivity, and high reproducibility (8–10). It can generate small droplets/aerosols with a narrow size distribution (6) by nebulizing the matrix solution at the capillary tip. It can also effectively handle liquid compositions from 100% aqueous to 100% organic (9). By controlling several parameters of OCN operation, the solvent content of the droplet approaching the sample surface can be manipulated (9) to reduce the analyte migration and enhance the matrix–analyte interaction. This feature demonstrates the strong potential of the OCN to improve the quality of data for imaging MALDI-MS. Furthermore, the OCN works well for both micro-flows (μl/min) and macro-flows (ml/min) with high transport efficiencies (5, 11), which can greatly minimize the time for matrix coating. This makes the OCN matrix application system very suitable for automated, high-throughput, and high-quality matrix deposition in imaging mass spectrometry of biological molecules.

This chapter describes the use of an oscillating capillary nebulizer (OCN) to spray small droplets of matrix aerosol onto the sample surface for improved matrix homogeneity, reduced crystal size, and controlled solvent effects. This protocol allows the visualization of many different lipid species and, of particular interest, sphingolipids in tissue slices of Tay-Sachs/Sandhoff disease by imaging MALDI-MS (12). The structures of these lipids were identified by analysis of tissue extracts using electrospray ionization in conjunction with tandem mass spectrometry (MS/MS and MS^3). These results illustrate the usefulness of tissue imaging MALDI-MS with matrix deposition by OCN for the molecular analysis in normal physiology and pathology.

2. Materials

2.1. Chemicals

1. 2,5-Dihydroxybenzoic acid (DHB).
2. Trifluoroacetic acid (TFA).

3. Sulfatides (Porcine Brain) (Avanti Polar Lipids, Inc., Alabaster, AL, USA).
4. Total ganglioside mixtures (Porcine Brain) (Avanti Polar Lipids, Inc., Alabaster, AL, USA).
5. Monosialogangliosides GM1, GM2, and GM3 (as NH_4^+ salts) (Matreya LLC, Pleasant Gap, PA, USA).
6. *Angiotensin* III. (Ang III) (Sigma Chemicals, St. Louis, MO, USA).
7. Fibrinopeptide B (*GluFib*) (Sigma Chemicals, St. Louis, MO, USA).
8. Adrenocorticotropic hormone (ACTH) 1–24 (Sigma Chemicals, St. Louis, MO, USA).
9. Hematoxylin–Eosin (H&E) Staining Solution (VWR, West Chester, PA, USA).
10. HPLC grade acetonitrile (ACN).
11. HPLC grade methanol (MeOH).
12. Diethylaminoethyl cellulose (DEAE)-Sephadex A-25 (Pharmacia LKB Biotechnology, Uppsala, Sweden).
13. Nanopure water (18 MΩ).

2.2. Experimental Animals

The hexb$^{+/-}$ and hexb$^{-/-}$ mice (strain: B6; 129S-Hexbtm1Rlp, Jackson Laboratory, Bar Harbor, ME, USA) were obtained by crossing homozygous males with heterozygous females or by heterozygous mating (13). The hexb genotype was determined by PCR on mouse-tail DNA with primers B3 (5′-ATGTGG ATGCAACTAACC-3′) and B4 (5′-AGGTTGTGCAGCTATT CC-3′) that flank the disrupting MC1NeopolyA cassette in exon 13. The hexb$^{-/-}$ and hexb$^{+/-}$ male mice were sacrificed and dissected at 18.5 and 23 weeks, respectively. The whole procedure was conducted using protocols approved under license by the UK Home Office (Animals Scientific Procedures Act, 1986).

2.3. Experimental Instruments

1. Cryomicrotome (Cryo-Star HM560, MICROM, Walldorf, Germany).
2. Leica autostainer XL (Leica Microsystems, Bannockburn, IL, USA).
3. Nikon Eclipse E600 microscope (Nikon, Melville, NY, USA).
4. Stainless steel-polished blank MALDI sample plate (Applied Biosystems, Foster City, CA, USA).
5. Syringe pump (KD Scientific, Holliston, MA, USA).
6. 25 ml TLC reagent sprayer with standard ground glass joint (Kimble/Kontes, Vineland, NJ, USA).
7. Scanning electron microscope (SEM) (Nova Nanolab 200 system, FEI company, Hillsboro, OR, USA).

8. Voyager DE-STR MALDI-TOF mass spectrometer with a 337 nm N_2 laser (3 Hz) (Applied Biosystems, Foster City, CA, USA).
9. API 4000 QTrap tandem mass spectrometer (Applied Biosystems, Foster City, CA, USA).

2.4. Oscillating Capillary Nebulizer Setup

A diagram of the design and operation of the oscillating capillary nebulizer (OCN) matrix application system is shown in **Fig. 7.1**. The OCN sprayer (5, 7) consists of two coaxial fused silica capillary tubes (Polymicro Technologies, LLC, Phoenix, AZ, USA) that are friction-fit mounted with PEEK Sleeves (Upchurch Scientific, Oak Harbor, WA, USA) housed in a 1/16″ stainless steel union tee (Swagelok, Solon, OH, USA). The inner capillary (i.d. 50 μm, o.d. 150 μm, length 80 mm) was used to transfer the matrix solution (signified by blue in **Fig. 7.1**) and the outer capillary (i.d. 250 μm, o.d. 350 μm, length 30 mm) allowed gas (signified by pink in **Fig. 7.1**) to pass through the annular space between the outer wall of inner capillary and the inner wall of the outer capillary to generate the oscillation of the inner capillary tip,

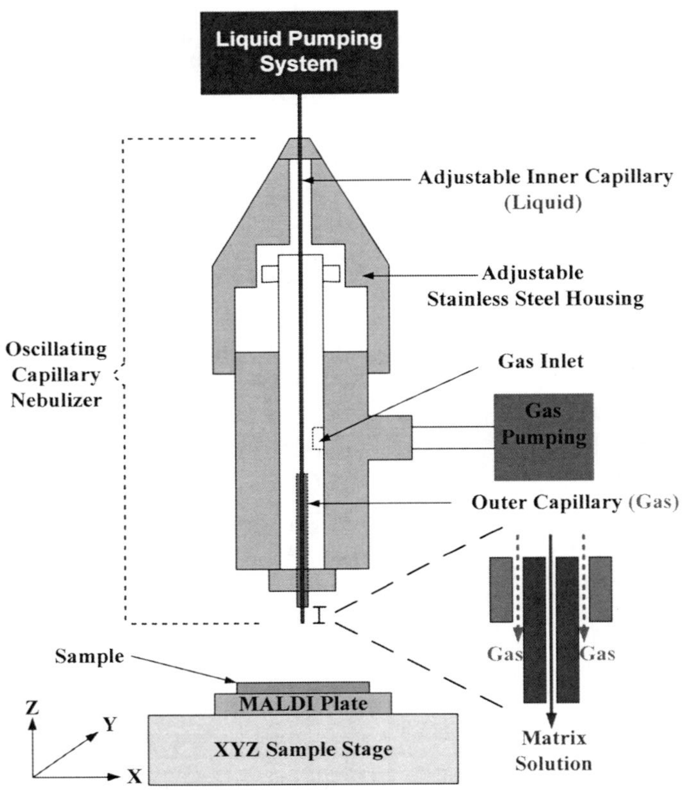

Fig. 7.1. Schematic of the oscillating capillary nebulizer (OCN) matrix application system with matrix solution and nebulizing gas.

which extends about 1 mm (R) from the outer capillary tip. The high-frequency oscillation induces the nebulization of the matrix solution and generates a fine and uniformly dispersed spray of matrix droplets/particles onto sample plates fixed on a xyz translation stage (Newport, Irvine, CA, USA) (*see* **Note 1**).

3. Methods

3.1. Preparation of Tissue Samples

1. The tissue samples were snap-frozen after dissection in liquid nitrogen and stored at –80°C.
2. Before tissue section, the tissues were first put into a sealed box with dry ice to equilibrate at respective temperature for 60 min.
3. The tissues were then transferred into cryomicrotome at –20°C for another 60 min.
4. Attach the tissues to the sample holder with small amount of optimal cutting temperature (OCT) compounds (Cryo-OCT Compound, Pittsburgh, PA, USA) at the back of tissue. No OCT compounds were used on the tissue area to be cut (*see* **Note 2**).
5. The tissue samples were sectioned as 10 μm slices at –18°C and thaw-mounted onto chilled MALDI plates (~5°C). The sample plates were sealed in Petri dishes and stored at –80°C.
6. Neighboring sections were also cut into 10 μm thickness under the same conditions and attached onto glass slides for histological staining.

3.2. Matrix Deposition

1. The tissue slices on the MALDI plates were taken out from the –80°C and brought into a desiccator at room temperature for 60 min before matrix coating.
2. Standard solutions were pipetted on different spots around the tissue slices for mass spectrometer calibration (*see* **Note 3**). For positive mode mass spectrometry operation, 1 μl of standard solution containing 5.0 pmol/μl Leu-enkephalin, 1.0 pmol/μl angiotensin II, 1.0 pmol/μl angiotensin I, 1.0 pmol/μl Glu-fibrinopeptide B, and 2.0 pmol/μl ACTH (18–39 clip) in acetonitrile:water (50:50, v:v) was deposited on each spot. For negative mode mass spectrometry operation, 1 μl of standard solution containing 1.0 pmol/μl sulfatides and 1.0 μg/μl total gangliosides in methanol solution was deposited on each spot. The mass spectrometer was calibrated using the m/z values of the standard molecules.

3. The sample plate was then fixed on a xyz stage under OCN for matrix application.
4. The concentration of matrix solution, flow rate of matrix solution, nebulizing gas pressure, the length of inner capillary tip extending from the outer capillary tip, OCN–sample distance, and the moving speeds of the xy stages can be adjusted to control the size of matrix droplet and the surface wetting on the sample plate for optimized matrix coating (*see* **Note 4**).
5. The characteristics of the matrix crystals were observed using a scanning electron microscope (SEM) (*see* **Note 5**). The example results of matrix crystal formation using different matrix application techniques were shown in **Fig. 7.2** (*see* **Note 6**).

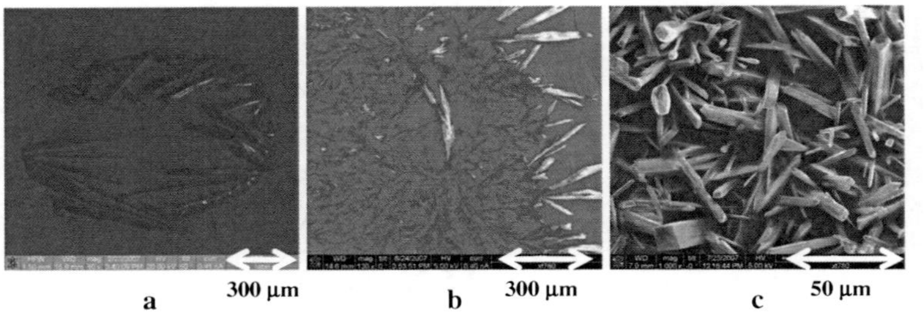

Fig. 7.2. SEM images of DHB crystals formed from matrix solution (30 mg/ml DHB in acetonitrile:water 50:50, v:v, with 0.1% TFA) using different matrix coating methods. (**a**) Direct drying of 1 µl matrix solution. (**b**) Matrix deposited using TLC sprayer (nitrogen pressure: 5 psi; sprayer–sample distance: 12 cm; spraying time: 10 s/cycle; drying time: 30 s/cycle; coating cycle: 30 times). (**c**) Matrix deposited using OCN sprayer (matrix solution flow rate: 60 µl/min; nitrogen pressure: 50 psi; sprayer–sample distance: 10 cm; coating time: 5 min).

6. The typical distance (L) between the OCN and the sample on the xyz translation stage is 8–20 cm depending on the flow rate of the matrix solution and gas pressure.
7. The matrix solution (30 mg/ml DHB in acetonitrile:water 50:50, v:v, with 0.1% TFA) was injected into the inner capillary of OCN through a 1 ml gastight syringe using a syringe pump with a flow rate ~60 µl/min.
8. The pressure of nebulizing gas (N_2) was adjusted to ~50 psi to generate a fine matrix aerosol which covers the selected sample area.
9. The sample plate was continually moved across the aerosol deposition area in the *x* direction (5 mm/s) and *y* direction (5 mm/s) to obtain an even matrix distribution throughout the sample surface.

10. The typical time of optimized OCN matrix coating for a 4 cm^2 sample is about 5 min with an estimated thickness of 10–20 μm (*see* **Note 7**).
11. After the matrix deposition is finished, the sample plate was further dried in a desiccator at room temperature for 10 min before being transferred into the imaging mass spectrometer.
12. The OCN system was flushed with nanopure water at a flow rate of 100 μl/min for 10–20 min to wash off any remaining residue (*see* **Note 8**).

3.3. MALDI Imaging Mass Spectrometry

1. The matrix-coated sample plate was loaded into the Voyager DE-STR MALDI-TOF mass spectrometer. The real-time video image of the tissue slice can be viewed via the video monitor. The tissue section boundaries were determined by their logical x and y coordinates on the MALDI plate.
2. Standard spots were used to optimize the instrument parameters for maximum sensitivity, resolution, and mass accuracy. The TOF mass spectrometer was operated in reflector mode with delayed extraction. The accelerating voltage, grid voltage, and delay time are typically 22 kV, 70%, and 400 ns, respectively. The laser intensity was checked daily to obtain the best signal-to-noise ratio. In negative mode, the laser intensity is usually a little bit higher (5–8%) than the positive mode. The acquisition mode was "manual" and the number of laser shots at each position was 10 for all acquisition methods (*see* **Note 9**).
3. Obtain an MALDI mass spectrum of the standard spot using the optimized acquisition method.
4. Load the *MALDI-MS Imaging Tool* software (free version at http://www.maildi-msi.org). Choose a calibrated mass spectrum of standard compounds as the "Data File" and select the optimized acquisition method as "Control File").
5. Define imaging area by setting the x and y coordinates of the tissue boundaries.
6. Set the step size of the laser rastering, which determines the pixel number of the MALDI image. Typically, a step size of 60 μm is used for brain tissue within the size range of 15–30 mm^2 (*see* **Note 10**).
7. Start the data acquisition and the *MALDI-MS Imaging Tool* creates an .img file.

3.4. Imaging Data Analysis

1. MALDI mass spectra of every individual pixel can be acquired by the *MALDI-MS Imaging Tool*. Using the "calculate image" function, an image of a selected m/z value can be visualized. The mass spectrum of a pixel can be displayed

by *Data Explorer* (factory-equipped software on Voyager DE-STR MALDI-TOF mass spectrometer) when the point of interest is selected. The example results were shown in **Fig. 7.3** (*see* **Note 11**).

2. MALDI images can be visualized using *Biomap* software package (free 3.7.4 version at http://www.maldi-msi.org) by loading the .img file (imaging MALDI-MS data).
3. MALDI images of all the *m/z* values can be viewed by the movie function (File/Export/Movie/). The MALDI images of selected *m/z* values were obtained by choosing the corresponding values of the data points (the *N* number) on the left panel.
4. The contrast of an ion image can be adjusted by changing the values of the slide bars on the left panel and the colors can be defined by clicking the "Set color table" button.
5. The display mode of the ion images was set to the default "interpolated" mode.
6. The ion images were saved using the export function of *Biomap* (File/Export/Image). The example ion images of sphingolipids in mouse brain of Tay-Sachs and Sandhoff disease model are shown in **Fig. 7.4** (*see* **Note 12**).

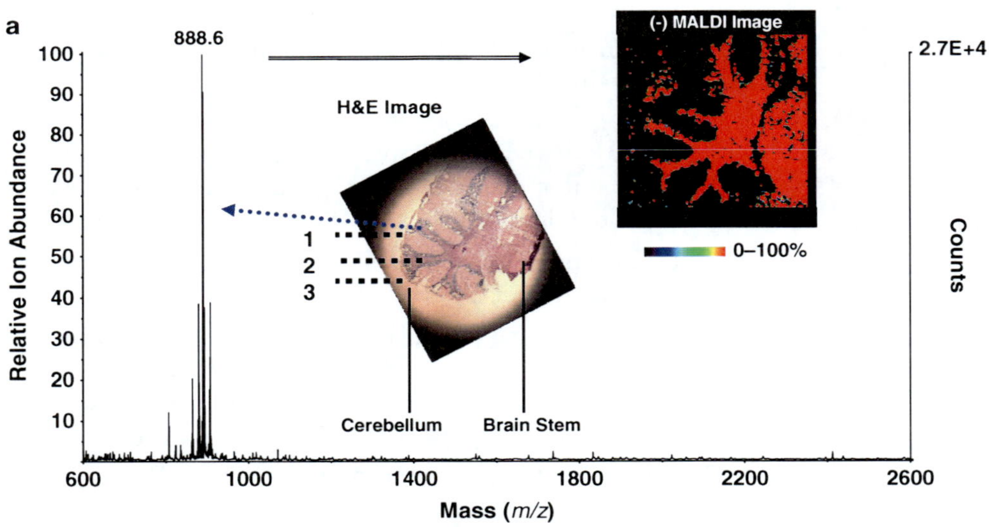

Fig. 7.3. Imaging MALDI-MS data from hexb$^{-/-}$ mouse brain (cerebellum and brain stem, 4.954 mm × 5.358 mm). The fine structures of cerebellum in the H&E-stained images are labeled as (1) molecular layer, (2) myelinated fiber (white matter), and (3) granular layer. The MALDI spectra present the ion yield from specific spots in (**a**) myelinated fiber (white matter) in negative ion mode, (**b**) granular layer region in negative ion mode, and (**c**) granular layer region in positive ion mode, respectively. The molecular distributions of *m/z* 888.6 ions, *m/z* 1,383 ions, and *m/z* 1,160 ions are shown in (**a**–**c**), respectively.

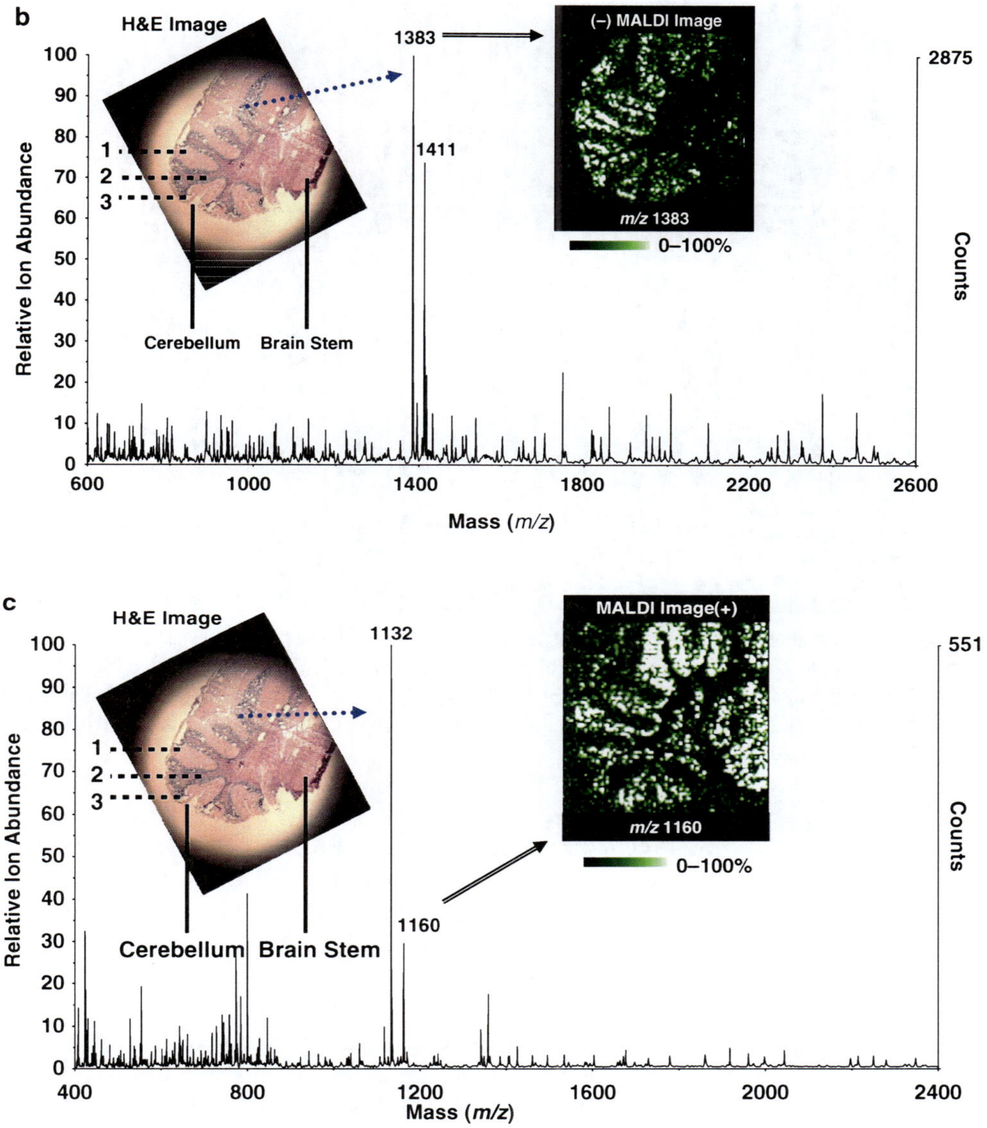

Fig. 7.3. (continued)

7. The summed MALDI mass spectrum of the whole sample can be obtained by the plot function (Analysis/Plot/Global/Scan).

3.5. ESI Mass Spectrometry

1. Brain tissues were homogenized (10 mg/ml) in water on ice. The lipids were extracted and the acidic glycolipids recovered by batch elution from a DEAE column (14).
2. The extracts were dissolved in 1.0 ml of MeOH and introduced via syringe infusion (0.6 ml/h) into an API 4000 QTrap tandem mass spectrometer.

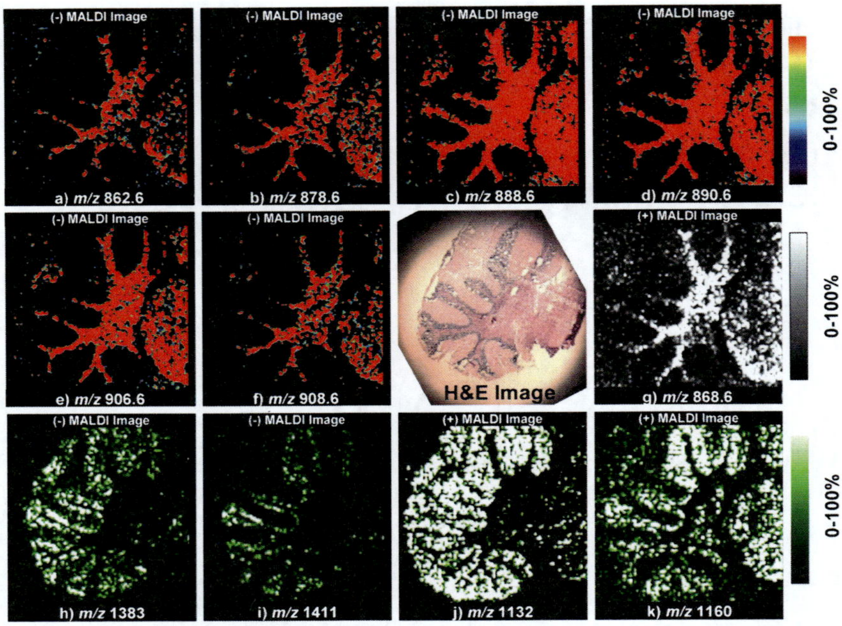

Fig. 7.4. Selected ion images of various sphingolipid species from hexb$^{-/-}$ mouse brain, which illustrate different histological localizations. (**a**) *m/z* 862.6 [ST d18:1/C22:0]; (**b**) *m/z* 878.6 [ST(OH) d18:1/h22:0]; (**c**) *m/z* 888.6 [ST d18:1/C24:1]; (**d**) *m/z* 890.6 [ST d18:1/C24:0]; (**e**) *m/z* 906.6 [ST(OH) d18:1/h24:0]; (**f**) *m/z* 908.6 [ST(OH) d18:0/h24:0]; (**g**) *m/z* 868.6 [unknown]; (**h**) *m/z* 1,383 [GM2 d18:1/C18:0]; (**i**) *m/z* 1,411 [GM2 d20:1/C18:0]; (**j**) *m/z* 1,132 [GA2 d18:1/C18:0+K]; (**k**) *m/z* 1,160 [GA2 d20:1/C18:0+K].

3. Precursor ion scans for *m/z* 96.9 in negative ion mode were used to determine the potential *N*-acyl chain length of sulfatide subspecies in each sample. The declustering potential (DP) was set to –220 eV and the collision energies were ranged from –100 to 120 eV. Precursor ion scans for *m/z* 290.1 in negative mode were used to identify the potential *N*-acyl chain length subspecies within each family of acidic gangliosides (i.e., GM1, GD1, GT1). These scans were performed with declustering potential of –70 to 100 eV (lower DP was required to reduce in-source fragmentation for species having multiple sialic acid residues). Collision energies ranged from –55 to 75 eV with lower collision energies used for species having increasing numbers of sialic acid residues because of the lability of these molecules toward fragmentation.

4. Ionization conditions were optimized for individual sulfatide or ganglioside subspecies and enhanced product ion (EPI) scans were selected to provide a greater diversity of product ions. EPI scans were performed with Q0 trapping set to "on," a linear ion trap fill time of 100 ms, and a scan rate of 1,000 amu/s. Example of ESI-MS/MS analysis of GM2 subspecies is shown in **Fig. 7.5a** (*see* **Note 13**).

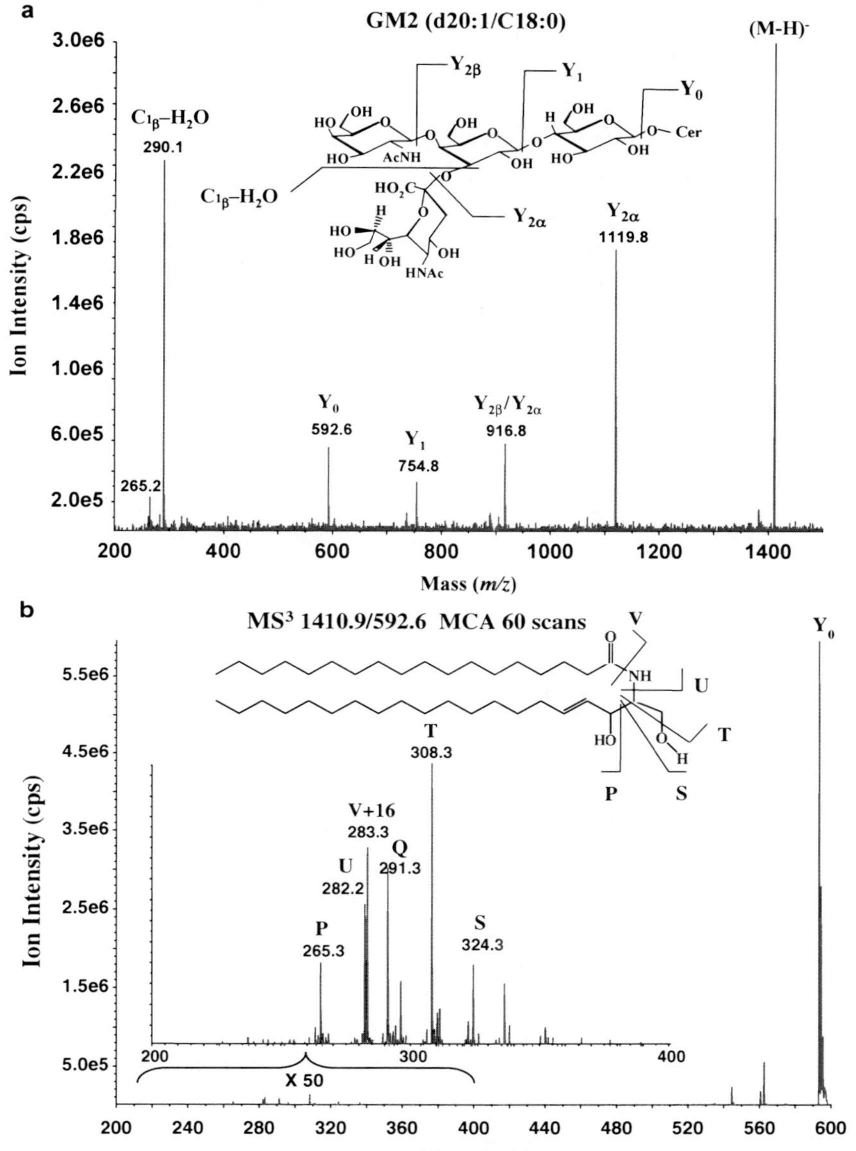

Fig. 7.5. (**a**) ESI-MS/MS spectrum of *m/z* 1,411 and (**b**) ESI-MS^3 spectrum of *m/z* 1,410.9/592.6 transition.

5. The MS^3 analysis is performed with the first mass analyzer (Q1) setting to "open" to pass a wide *m/z* window around the precursor ion of interest. This is transmitted to Q2 where it collides with a neutral gas and dissociates to various fragment ions. The linear ion trap (LIT) is then set to trap and hold a 2 *m/z* unit window centered on the product ion of interest. The selected *m/z* ions were fragmented further to secondary product ions, which are then scanned

out of the LIT. The sphingoid base and fatty acid composition of each ganglioside can be successfully identified by MS^3 analysis. An example of ESI-MS^3 analysis of GM2 subspecies (1,410.9/592.60 transition) is shown in **Fig. 7.5b** (*see* **Note 14**).

6. Neutral glycosphingolipids were analyzed in positive ion mode as both $(M+H)^+$ and $(M+Na)^+$ species. Neutral glycosphingolipids fragment primarily via cleavage of carbohydrate groups. Potential subspecies were identified via neutral loss scans for hexose and *N*-acetylhexosamine (162 and 203 units, respectively). The parameters of EPI and MS^3 scans for neutral glycosphingolipids were kept the same as those for acidic gangliosides.

4. Notes

1. An important feature of the OCN system is the ability to minimize the amount of solvent that comes into contact with the tissue. This serves to reduce analyte migration and matrix crystal size to minimize the loss of molecular spatial information.
2. OCT compounds are not compatible with imaging MALDI-MS. Even small amounts of these compounds residing on a cutting blade may cause poor results. Precautions should be made to avoid the contamination of OCT compounds on studied specimens.
3. The diffusion of the standard solution on sample plate should be tested first to determine the positions of standard spots to avoid the spreading of standard solution onto the tissue slices.
4. It was determined that 30 mg/ml of DHB in acetonitrile:water (50:50, v:v, with 0.1% TFA) can form evenly distributed small (0.5–20 μm mean diameter) matrix crystals on the sample surface. Several different combinations of matrix solution flow rate, nebulizing gas pressure, and OCN–sample distance were evaluated for matrix deposition and have been reported (12). Briefly, the matrix molecules formed irregular clusters without obvious crystal boundaries when using a high flow rate with a low gas pressure and short sprayer–sample distance. This was anticipated because these conditions would likely produce big droplets with a high solvent content (9). A reduction of the matrix solution flow rate and an increase of gas pressure and sprayer–sample distance resulted in the matrix crystals

becoming flat and rectangular in shape and having a size of several micrometers. This presumably occurs because the droplets are smaller and the solvent has more time to evaporate and begin to form DHB crystals in flight. At an even slower flow rate and higher gas pressure and sprayer–sample distance, the smallest needle-like matrix crystals were observed. A longer sprayer–sample distance (> 8 cm) was recommended to produce dryer DHB particles, which can minimize the analyte migration induced by the solvent and yield much greater signal response (15).

5. The accelerator voltage of the SEM system was typically 5 kV. No carbon or gold was further coated to the matrix surface prior to the SEM analysis.
6. The formation of matrix crystals on the sample surface can greatly affect the results of imaging MALDI-MS. In **Fig. 7.2**, it is clearly shown that the size and surface distribution of DHB crystals are dramatically different when using various matrix deposition protocols. SEM characterization of matrix crystals can be used to assist the optimization of matrix coating and improve the quality of the MALDI images.
7. It was observed that a matrix thickness between 5 and 50 μm was sufficient for reasonable signal-to-noise ratio (s/n) of lipid ions in mouse brain tissue and other samples.
8. To avoid a clogging problem with the OCN, it is suggested to pump the matrix solution for 2–3 min before turning on the nebulizing gas. If the flow of matrix solution needs to be stopped because of changing the syringe, adding more matrix solution, or switching to nanopure water when the matrix application is finished, turn off the nebulizing gas first and keep pumping the matrix solution for another 2–3 min after there is no gas flow through the OCN.
9. The mass spectra and resulting ion images can be obtained with only 8–10 laser shots per spot in either the positive or the negative ionization mode. This suggests that the OCN matrix deposition system is able to generate a good matrix–analyte interaction, which promotes efficient laser desorption and subsequent ionization of lipids. Since the OCN system can yield mass spectra with high s/n ions via fewer laser shots, the data acquisition time was also greatly reduced, which is a considerable benefit for time-consuming imaging mass spectrometry experiments.
10. The step size should be carefully selected. Although decreasing the step size may provide more details about the sample, it also causes longer data acquisition time,

larger data files, and the possibility of matrix subliming off the sample in ultra-high vacuum (UHV) chamber, especially for samples prepared using dryer matrix coating conditions.

11. MALDI-MS spectra acquired from the hexb$^{-/-}$ mouse brain slices prepared by OCN matrix coating system showed several prominent ions of *m/z* 888.6, 1,132 and 1,383 (**Fig. 7.3**) localized in different regions of the brain.
12. Sphingolipids (sulfatide, ganglioside GM2, and asialo-GM2 (GA2)) were distinctly visible in hexb$^{-/-}$ mouse brain samples by using OCN for matrix application. These ion images clearly demonstrate that the OCN system is useful for sample preparation for imaging MALDI-MS of lipids. The spatial distribution of sulfide subspecies ST d18:1/C22:0 (*m/z* 862.6), ST(OH) d18:1/h22:0 (*m/z* 878.6), ST d18:1/C24:1 (*m/z* 888.6), ST d18:1/C24:0 (*m/z* 890.6), ST(OH) d18:1/h24:0 (*m/z* 906.6), ST(OH) d18:0/h24:0 (*m/z* 908.6), and an unknown ion (*m/z* 868.6) displayed a remarkably similar pattern to the myelinated fiber (white matter) region of the cerebellum and a relatively even distribution in brain stem (c.f., H&E staining) (**Fig. 7.4a–g**). The localizations of potassiated ganglioside GA2 (d18:1/C18:0) (*m/z* 1,132), potassiated ganglioside GA2 (d20:1/C18:0) (*m/z* 1,160), ganglioside GM2 (d18:1/C18:0) (*m/z* 1,383), and ganglioside GM2 (d20:1/C18:0) (*m/z* 1,411), which are also known to accumulate in mice with this genetic defect (16), closely matched the granular cell region in cerebellum and produced no detectable ions in the brain stem region (**Fig. 7.4 h–k**). The imaging MALDI-MS results in **Fig. 7.4** illustrate that various subcategories of sphingolipids are localized to specific regions of the brain. Therefore, this technology is a valuable complement to other types of "lipidomic" analysis, which uses homogenized extracts of the entire tissue which may miss potentially important regional changes in both the types and the amounts of the lipids present.
13. ESI-MS/MS analysis of the lipid extracts from the mouse brains was performed to confirm the structure of sulfatide, GM2, and GA2. For example, in negative ion mode, MS/MS of *m/z* 1,410.9 generates five major fragment ions corresponding to losses of different sugar moieties in the head group (**Fig. 7.5a**). The product ions of *m/z* 1,119.8, 916.8, 754.8, and 592.6 correspond to the Y-type glycosidic bond cleavage involving loss of NeuAc, NeuAc/GalNac, NeuAc/GalNac/Gal, and NeuAc/GalNac/Gal/Glc, respectively. The *m/z* 290.1

ions were produced by C-type cleavage and charge retention on the sialic acid with subsequent dehydration, which confirms the existence of a sialic acid moiety.

14. An MS^3 experiment was performed on the Y_0 fragment ion of *m/z* 592.6 to determine the ceramide backbone of the *m/z* 1,410.9 ion. The resulting MS^3 spectra (**Fig. 7.5b**) showed secondary fragment ions of *m/z* 324, 308, 282, and 283, corresponding to S, T, U, and V + 16 fragments, respectively (17), revealing that the amide-linked fatty acid is stearate (C18:0). The ions of *m/z* 265 and 291 correspond to complimentary P and Q fragments, respectively (17), showing that the sphingoid base backbone is d20:1. Thus, this major species in $hexb^{-/-}$ mouse brain is ganglioside GM2 (d20:1/C18:0).

Acknowledgments

The authors would like to thank Drs. Richard Browner and Facundo Fernandez for providing the OCN sprayer, Dr. Markus Stoeckli for sharing the modified MMIST software, and Lan Sun for SEM analysis. This work is supported by NIH GM069338 (Lipid MAPS) and seed funding from Georgia Institute of Technology for the Mass Spectrometry Bio-Imaging Center.

References

1. Caprioli, R. M., Farmer, T. B., Gile, J. (1997) Molecular imaging of biological samples: localization of peptides and proteins using MALDI-TOF MS. *Anal Chem*, **69**, 4751–4760.
2. Chaurand, P., Schwartz, S. A., Caprioli, R. M. (2004) Profiling and imaging proteins in tissue sections by MS. *Anal Chem*, **76**, 86A–93A.
3. Rubakhin, S. S., Greenough, W. T., Sweedler, J. V. (2003) Spatial profiling with MALDI MS: distribution of neuropeptides within single neurons. *Anal Chem*, **75**, 5374–5380.
4. McDonnell, L. A., Heeren, R. M. A. (2007) Imaging mass spectrometry. *Mass Spectrom Rev*, **26**, 606–643.
5. Wang, L., May, S. W., Browner, R. F. (1996) Low-flow interface for liquid chromatography-inductively coupled plasma mass spectrometry speciation using an oscillating capillary nebulizer. *J Anal Atomic Spectrom*, **11**, 1137–1146.
6. Reyderman, L., Stavchansky, S. (1996) Novel methods of microparticulate production: application to drug delivery. *Pharm Dev Technol*, **1**, 223–229.
7. Perez, J., Petzold, C. J., Watkins, M. A., Vaughn, W. E., Kenttamaa, H. I. (1999) Laser desorption in transmission geometry inside a Fourier-transform ion cyclotron resonance mass spectrometer. *J Am Soc Mass Spectrom*, **10**, 1105–1110.
8. Lake, D. A., Johnson, M. V., McEwen, C. N., Larsen, B. S. (2000) Sample preparation for high throughput accurate mass analysis by matrix-assisted laser desorption/ionization time-of-flight mass spectrometry. *Rapid Commun Mass Spectrom*, **14**, 1008–1013.
9. Basile, F., Kassalainen, G. E., Williams, S. K. R. (2005) Interface for direct and continuous sample-matrix deposition onto a MALDI probe for polymer analysis by thermal field flow fractionation and off-line MALDI-MS. *Anal Chem*, 77, 3008–3012.

10. Fung, K. Y. C., Askovic, S., Basile, F., Duncan, M. W. (2004) A simple and inexpensive approach to interfacing high-performance liquid chromatography and matrix-assisted laser desorption/ionization-time of flight-mass spectrometry. *Proteomics*, **4**, 3121–3127.
11. Kirlew, P. W., Caruso, J. A. (1998) Investigation of a modified oscillating capillary nebulizer design as an interface for CE-ICP-MS. *Appl Spectrosc*, **52**, 770–772.
12. Chen, Y., Allegood, J., Liu, Y., Wang, E., Cachon-Gonzalez, B., Cox, T. M., Merrill, A. H., Jr., Sullards, M. C. (2008) Imaging MALDI mass spectrometry using an oscillating capillary nebulizer matrix coating system and its application to analysis of lipids in brain from a mouse model of Tay-Sachs/Sandhoff disease. *Anal Chem*, **80**, 2780–2788.
13. Cachon-Gonzalez, M. B., Wang, S. Z., Lynch, A., Ziegler, R., Cheng, S. H., Cox, T. M. (2006) Effective gene therapy in an authentic model of Tay-Sachs-related diseases. *Proc Natl Acad Sci U S A*, **103**, 10373–10378.
14. van Echten-Deckert, G. (2000) *Sphingolipid Metabolism and Cell Signaling*, Pt B, **Vol. 312**, Academic Press Inc, San Diego, CA, 64–79.
15. Hankin, J. A., Barkley, R. M., Murphy, R. C. (2007) Sublimation as a method of matrix application for mass spectrometric imaging. *J Am Soc Mass Spectrom*, **18**, 1646–1652.
16. Conzelmann, E., Sandhoff, K. (1978) Ab-variant of infantile Gm2 gangliosidosis - deficiency of a factor necessary for stimulation of hexosaminidase a-catalyzed degradation of ganglioside Gm2 and glycolipid Ga2. *Proc Natl Acad Sci U S A*, **75**, 3979–3983.
17. Merrill, A. H., Sullards, M. C., Allegood, J. C., Kelly, S., Wang, E. (2005) Sphingolipidomics: high-throughput, structure-specific, and quantitative analysis of sphingolipids by liquid chromatography tandem mass spectrometry. *Methods*, **36**, 207–224.

Chapter 8

Mapping Pharmaceuticals in Rat Brain Sections Using MALDI Imaging Mass Spectrometry

Yunsheng Hsieh, Fangbiao Li, and Walter A. Korfmacher

Abstract

Matrix-assisted laser desorption/ionization-tandem mass spectrometric method (MALDI-MS/MS) has proven to be a reliable tool for direct measurement of the disposition of small molecules in animal tissue sections. As example, MALDI-MS/MS imaging system was employed for visualizing the spatial distribution of astemizole and its primary metabolite in rat brain tissues. Astemizole is a second-generation antihistamine, a block peripheral H1 receptor, which was introduced to provide comparable therapeutic benefit but was withdrawn in most countries due to toxicity risks. Astemizole was observed to be heterogeneously distributed to most parts of brain tissue slices including cortex, hippocampus, hypothalamic, thalamus, and ventricle regions while its major metabolite, desmethylastemizole, was only found around ventricle sites. We have shown that astemizole alone is likely to be responsible for the central nervous system (CNS) side effects when its exposures became elevated.

Key words: MALDI, tandem mass spectrometry, astemizole, rat brain tissues, drug localization.

1. Introduction

First-generation antihistamines provide symptomatic relief from allergies and the common cold to patients. However, their therapeutic potential is hampered by the sedation caused by their effects on histamine receptors in the brain (1–3). Second-generation antihistamine (astemizole, as an example) block peripheral H1 receptors were introduced to provide comparable therapeutic benefit without the CNS side effects under manufactures' recommended doses (1–4). It was reported that astemizole was found to cause arrhythmias when drug exposures became

S.S. Rubakhin, J.V. Sweedler (eds.), *Mass Spectrometry Imaging*, Methods in Molecular Biology 656,
DOI 10.1007/978-1-60761-746-4_8,

elevated and the cardiac toxicity was mainly due to the parent drug (5). As the site of drug administration is often distinct from the site of the drug target, in vivo drug distribution is a critical aspect of toxicological and pharmacological action. Autoradiography (6–8) has been a common method for measuring drug distribution in animal tissue sections. However, the major disadvantage in this radioautographic technique is that it detects total drug-related materials but does not distinguish between the administered drug and its metabolites. This is because the measured radioactivity signals come from any drug-related material including the administered compound and its metabolites, which may confound data interpretation. In this work, the uptake and the retention of astemizole into rat brains with and without perfusion with saline solution using MALDI-MS/MS imaging technique (9–19) was investigated in order to resolve the possible causes of its CNS side effects.

2. Materials

2.1. Chemical Reagents

1. 2, 5-Dihydroxybenzoic acid (DHB) used as matrix substances for MALDI (Aldrich, Milwaukee, WI, USA) is dissolved in 50:50 acetonitrile in water as a 20 mg/ml matrix solution.
2. Astemizole used as an analyte (Sigma, St. Louis, MO, USA) is dissolved in acetonitrile at a concentration of 1 mg/ml and then serially diluted with 50:50 acetonitrile:water to the appropriate concentrations or is formulated with 0.4% methylcellulose dosing solution at a concentration of 10 mg/ml.
3. Acetonitrile (HPLC grade) (Fisher Scientific, Pittsburg, PA, USA) is used as organic solvent.
4. Methylcellulose (The Dow Chemicals, Piscataway, NJ, USA) is dissolved in water as 0.4% dosing solution.
5. Dulbecco's Phosphate-Buffered Saline (D-PBS) (Invitrogen, California, CA, USA) for perfusion experiment.

2.2. Equipment

1. Water purification system (Millipore, Billerica, MA, USA).
2. Multistep pipette with 1 μl pipette tips (Eppendorf, Westbury, NY, USA).
3. A CM3050 cryostat (Leica Microsystems Inc., Bannockburn, IL, USA) designed for rapid freezing and sectioning of tissue samples was used to produce the high-quality frozen sections.

4. A Nikon microscope (Micron Optics, Cedar Knolls, NJ, USA) was used to record optical images to define the outline of tissue sections.
5. A glass reagent sprayer (Kontes Glass Company, Vineland, NJ, USA) was used to coat the matrix solution over the tissue sections.
6. A vacuum desiccator (Fisher Scientific, Pittsburg, PA, USA).
7. Opti-TOF™ MALDI plate system (Applied Biosystems, Foster City, CA, USA) provides hydrophobically coated MALDI sample plates (5×5 cm) for spotting and tissue imaging applications which include a reusable holder and insert and are compatible with all Applied Biosystems MALDI plateforms.
8. A QStar Pulsar (Applied Biosystems, Foster City, CA, USA) hybrid QqTOF mass spectrometer equipped with an oMALDI™ (orthogonal MALDI) ion source and a nitrogen laser (337 nm) was used to generate high-quality MS and MS/ MS data.
9. The oMALDI™ Server software (Applied Biosystems, Foster City, CA, USA), a Windows-based program consisting of a main window, a set of drop-down menus, and a series of dialog boxes, controls the oMALDI™ ion source to operate the motor controller and the UV laser and directs the acquisition software (Analyst QS, Applied Biosystems) to acquire mass spectra. The oMALDI™ Server software with Imaging option allows the user to select a rectangular region of the user-defined area anywhere on the MALDI plate, to build a search pattern covering the defined region according to the spot width and height parameters, to acquire MS or MS/MS spectra from each pixel, and to rapidly obtain two-dimension mode (2D) profile information for small molecules from intact tissue sections.

3. Methods

3.1. Drug Administration and Tissue Sampling

1. Treat two Spray-Dawley rats with astemizole orally at 100 mg per kg of body weight in 0.4% methylcellulose as the dosing vehicle (*see* **Note 1**).
2. Attach perfusion needle (20 gauge) with a clamp closed.
3. Fill perfusion reservoir with 50 ml ice cold D-PBS.
4. Anesthetize the rat using CO_2 and open the abdominal cavity exposing the heart.

5. Insert the needle into the left ventricle of the heart and open the clamp to allow the flow of D-PBS. Make a cut in the right atrium with sharp scissors to allow for the escape of blood from the animal.
6. Stop the perfusion when blood has been cleared from body (liver has turned pale in color).
7. Excise the brain from the skull, snap freeze it, and allow the brain immediately to freeze in dry ice to prevent drug diffusion and to keep the integrity of the tissue.
8. Store the frozen tissue in sealed container at –20°C until further analysis to prevent the water evaporation from the tissue.

3.2. Tissue Preparation

1. Use ice slush made from distilled water to attach the rat brains to the cryostat sample stages (*see* **Notes 2** and **3**).
2. Adjust the chamber temperature of cryostat to –20°C.
3. Clean a new slicing blade with methanol.
4. Install the blade on Cryostat.
5. Set the section thickness to 16 μm.
6. Equilibrate steel MALDI sample plate to the chamber temperature of cryostat.
7. Slice the rat brain tissue samples along coronal direction to observe cortex and hypothalamic areas, main binding sites for histamine H-1 receptors, and three brain ventricles, choroids plexus, dorsal third ventricle, and lateral ventricle. These ventricles are important for us to evaluate drug penetration of the brain–blood barrier. Other brain substructures, such as hippocampus, cingulum, and external capsule, were also exposed.
8. Keep slicing the brain tissue until the desired position is reached.
9. Transfer the tissue slice on a cold MALDI plate with an artist brush and move it to the desired location (*see* **Note 4**).
10. Put one finger against the opposite side of the MALDI plate to thaw the tissue slice on the MALDI sample plate.
11. Dehydrate the rat tissue sections in a vacuum desiccator at the room temperature overnight before matrix application.

3.3. Matrix Application

1. Place ~10 ml of a matrix solution into a glass reagent sprayer.
2. Hold the sample plate vertically about 20–30 cm from the sprayer nozzle.

3. Spray multiple coats of matrix across the surface of the tissue. One coating cycle consisted of passing the sprayer two times across the surface of the tissue.
4. Allow the sample to dry for at least 2–5 min before the next coating cycle to avoid deposition of a large quantity of solvent in any one region of the tissue. The combination of the spray rate and spray distance should be adjusted to avoid excessive wetting of the tissue which could lead to analyte dispersion (*see* **Notes 5, 6,** and 7).
5. Iterates this process to redissolve and recrystallize the matrix enhancing the incorporation efficiency of analytes in the crystals until there was a relatively homogeneous layer of matrix crystals over the surface of the tissue, usually about 15 cycles.
6. Dehydrate the processed tissue sections under the hood for 15 min at room temperature.

3.4. MALDI-MS Method Development

1. Deposit a few drops of matrix solution containing 10 ng/μl astemizole on the MALDI plates.
2. Allow all spots to dry completely under the hood prior to acquiring MS and MS/MS information of astemizole with the oMALDI™ Server software (**Fig. 8.1**) (*see* **Notes 8–10**).
3. Load the MALDI plate into the MALDI chamber of a QStar Pulsar hybrid QqTOF mass spectrometer.
4. Operate repetition rates and the pulse energy of the laser at 20 Hz and 80 μJ.

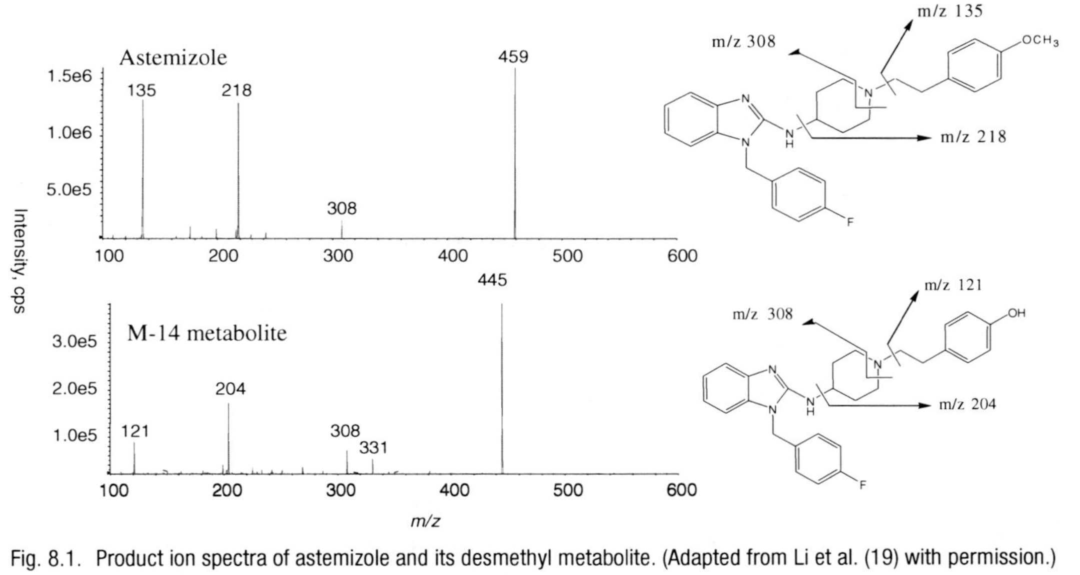

Fig. 8.1. Product ion spectra of astemizole and its desmethyl metabolite. (Adapted from Li et al. (19) with permission.)

5. Set argon used as the collision gas with a collision-induced dissociation (CAD) gas pressure setting of 5 (~3–4 × 10^{-5} torr).
6. Adjust collision energy for dissociation of the protonated astemizole molecular ion $[M+H]^+$ (parent ion *m/z* 459 shown in **Fig. 8.1**) (*see* **Note 11**).
7. Acquire *m/z* 218 product ion scans for astemizole and *m/z* 204 product ion scans for its demethyl metabolite (**Fig. 8.1**) under the "enhanced mode" as suggested by high performance liquid chromatography/quadrupole linear ion trap mass spectrometric experiments (**Scheme 8.1**) to build a MALDI-IMS method for further imaging mass spectrometry studies. The predominant fragments for both astemizole and its primary metabolite were then "enhanced," using a feature of the Pulsar that decelerates and traps ions prior to accelerating them into the orthogonal time-of-flight region. Sensitivity could be further increased by allowing a wide mass range into the collision quadrupole (~5 *m/z*), by using the maximum pulsing rate into the TOF region, and by detecting only a small mass range surrounding the fragment of interest.

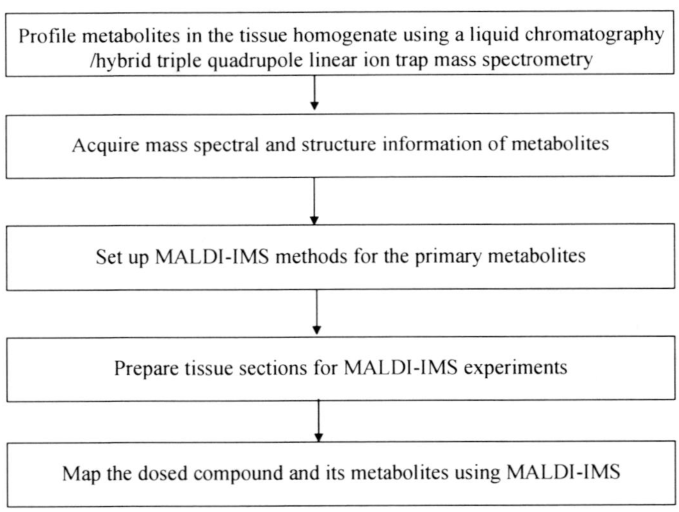

Scheme 8.1 General strategy of mapping pharmaceuticals and their metabolites.

8. Deposit a few 0.2 μl drops of the matrix solution containing the analyte within the various regions of the blank brain section.
9. Allow all spots to dry under the hood for 1 h.
10. Load the MALDI plate into the MALDI chamber of a QStar Pulsar hybrid QqTOF mass spectrometer.

11. Measure the MALDI signals of the analyte from the spotted regions under the established instrumental conditions to evaluate the degree of the ionization suppression due to the endogenous materials from the blank tissues (*see* **Note 12**).
12. Record the optical image of a sagittal section of the brain (**Fig. 8.2a**) to indicate the presence of the cortex, limbic system, cerebellum, brain stem, and ventricles regions and to define the outline of a given tissue section.

3.5. Imaging Astemizole and Its Metabolite in the Rat Brain Sections

1. Load the MALDI plate with the study brain slices coated with DHB into the MALDI chamber of a QStar Pulsar hybrid QqTOF mass spectrometer.
2. Initiate the image process with the oMALDI™ Server software. The sample MALDI plate will be moved from one spot to the other spot under a stationary laser automatically, creating a raster of desorbed areas over the tissue surface.

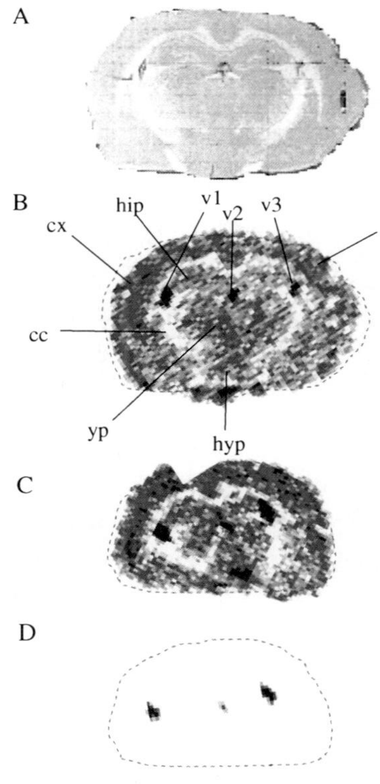

Fig. 8.2. (**a**) The optical image of a rat brain from a coronal section. (**b**) MALDI-MS/MS images of astemizole in the rat brain slice without perfusion and (**c**) with perfusion; cortex (cx), hippocampus (hip), corpus callosum (cc), hypothalamic region (hyp), thalamus region (yp), choroid plexus (v1), dorsal third ventricle (v2), and lateral ventricle (v3) are indicated by *arrows*. (**d**) MALDI-MS/MS images of M-14 metabolite of astemizole in the rat brain slice. (Adapted from Li et al. (19) with permission.)

The oMALDI server directs the acquisition software to acquire mass spectra from each point, treating each laser spot as one sample in a batch run. The laser was turned off for 2 s while the MALDI plate was repositioned for the next acquisition location in a rectangular pattern with a spatial resolution of 100–150 μm (the size of the laser beam). During an imaging mass spectrometric experiment, the resulting mass spectra data are first obtained as a function of the acquisition times which are associated with the location in an array of pixels (*see* **Notes 13, 14, 15,** and **16**).

3.6. Data Analysis

1. Construct the ion density image maps after data acquisition has completed using the oMALDI Server 4.0 imaging software. Two-dimensional images are obtained by plotting the spatial dimensions of *x* and *y* versus the signal amplitude of selected ion range as a function of the location on the tissue surface.
2. Construct two-dimensional ion maps for astemizole based on a given product ion *m/z* 218 value that was monitored in each mass spectrum. As shown in **Fig. 8.2**, the color of each pixel represents the intensity of the selected ion contributed from astemizole. In order to detect small molecules in a complex biological tissue section using MS/MS was normally required in order to generate signals from the compound of interest that could easily be distinguished from the background interference produced by the matrix. Analysis of the resulting product ions by the second analyzer generates a mass spectrum of the product ions that can be used to provide structural information. Therefore, by monitoring the transition of a selected precursor ion to its product ions, tandem mass spectrometry systems offer a tool for distinguishing between isobaric compounds such as matrix ions and the analytes (*see* **Notes 17** and **18**).
3. Reconstruct two-dimensional ion maps for the demethyl metabolite based on a given product ion *m/z* 204 value that was monitored in each mass spectrum. As shown in **Fig. 8.2d**, the color of each pixel represents the intensity of the selected ion contributed from the metabolite (*see* **Note 19**).

4. Notes

1. The animal dosing experiments were carried out in accordance to the US National Institutes of Health Guide to the Care and Use of Laboratory Animals and the Animal Welfare Act.

2. Tissue sample preparation is one of the critical steps for the success of detecting small molecules in biological tissues using MALDI-IMS. Inappropriately handling tissue samples in the sample preparation steps may cause delocalization and degradation of the analytes. Tissue containing drugs subject to photodegradation should be maintained in a dark container as much as possible during the processing.
3. Cryostats are normally used as a standard tool for slicing frozen tissue to reducing sample contamination. Contamination of the tissue surface due to the use of OCT (optimal cutting temperature polymer) as an embedding medium for stabilizing the organ specimens should be avoided because it would lead to ionization suppression in MALDI-MS analysis.
4. There are several other ways available to transfer tissue slices to the sample plate. First, the frozen section could adhere to the MALDI plate held at room temperature by placing the plate over the section. Second, the section is stuck to a double-sided transparent tape which is then stuck on the sample plate.
5. Once the section is mounted to the sample plate with the desired orientation, matrix solution could be also deposited on the tissue surface by electrospray deposition or using robotics to deposit small matrix droplets across the tissue surface before MALDI analysis.
6. High-resolution images with MALDI-IMS necessitate uniform coating by the matrix, no redistribution of surface analytes and linearity of MALDI signals after matrix application. In order to produce a homogeneous crystal layer over the tissue surface, spraying multiple coats of matrix across the surface of the tissue is recommended.
7. For certain analytes, contaminants such as salts and phospholipids in tissues may significantly inhibit co-crystallation of analytes with MALDI matrix resulting in low MALDI intensities of the analytes. Washing of tissue sections with weak organic solvents such as 70% ethanol prior to matrix application might eliminate endogenous materials from the native tissue and further to improve the MALDI matrix crystallization process. However, for small molecule applications, there is a potential risk of either losing the analytes or altering the spatial integrity of localization of drug-related components on tissues.
8. MALDI-MS employs a matrix solution to mix with the analyte to allow drying and to form co-crystals form. These crystals are subject to absorb energy at the wavelength of

the laser beam resulting in desorption and ionization of the analytes that were included in the crystals. In this process, matrix is uniformly deposited over various tissue sections to extract analytes into the surface of the tissue and to produce crystals.

9. The selection of matrix and matrix solution conditions such as solvent compositions, pH, and the rates of co-crystals growth can affect the quality of mass spectra for small molecules. The success of the MALDI-IMS applications to an analyte in tissue is strongly dependent on the choice of appropriate matrix materials.
10. The common UV-absorbing molecules used as matrices for MALDI analysis are benzoic acid-based components with low molecular weights (< 500 Da) which dominate the low-mass range background for a typical MALDI-MS spectrum further challenging the advancement of MALDI for the analysis of small molecules. DHB was chosen in this work over sinapinic acid (SA, 3,5-dimethoxy, 4-hydroxy cinnamic acid) commonly used as the matrix because it generated fewer background signals and greater analyte signals for astemizole.
11. The instrument parameters for the analytes are first optimized prior to MALDI-IMS experiments.
12. In general, it is possible for MALDI to suffer some degrees of signal irreproducibility due to crystallization behavior and laser properties such as energy profile and firing repetition rate. Variations in peak intensity of analytes may be seen when the laser focuses on different regions of the same tissue section. To investigate the potential of the ionization reproducibility of astemizole from different regions of rat brain tissue sections, seven drops (0.1 μl each) containing 10 ng of both analytes were deposited on different locations of a blank rat brain slice. The MALDI responses of astemizole from the spiked areas of rat brain sections were found to be comparable. The results confirmed MALDI-MS to be a feasible technique in providing spatial resolution with a high degree of consistency to reveal the density and distribution of astemizole and its primary metabolite in rat brain tissue.
13. A raster of organ sections from a rodent species containing the compounds of interest under a stationary laser beam is performed over a predetermined two-dimensional array to generate ion plumes directly from the tissue sections in a MALDI plate array.
14. The movement of the sample stages is automatically accomplished in the *x* and *y* directions to locate the edges of

the tissue sample and to define the exact region of interest. The localization of the laser beam on the sample is accurate to within 5 μm.

15. MALDI imaging resolution is governed by both crystal size and laser diameter. Generally smaller crystal sizes yield better imaging spatial resolution. With crystal diameters smaller than the laser beam, typically 50–200 μm depending on the instrument, imaging resolution is generally limited to the laser diameter.
16. Acquisition times for tissue imaging relies on several instrumental parameters such as spatial resolution requirements, the laser repetition rate, spot-to-spot sample repositioning transfer time, and data processing. As laser with fast repetition rates and improved electronics become available, one could reduce the acquisition times from hours to minutes.
17. **Figure 8.2b** shows the MALDI-MS/MS image of astemizole in a rat brain slice. The product ion spectrum of astemizole from the rat brain tissue was found to be consistent with that from the authentic standard material deposited into the blank rat brain section. The MALDI-MS/MS imaging results as given in **Fig. 8.2b** revealed that the astemizole was readily detected in the entire brain and the most intense signal was observed in the ventricle area. A significant amount of astemizole appeared to be distributed and localized at the cortex, hippocampus, hypothalamic, and thalamus regions over the entire thickness of the section. Astemizole was found to be low in the corpus callosum which contains nerve fibers. The product ion spectrum of astemizole clearly indicated the presence of the drug with variable concentrations as a function of the location within the tissue section monitored.
18. The MALDI-MS/MS imaging results of astemizole shown in **Fig. 8.2c** indicated that the relative distribution patterns between the perfused and untreated rat brain sections remained consistent over the entire investigation period. The fact that the images were similar despite the sample preparation differences suggested that the drug level in the slices was likely due to tissue binding activities rather than being due to residual biological fluids from the systemic circulation.
19. The MALDI-MS/MS imaging results as given in **Fig. 8.2d** revealed that the M-14 metabolite was also measured by MS primarily around the three ventricle sites. The product ion spectrum of M-14 metabolite from the rat brain tissue was found to be consistent with that from brain homogenates by HPLC-MS/MS (data not shown). The

signals of the M-14 metabolite cannot be confirmed in other parts of the brain section. The results implied that the metabolite tended to stay in cerebrospinal fluid than to cross the brain–blood barrier (BBB). Therefore, astemizole was likely the major cause of efficacy and CNS side effects following drug administration in rats.

References

1. Simons, F. E. (1994) The therapeutic index of newer H1-receptor antagonists. *Clin Exp Allergy*, **24**, 707–723.
2. Simons, F. E. (1999) Prospective, long-term safety evaluation of the H1-receptor antagonist cetirizine in very young children with atopic dermatitis. ETAC Study Group. Early treatment of the atopic child. *J Allergy Clin Immunol*, **104**, 433–440.
3. Passalacqua, G., Bousquet, J., Bachert, C., Church, M. K., Bindsley-Jensen, C., Nagy, L., Szemere, P., Davies, R. J., Durham, S. R., Horak, F., Kontou-Fili, K., Malling, H. J., Cauwenberge, P., Canonica, G. W. (1996) The clinical safety of H1-receptor antagonists. *Allergy*, **51**, 666–675.
4. Estelle, F., Simons, R. (1999) H1-receptor antagonists: safety issues. *Ann Allergy Asthma Immunol*, **83**, 481–488.
5. Horak, F., Stubner, U. P. (1999) Comparative tolerability of second generation antihistamines. *Drug Saf*, **20**, 385–401.
6. Jansen, F. P., Wu, T. S., Voss, H. P., Steinbusch, H. W., Vollinga, R. C., Rademaker, B., Bast, A., Timmerman, H. (1994) Characterization of the binding of the first selective radiolabelled histamine H3-receptor antagonist, [125I]-iodophenpropit, to rat brain. *Br J Pharmacol*, **113**, 355–362.
7. Alves-Rodrigues, A., Timmerman, H., Willems, E., Lemstra, S., Zuiderveld, O. P., Leurs, R. (1998) Pharmacological characterisation of the histamine H3 receptor in the rat hippocampus. *Brain Res*, **788**, 179–186.
8. Khatib-Shahidi, S., Andersson, M., Herman, J. L., Gillespie, T. A. Caprioli, R. M. (2006) Direct molecular analysis of whole-body animal tissue sections by imaging MALDI mass spectrometry. *Anal Chem*, **78**, 6448–6456.
9. Hsieh, Y., Casale, R., Fukuda, E., Chen, J., Knemeyer, I., Wingate, J., Morrison, R., Korfmacher, W. A. (2006) Matrix-assisted laser desorption/ionization imaging mass spectrometry for direct measurement of clozapine in rat brain tissue. *Rapid Commun Mass Spectrom*, **20**, 965–972.
10. Hsieh, Y., Chen, J., Korfmacher, W. A. (2007) Mapping pharmaceuticals in tissues using MALDI imaging mass spectrometry. *J Pharmacol Toxicol Methods*, **55**, 193–200.
11. Reyzer, M. L., Hsieh, Y., Ng, K., Korfmacher, W. A., Caprioli, R. M. (2003) Direct analysis of drug candidates in tissue by matrix-assisted laser desorption/ionization mass spectrometry. *J Mass Spectrom*, **38**, 1081–1092.
12. Stoeckli, M., Staab, D., Schweitzer, A., Gardiner, J., Seebach, D. (2007) Imaging of a beta-peptide distribution in whole-body mice sections by MALDI mass spectrometry. *J Am Soc Mass Spectrom*, **18**, 1921–1924.
13. Chaurand, P., Norris, J. L., Cornett, D. S., Mobley, J. A., Caprioli, R. M. (2006) New developments in profiling and imaging of proteins from tissue sections by MALDI mass spectrometry. *J Proteome Res*, **5**, 2889–2900.
14. Chaurand, P., Sanders, M. E., Jensen, R. A., Caprioli, R. M. (2004) Proteomics in diagnostic pathology: profiling and imaging proteins directly in tissue sections. *Am J Pathol*, **165**, 1057–1068.
15. Chaurand, P., Schwartz, S. A., Billheimer, D., Xu, B. J. Crecelius, A. Caprioli, R. M. (2004) Integrating histology and imaging mass spectrometry. *Anal Chem*, **76**, 1145–1155.
16. Chaurand, P., Schwartz, S. A., Caprioli, R. M. (2002) Imaging mass spectrometry: a new tool to investigate the spatial organization of peptides and proteins in mammalian tissue sections. *Curr Opin Chem Biol*, **6**, 676–681.
17. Chaurand, P., Schwartz, S. A., Caprioli, R. M. (2004) Assessing protein patterns in disease using imaging mass spectrometry. *J Proteome Res*, **3**, 245–252.
18. Chen, J., Hsieh, Y., Crossman, L., Knemeyer, I., Korfmacher, W. A. (2008) Visualization of first-pass drug metabolism of terfenadine by MALDI-imaging mass spectrometry. *Drug Metab Lett*, **2**, 1–4.
19. Li, F., Hsieh, Y., Kang, L., Sondey, C., Lachowicz, J., Korfmacher, W. A. (2009) MALDI-tandem mass spectrometry imaging of astemizole and its primary metabolite in rat brain sections. *Bioanalysis*, **1**, 299–307.

Chapter 9

Laser Ablation Electrospray Ionization for Atmospheric Pressure Molecular Imaging Mass Spectrometry

Peter Nemes and Akos Vertes

Abstract

Laser ablation electrospray ionization (LAESI) is a novel method for the direct imaging of biological tissues by mass spectrometry. By performing ionization in the ambient environment, this technique enables in vivo studies with potential for single-cell analysis. A unique aspect of LAESI mass spectrometric imaging (MSI) is depth profiling that, in combination with lateral imaging, permits 3D molecular imaging for the first time under native conditions. With current lateral and depth resolutions of ~100 and ~40 μm, respectively, LAESI MSI helps to explore the molecular architecture of live tissues.

Key words: Mass spectrometry, imaging, ambient, direct analysis, depth profiling, three-dimensional, in vivo, tissue imaging.

1. Introduction

Traditional mass spectrometric imaging (MSI) methods, such as matrix-assisted laser desorption ionization (MALDI) and secondary ion mass spectrometry (SIMS), have become important tools for the investigation of molecular distributions in tissues due to their high ionization efficiencies and excellent lateral and depth resolutions. Invasive sample preparation and the need for vacuum conditions, however, are incompatible with the analysis of live samples.

Novel ionization methods in ambient mass spectrometry (1) overcome these limitations by performing imaging under native conditions. Desorption electrospray ionization (2), atmospheric pressure (AP) mid-infrared (mid-IR) MALDI (3), laser ablation

S.S. Rubakhin, J.V. Sweedler (eds.), *Mass Spectrometry Imaging*, Methods in Molecular Biology 656, DOI 10.1007/978-1-60761-746-4_9, © Springer Science+Business Media, LLC 2010 electrospray ionization (LAESI) (4), and most recently laser ablation coupled to flowing AP afterglow MS (5) have demonstrated imaging capabilities with lateral resolutions 20–400 μm while obtaining low limits of detection. **Figure 9.1a** shows the schematics of the LAESI ion source. **Figure 9.1b** depicts the fast imaging of the entrainment of laser-ablated particulates into the electrospray plume. The interaction of the sprayed droplets with the particulates and neutrals emerging from the laser ablation (LA) produces coalesced charged particles that are thought to be the basis of the LAESI signal. In this chapter, we describe the protocols for lateral and 3D MSI of biological tissues using LAESI.

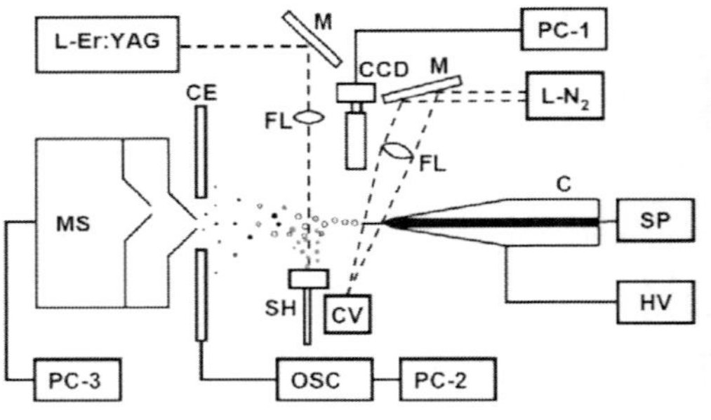

(A)

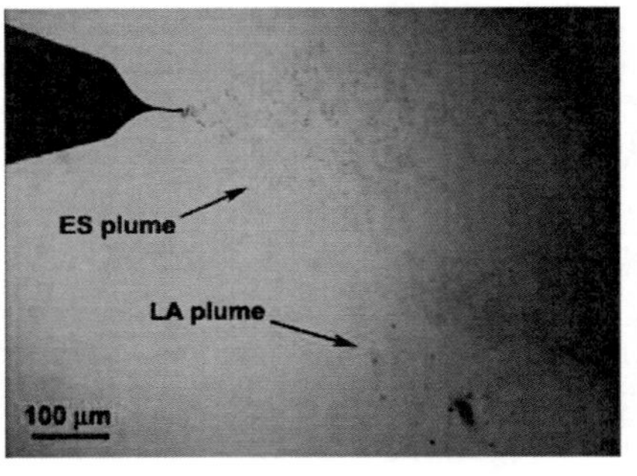

(B)

Fig. 9.1. (**a**) Schematics of the LAESI setup with (optional) spray current measurement and fast imaging system (C, capillary emitter; SP, syringe pump; HV, high-voltage power supply; L-N_2, nitrogen laser; M, mirrors; FL, focusing lenses; CV, cuvette; CE, counter electrode; OSC, digital oscilloscope; SH, sample holder; L-Er:YAG, Er:YAG laser; MS, mass spectrometer; PC-1 to PC-3, personal computers). *Open black dots* represent the droplets formed by the electrospray. Their interaction with the particulates and neutrals (*solid gray dots*) emerging from the laser ablation produces some fused particles (*solid black dots*) that are thought to be the basis of the LAESI signal. (Adapted with permission from (6). Copyright 2007 American Chemical Society.) (**b**) Our fast imaging experiments supported this scenario. The image captured with ~10 ns exposure time shows the interaction of the laser ablation plume (LA) and the electrospray plume (ES). (Adapted with permission from (6). Copyright 2007 American Chemical Society.)

2. Materials

2.1. Reagents and Sample

1. Electrospray solution for positive ion mode analysis: 50% methanol or acetonitrile, 1% acetic acid or formic acid in water. ▼CAUTION: Glacial acetic acid is extremely harmful when inhaled; causes burns to the skin and eyes; and avoid contact with skin, eyes, and the respiratory system.
2. Electrospray solution for negative ion mode analysis: 50% methanol or acetonitrile, 1% ammonium acetate or ammonium hydroxide in water. ▼CAUTION: May cause irritation to skin, eyes, and the respiratory tract; avoid direct contact and inhalation.

3. Electrospray solution for reactive LAESI experiments: reagents, e.g., ~1 μM–1 mM lithium sulfate.
4. Pre-cleaned glass slides (e.g., Thermo Fisher Scientific, Inc., Waltham, MA, USA).
5. Double-sided tape.
6. 1-Methyl-2-pyrrolidinone and an electric heater for the removal of the optical fiber jacket and 1% nitric acid (reagent grade) for the etching of the germanium oxide fiber. ▼CAUTION: Combustible, causes irritation to skin and eyes; and avoid contact with skin and eyes.
7. For plant studies, growth chamber and plant growing protocol (e.g., 14 h photoperiod, 24°C).
8. Cryomicrotome with microtome blades (e.g., Shandon Cryostat, Thermo Fisher Scientific), frozen specimen embedding medium (FSEM, Shandon Cryomatrix, Thermo Fisher Scientific, Inc., Waltham, MA, USA), and liquid N_2. ▼CAUTION: It causes suffocation when present at amounts sufficient to reduce oxygen concentration below 19.5%. Contact with tissue can cause severe cryogenic burns. Always handle with protective gloves.
9. –80°C freezer (e.g., Revco freezers, Thermo Fisher Scientific, Inc., Waltham, MA, USA).
10. Aluminum foil.

2.2. Mid-IR Ablation on Sample

1. Q-switched mid-IR laser emitting light at 2.94 μm wavelength or an optical parametric oscillator tunable in the vicinity of 2.94 μm wavelength pumped by a Q-switched Nd:YAG laser (e.g., Opotek, Inc., Carlsbad, CA, USA). ▼CAUTION: Class IV laser; Direct exposure of the eye to the laser beam can cause permanent eye damage. Always wear laser protective eyewear of sufficient optical density at the operating wavelengths.
2. Mirrors for mid-IR light (e.g., gold-coated mirrors, Thorlabs, Newton, NJ, USA).
3. Focusing lens for mid-IR light, e.g., plano-convex CaF_2 or antireflection-coated ZnSe lens (Infrared Optical Products, Farmingdale, NY, USA); or an aspherical lens; or a reflective microscope objective (Newport Corp., Irvine, CA, USA); or a sharpened optical fiber for mid-IR light delivery, e.g., 450 μm core germanium oxide-based optical fiber (Infrared Fiber Systems, Silver Spring, MD, USA) with chucks and positioners (e.g., Newport Corp., Irvine, CA, USA) and a three-axis translation stage (e.g., Thorlabs, Inc., Newton, NJ, USA).

4. Thermo- and photosensitive papers for beam alignment and core spot size optimization (e.g., R30C5W, Liquid Crystal Resources L.L.C., Glenview, IL, and multigrade IV, Ilford Imaging Ltd, UK), respectively.
5. Plate holders (e.g., FP02 or FP02, Thorlabs, Inc., Newton, NJ, USA) for sample mounting.
6. Enclosure for the ion source including the laser ablation plume. ▼CAUTION: Ablated particles may become airborne and pose health hazard upon inhalation or contact with skin and eye. Please always be aware of related health risks and take proper measures for protection.
7. For frozen samples Peltier cooling stage (e.g., Ferrotec Corp., Bedford, NH, USA) with heat fan (e.g., Allied Electronics, Inc., Fort Worth, TX, USA), and DC power supply.
8. Optical microscope for measurement of ablation crater dimensions.

2.3. Post-ionization with Charged Droplets

1. Electrospray emitters (e.g., MT320-50-5-5 or FS360-75-30-N-5, New Objective, Inc., Woburn, MA, USA).
2. Metal union, conductive perfluoroelastomer ferrule, fittings, tubing sleeve, fused silica capillary, needle port (e.g., IDEX Health & Sciences, Oak Harbor, WA, USA, or Waters Corp., Milford, MA, USA).
3. Syringe pump (e.g., Harvard Apparatus, Holliston, MA, USA) or an LC solvent-delivery system.
4. For experiments with controlled spaying mode, stainless steel electrode with oscilloscope to perform spray current measurements (e.g., WaveSurfer 452, LeCroy, Chestnut Ridge, NY, USA).
5. High-voltage power supply (e.g., Stanford Research Systems, Inc., Sunnyvale, CA, USA). ▼CAUTION: Electric shock hazard. Please make sure that all electric connections are properly shielded.

2.4. Molecular Imaging and Data Analysis

1. Three-axis translation stage (e.g., Newport Corp., Irvine, CA, USA) with motorized actuators and controller (e.g., with LTA-HS, Newport Corp., Irvine, CA, USA).
2. Mass spectrometer (e.g., Q-TOF Premier, Waters Corp., Milford, MA, USA).
3. Software for correlated positioning of the three-axis translation stage with laser ablation and mass spectrometric data acquisition (written in, e.g., LabView, National Instruments, Austin, TX, USA).

4. Software for data analysis and scientific visualization package for molecular image generation (e.g., Origin, OriginLab Corp., Northampton, MA, USA); ImageJ (NIH, available at http://rsb.info.nih.gov), Biomap (available at http://www.maldi-msi.org).

3. Methods

Figure 9.2a shows the workflow of LAESI imaging in an MSI setup. In a 2D imaging experiment, the sample surface is scanned in the focal plane of the mid-infrared laser light, the ablated neutrals and particulate matter are post-ionized with a cloud of charged droplets, and the resulting ions are mass analyzed and recorded. Mass-selected molecular images are reconstructed by correlating the intensity of ion signal for an *m/z* of interest with the absolute lateral coordinate of analysis for each pixel of the interrogated area (*see* **Fig. 9.2b** for an *Aphelandra squarrosa* leaf and **Fig. 9.4b,c** for rat brain tissue section). For 3D molecular imaging, the chemical depth profile of the sample is acquired with a selected number of laser pulses delivered at each pixel. The 3D molecular image is represented by a set of 2D images correlated with the absolute depth of analysis (*see* **Fig. 9.2c**).

The imaging resolution of LAESI MSI laterally is characterized by the diameter of the ablation crater along with the ablation depth in the third dimension. In regular experiments, lateral step sizes, larger than or equivalent to the diameter of the ablation crater, are applied. The molecular imaging resolution is therefore limited by the divergence of the laser and the properties of the optical focusing component. High-resolution LAESI MSI is also dependent on reliable and effective ion generation at each pixel of the image. Governing factors are the efficiency of the interaction between the electrospray and the laser ablation plumes and the success of entrainment of the resulting droplets and ions into the mass spectrometer. **Figure 9.3a** summarizes the major variables that require optimization for a robust operation. The following section is guidance for setting up a LAESI MSI experiment. For best results, please consider the average ion count for at least three pixels each time adjustment is made to a variable.

3.1. Preparation and Mounting of Tissue

1. Mount sample directly on sample holder with the surface of interest exposed on top and adjust $d_{OR\text{-}FP}$ to ~10 mm (*see* **Fig. 9.3a**). Small clamps and single- and double-sided adhesion tape often enable easy mounting. ▼CAUTION:

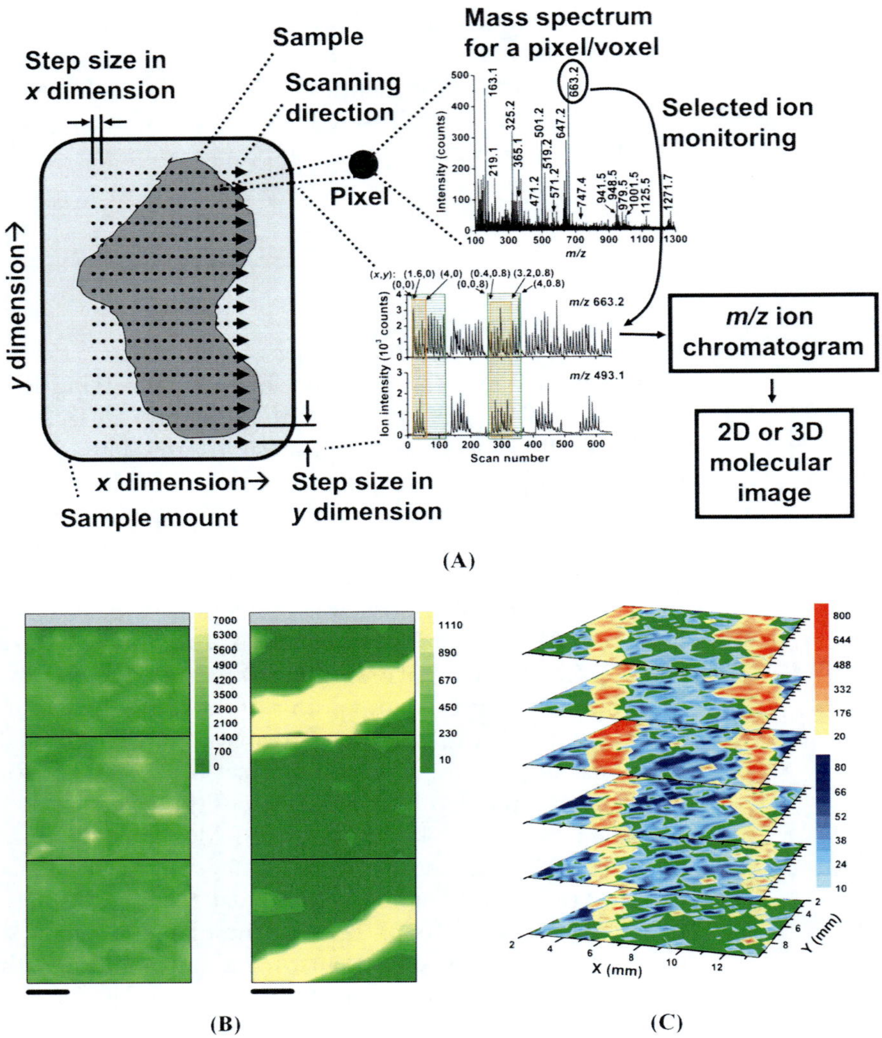

Fig. 9.2. (**a**) Schematics of LAESI MSI in two and three dimensions. Molecular images are obtained by correlating the coordinates of the analyzed pixels with the selected ion abundances. (**b**) Examples for metabolites with uniform (*m/z* 663.16 assigned as kaempferol-(diacetyl coumaryl-rhamnoside)) and heterogeneous (*m/z* 493.10 assigned as methoxy kaempferol glucoronide) lateral distributions in *A. squarrosa* leaves. (Reprinted with permission from (4). Copyright 2008 American Chemical Society.) (**c**) 3D imaging revealed that kaempferol or luteolin detected at *m/z* 287.06 (*yellow* scale) followed the *yellow* variegation sectors whereas chlorophyll a with *m/z* 893.54 (*blue* scale) accumulated in the third and fourth layers of the leaf. (Reprinted with permission from (11). Copyright 2009 American Chemical Society.)

Native water content of sample must be retained for mid-IR light coupling (*see* **Note 4.1.1**). (Optional) Evaporative loss of water can be prevented by thaw mounting, whereby the sample is directly frozen onto the sample holder or a glass slide and is kept at low temperatures during imaging (*see* **Fig. 9.4a**).

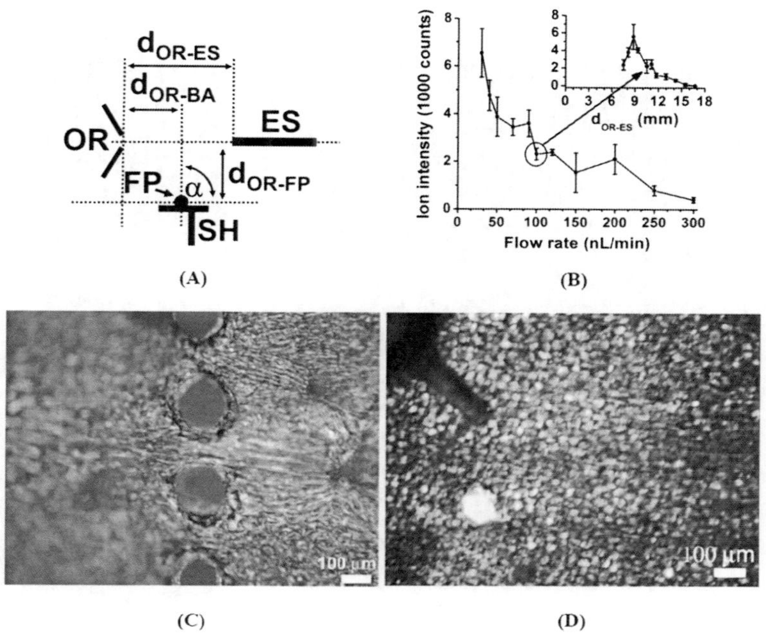

Fig. 9.3. (**a**) Geometry parameters in a LAESI MSI experiment used to optimize signal-to-noise ratio and lateral and depth resolution of molecular images. The relative position of the electrospray capillary, ES, and the focal point, FP, with respect to the mass spectrometer orifice, OR, and the incidence angle, α, of the laser beam are the essential factors that determine the efficiency of interaction between the ablation plume and the electrospray. (**b**) The smaller droplets generated by a low flow rate nanospray source (compare the ion intensities at 25 and 300 nl/min) and careful alignment of the ablation plume improved the sensitivity of LAESI MSI. (**c**) Differential interference contrast microscope image of the adaxial surface of *Arabidopsis thaliana* leaf ablated using a plano-convex ZnSe focusing lens showed tissue removal in an area of ~200 μm diameter suggesting a similar lateral resolution for the MSI experiment. (**d**) Online visualization of tissue sampling with a fiber sharpened to ~50 μm (see the top left corner of the image) further improved the lateral resolution by yielding ablation marks of ~100 μm in diameter.

2. (Optional) Sectioning may be required for certain tissue types such as animal or human organs. Wrap sample in aluminum foil and dip it into liquid N_2 for ~20 s. ▼CAUTION: Improper timing can lead to tissue fracturing. Section tissue into 10–100 μm thick slices with a cryomicrotome at –10 to –20°C. Note that temperature and time requirements might depend on the tissue type. Thaw-mount slices onto microscope slides or directly onto sample holder.
3. (Optional) Store tissues and sections wrapped in aluminum foil at –80°C. This maintains sample integrity for up to a few months. ▼CAUTION: Improper storage conditions can result in postmortem tissue degradation (*see* **Note 4.1.2**).

3.2. Mid-IR Ablation of Sample

1. For lateral imaging experiment, operate a mid-IR laser at 2.94 μm wavelength and 10 Hz repetition rate. (Optional)

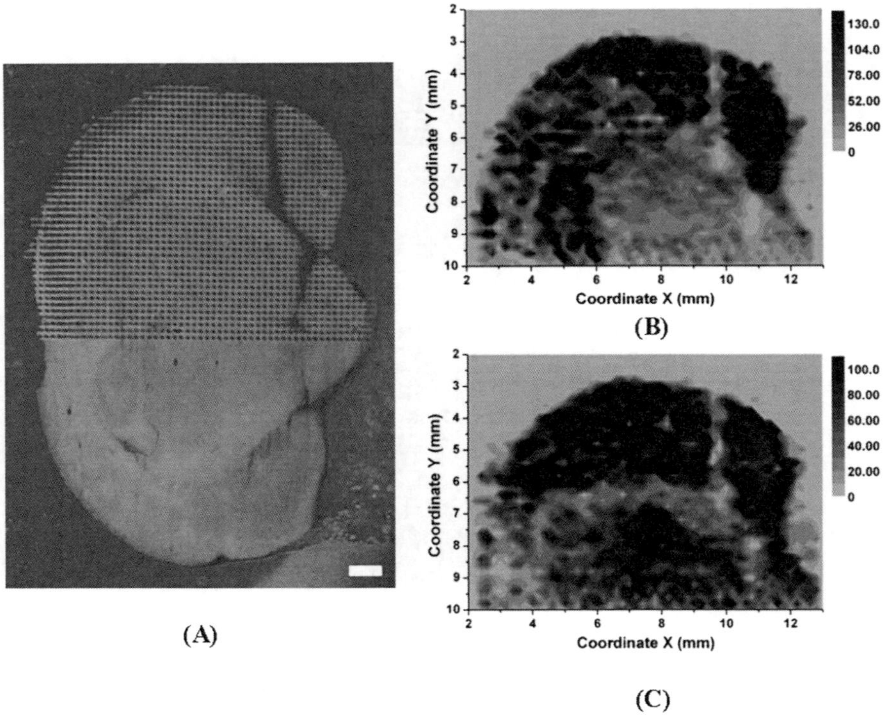

Fig. 9.4. LAESI MSI experiment on ~100 μm thick coronal section of a rat (*Rattus norvegicus*) brain. The sample was mounted on a Peltier cooling stage and was kept frozen during imaging to avoid postmortem tissue degradation. (**a**) The imaged area is shown by an array of ablation marks located 200 μm apart in the *x* and *y* directions. Scale bar corresponds to 1 mm. (**b**) Glycerophosphocholine observed at *m/z* 258.11 exhibited heterogeneous distribution in the tissue. (**c**) Glycerophosphocholine (38:6) measured at *m/z* 806.56 was present throughout the section and appeared especially abundant in the caudate putamen (striatum) and cerebral cortex regions of the brain section.

For samples with low water content (e.g., skin) increase the incident pulse energy. For very low water content samples (such as bone and tooth), tune the laser wavelength to a strong absorption band (*see* **Note 4.2.1**).

2. Use gold mirrors and a focusing element (e.g., a plano-convex ZnSe lens) to couple the mid-IR laser pulse into the sample at right angle ($\alpha=90°$ in **Fig. 9.3a**) (*see* **Note 4.2.2**). Position the focal point below the orifice axis and set $d_{OR\text{-}FP}$ to 5–8 mm (*see* **Fig. 9.3a**) (*see* **Note 4.3.3**). ▼CAUTION: Protective enclosure must be used to avoid inhalation of airborne ablated particles.
3. Optimize the position of the focusing element and the pulse energy of the laser beam to achieve tissue removal in desired dimensions – this determines the pixel size (*see* **Note 4.2.2**). For example, **Fig. 9.3c** shows that a ZnSe lens allowed the sampling of *Arabidopsis thaliana* with 200 μm diameter ablation mark at 1 mJ/pulse energy. (Optional) Fine focusing can also be achieved by vertical positioning of the

sample holder (*see* SH in **Fig. 9.3a**). Please remain within 5 < $d_{OR\text{-}FP}$ < 30 mm range (*see* **Note 4.3.3**). At distances less than 5 mm the expansion of the ablation plume can destabilize the electrospray (6). Most of the ablated particulate matter does not travel beyond 30 mm due to the drag force in the atmosphere (7).

4. (Optional) Sharpened optical fiber delivery can offer superior lateral resolutions by producing ablation craters with decreased dimensions. For example, **Fig. 9.3d** shows a 100-μm diameter ablation crater. For a GeO_2-based fiber delivery system, first place ~1 cm of both tips of the fiber in boiling 1-methyl-2-pyrrolidinone for ~5 min until the polyimide coating is dissolved. ▼CAUTION: Work under hood with good ventilation to avoid inhalation of vapor. Wash residues off the glass fiber tip with water and methanol. Cleave both ends for flat surfaces with a diamond scribe and etch one end to desired diameter in 0.1% HNO_3 solution at room temperature (e.g., 50 μm in ~15 min). Clean fiber tip with water. Position fiber tip to surface using a three-axis translation stage. ▼CAUTION: Etched tip poses sharp object hazard; handle with care.

3.3. Post-ionization with Charged Droplets

1. Position a blunt-tip nanospray emitter (e.g., TaperTip™ with 100 μm inner diameter) in line with the inlet axis of the mass spectrometer and at orifice-to-emitter tip distance, $d_{OR\text{-}ES}$, of ~10 mm (*see* **Fig. 9.3a**).
2. Feed 50% methanol solution containing 0.1% acetic acid or 0.1% ammonium acetate for positive and negative ion mode, respectively, through the metal emitter at <300 nl/min. Initiate electrospray by directly applying high voltage to the emitter (*see* **Notes 4.3.1** and **4.3.2**). **Figure 9.3b** reveals that lower flow rates, down to 30 nl/min, give higher kaempferol-(diacetyl coumaryl-rhyamnoside) counts when leaf tissues of *A. squarrosa* are studied. The optimum flow rate may be sample dependent as different ablated particulate size distributions may require different electrospray droplet size distributions for sufficient interaction. (Optional) For emitters with nonconductive surface, charge the solution through electrifying a metal union. (Optional) Other organic solvents, such as acetonitrile and isopropanol, may replace methanol. For reactive LAESI imaging, the electrosprayed solution may contain additives (8).
3. While sampling intact sample areas, optimize the spray voltage for LAESI ion yield. (Optional) Follow the temporal behavior of the spray current on a counter electrode with an oscilloscope to determine the flow rate and spray voltage conditions for the cone-jet spraying mode (*see* **Notes 4.3.1**

and **4.3.2**). This operating regime, characterized by elevated steady spray current, is reported to be efficient for the post-ionization of ablated particles (6). ▼CAUTION: Avoid electrical breakdown (*see* **Note 4.3.3**). Sudden rise in the spray current, measured either on the counter electrode or on the orifice of the mass spectrometer, can be an indication of corona discharge. It is usually observed at 3,500–4,000 V for the above-recommended solution, electrosprayed through 100 μm inner diameter emitter.

4. Adjust $d_{OR\text{-}BA}$ (*see* **Fig. 9.3a**) to optimize horizontally the overlap between the laser plume and the electrospray, thereby improving the LAESI ion yield (*see* **Note 4.3.3**). Please note that the laser beam, the emitter, and the orifice axis should remain in the same plane with the latter two aligned on the same axis (*see* **Note 4.2.2**). The inset in **Fig. 9.3b** shows that an order of magnitude improvement can be achieved in the ion counts by moving the emitter by few millimeters.
5. Optimize $d_{OR\text{-}FP}$ (*see* **Fig. 9.3a**) while refocusing the lens to retain the pixel size obtained in **Section 3.2**. The aim of this step is to optimize vertically the overlap between the laser plume and the electrospray, thereby improving the LAESI ion yield. Please note that the laser beam, the emitter, and the orifice axis should remain in the same plane (*see* **Note 4.2.2**).
6. (Optional) Optimal ion transfer is inherently dependent on the ion source geometry of the mass spectrometer. Fine-tune the position the electrospray emitter perpendicularly to the orifice axis to optimize for ion transfer and repeat Steps 2–6 after each time an adjustment is made.
7. Repeat Steps 2 through 6 for optimum LAESI ion yield.
8. Measure the dimensions of the ablation crater on the sample to be imaged. (Optional) For 3D MSI experiment, perform ablation with individual laser pulses and determine the depth of a voxel.

3.4. LAESI MSI in 2D and 3D

1. Establish software control over the three-axis translation stage and select a gridding algorithm (e.g., adaptive grid, selected region imaging, rectangular grid, spiral pattern, Z scanning) with which to raster the sample surface with selected dwell time at each pixel over the area to be imaged. In our experiments, for example, a LabView program was written in-house to position the sample in a rectangular pattern (e.g., based on left–right scanning shown in **Fig. 9.2a**) with lateral step sizes in x and y directions equal to or slightly larger than the ablation spot diameter (*see* **Fig. 9.4a**) (4). Alternatively, molecular imaging can be

performed with lateral step sizes smaller than the diameter of the ablation spot. This approach, also known as oversampling, has been successfully used in combination with classical vacuum and AP IR-MALDI imaging (9, 10). Calculate the total time required for imaging. Wait for START signal to initiate scanning sequence.

2. Operate the mid-IR laser source at a repetition rate sufficient to produce acceptable signal-to-noise ratio in the mass spectrum within the dwell time at each pixel to perform a LAESI 2D MSI experiment. (Optional) For molecular imaging in 3D, use a spectrum acquisition rate higher than the laser source repetition rate to successfully mass analyze the ions generated within a single laser pulse. Wait for START signal to initiate ablation sequence (11).
3. Load the tuning conditions for the ion source of the mass spectrometer and define acquisition parameters (mass range, scan rate, etc.) for the calculated imaging time (*see* **Note 4.2.2**). Wait for START signal to start the collection of mass spectra.
4. Start the electrospray source at experimental conditions optimized in **Section 3.3**. Please make sure that there is enough electrospray solution in the syringe for the total imaging time.
5. Simultaneously START acquisition of mass spectra, mid-IR ablation, and surface scanning (*see* **Note 4.3.3**).
6. When imaging has finished, STOP surface scanning, mid-IR lasing, and data acquisition (*see* **Note 4.3.3**).
7. Turn off spray voltage and stop the flow through the electrospray emitter.

3.5. Data Analysis

1. Correlate absolute coordinates of pixels in 2D or voxels in 3D with the mass spectra to obtain 2D and 3D molecular images (*see* **Fig. 9.2**).
2. Repeat **Sections 3.1–3.5** to obtain optimal results for a particular experiment. Please review **Section 4** for a list of possible sources of problems, their identification, and prevention during the LAESI MSI experiments.

4. Notes

4.1. Preparation and Mounting of Tissue

1. Evaporative water loss can be a significant source of problems during LAESI MSI experiments. Unsuccessful tissue removal may be an indication of insufficient water content

in the sample. Freeze–thaw mounting and active temperature and/or humidity control during experiments can mitigate these effects.

2. Like in many imaging experiments, improper handing of tissue may result in postmortem degradation. Changing in the chemical composition can yield inconsistent results. Carefully scrutinize sample preparation steps and mounting and storage conditions (*see* **Section 3.1**).

4.2. Mid-IR Ablation of Sample

1. The water content and the mechanical properties, primarily the tensile strength, of the sample are critical factors during mid-IR ablation of samples (12) in LAESI experiments. Although significant changes in these properties can detrimentally affect the imaging results, they are unlikely to occur within a tissue type. These effects would take place, for example, when moving from soft tissue to bone or tooth within a sample. Unsuccessful or incomplete ablation might be an indication of changing water content or tensile strength. Monitor tissue removal at the imaged area and adjust the pulse energy where required.

2. As the LAESI MSI experiments correlate the peak intensity of a particular ion in the mass spectrum with the concentration of the compound at a particular location in the sample (6), it is crucial to ensure that equal amounts of material be ablated from pixel to pixel during the course of the molecular imaging experiment. Significant variations in the diameter and depth of the ablation craters may yield artifacts during the experiment. Likely causes are uneven height of the sample surface, changing angle between the sample holder and the laser beam (α in **Fig. 9.3a**), or temporal variations in the laser pulse stability. Varying water content and tensile strength might further contribute to these effects. Optimize the sample elevation and the relative alignment of the sample holder and the infrared beam axis in **Section 3.2**. Alternatively, adjust the pulse energy to correct for changing ablation dimensions.

4.3. Post-ionization with Charged Droplets

1. Stability of charged droplet generation with the electrospray source is critical for successful ion production with the LAESI source. Sudden changes in the ion or spray current levels usually point to unstable liquid dispersion. Inspect connections and look for material deposition and oxidation on the emitter tip. Clean the emitter tip with the electrospray solvent. Alternatively, replace emitter if required.

2. Liquid filament formation may occur in the negative ion mode and can significantly lower the efficiency of liquid dispersion. In turn, the interception volume between the laser

ablation and the charged droplets is dramatically decreased, causing reduced ion yields. Low LAESI ion yield in the negative ion mode often indicates filament formation. Alter the flow rate and/or the high voltage on the emitter or use a sheath gas.

3. Some ablated particulates may deposit on the electrospray emitter and contaminate the electrospray solvent resulting in regular electrospray ionization. Prolonged presence of ions related to an analyte is often an indication of cross contamination between pixels due to material deposition. Inspect the emitter under optical microscope for significant material deposition. Optimize **Sections 3.2** and **3.3**. for larger $d_{OR\text{-}FP}$ (*see* **Fig. 9.3a**) and/or increase the ($d_{OR\text{-}ES} - d_{OR\text{-}BA}$) difference by moving the focal point away from the emitter tip. Please note that too large distances may significantly lower the ion yields. Clean emitter tip with the electrospray solution. Replace the emitter if required. Alternatively, use a protective sleeve around the emitter to prevent material deposition on it.

References

1. Cooks, R. G., Ouyang, Z., Takats, Z., Wiseman, J. M. (2006) Ambient mass spectrometry. *Science*, **311**, 1566–1570.
2. Takats, Z., Wiseman, J. M., Gologan, B., Cooks, R. G. (2004) Mass spectrometry sampling under ambient conditions with desorption electrospray ionization. *Science*, **306**, 471–473.
3. Li, Y., Shrestha, B., Vertes, A. (2008) Atmospheric pressure infrared MALDI imaging mass spectrometry for plant metabolomics. *Anal Chem*, **80**, 407–420.
4. Nemes, P., Barton, A. A., Li, Y., Vertes, A. (2008) Ambient molecular imaging and depth profiling of live tissue by infrared laser ablation electrospray ionization mass spectrometry. *Anal Chem*, **80**, 4575–4582.
5. Shelley, J. T., Ray, S. J., Hieftje, G. M. (2008) Laser ablation coupled to a flowing atmospheric pressure afterglow for ambient mass spectral imaging. *Anal Chem*, **80**, 8308–8313.
6. Nemes, P., Vertes, A. (2007) Laser ablation electrospray ionization for atmospheric pressure, *in vivo*, and imaging mass spectrometry. *Anal Chem*, **79**, 8098–8106.
7. Vertes, A., Nemes, P., Shrestha, B., Barton, A. A., Chen, Z. Y., Li, Y. (2008) Molecular imaging by Mid-IR laser ablation mass spectrometry *Appl Phys Mater Sci Process*, **93**, 885–891.
8. Shrestha, B., Nemes, P., Nazarian, J., Hathout, Y., Hoffman, E., Vertes, A. (2010) Direct analysis of lipids and small metabolites in mouse brain tissue by AP IR-MALDI and reactive LAESI mass spectrometry. *Analyst*, **135,** 751–758.
9. Jurchen, J. C., Rubakhin, S. S., Sweedler, J. V. (2005) MALDI-MS imaging of features smaller than the size of the laser beam. *J Am Soc Mass Spectrom*, **16**, 1654–1659.
10. Li, Y., Shrestha, B., Vertes, A. (2007) Atmospheric pressure molecular imaging by infrared MALDI mass spectrometry. *Anal Chem*, **79**, 523–532.
11. Nemes, P., Barton, A. A., Vertes, A. (2009) Three-dimensional imaging of metabolites in tissues under native conditions by laser ablation electrospray ionization mass spectrometry. *Anal Chem*, **81,** 6668–6675.
12. Apitz, I., Vogel, A. (2005) Material ejection in nanosecond Er:YAG laser ablation of water, liver, and skin. *Appl Phys Mater Sci Process*, **81**, 329–338.

Chapter 10

Matrix-Assisted Laser Desorption/Ionization and Nanoparticle-Based Imaging Mass Spectrometry for Small Metabolites: A Practical Protocol

Yuki Sugiura and Mitsutoshi Setou

Abstract

Matrix-assisted laser desorption/ionization (MALDI)-imaging mass spectrometry (IMS, also referred to as mass spectrometry imaging [MSI]) enables visualization of the distribution of biomolecules with varied and vast structures in tissue sections. This emerging imaging technique was initially developed as a tool for protein imaging; however, the number of studies reporting imaging of small organic molecules has recently increased. IMS is an effective technique for the visualization of endogenous small metabolites, especially lipids, facilitated by the unique advantages of mass spectrometry-based molecular detection. Despite the promising capability of MALDI-IMS for imaging small metabolites, this technique still has several issues, especially in spatial resolution. One of the critical limitations of the spatial resolution of MALDI-IMS is the size of the organic matrix crystal and the analyte migration during the matrix application process. To overcome these problems, we reported a nanoparticle (NP)-assisted laser desorption/ionization (nano-PALDI)-based IMS, in which the matrix crystallization process is eliminated. In this chapter, a practical protocol for MALDI-IMS of lipids is outlined. In addition, as an attractive alternative to MALDI-based IMS, we also present nanoparticle-based IMS that improves spatial resolution.

Key words: Imaging mass spectrometry, matrix-assisted laser desorption/ionization, MALDI, small molecules, lipids, nanoparticles.

1. Introduction

Traditional matrix-assisted laser desorption/ionization mass spectrometry (MALDI-MS) can be used to analyze a wide range of molecules that have varied physical and chemical features. Therefore, in the biological and medical fields, MALDI-IMS can be used to visualize the distribution of vast numbers of biomolecules,

S.S. Rubakhin, J.V. Sweedler (eds.), *Mass Spectrometry Imaging*, Methods in Molecular Biology 656,
DOI 10.1007/978-1-60761-746-4_10, © Springer Science+Business Media, LLC 2010

including small metabolite molecules (1, 2) and large proteins (3, 4), in cells and tissues.

IMS is an effective imaging technique particularly for small metabolite distributions. Facilitated by the mass spectrometry-based molecular detection, IMS stands out as a tool for imaging metabolites in tissues and cells because of several advantages. First, IMS does not require any labels or specific probes. Second, IMS is a non-targeted imaging method. Thus, unexpected metabolites can be imaged. Finally, many types of metabolites can be simultaneously imaged. Because of the enormous molecular diversity of metabolite species, of which large numbers are still unknown, these features are necessary for localizing metabolites. In addition, by using tandem MS analysis (MS^n), the detailed structure of metabolites can be identified. Thus, whether observed mass signals are from the molecule of interest can be determined (1, 5).

The emergence of IMS as a tool for imaging metabolites has a large impact, because there was previously no established technique for metabolite imaging; RNA transcripts can be visualized with oligonucleotide probes using in situ hybridization and proteins can be visualized using immunohistochemistry with appropriate antibodies (**Table 10.1**). In this context, metabolic imaging by IMS is critical for interpretations of various aspects of life sciences. Today, the application of IMS to visualizing small organic compounds ($m/z < 1,000$) can be subdivided into two distinct research areas, namely measurement of endogenous compounds and of exogenous compounds such as drugs. For endogenous metabolites, a number of reports regarding lipid metabolites have been intensively reported (**Table 10.2**).

Table 10.1
Representative molecular imaging methods in tissues or cells

	Representative molecular imaging methods	probe
DNA	FISH (fluorescence in situ hybridization)	Oligo nucleotide probe
RNA	In situ hybridization	Oligo nucleotide probe
Protein	Immunohistochemistry, green fluorescent protein-fused protein	Antibody
Metabolites (especially lipids)	Imaging mass spectrometry	-

Despite the promising capability of IMS for imaging small molecules, this technique still has several unresolved issues. An important challenge is the improvement of spatial resolution in the range of subcellular resolution (>10 μm). Metabolite imaging within cellular organelles, which requires resolution at submicrometer level, will provide researchers with valuable information about cellular metabolism. In this regard, imaging with SIMS has achieved submicron spatial resolution and has successfully visualized subcellular structures (6, 7) and dynamic changes of metabolite molecules during *Tetrahymena* mating (6).

Table 10.2
Current IMS for lipids

	Class	Ion detection mode	References
Glycerophospholipids	PCs	Positive	(1, 12, 27, 34, 43)
	PEs	Negative	(27, 43, 44)
	PIs	Negative	(27, 43, 44)
	PSs	Negative	(27, 43, 44)
	PGs	Negative	(27, 43, 44)
	Cardiolipins	Negative	(27, 43, 44)
Neutral lipids	Triacylglycerols	Positive	(43)
	Diacylglycerols	Positive	(43)
	Cholesterol	Positive	(27, 46)
Sphingolipids	Gangliosides	Negative	(12, 28)
	Sulfatides	Negative	(28, 32, 44)
	Galactosyl-ceramide	Positive	(9, 47)
Fatty acids	–	Negative	(48)

Note: PC=phosphatidylcholine, PE=phosphatidylethanolamine, PI=phosphatidylinositol, PS=phosphatidylserine, PG=phosphatidylglycerol.

However, both soft ionization of analytes and tandem MS are difficult to achieve with typical SIMS technique (8). In contrast, spatial resolution of MALDI-IMS is lower than that of SIMS. The spatial resolution depends on the experimental conditions and the instrument used but is typically 20–100 μm. Limitations of the spatial resolution of MALDI-IMS include the size of the organic matrix crystal and the analyte migration during the matrix application process. To overcome these problems, Taira and colleagues reported a nanoparticle (NP)-assisted laser desorption/ionization (nano-PALDI)-based IMS, in which the matrix crystallization process is eliminated (9). The use of nano-PALDI has enabled researchers to image compounds with spatial resolution at the cellular level (15 μm; almost equal to the size of the diameter of a laser spot).

Here, practical experimental procedures using MALDI-IMS for lipid metabolites are presented. The technical problems specific to IMS of small metabolites are introduced and the techniques for mitigating these problems are explained. In addition, as an attractive alternative to MALDI-IMS, nanoparticle-based IMS, which improves spatial resolution, is discussed. A protocol for the preparation of extremely small nanoparticles ($d = 3.7$ nm) developed by our group (10) and its application method is presented.

Before describing the detailed experimental procedure, the structure of glycerophospholipids (GPLs) should be briefly mentioned. GPLs are the most abundant type of lipids, especially in the brain, and form a large molecular family in which phosphoric acid in the ester form is bound to a glycerolipid. They are subdivided into distinct classes (e.g., phosphatidylcholines [PCs], phosphatidylethanolamines, and phosphatidylinositols) based on the structure of the head group linked to the phosphate, which is attached at the *sn*-3 position of the glycerol backbone. GPLs are further subdivided into numerous molecular species based on the composition of the fatty acids linked to the *sn*-1 and *sn*-2 positions of the glycerol backbone (**Fig. 10.1**).

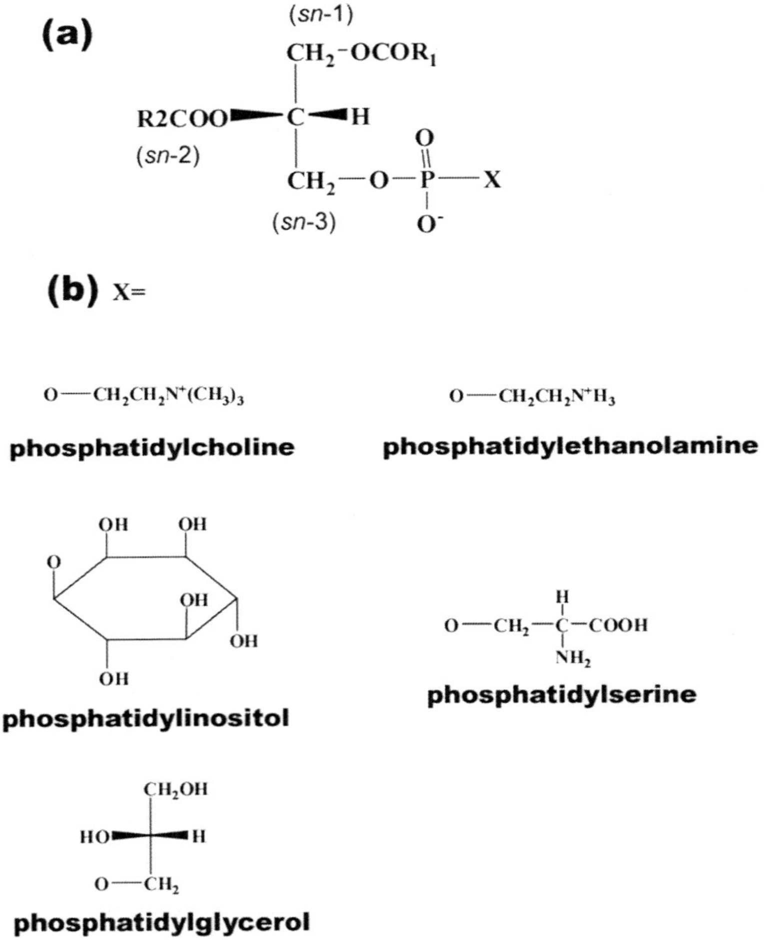

Fig. 10.1. Structure of phospholipids. Structure of the glycerol backbone of glycerophospholipids (GPLs) (**a**) and their head group (**b**) is shown. GPLs are subdivided into distinct classes (e.g., phosphatidylcholines, phosphatidylethanolamines, and phosphatidylinositols) based on the structure of the head group linked to the phosphate, which is attached at the *sn*-3 position of the glycerol backbone (**b**). GPLs are further subdivided into numerous molecular species based on the composition of fatty acids linked to the *sn*-1 and *sn*-2 positions of the glycerol backbone.

IMS allows the visualization of these multiple molecular species in parallel. The distinct localization of GPL molecular species can be determined; in other words, the distinct fatty acid composition of biological membranes in different tissue locations can be determined.

2. Materials

2.1. Chemicals

1. Trifluoroacetic acid (TFA).
2. 1,2-Dihexanoyl-*sn*-glycero-3-phosphocholine (MW: 453.5, monomer and dimer were used for calibration) as a calibration standard.
3. 2,5-Dihydroxybenzoic acid (DHB).

2.2. Matrix Solution for MALDI-IMS of PCs

1. For measurement of PC, a DHB solution (40 mg/ml DHB, 10 mM potassium acetate, 70% methanol) (*see* **Notes 1** and **2**) was used as the matrix solution (11).

2.3. Matrix Solution for MALDI-IMS of Gangliosides

1. For measurement of gangliosides, a matrix solution without potassium salt (40 mg/ml DHB, 70% methanol) was used as the matrix solution (12).

2.4. Chemicals for Nanoparticle Preparation

1. $FeCl_2{\cdot}4H_2O$ (99.9%).
2. γ-Aminopropyltriethoxysilane (γ-APTES).

All of the chemicals used in this study were of the highest purity available.

3. Methods

3.1. Animal Sacrifice and Tissue Extraction

All experiments involving mice were conducted in accordance with the protocols approved by the animal care and use committee at the participating research institute.

1. The brains of 8-week-old male C57BL/6 J Cr mice were used.
2. The brains were extracted within 1 min (typically 40 s) after sacrifice (*see* **Note 3**).
3. The trimmed tissue blocks were immediately frozen in powdered dry ice, which allows tissues to be frozen without cracks, and stored at –80°C until use.

3.2. Preparation of Tissue Sections

1. Tissues blocks were sectioned at −16°C using a cryostat (CM 3050; Leica, Germany) to a thickness of 5 μm (for detailed discussion about slice thickness, see Refs. (13, 14)).
2. Tissue blocks were held by an optimum cutting temperature (OCT) polymer, but they were not embedded into the polymer because it was thought that any residual polymer on the tissue slices might degrade the mass spectra (14) (*see* **Note 4** and Ref. (14)).
3. The frozen sections were thaw-mounted on indium-tin-oxide (ITO)-coated glass slides (Bruker Daltonics) and ITO-coated sheets (Tobi Co., Ltd., Kyoto, Japan). The slides were used for tandem time-of-flight (TOF/TOF) measurements and the sheets were used for quadrupole ion trap (QIT)-TOF measurements.
4. The prepared sections were subjected to matrix application within 5 min (*see* **Note 5**).

3.3. Spray Coating of the Matrix Solution

1. The matrix solution was sprayed over the tissue surface using a 0.2-mm nozzle caliber airbrush (Procon Boy FWA Platinum; Mr. Hobby, Tokyo, Japan).
2. The distance between the nozzle tip and the tissue surface was held at 10 cm and the spraying period was fixed at 5 min. Approximately 100 μl of matrix solution was sprayed onto each brain section.

3.4. Production of Nanoparticles

200 ml is the best quantity for the preparation of nanoparticles.

Day 1

1. Cool water and ethanol at 4°C temperature.
2. Dissolve 2 g of $FeCl_2 \cdot 4H_2O$ in 100 ml water.
3. Decant 100 ml γ-APTES directly into another beaker. γ-APTES dissolves in both water and organic solvents.
4. Agitate γ-APTES using a stir bar.
5. Pour all the solution of $FeCl_2$ into the γ-APTES solution with agitation. The mixing ratio of γ-APTES to $FeCl_2 \cdot 4H_2O$ should be 45:1.
6. Put an aluminum foil lid on the beaker and allow it to cool for 30 min.
7. Separate the solution into 35 ml portions in 50-ml Falcon tubes and centrifuge the solution at 8,000×*g* force at 4°C for 20 min.
8. Remove the supernatant.
9. Add chilled water so that the total volume is 35 ml.

10. Break the precipitate with supersonic waves and vortex it to suspend the precipitate.
11. Centrifuge the solution at 8,000×g force at 4°C for 20 min.
12. Remove the supernatant.
13. Repeat procedures 9 through 12, three times.
14. Add ethanol, centrifuge the solution, and remove the supernatant (**Fig. 10.2a**).
15. Leave the tube overnight at 80°C to completely evaporate the ethanol.

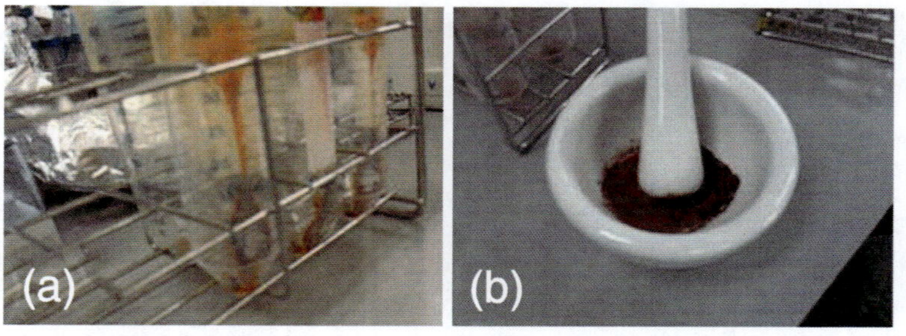

Fig. 10.2. Precipitated nanoparticles by centrifugation (**a**) and after drying/crushing in a mortar (**b**).

Day 2

1. Crush the precipitate using a mortar as shown in **Fig. 10.2b**.
2. After crushing, collect the powder from the crushed precipitate and put it in Eppendorf tubes.
3. The prepared nanoparticles can be preserved for 1 month in a dark room at 4°C.

3.5. Use of Nanoparticles for Imaging Mass Spectrometry

1. Put 10 mg each of nanoparticles into two Eppendorf tubes.
2. Put 1 ml of 100% methanol with 10 mM sodium acetate into each tube.
3. Vortex the contents for 1 min with supersonic waves.
4. Centrifuge the tubes at 9,000×g at 4°C for 10–30 s.
5. Adjust the centrifuging time so that the ideal color of the fluid can be obtained (**Fig. 10.3**).
6. Transfer the supernatant, which contains the nanoparticles, to fresh Eppendorf tubes.
7. Spray the supernatant onto the tissue sections using an airbrush. Adjust the distance between the airbrush and the tissue samples so that the sprayed droplets of methanol evaporate during spraying or so that they dry immediately on the tissue surface.

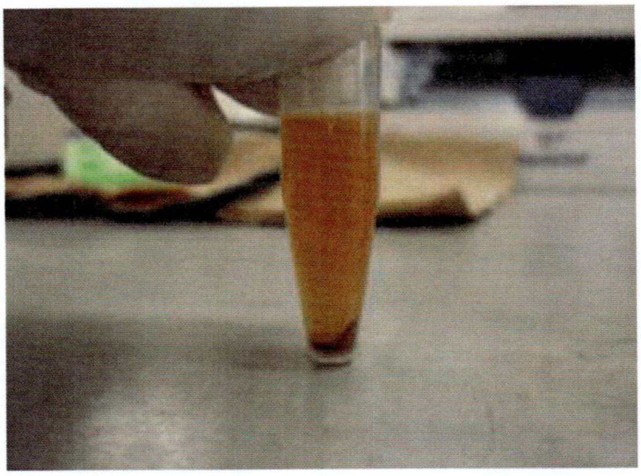

Fig. 10.3. A suspension of nanoparticles.

3.6. Instrument Parameter Settings

IMS was performed using a MALDI TOF/TOF instrument (Ultraflex 2 TOF/TOF; Bruker Daltonics), which was equipped with a 355 nm Nd:YAG laser.

1. The number of laser irradiations was 100 in each spot.
2. Raster scans on tissue surfaces were performed automatically using FlexControl and FlexImaging 2.0 software (Bruker Daltonics).
3. Image reconstruction was performed using FlexImaging 2.0 software.
4. For measurement of PCs, the data were acquired in the positive reflectron mode under an accelerating potential of 20 kV using an external calibration method.
5. For measurement of gangliosides, experiments were performed in the negative reflectron mode under an accelerating potential of –20 kV.

3.7. Spectrum Normalization

The variation in ionization efficiency in the IMS results, caused by heterogeneous distribution of organic matrix crystals and their sublimation during the measurement, was eliminated for each data point by equalizing the total ion current for each mass spectrum, using the "normalize spectra" function of the FlexImaging 2.0 software (11) (*see* **Note 6**).

3.8. Tandem Mass Spectrometry for Molecular Identification

Molecular identification was performed with tandem mass spectrometry using a QIT-TOF mass spectrometer (AXIMA-QIT; Shimadzu, Kyoto, Japan) to ensure the molecular assignment which was performed using only mass. The MS^n analysis was performed directly on the sections of tissue sections in the mid-mass range mode.

3.9. Results of Imaging Mass Spectrometry for Phosphatidylcholines

As a number of studies have shown, MALDI-IMS is effective for profiling and visualizing the distribution of GPLs because of easy detection. GPLs are ionized efficiently because of the following reasons. First, large amounts (more than 60% dry weight) of mouse brain are composed of lipids (high expression). Second, GPLs have an easily ionized structure. Phospholipids, in particular PCs, contain a phosphate group and a trimethylamine that are charged easily (15).

Figure 10.4a shows representative mass spectrum obtained by IMS of mouse brain sections. Nine intense mass peaks were assigned to eight abundant PC molecular species and one sphingomyelin based on *m/z* values. All of these contain a trimethylamine head group. In **Fig. 10.4b**, the tissue distributions of the five major PC molecular species in sections of the sagittal brain are shown. Although the most abundant molecular PC species, PC (diacyl-16:0/18:1), was uniformly distributed across the entire gray matter region of the section, other PC molecular species showed rather heterogeneous distribution patterns.

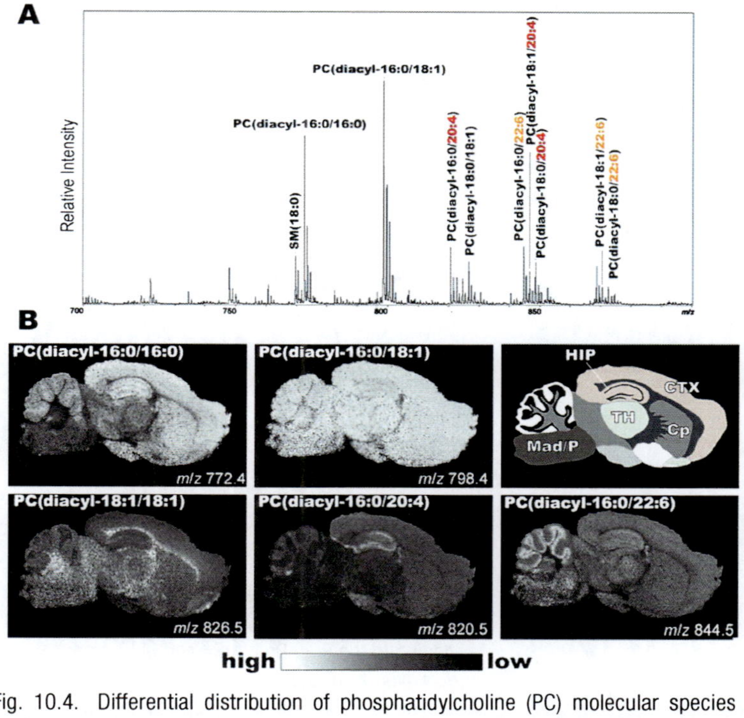

Fig. 10.4. Differential distribution of phosphatidylcholine (PC) molecular species in sagittal mouse brain sections. (**a**) An averaged mass spectrum obtained from an entire mouse brain section. In the spectrum, intense mass peaks corresponding to nine abundant PCs were assigned according to mass. (**b**). MALDI-IMS of a brain section simultaneously identified the heterogeneous distributions of several PCs. Schema of the mouse brain sagittal section and ion images of PCs obtained by IMS are shown.

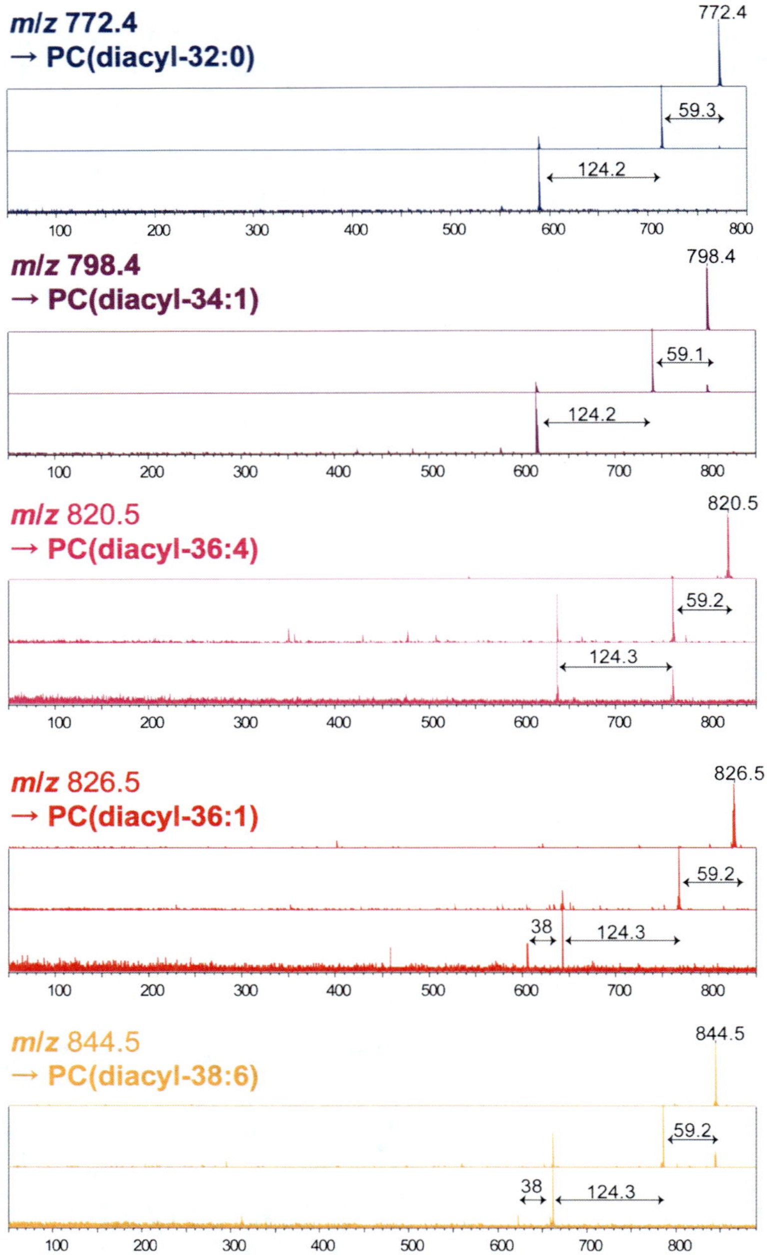

Fig. 10.5. Tandem mass spectrometry (MS) allows molecular identification of glycerophospholipids on the tissue surface. Results of MS^n structural analysis of ions corresponding to the PCs. Both MS^2 and MS^3 product ion spectra show that the mass peaks are derived from the PCs. Neutral losses of 59 and 124 u from precursor ions, corresponding to trimethylamine and cyclophosphate, respectively, were used as diagnostic ions.

The molecular assignments were verified by structural analysis of each peak using MS^n (**Fig. 10.5**). A QIT-TOF mass spectrometer was used for this purpose. This instrument can identify molecules using a highly selective/sensitive MS^n from mixture ions generated on the tissue surface (5). In each mass peak, the presence of a trimethylamine head group and a phosphate was confirmed (neutral losses of 59 and 124 u from precursor ions corresponding to trimethylamine and cyclophosphate, respectively), which are used as diagnostic ions in product ion mass spectra (5, 16, 17).

3.10. Results of Imaging Mass Spectrometry for Gangliosides

Gangliosides are glycosphingolipids consisting of mono- to poly-sialylated oligosaccharide chains of variable lengths attached to a ceramide unit. Gangliosides are inserted in the outer layer of plasma membranes with the hydrophobic ceramide moiety acting as an anchor. Their oligosaccharide moiety is exposed to the external medium (18). Gangliosides also form a large family; their constituent oligosaccharides differ in the glycosidic linkage position, sugar configuration, and the content of neutral sugars and sialic acid. In particular, based on the number of sialic acids, they can be subdivided into GM (i.e., mono-sialylated), GD (di-sialylated), GT (tri-sialylated), and GQ (quadra-sialylated) groups. As well as the oligosaccharide unit, the ceramide moiety also varies with respect to the type of long-chain base (LCB; sphingosine base) and fatty acid moiety (**Fig. 10.6**).

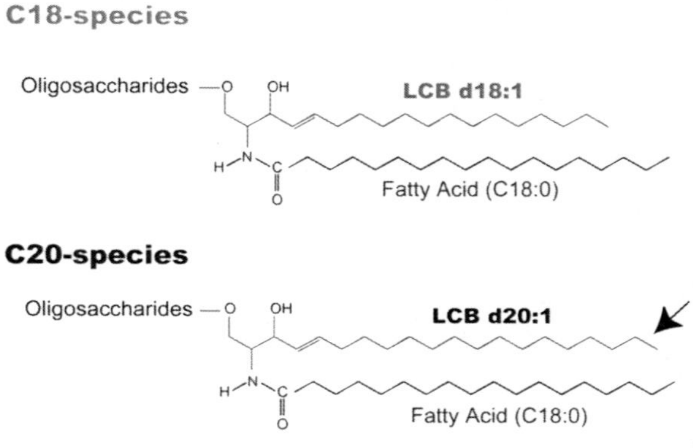

Fig. 10.6. Structures of ganglioside molecular species containing C18-long-chain backbone (LCB) and C20-LCB. The C20 species has two more carbons in its LCB moiety than does the C18 species (*arrow*).

Previous biochemical studies have shown that the LCB of brain gangliosides has either 18 or 20 carbons (i.e., C18- or C20-sphingosine). The C20-sphingosine (C20-LCB species) is present in significant amounts only in the central nervous system (19–22). Its content increases significantly throughout life in rodents and in human (23–25). Because of its characteristic brain specificity and the dramatic increase during the course of life, the

C20-LCB ganglioside was of great interest. However, a lack of visualization technology for the specific detection and visualization of C18 and C20 gangliosides has prevented researchers from determining their precise tissue distribution. Antibodies against some oligosaccharide moieties are available for visualizing molecular species with different constituent oligosaccharides (26); however, such immunological methods cannot detect differences in the ceramide structure, which is hidden in the lipid bilayer.

Because of the negatively charged sialic acids and their rich abundance in the brain, gangliosides are strongly detected in the $1,500 < m/z < 2,500$ range in the negative ion detection mode (12, 27, 28) (**Table 10.3**). In addition, as well as structural differences in oligosaccharides, IMS can discriminate structural differences in lipid moieties and has successfully revealed the specific distribution of C20-LCB species in mouse brain. Although C18 species are widely distributed throughout the frontal brain, C20 species are selectively localized in specific brain regions, namely in the molecular layer of the dentate gyrus (**Fig. 10.7A**).

Table 10.3
Detection of gangliosides in the mouse hippocampus

	Negative ions					
	$[M-H]^-$	$[M+Na-2H]^-$	$[M+K-2H]^-$	$[M+Na-3H]^-$	$[M+Na+K-3H]^-$	$[M+2K-3H]^-$
GM1 (d18:1/18:0)	1,544	–	–	–	–	–
GM1 (d20:1/18:0)	1,572	–	–	–	–	–
GD1 (d18:1/18:0)	–	1,858	1,874	–	–	–
GD1 (d20:1/18:0)	–	1,886	1,902	–	–	–
GT1 (d18:1/18:0)	–	–	–	2,170	2,186	2,202
GT1 (d20:1/18:0)	–	–	–	2,198	2,214	2,230

Fig. 10.7. (continued) spectra of m/z 888.3 and 916.3 were obtained to determine the different structural constituents in the ceramide moieties. Because of the detection of m/z 283.0 (fatty acid-related ion) in both spectra, the 28 u difference between m/z 1,544 and m/z 1,572 was attributed to the difference in the sphingosine constituent (m/z 1,544 had C18 sphingosine and m/z 1,572 had C20 sphingosine).

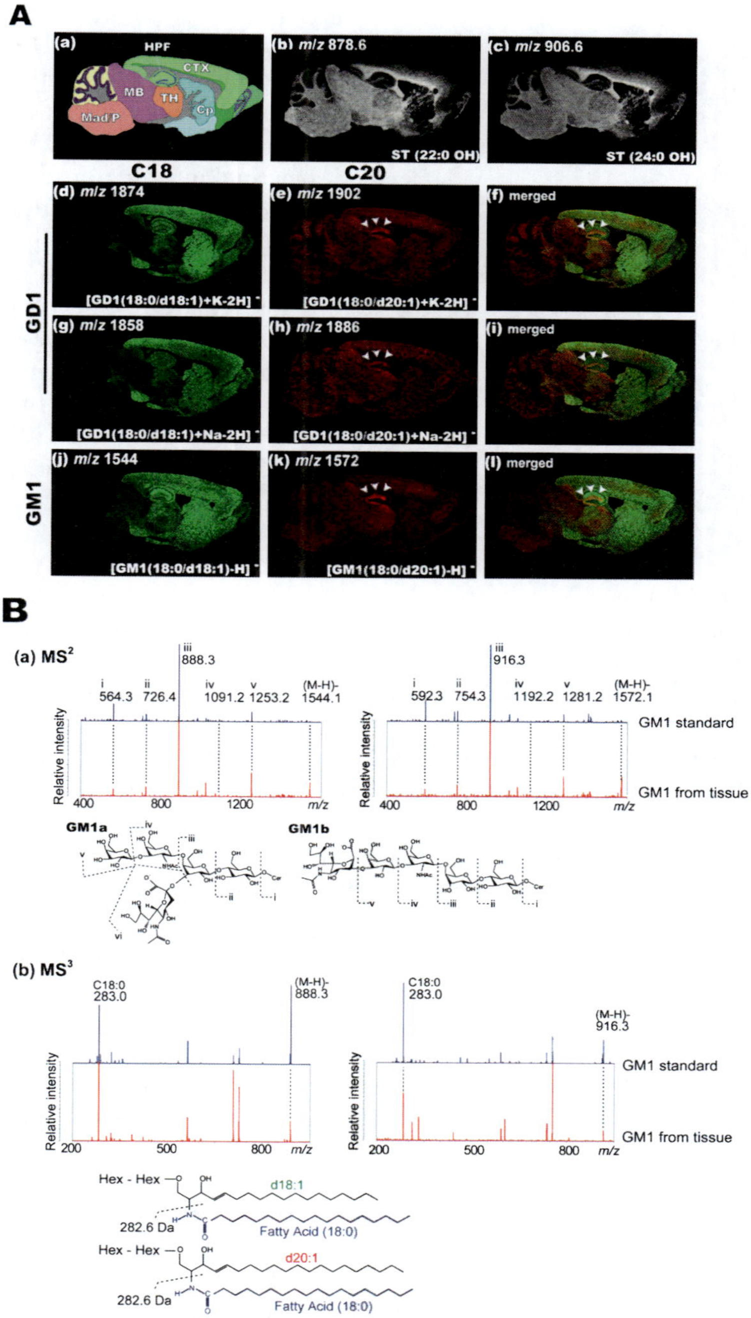

Fig. 10.7. Imaging mass spectrometry (IMS) and direct MS^n allow specific visualization and detection of ganglioside molecular species. (**A**) IMS at 50 μm raster step size was used to gain an overview of ganglioside distribution in different brain regions. Schematic diagram of the brain section (**a**) and ion images of sulfatides (**b–c**) are shown. Ions corresponding to gangliosides, namely GD1 (**d–i**) and GM1 (**j–l**), were visualized. (B) (**a**) The MS^2 product ion spectra show that the ions at *m/z* 1,544 and 1,572 had the same oligosaccharide structure (i.e., they contained a sialic acid moiety) but the ceramide mass peaks were observed at different *m/z* values. (**b**) MS^3 product ion mass

To confirm that the difference of 28 u, which corresponds to a $(CH_2)_2$ unit, observed between C18 and C20 species (**Table 10.3**) can be attributed to differences in LCB chain lengths, structural analysis of ions corresponding to GM1 gangliosides was performed using MS^n (**Fig. 10.7B**). This technique can provide detailed structural information about the ion of interest. The MS^2 results for both *m/z* 1,544 and 1,572 showed a ceramide peak and peaks corresponding to oligosaccharides containing a sialic acid (**Fig. 10.7B**). This **10.7B**, a). The peaks in the MS^2 spectra for oligosaccharides of *m/z* 1,544 and 1,572 were exactly the same. These gangliosides have the same oligosaccharide moiety. Next, MS^3 was performed to determine the detailed structure of the ceramide. In the MS^3 spectra, a common peak was observed at *m/z* 283.0, which corresponded to $(C_{17}H_{35}COOH)^-$, a fatty acid (**Fig. 10.7B**, b). Thus the difference was because of the difference in the chain length of the LCBs, namely the C18 and C20 sphingosines.

3.11. Results of Nanoparticle-Assisted Laser Desorption/Ionization Imaging Mass Spectrometry

In the early study of the soft ionization, NPs ($d = \sim 30$ nm) with metal oxide cores were found to assist laser desorption/ionization of analyte molecules in the presence of diluted glycerol (29). Recently, gold-NPs ($d = \sim 5.5$ nm) have been used in MS (30) and IMS (31). Compared to the traditional MALDI-IMS, nanoparticle-based IMS has several advantages. Gold-NPs ionize biomolecules that are different from those by traditional organic matrices (31) (in the results shown here, NPs more easily ionize sphingolipids than GPLs). Also, elimination of matrix-derived signals is important, especially for analysis of small molecules (9, 32).

Elimination of the matrix–analyte co-crystallization process is another important advantage of NP-based IMS. Spatial resolution is not restricted by the crystal size but by the diameter of the laser spot. **Figure 10.8** shows the results of NP and MALDI-IMS of lipids in rat cerebellum. As can be seen, spraying NPs on the tissue surface did not alter the optical image of the tissue structure (**Fig. 10.8**, upper panel), and it enhanced soft ionization of lipids to visualize them (**Fig. 10.8**, middle panel). In contrast, when the sections were sprayed with a DHB solution, although shown is a poor example, non-homogeneous crystals were observed on the section, which obscured the optical view of the sample surface (**Fig. 10.8**, upper panel). This resulted in blurred images of ions in this section (**Fig. 10.8**, middle panel). Furthermore, soft ionization with NP achieves sufficient ion yields to perform MS/MS on the tissue surface (**Fig. 10.8**, bottom).

Because of the extremely small particle size, it is possible to localize lipids in fine tissue structures (within several tens of micrometers). **Figure 10.9** shows another example of nanoparticle-IMS used to visualize the distribution of lipids

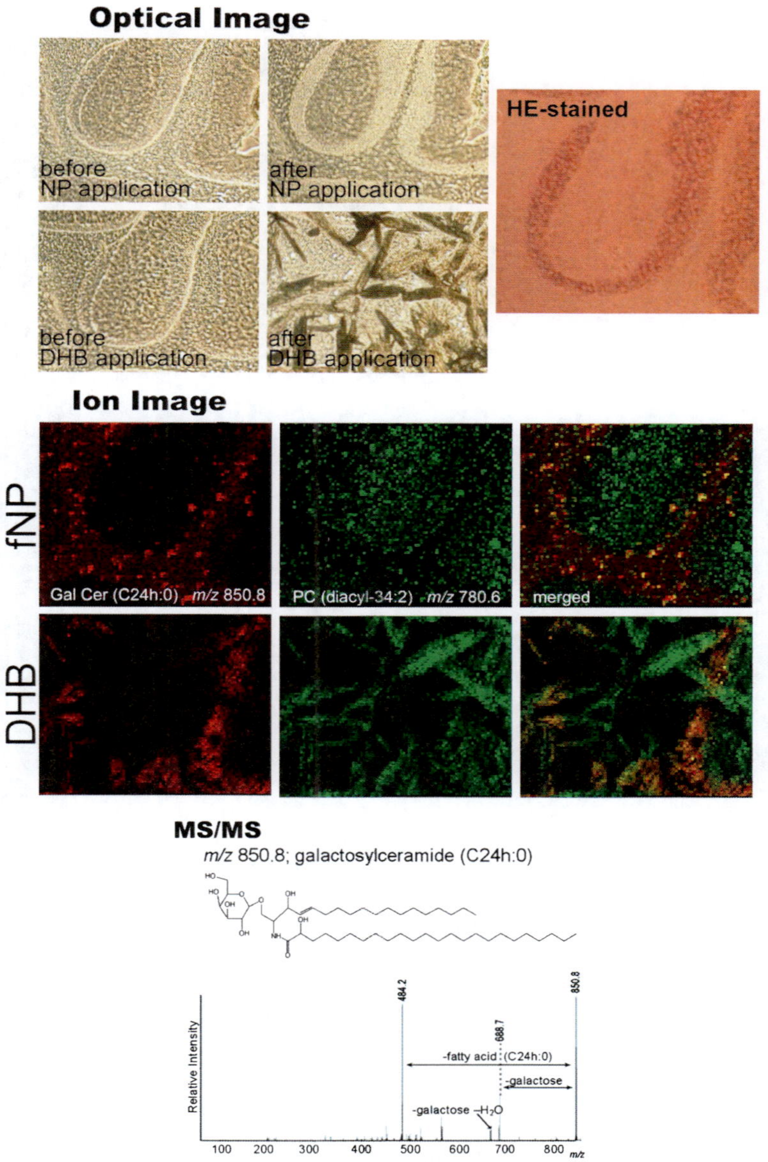

Fig. 10.8. Nanoparticle-assisted laser desorption/ionization imaging of lipids. (*Upper panel*) Optical images of rat cerebellum tissue before/after being sprayed with nanoparticles (NPs) and 2,5-dihydroxybenzoic acid (DHB) solution were shown. Successive brain section stained with hematoxylin–eosin (H&E) is also presented. (*Middle panel*) Ion images obtained with NPs and DHB are shown. Visualized ions were identified as galactosylceramide (C24h:0) and PC (diacyl-34:2) by tandem mass spectrometry on both DHB- and NP-coated sections. (*Bottom panel*) Example representation of MS/MS result on the tissue sprayed with NP, for ion at *m/z* 850.8. Product ion spectrum indicates that the ion was derived from galactosylceramide(C24h:0).

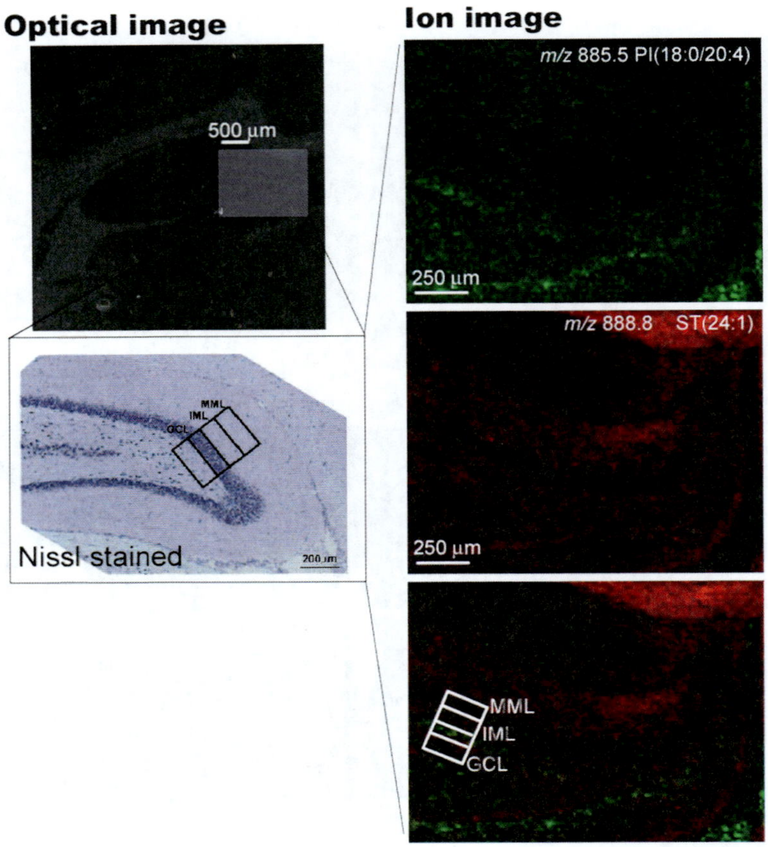

Fig. 10.9. Nanoparticle-assisted laser desorption/ionization improves spatial resolution in imaging mass spectrometry. (*Left panel*) Optical image of rat hippocampus indicating measurement area for nanoparticle-based IMS. Nissl-stained section indicating fine layer structure of rat hippocampus is also shown. (*Right panel*) Ion images which reveal hippocampal layer-specific distribution of phosphatidylinositol (18:0/20:4) at *m/z* 885.5 and sulfatide (24:1) at *m/z* 888.8 were presented. GCL = granular cell layer, IML = inner molecular layers, MML = middle molecular layers. Reprinted from Ref. (9).

within the layer structure of the rat hippocampus. In this case, hippocampal layer-specific localization of sulfatides and phosphatidylinositol was clearly observed (33).

4. Notes

1. The composition of organic solvents (methanol) used in matrix solutions influences the signal detection sensitivity for lipids and peptides (**Fig. 10.10**). Higher composition of methanol enhances lipid detection by efficient extraction of lipid molecules from tissue sections. In contrast, hydrophilic

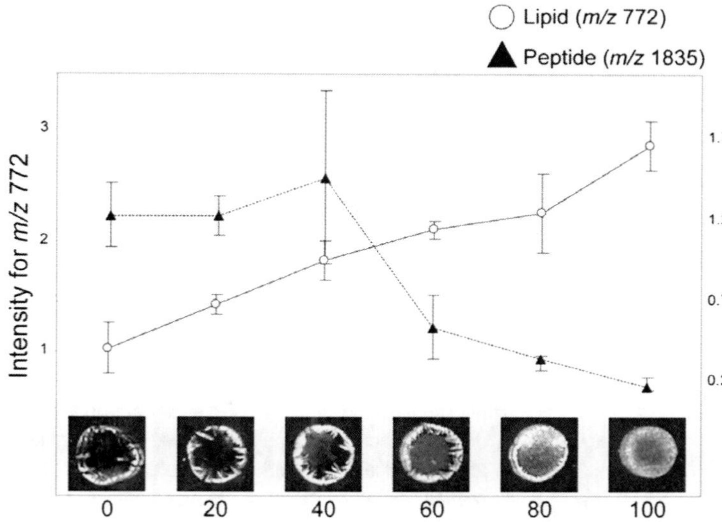

Fig. 10.10. Composition of organic solvents (methanol) in the matrix solution influences signal detection of lipids and peptides. Matrix solutions (containing 35 mg/ml DHB and 0.1% TFA) of different ratios of water and methanol were prepared. Next, 0.5 μl of each solution was applied to the section of mouse brain homogenate, which has homogeneous molecular distribution at any location of the section ($n = 3$). By increasing the methanol concentration, crystal form was also changed; needle-like crystal, from which peptides were detected, was changed to the aggregate of smaller crystals from which lipids were detected.

solutions enhance peptide detection for the same reason. Considering the easier application of methanol by spraying, a solution of 70% methanol was used in this study.

2. GPLs, especially PCs, preferentially cationize as their alkali metal-adduct molecules (34–36). Because tissue sections contain rich sodium and potassium salts, such alkali metal-adducted GPLs, rather than protonated molecules, are preferentially generated. Molecular ionization with such multiple ion forms from a single species often hampers IMS experiments, because GPLs have many molecular species and a single peak might contain multiple types of ions. Generation of such multiple molecular ions from a single PC molecular species can be suppressed by adding potassium acetate to the matrix solution (1). In this study, potassium salt (20 mM potassium acetate) was added to the matrix solution. Thus, the molecular ion forms were limited to potassium-adducted molecules, and the spectra were simplified.
3. As shown in **Fig. 10.11** postmortem degradation of GPLs was observed by IMS within 15 min in a series of mouse brains extracted at different times (15, 30, 60, and 120 min).

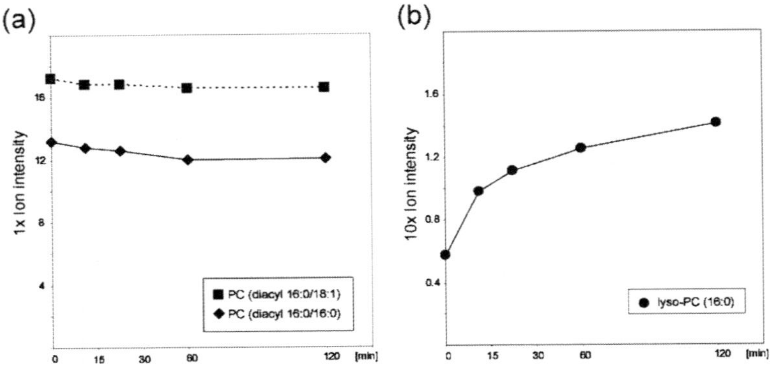

Fig. 10.11. Postmortem degradation of phosphatidylcholines and increase in lysophosphatidylcholines. An IMS series was performed on mouse brains extracted after different amounts of time had elapsed after sacrifice (within 1, 15, 30, 60, and 120 min after sacrifice). After IMS, the ion intensities of the PCs were averaged over the entire section. As postmortem events, degradation of PCs and an increase in lyso-PCs were observed within 15 min, presumably because of the stimulation of phospholipase A under ischemic conditions (37, 38). In this study, mouse brains were extracted within 1 min (typically within 40 s) after sacrifice. Reprinted from Ref. (11).

This is presumably because of stimulation of phospholipase enzymes under ischemic conditions (37, 38).

4. In general, embedded tissues are cut into thin slices. Embedding enables the sample to retain its shape and makes the cutting process easier. However, in IMS experiments, attachment and penetration of the embedding agents (e.g., OTCs) in the sample lead to a deterioration of the MS signal (4, 14). In particular, when analyzing small molecules with an *m/z* of 800–2,000, contamination with OTC leads to the presence of extremely high polymer peaks in the mass spectra of positive ions. This virtually hides all of the smaller peaks (**Fig. 10.12**). For this reason, when preparing sections for IMS, OTC is used only to "support" the tissue blocks, and thus, OTC does not directly attach to the tissue being analyzed. As an alternative, a precooled semiliquid gel of 2% sodium carboxymethyl cellulose (CMC) can be used as an embedding compound that does not interfere with mass spectrometry (39).
5. Dehydration of tissue sections for long times can lead to altered signals (40). Goodwin and colleagues demonstrated that, even within 1 min, signals were altered, both increasing and decreasing. Therefore, tissue slices should be moved to the next step (matrix application) as quickly as possible. Considerable care is required at these stages in order to facilitate a comparison between the biomarkers in independent IMS experiments.

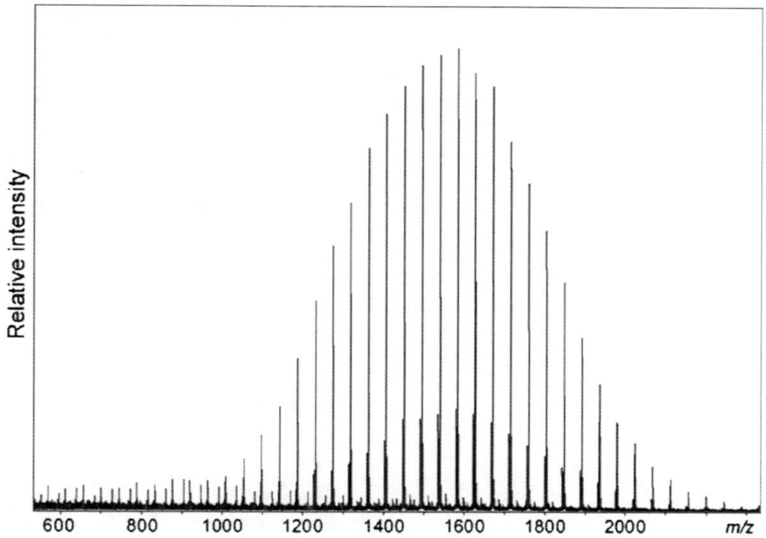

Fig. 10.12. Mass spectra of small molecules were obtained from a representative section that was completely embedded in optimal cutting temperature compound. Reprinted from Ref. (11).

6. The matrix–analyte crystallization process, particularly when using salt-added matrix solution to reduce the molecular ion forms of GPLs, leads to the development of heterogeneous crystals which in turn results in spot-to-spot variance of signal intensities (41, 42). This problem was solved using a spectrum normalization procedure with TIC (**Fig. 10.13**).

We performed spectrum normalization with TIC; the obtained spectra were multiplied with arbitrary variables such that all spectra had equal TIC values (i.e., equal integral values of the measured *m/z* region [400 < *m/z* < 900]). Such TIC normalization is available with the "normalize spectra" function of FlexImaging 2.0 software with filter function to exclude a number of noise spectra from the normalization process (see details in the software manual).

To evaluate the effect of the normalization procedure, we prepared a section of mouse brain homogenate that had a uniform distribution of biomolecules. **Figure 10.13a** shows the ion images for *m/z* 772.6 corresponding to PC (diacyl-16:0/16:0), with and without spectrum normalization. After the normalization procedure, the image was corrected such that the ion distribution was uniform throughout the section. The signal intensity was then plotted and found to have a Gaussian distribution. Spectrum normalization with TIC improved the results of the IMS of mouse brain sections. **Figure 10.13b** shows the ion images of a mouse brain section for PC (diacyl-16:0/16:0),

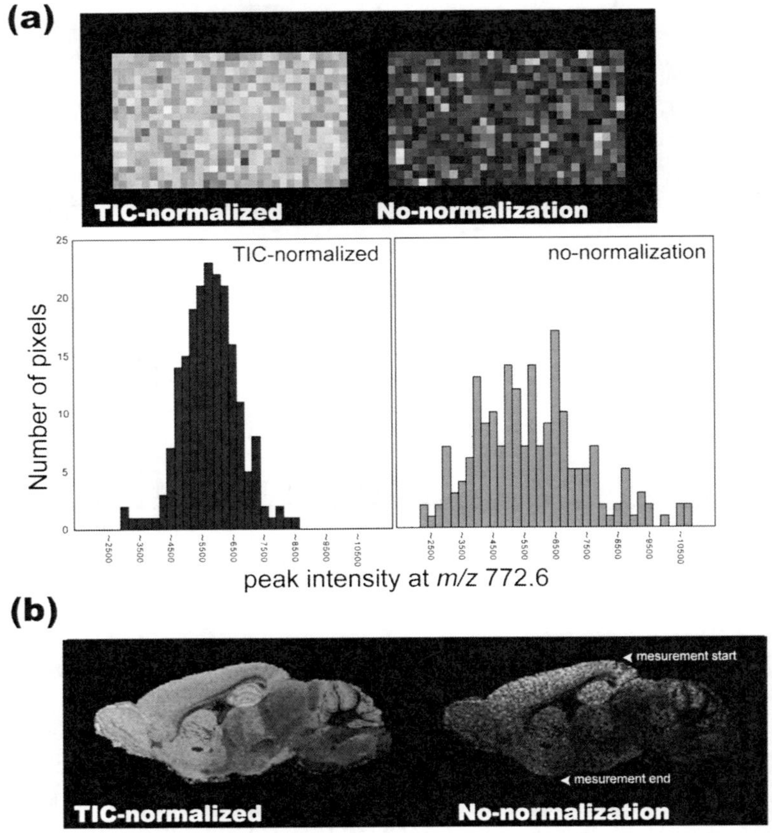

Fig. 10.13. Spectrum normalization using TIC improves both the quantitative ability and the visualization quality of IMS. (**a**) IMS results for PC (diacyl-16:0/16:0) on a section of mouse brain homogenate, processed with/without TIC normalization (*upper panel*), and plot of ion-intensity distribution for PC (diacyl-16:0/16:0) obtained from a brain homogenate section, with/without TIC normalization (*lower panel*). (**b**) Ion images of PC (diacyl-16:0/16:0) on an adult mouse brain section, in which spectra were processed with/without TIC normalization. Reprinted from Ref. (11).

with and without spectrum normalization. In the ion image without normalization, the ion distribution was heterogeneous, even between adjacent pixels. Furthermore, the signal intensity was found to decrease with time (arrowhead). In contrast, when the normalization procedure was used, a clear ion-distribution pattern that correlated well with the anatomical features of the brain section was obtained (11).

Acknowledgments

The authors would like to thank Dr. S. Taira and Dr. H. Ageta for their advice and fruitful discussion. This work was supported by the SENTAN program of the Japan Science and Technology Agency.

References

1. Garrett, T. J., Prieto-Conaway, M. C., Kovtoun, V., Bui, H., Izgarian, N., Stafford, G. Yost, R. A. (2006). Imaging of small molecules in tissue sections with a new intermediate-pressure MALDI linear ion trap mass spectrometer. *Int J Mass* Spectrom, **260**, 11.
2. Khatib-Shahidi, S., Andersson, M., Herman, J. L., Gillespie, T. A., Caprioli, R. M. (2006). Direct molecular analysis of whole-body animal tissue sections by imaging MALDI mass spectrometry. *Anal Chem*, **78**, 6448–6456.
3. Stoeckli, M., Staab, D., Staufenbiel, M., Wiederhold, K. H., Signor, L. (2002). Molecular imaging of amyloid beta peptides in mouse brain sections using mass spectrometry. *Anal Biochem*, **311**, 33–39.
4. Chaurand, P., Norris, J. L., Cornett, D. S., Mobley, J. A., Caprioli, R. M. (2006). New developments in profiling and imaging of proteins from tissue sections by MALDI mass spectrometry. *J Proteome Res*, **5**, 2889–2900.
5. Shimma, S., Sugiura, Y., Hayasaka, T., Zaima, N., Matsumoto, M., Setou, M. (2008). Mass imaging and identification of biomolecules with MALDI-QIT-TOF-based system. *Anal Chem*, **80**, 875–885.
6. Ostrowski, S. G., Van Bell, C. T., Winograd, N., Ewing, A. G. (2004). Mass spectrometric imaging of highly curved membranes during Tetrahymena mating. *Science*, **305**, 71–73.
7. Monroe, E. B., Jurchen, J. C., Lee, J., Rubakhin, S. S., Sweedler, J. V. (2005). Vitamin E imaging and localization in the neuronal membrane. *J Am Chem* Soc, **127**, 12152–12153.
8. Kraft, M. L., Weber, P. K., Longo, M. L., Hutcheon, I. D., Boxer, S. G. (2006). Phase separation of lipid membranes analyzed with high-resolution secondary ion mass spectrometry. *Science*, **313**, 1948–1951.
9. Taira, S., Sugiura, Y., Moritake, S., Shimma, S., Ichiyanagi, Y., Setou, M. (2008). Nanoparticle-assisted laser desorption/ionization based mass imaging with cellular resolution. *Anal Chem*, **80**, 4761–4766.
10. Moritake, S., Taira, S., Sugiura, Y., Setou, M., Ichiyanagi, Y. (2009). Magnetic nanoparticle-based mass spectrometry for the detection of biomolecules in cultured cells. *J Nanosci Nanotechnol*, **9**, 169–176.
11. Sugiura, Y., Konishi, Y., Zaima, N., Kajihara, S., Nakanishi, H., Taguchi, R., Setou, M. (2009). Visualization of the cell-selective distribution of PUFA-containing phosphatidylcholines in mouse brain by imaging mass spectrometry. *J Lipid Res*, **50**, 1776–1788
12. Hayasaka, T., Goto-Inoue, N., Sugiura, Y., Zaima, N., Nakanishi, H., Ohishi, K., Nakanishi, S., Naito, T., Taguchi, R., Setou, M. (2008). Matrix-assisted laser desorption/ionization quadrupole ion trap time-of-flight (MALDI-QIT-TOF)-based imaging mass spectrometry reveals a layered distribution of phospholipid molecular species in the mouse retina. *Rapid Commun Mass* Spectrom, **22**, 3415–3426.
13. Sugiura, Y., Shimma, S., Setou, M. (2006). Thin sectioning improves the peak intensity and signal-to-noise ratio in direct tissue mass spectrometry. *J. Mass Spectrom Soc Jpn*, **54**, 4.
14. Schwartz, S. A., Reyzer, M. L., Caprioli, R. M. (2003). Direct tissue analysis using matrix-assisted laser desorption/ionization mass spectrometry: practical aspects of sample preparation. *J Mass Spectrom*, **38**, 699–708.
15. Pulfer, M., Murphy, R. C. (2003). Electrospray mass spectrometry of phospholipids. *Mass Spectrom Rev*, **22**, 332–364.
16. Stubiger, G., Pittenauer, E., Allmaier, G. (2008). MALDI seamless postsource decay fragment ion analysis of sodiated and lithiated phospholipids. *Anal Chem*, **80**, 1664–1678.
17. Touboul, D., Piednoel, H., Voisin, V., De La Porte, S., Brunelle, A., Halgand, F., Laprevote, O. (2004). Changes of phospholipid composition within the dystrophic muscle by matrix-assisted laser desorption/ionization mass spectrometry and mass spectrometry imaging. *Eur J Mass Spectrom (Chichester, Eng)*, **10**, 657–664.
18. van Echten, G., Sandhoff, K. (1993). Ganglioside metabolism. enzymology, topology, and regulation. *J Biol Chem*, **268**, 5341–5344.
19. Sonnino, S., Chigorno, V. (2000). Ganglioside molecular species containing C18- and C20-sphingosine in mammalian nervous tissues and neuronal cell cultures. *Biochim Biophys Acta*, **1469**, 63–77.
20. Sambasivarao, K., McCluer, R. H. (1964). Lipid components of gangliosides. *J Lipid* Res, **15**, 103–108.
21. Schwarz, H. P., Kostyk, I., Marmolejo, A., Sarappa, C. (1967). Long-chain bases of brain and spinal cord of rabbits. *J Neurochem*, **14**, 91–97.
22. Jungalwala, F. B., Hayssen, V., Pasquini, J. M., McCluer, R. H. (1979). Separation of molecular species of sphingomyelin

by reversed-phase high-performance liquid chromatography. *J Lipid Res*, **20**, 579–587.

23. Palestini, P., Sonnino, S., Tettamanti, G. (1991). Lack of the ganglioside molecular species containing the C20-long-chain bases in human, rat, mouse, rabbit, cat, dog, and chicken brains during prenatal life. *J Neurochem*, **56**, 2048–2050.
24. Palestini, P., Masserini, M., Sonnino, S., Giuliani, A., Tettamanti, G. (1990). Changes in the ceramide composition of rat forebrain gangliosides with age. *J Neurochem*, **54**, 230–235.
25. Mansson, J. E., Vanier, M. T., Svennerholm, L. (1978). Changes in the fatty acid and sphingosine composition of the major gangliosides of human brain with age. *J Neurochem*, **30**, 273–275.
26. Kotani, M., Kawashima, I., Ozawa, H., Terashima, T., Tai, T. (1993). Differential distribution of major gangliosides in rat central nervous system detected by specific monoclonal antibodies. *Glycobiology*, **3**, 137–146.
27. Jackson, S. N., Wang, H. Y., Woods, A. S. (2005). Direct profiling of lipid distribution in brain tissue using MALDI-TOFMS. *Anal Chem*, 77, 4523–4527.
28. Chen, Y., Allegood, J., Liu, Y., Wang, E., Cachon-Gonzalez, B., Cox, T. M., Merrill, A. H., Jr., Sullards, M. C. (2008). Imaging MALDI mass spectrometry using an oscillating capillary nebulizer matrix coating system and its application to analysis of lipids in brain from a mouse model of Tay-Sachs/Sandhoff disease. *Anal Chem*, **80**, 2780–2788.
29. Tanaka, K., Ido, Y., Akita, S., Yoshida, Y., Yoshida, T. (1987). *Proceedings of the 2nd Japan–China Joint Symposium on Mass spectrometry* , Osaka, Japan, 185–187.
30. McLean, J. A., Stumpo, K. A., Russell, D. H. (2005). Size-selected (2–10 nm) gold nanoparticles for matrix assisted laser desorption ionization of peptides. *J Am Chem Soc*, **127**, 5304–5305.
31. Jackson, S. N., Ugarov, M., Egan, T., Post, J. D., Langlais, D., Albert Schultz, J., Woods, A. S. (2007). MALDI-ion mobility-TOFMS imaging of lipids in rat brain tissue. *J Mass Spectrom*, **42**, 1093–1098.
32. Ageta, H., Asai, S., Sugiura, Y., Goto-Inoue, N., Zaima, N., Setou, M. (2009). Layer-specific sulfatide localization in rat hippocampus middle molecular layer is revealed by nanoparticle-assisted laser desorption/ionization imaging mass spectrometry. *Med Mol Morphol*, **42**, 16–23.
33. Ageta, H., Asai, S., Sugiura, Y., Goto-Inoue, N., Zaima, N., Setou, M. (2009). Layer-specific sulfatide localization in rat hippocampus middle molecular layer is revealed by nanoparticle-assisted laser desorption/ionization imaging mass spectrometry. *Med Mol Morphol*, **42**, 16–23.
34. Jackson, S. N., Wang, H. Y., Woods, A. S. (2005). *In situ* structural characterization of phosphatidylcholines in brain tissue using MALDI-MS/MS. *J Am Soc Mass Spectrom*, **16**, 2052–2056.
35. Han, X., Gross, R. W. (2001). Quantitative analysis and molecular species fingerprinting of triacylglyceride molecular species directly from lipid extracts of biological samples by electrospray ionization tandem mass spectrometry. *Anal Biochem*, **295**, 88–100.
36. Hsu, F. F., Turk, J. (2001). Structural determination of glycosphingolipids as lithiated adducts by electrospray ionization mass spectrometry using low-energy collisional-activated dissociation on a triple stage quadrupole instrument. *J Am Soc Mass Spectrom*, **12**, 61–79.
37. Umemura, A., Mabe, H., Nagai, H., Sugino, F. (1992). Action of phospholipases A2 and C on free fatty acid release during complete ischemia in rat neocortex. Effect of phospholipase C inhibitor and *N*-methyl-D-aspartate antagonist. *J Neurosurg*, **76**, 648–651.
38. Rehncrona, S., Westerberg, E., Akesson, B., Siesjo, B. K. (1982). Brain cortical fatty acids and phospholipids during and following complete and severe incomplete ischemia. *J Neurochem*, **38**, 84–93.
39. Stoeckli, M., Staab, D., Schweitzer, A. (2006). Compound and metabolite distribution measured by MALDI mass spectrometric imaging in whole-body tissue sections. *Int J Mass* Spectrom, **260**, 195–202.
40. Goodwin, R. J., Dungworth, J. C., Cobb, S. R., Pitt, A. R. (2008). Time-dependent evolution of tissue markers by MALDI-MS imaging. *Proteomics*, **8**, 3801–3808.
41. Sugiura, Y., Shimma, S., Setou, M. (2006). Two-step matrix application technique to improve ionization efficiency for matrix-assisted laser desorption/ionization in imaging mass spectrometry. *Anal Chem*, **78**, 8227–8235.
42. Mock, K. K., Sutton, C. W., Cottrell, J. S. (1992). Sample immobilization protocols for matrix-assisted laser-desorption mass spectrometry. *Rapid Commun Mass Spectrom*, **6**, 233–238.
43. Astigarraga, E., Barreda-Gomez, G., Lombardero, L., Fresnedo, O., Castano, F., Giralt, M. T., Ochoa, B., Rodriguez-Puertas, R., Fernandez, J. A. (2008). Profiling and imaging of lipids on brain and

liver tissue by matrix-assisted laser desorption/ionization mass spectrometry using 2-mercaptobenzothiazole as a matrix. *Anal Chem*, **80**, 9105–9114.
44. Jackson, S. N., Wang, H. Y., Woods, A. S. (2007). *In situ* structural characterization of glycerophospholipids and sulfatides in brain tissue using MALDI-MS/MS. *J Am Soc Mass Spectrom*, **18**, 17–26.
45. Wang, H. Y., Jackson, S. N., Woods, A. S. (2007). Direct MALDI-MS analysis of cardiolipin from rat organs sections. *J Am Soc Mass Spectrom*, **18**, 567–577.
46. Altelaar, A. F., Klinkert, I., Jalink, K., de Lange, R. P., Adan, R. A., Heeren, R. M., Piersma, S. R. (2006). Gold-enhanced biomolecular surface imaging of cells and tissue by SIMS and MALDI mass spectrometry. *Anal Chem*, **78**, 734–742.
47. Cha, S., Yeung, E. S. (2007). Colloidal graphite-assisted laser desorption/ionization mass spectrometry and MSn of small molecules. 1. Imaging of cerebrosides directly from rat brain tissue. *Anal Chem*, **79**, 2373–2385.
48. Zhang, H., Cha, S., Yeung, E. S. (2007). Colloidal graphite-assisted laser desorption/ionization MS and MS(n) of small molecules. 2. Direct profiling and MS imaging of small metabolites from fruits. *Anal Chem*, **79**, 6575–6584.

Chapter 11

Cellular Imaging Using Matrix-Enhanced and Metal-Assisted SIMS

A.F. Maarten Altelaar and Sander R. Piersma

Abstract

Imaging mass spectrometry (IMS) allows the direct investigation of both the identity and the spatial distribution of the entire molecular content directly in tissue sections, single cells, and many other biological surfaces. We describe here the steps required to retrieve the molecular information from tissue sections using matrix-enhanced (ME) and metal-assisted (MetA) secondary ion mass spectrometry (SIMS). Surface metallization by plasma coating enhances desorption/ionization of membrane components such as lipids and sterols in imaging time-of-flight (ToF) SIMS of tissues and cells. High-resolution images of cholesterol and other membrane components can be obtained for single neuroblastoma cells and reveal subcellular details. Alternatively, in ME-SIMS, 2,5-dihydroxybenzoic acid electrosprayed on neuroblastoma cells allows intact molecular ion imaging of phosphatidylcholine (PC) and sphingomyelin (SM) at the cellular level.

Key words: Imaging mass spectrometry, lipids, neuroblastoma cells, gold coating.

1. Introduction

In SIMS (1) the sample surface is bombarded with a high-energy primary ion beam between 1 and 25 kiloelectronvolts (keV). Typical primary ions used in SIMS include Ga^+, Cs^+, and In^+, with Ga^+ being able to provide the smallest probe size (less than 10 nm). Although SIMS does not routinely yield intact protein and peptide signals directly from tissue sections, it does have several advantages compared to MALDI. The most important advantage of SIMS over MALDI is the chemical imaging capabilities routinely delivering submicron spatial resolution (2). Furthermore, the SIMS technique is very sensitive and remarkably

S.S. Rubakhin, J.V. Sweedler (eds.), *Mass Spectrometry Imaging*, Methods in Molecular Biology 656,
DOI 10.1007/978-1-60761-746-4_11, © Springer Science+Business Media, LLC 2010

versatile since it can analyze almost any kind of solid surface (3). The bombardment of these solid surfaces with high-energy primary ions will induce damage over a certain depth in the sample, resulting in changes in the molecular structure of the constituents in this area. To prevent imaging of the induced damage, SIMS imaging experiments are conducted in two different regimes. In the dynamic regime the entire sample surface is eroded in time and the complete top monolayer is removed. Dynamic SIMS is primarily used in quantitative elemental imaging (4–6) (not the topic of this chapter). In the static SIMS regime a much lower primary ion dose is used compared to dynamic SIMS resulting in less than 1% of the top surface atoms and molecules to interact with the primary ion beam and no primary ion strikes again the damaged region. Consequently, in static SIMS significantly less fragmentation of the molecular content occurs, which allows the technique to be used in imaging of small organic components (7–9). In order to enhance the ionization yield for large intact molecular ions by SIMS, different kinds of surface modifications (MALDI matrices (10–14), silver (15), and gold (16–20)) as well as the use of polyatomic primary ion beams (21–26) have been suggested. Although these methods have shown to be able to desorb and ionize peptide and proteins from model samples, in direct tissue analysis they are highly biased toward lipids and steroids. One explanation for this phenomenon is the surface sensitivity of the technique. Since with SIMS only the top few monolayers are sampled, the technique favors the ionization of compounds with surface propensity like cholesterol and lipids, which are highly abundant in tissue sections.

In this chapter surface modifications in SIMS, such as metal-assisted (MetA) and matrix-enhanced (ME) SIMS, are described for the ionization of intact biomolecular ions, increasing the applicability of SIMS to real-world biological problems. In MetA-SIMS a very thin layer (~1 nm) of a metal (e.g., gold) is deposited on the sample surface to assist in the desorption/ionization process (3, 17, 18). One crucial factor in this method seems the migration of the analytes on the gold surface. In a recent study we have shown that SIMS signals for both the cholesterol and the lipid phosphatidylcholine (PC) increased when these species were deposited on a thin layer of gold. Increased signals for cholesterol were exclusively obtained when the layer of gold was deposited on top of the cholesterol and PC sample (27). The same effect was observed in direct MetA-SIMS tissue (27) analysis as well as in a MetA-SIMS study of dyes by Adriaensen et al. (16).

Using ME-SIMS we have demonstrated the possibility of obtaining peptide signals, from a nervous tissue extract from the pond snail *Lymnaea stagnalis*, identical to those obtained with MALDI-MS, up to *m/z* 2,590 (10). The analysis readily identified five known peptides with ME-SIMS using 2,5-dihydroxybenzoic acid as matrix. In ME-SIMS one is not dependent on the migration of the analyte molecules onto the matrix-covered surface, like in MetA-SIMS. The matrix is deposited in a wet environment, and the analyte molecules are extracted from the tissue surface. Still the biases for species with surface propensity remain since during drying these species are pushed to the outside of the forming matrix crystals. Furthermore, sample preparation (described in the protocol) is more stringent since the size of the matrix crystals determines the obtainable spatial resolution.

2. Materials

2.1. Chemicals

1. α-Cyano-4-hydroxycinnamic acid.
2. 2,5-Dihydroxybenzoic acid.
3. Trifluoracetic acid.
4. HPLC-grade water.
5. Ethanol.
6. Gelatine.
7. Liquid isopentane.
8. Dry ice.

2.2. Instruments and Materials

1. Physical Electronics (Eden Prairie, MN, USA) TRIFTII time-of-flight SIMS (ToF-SIMS) equipped with an $^{115}In^{+}$ liquid metal ion gun.
2. Dissection microscope allowing low magnification (10×/40×) (CETI, Antwerpen, Belgium).
3. Cryomicrotome; Leica CM 3000 cryostat (Leica Microsystems, Nussloch, Germany).
4. −80°C freezer.
5. Desiccator containing a silica gel canister.
6. Optical microscope (Leica DMRX) equipped with a digital camera (Nikon DXM1200).
7. Chromatography sprayer, 10 ml (Sigma-Aldrich; Z529710).
8. 9 mm screw top vials, 12×32 mm (Waters, P/N 186000272).
9. Quorum Technologies (Newhaven, East Sussex, UK) SC7640 sputter coater equipped with a gold target (SC510–314A), a FT7607 quartz crystal microbalance stage, and a FT7690 film thickness monitor.

10. Conductive glass slides; 25×50×1.1 mm unpolished float glass, SiO_2 passivated/indium-tin-oxide coated, Rs = 6±2 Ω (Delta Technologies; CG-40IN-1115).
11. Male Wistar rats (Crl:WU) weighing 350 g (Charles-River, Maastricht, The Netherlands).
12. Freshwater snails *Lymnaea stagnalis*, raised under laboratory conditions.

2.3. Software

1. Both TRIFT systems are operated by WinCadence software (version 3.7.1.5) and controlled by the vacuum watcher (Physical Electronics, Watcher 2.1.2.140).
2. AcqirisLive 2.11 controls the acqiris settings and data acquisition.
3. LaVision, DaVis 6.2.3. controls the CCD camera settings and data acquisition.
4. Mass spectral data analysis is performed with WinCadence 3.7.1.5, MatLab 7.0.4 (PCA), AWE3D 1.5.2.0, and tofToCsv (tool to convert entire *m/z* data file to a comma-separated file (csv)).
5. Image data analysis is performed with WinCadence 3.7.1.5, MatLab 7.0.4 (PCA) and PCA-based in-house-developed algorithms.

3. Method

The described protocol for IMS allows the mapping of molecular distributions directly in tissue sections and on cell surfaces. The different approaches to SIMS are able to deliver images of distributions of small organic compounds like lipids and steroids. Since ME-SIMS and especially MetA-SIMS are capable of imaging biological surfaces with very high spatial resolution, these techniques could be very well suited for direct analysis of lipid distributions on single-cell surfaces.

3.1. Tissue Sections

All experimental procedures should be in accordance with the European directives (86/609/EEC) and approved by the Commission on Laboratory Animal Experiments.

Male Wistar rats (Crl:WU) weighing 350 g are decapitated without prior anesthesia, and brains are dissected and frozen in liquid isopentane, cooled to −50°C on dry ice, and then stored at −80°C until sectioning.

Freshwater snails (*Lymnaea stagnalis*) raised under laboratory conditions; 20±1°C water temperature, 12 h light/12 h dark cycle, and fed lettuce ad libitum. Adult specimens are decapitated, and the cerebral ganglia are dissected, directly embedded in 10% gelatine at 30°C, and frozen at −80°C. Gelatine embedding allows cryostat sectioning down to 5 μm thickness and does not interfere with ME-SIMS analysis.

Sections are thaw-mounted on indium-tin-oxide (ITO, 4–8 Ω resistance)-coated glass slides and are stored at −80°C until use. Conservation of morphology was checked by optical microscopy during the sectioning, drying, and storage process. No obvious ice crystal damage was observed after freezing at −80°C in gelatine. Light microscopy images were acquired using a microscope with a digital camera. Prior to mass spectrometry, frozen tissue sections were brought to room temperature in a desiccator over a silica gel canister.

1. Freeze tissue directly in liquid isopentane after dissection, cool on dry ice, and store at −80°C until use. Postmortem changes are known to occur in the proteome of susceptible peptides and proteins within minutes. To prevent these alterations of the sample, the dissected tissue must be snap-frozen and defrosting of the sample should be done only just before sample preparation starts (*see* **Note 1**).
2. Place the dissected tissue on the sample holder in the cryostat. Tissue section can be attached to the sample holder by a very small amount of Tissue-Tek. Great care has to be taken that the Tissue-Tek does not come into contact with the tissue area of interest (*see* **Note 2**).
3. For very small organs (like the pituitary gland or a mollusk central nervous system) embedding in non-polymer containing solutions like 10% gelatine helps to prevent damaging of the tissue and assist in cutting. Tissue is embedded in 10% gelatine at 30°C directly after dissection and frozen at −80°C. Gelatine embedding allows sectioning down to 5 μm thickness results in no tissue damage during freezing and is compatible with MS.
4. Cut tissue section ~10-μm thick using a cryomicrotome at −20°C. 10-μm thickness is optimal for IMS; enough analyte molecules available for extraction and no problems with conductivity are observed (*see* **Note 3**).
5. Pick up the tissue sections with microscope glass slides by thaw-mounting, place them in a closed container (e.g., plastic Petri dishes sealed with parafilm) on dry ice, and store at −80°C until use.
6. Take a tissue section in its closed container from the −80°C storage and allow it to come to room temperature in a dry box with silica gel canisters before analysis.

3.2. Cell Cultures

1. Cells are grown directly on conductive ITO-coated glass slides by seeding (neuroblastoma) cells in six-well plates at ~25,000 cells/well and cultured in 3 ml of DMEM supplemented with 10% serum and antibiotics.
2. The cells were washed in 300 mM sucrose solution (to prevent hypotonic shock).
3. Next the sucrose is removed by washing with Milli-Q water, after which the cells are rapidly frozen on dry ice.
4. The snap-frozen cells are freeze-dried for 30 min, to remove any ice crystals, and stored at −80°C until use.
5. Prior to mass spectrometry, the cell cultures are brought to room temperature in a desiccator over a silica gel canister (1 h).
6. Conservation of cell morphology is checked by optical microscopy, using a Leica DMRX microscope with a Nikon DXM1200 digital camera.

3.3. Matrix Deposition for ME-SIMS

The matrix deposition method of choice depends on the spatial resolution required. For high spatial resolution IMS, electrospray deposition (ESD) is preferred since it results in considerably smaller crystal sizes than pneumatically assisted nebulization (e.g., by a TLC sprayer). Key issues in development of a matrix deposition method are optimal incorporation of analyte into the matrix crystals and minimal lateral diffusion. These two requirements can be met if the matrix arrives at the tissue surface in very small droplets before all solvent has evaporated.

1. For ESD, a syringe pump pumps matrix solution (15 mg/ml 2,5-DHB in 50% MeOH/0.1% TFA (V/V)) from a gastight syringe through a stainless steel electrospray capillary (o.d. 220 μm, i.d. 100 μm) maintained at 3–5 kV (*see* **Note 4**).
2. The capillary is mounted on an electrically insulated manual translation stage in a vertical orientation. The stage is fitted with a digital micrometer for accurate positioning of the needle tip with respect to the grounded sample plate.
3. The sample plate is mounted on an X–Y moveable table.
4. Matrix deposition was performed by spraying for 10 min at a flow rate of 12 μl/h, a voltage of 4.7 kV, and a needle to sample plate distance of 5.0 mm.
5. For ME-SIMS analysis, samples are covered with a thin layer of matrix by electrospray deposition yielding small (0.3–1 μm) matrix crystals (*see* **Note 5**).
6. After matrix deposition the matrix coverage is checked using an optical microscope and the tissue sections are left to dry for 30 min.

3.4. Gold Coating for MetA-SIMS

1. Place the ITO slide with the sample in the Quorum Technologies SC7640 sputter coater. Press the start sequence button to pump down, flush with argon, further pump down the vacuum chamber, and leak in argon until the pressure reaches 0.1 mbar (all done automatically).
2. Enter on the film thickness monitor the density of the metal used (19.30 g/cm^3 for gold) and the desired thickness of the sputtered metal layer.
3. Coat the tissue sections with 1 nm of gold directly on the tissue.
4. Put the discharge voltage on 1 kV and press start on the film thickness monitor. Adjust the plasma current to 25 mA for homogenous coverage.

IMS experiments can be performed using ME-SIMS or MetA-SIMS. In the cases described here, these experiments are conducted on a ToF-SIMS equipped with an $^{115}In^+$ liquid metal ion gun. The experimental procedure for ME-SIMS and MetA-SIMS data acquisition is identical.

3.5. ME-SIMS and MetA-SIMS Imaging Mass Spectrometry

All static SIMS experiments were performed on a time-of-flight SIMS (ToF-SIMS) equipped with an $^{115}In^+$ liquid metal ion gun. In short, the TRIFT is a stigmatic imaging ToF analyzer incorporating a 2-m flight path and three quasi-hemispherical electrostatic sector analyzers (ESAs) as an integral part of the ToF analyzer. These ESAs compensate for the differences in ion flight times due to variations in the secondary ion's initial kinetic energies and ion emission angles. The secondary ions were extracted through a 3.2-kV electric field into the ToF analyzer and post-accelerated by an additional 5 kV prior to detection on a dual multichannel plate/phosphor screen detector. A multistop time-to-digital converter with 138-ps time resolution was used to acquire the detector signals. All experiments were performed with a primary ion beam current of 450 pA, a primary pulse length of 30 ns, a spot diameter of 500 nm, and a primary ion energy of 15 kV. The experimental conditions were chosen in such a way that all the analyses were conducted in the static SIMS regime. For each chemical image, the primary beam was rastered over a 150 × 150 μm sample area, divided into 256 × 256 square pixels. The positive ion mode mass spectra and chemical images were taken from the same 150 × 150 μm areas within the sample. For each ion detected, the mass (ToF) and primary ion beam position were recorded, allowing post-processing of the data. The chemical images were compared to optical images of the corresponding sample areas.

1. Before conducting SIMS imaging experiments optimize the setup for image quality, using a copper grid with a 25 μm

repeat. In the vacuum watcher close spectro gate valve (V5). In WinCadance software go to hardware, start the DC beam, and raise the gain of the electron multiplier until the copper grid becomes visible on the secondary electron detector (SED). Select lens 1 and wobble. Use the multiple variable aperture (MVA) on the side of the instrument to improve the image quality. When best result is achieved stop wobble, select lens 2, and start wobble again. Adjust beam steering (x and y) to improve the image quality when needed. At optimal image quality stop wobble, select blanker, and start wobble again. This time adjust lens 2 to fix the image and lens 1 to refocus. After refocusing the procedure is repeated until a clear fixed and focused image of the copper grid can be seen on the SED.

2. In hardware make sure that there is no voltage on the bunching parameter (perform the imaging measurements in unbunched mode for optimal image quality). Before the start of the experiment select under acquisition setup/advanced settings: save as raw file, in order to be able to post-process the raw data after the measurement is completed.
3. Perform the ME-SIMS experiment in such a way that the analysis is conducted in the static SIMS regime. This can be achieved with a primary ion beam current of ~450 pA, a primary pulse length of 30 ns, a spot diameter of 500 nm, and a primary ion energy of 15 kV. At 3 min per experiment this results in a primary ion dose of 4.9×10^{11} ions/cm^2 (*see* **Note 6**).
4. For each chemical image, the primary beam is rastered over a 150×150 μm sample area, divided into 256×256 square pixels (larger or smaller areas can also be chosen). To image a significant larger surface, like a whole tissue section, analyze multiple 150×150 μm areas by stepping the sample stage in a mosaic pattern. To compensate for small deviations on the sample stage positioning a 10 μm overlap with the previous acquired sample is taken (the sample stage is moved by 140 μm).

Each individual experiment is saved as .raw file to allow post-processing of the data. In "spectra" choose specific m/z ranges and select "image" for each range. Now in "acquisition","setup/advanced settings" select "acquire from raw file" and under the tab "image" the selected distributions can be seen. For large tissue sections two approaches of image stitching are available. First 150 μm^2 images are stitched together manually in image handling software like Adobe Photoshop. Second a feature-based image alignment algorithm using PCA was developed for the visualization of the imaging data (28).

3.6. Discussion and Conclusions

Following the procedures discussed in **Section 3** images were acquired as shown in **Fig. 11.1**. In panel A ME-SIMS images of *lymnaea Stagnalis* brain tissue are shown and in panel B MetA-SIMS images of single neuroblastoma cells are shown. The plasma coating procedure in MetA SIMS enhances desorption/ionization of membrane components such as lipids and sterols in SIMS imaging of tissues and cells. As shown

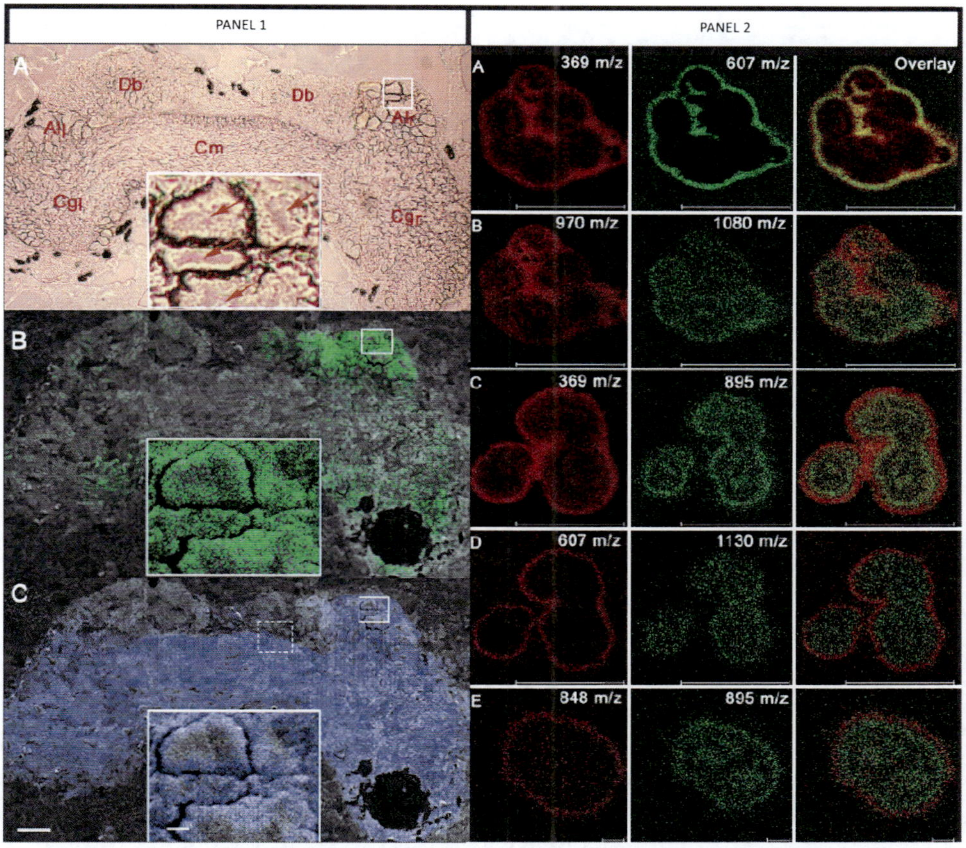

Fig. 11.1. (Panel 1) Direct molecular imaging of *Lymnaea stagnalis* nervous tissue by ME-SIMS. (**a**) Optical image of the *Lymnaea* cerebral ganglia, inset shows high magnification image of neurons in the anterior lobe (*solid box*), *arrows* indicate nuclei. Different regions in the section are right and left cerebral ganglia (Cgr and Cgl), anterior lobe (Al), commissure (Cm), and dorsal bodies (Db). (**b**) ME-SIMS image of APGWamide (429.0–433.2 *m/z*; *green*, 0–3 counts) distribution. (**c**) ME-SIMS image of cholesterol (368.2–371.3 *m/z*; *blue*, 0–4 counts) distribution. Scale bar: 200 μm; scale bar inset: 10 μm. Molecular images (**b** and **c**) are presented as colored overlays on top of the gray-scale TIC (total ion count) image (mass range: 1.0–5,000 *m/z*, 0–140 counts). The ME-SIMS measurements used indium primary ions (total ion dose 4.9×10^{11} ions/cm^2). Reprinted from Ref. (10). (Panel 2) Cellular localization of MetA-SIMS selected ion count signals from neuroblastoma cells. Cells were imaged after deposition of 1-nm gold. (**a**) *m/z* 369 (cholesterol $[M-OH]^+$, 0–14 counts) and 607 (DAG, 0–6 counts). (**b**) *m/z* 970 (cholesterol $[2\ M+Au]^+$, 0–4 counts) and 1,080 (0–1 counts). (**c**) *m/z* 369 (cholesterol $[M-OH]^+$, 0–12 counts) and 895 (0–2 counts). (**d**) *m/z* 607 (DAG, 0–4 counts) and 1,130 (0–1 counts). (**e**) *m/z* 848 (0–1 counts) and 895 (0–1 counts). Scale bar: 100 μm in panels (**a–d**) and 10 μm in panel E. Total ion dose 1.18×10^{12} ions/cm^2 in panels (**a–d**) and 4.7×10^{12} ions/cm^2 in panel (**e**). Reprinted from Ref. (27).

in **Fig. 11.1** High-resolution images of cholesterol and other membrane components can be obtained for single neuroblastoma cells and reveal subcellular details. Alternatively, in ME-SIMS, 2,5-dihydroxybenzoic acid electrosprayed on neuroblastoma cells and tissues allows intact molecular ion imaging of phosphatidylcholine (PC) and sphingomyelin (SM) at the cellular level.

4. Notes

1. Recent observations by Svensson et al. (29) point toward post-mortem changes in the proteome of susceptible peptides and proteins within minutes. To prevent these alterations of the sample Svensson et al. use focused microwave irradiation to sacrifice the animals. Since focused microwave irradiation is not available in every laboratory, an alternative way to minimizing alterations is snap-freezing of the dissected tissue and defrosting only just before sample preparation starts.
2. Be very careful not to make any contact between the Tissue-Tek and parts of the tissue used for sectioning. The Tissue-Tek contains a large amount of polymer substance, which smears over the tissue surface upon cutting. These polymers will dominate the resulting mass spectra.
3. After defrosting of the samples the degenerative processes in the sample continue by the reactivations of multiple proteases. For this reason the defrosting should be done just before sample preparation starts and tissues should not be left untreated for a sustained period of time. Furthermore, sample preparation should be done delicate but fast to prevent further degeneration and allow intact molecular species to be incorporated in the MALDI matrix.
4. The ESD needle can get clocked when too high matrix concentrations are used. Adjust the concentration of the matrix solution if needed.
5. Since in SIMS imaging experiments only the top layer is sampled, the matrix layer should not be too thick and the analytes have to be able to migrate to the top surface layer. This combination is critical for a successful ME-SIMS experiment and therefore the ESD conditions have to be optimized carefully, in order to produce a fine mist of matrix droplets so the matrix arrives onto the sample surface in a wet environment.

6. In static SIMS the same area is never sampled twice in order to prevent imaging of induced damage. To do so only 1% of the sample surface is analyzed, which converts to a primary ion dose of 10^{13} ions/cm^2.

References

1. Vickerman, J. C., Briggs, D. (Eds.) (2001) *ToF-SIMS: Surface Analysis by Mass Spectrometry*, IM Publications and SurfaceSpectra Limited, Chichester.
2. Todd, P. J., McMahon, J. M., Short, R. T., McCandlish, C. A. (1997) Organic SIMS of biologic tissue. *Anal Chem*, **69**, 529A–535A.
3. Delcorte, A., Garrison, B. J. (2000) High yield events of molecular emission induced by kiloelectronvolt particle bombardment. *J Phys ChemB*,**104**, 6785–6800.
4. Chandra, S., Morrison, G. H. (1995) Imaging ion and molecular-transport at subcellular resolution by secondary-ion mass-spectrometry. *Int J Mass Spectrom Ion Process*, **143**, 161–176.
5. Chandra, S., Smith, D. R., Morrison, G. H. (2000) Subcellular imaging by dynamic SIMS ion microscopy. *Anal Chem*, **72**, 104A–114A.
6. Strick, R., Strissel, P. L., Gavrilov, K., Levi-Setti, R. (2001) Cation–chromatin binding as shown by ion microscopy is essential for the structural integrity of chromosomes. *J Cell Biol*, **155**, 899–910.
7. Colliver, T. L., Brummel, C. L., Pacholski, M. L., Swanek, F. D., Ewing, A. G., Winograd, N. (1997) Atomic and molecular imaging at the single-cell level with TOF-SIMS. *Anal Chem*, **69**, 2225–2231.
8. Pacholski, M. L., Cannon, D. M., Ewing, A. G., Winograd, N. (1998) Static time-of-flight secondary ion mass spectrometry imaging of freeze-fractured, frozen-hydrated biological membranes. *Rapid Commun Mass Spectrom*, **12**, 1232.
9. Todd, P. J., Schaaff, T. G., Chaurand, P., Caprioli, R. M. (2001) Organic ion imaging of biological tissue with secondary ion mass spectrometry and matrix-assisted laser desorption/ionization. *J Mass Spectrom* **36**, 355–369.
10. Altelaar, A. F. M., van Minnen, J., Jimenez, C. R., Heeren, R. M. A., Piersma, S. R. (2005) Direct molecular imaging of *Lymnaea stagnalis* nervous tissue at subcellular spatial resolution by mass spectrometry. *Anal Chem*, 77, 735–741.
11. Luxembourg, S. L., McDonnell, L. A., Duursma, M. C., Guo, X. H., Heeren, R. M. A. (2003) Effect of local matrix crystal variations in matrix-assisted ionization techniques for mass spectrometry. *Anal Chem*, **75**, 2333–2341.
12. McDonnell, L. A., Piersma, S. R., Altelaar, A. F. M., Mize, T. H., Luxembourg, S. L., Verhaert, P., van Minnen, J., Heeren, R. M. A. (2005) Subcellular imaging mass spectrometry of brain tissue. *J Mass Spectrom*, **40**, 160–168.
13. Wittmaack, K., Szymczak, W., Hoheisel, G., Tuszynski, W. (2000) Time-of-flight secondary ion mass spectrometry of matrix-diluted oligo- and polypeptides bombarded with slow and fast projectiles: positive and negative matrix and analyte ion yields, background signals, and sample aging. *J Am Soc Mass Spectrom*, **11**, 553–563.
14. Wu, K. J., Odom, R. W. (1996) Matrix-enhanced secondary ion mass spectrometry: a method for molecular analysis of solid surfaces. *Anal Chem*, **68**, 873–882.
15. Nygren, H., Johansson, B. R., Malmberg, P. (2004) Bioimaging TOF-SIMS of tissues by gold ion bombardment of a silver-coated thin section. *Microsc Res Tech*, **65**, 282–286.
16. Adriaensen, L., Vangaever, F., Gijbels, R. (2004) Metal-assisted secondary ion mass spectrometry: influence of Ag and Au deposition on molecular ion yields. *Anal Chem*, **76**, 6777–6785.
17. Delcorte, A., Bertrand, P. (2004) *Interest of Silver and Gold Metallization for Molecular SIMS and SIMS Imaging*, 250–255, Elsevier Science, London.
18. Delcorte, A., Bour, J., Aubriet, F., Muller, J. F., Bertrand, P. (2003) Sample metallization for performance improvement in desorption/ionization of kilodalton molecules: quantitative evaluation, imaging secondary ion MS, and laser ablation. *Anal Chem*, **75**, 6875–6885.
19. Delcorte, A., Medard, N., Bertrand, P. (2002) Organic secondary ion mass spectrometry: sensitivity enhancement by gold deposition. *Anal Chem*, **74**, 4955–4968.
20. Keune, K., Boon, J. J. (2004) Enhancement of the static SIMS secondary ion yields of lipid moieties by ultrathin gold coating of

aged oil paint surfaces. *Surf Interface Anal*, **36**, 1620–1628.
21. Nguyen, T. C., Ward, D. W., Townes, J. A., White, A. K., Krantzman, K. D., Garrison, B. J. (2000) A theoretical investigation of the yield-to-damage enhancement with polyatomic projectiles in organic SIMS. *J Phys Chem B*, **104**, 8221–8228.
22. Sjovall, P., Lausmaa, J., Johansson, B. (2004) Mass spectrometric imaging of lipids in brain tissue. Anal Chem, **76**, 4271–4278.
23. Todd, P. J., McMahon, J. M., McCandlish, C. A. (2004) Secondary ion images of the developing rat brain. *J Am Soc Mass Spectrom*, **15**, 1116–1122.
24. Touboul, D., Halgand, F., Brunelle, A., Kersting, R., Tallarek, E., Hagenhoff, B., Laprevote, O. (2004) Tissue molecular ion imaging by gold cluster ion bombardment. *Anal Chem*, **76**, 1550–1559.
25. Townes, J. A., White, A. K., Wiggins, E. N., Krantzman, K. D., Garrison, B. J., Winograd, N. (1999) Mechanism for increased yield with SF5+ projectiles in organic SIMS: the substrate effect. *J Phys Chem A*, **103**, 4587–4589.
26. Weibel, D., Wong, S., Lockyer, N., Blenkinsopp, P., Hill, R., Vickerman, J. C. (2003) A C-60 primary ion beam system for time of flight secondary ion mass spectrometry: its development and secondary ion yield characteristics. *Anal Chem*, **75**, 1754–1764.
27. Altelaar, A. F. M., Klinkert, I., Jalink, K., de Lange, R. P. J., Adan, R. A. H., Heeren, R. M. A., Piersma, S. R. (2006) Gold-enhanced biomolecular surface imaging of cells and tissue by SIMS and MALDI mass spectrometry. *Anal Chem*, **78**, 734–742.
28. Broersen, A., van Liere, R., Altelaar, A. F. M., Heeren, R. M. A., McDonnell, L. A. (2008) Automated, feature-based image alignment for high-resolution imaging mass spectrometry of large biological samples. *J Am Soc Mass Spectrom*, **19**, 823–832.
29. Svensson, M., Skold, K., Svenningsson, P., Andren, P. E. (2003) Peptidomics-based discovery of novel neuropeptides. *J Proteome Res*, **2**, 213–219.

Chapter 12

Tandem Mass Spectrometric Methods for Phospholipid Analysis from Brain Tissue

Timothy J. Garrett and Richard A. Yost

Abstract

We describe the utility of intermediate-pressure MALDI and tandem mass spectrometry (MS/MS and MS^n) for the characterization and imaging of phospholipids in brain tissue sections. The use of both MS/MS spectra and MS/MS images allows for identification of isobaric compounds. The structural characterization of phosphatidylcholines, phosphatidylserines, phosphatidylethanolamines, and sphingomyelins directly from tissue sections is described.

Key words: MALDI, tandem mass spectrometry, phospholipids, brain.

1. Introduction

Imaging tissue sections by matrix-assisted laser desorption/ionization mass spectrometry (MALDI-MS) is a growing field in mass spectrometry. This is due in part to the prospects of analyzing for very specific molecular ions (1–3) directly from sectioned tissue and correlating changes in those specific ions to diseases (4, 5). Critical to this comparison is the role of tandem mass spectrometry (MS/MS or MS^n) in the detection and identification of known and unknown compounds desorbed from the tissue (3, 6–9). The field of imaging mass spectrometry pertains to the direct analysis of surfaces, primarily tissue sections, using a focused ion beam (secondary ion mass spectrometry or SIMS) (10) or a focused laser beam with a highly absorbing matrix (MALDI) (11). In order to create an image, the laser or ion beam is rastered in a discrete pattern across the tissue surface. Each spot

S.S. Rubakhin, J.V. Sweedler (eds.), *Mass Spectrometry Imaging*, Methods in Molecular Biology 656,
DOI 10.1007/978-1-60761-746-4_12,

that the laser interrogates generates a mass spectrum and is considered a pixel with dimensions related to the spot size of the laser or ion beam (12).

A majority of the work in imaging MS has been focused on determining the distribution of proteins and peptides in tissue sections. Smaller molecules such as drugs (3, 8, 13, 14) and phospholipids (6, 7, 15–18) have also been analyzed, showing that a wide variety of compounds can be localized in tissue sections with MALDI-MS and identified by tandem MS (3, 6, 7, 9, 18).

Phospholipid (PL) ions have been analyzed directly from tissue sections by mass spectrometry for several years. Secondary ion mass spectrometry (SIMS) offered images for the fragment ion at *m/z* 184, which corresponds to the head group for all phosphatidylcholines and sphingomyelins, showing distribution primarily in the gray matter of the brain (19). With the use of MALDI, researchers have been able to generate images of PLs from both molecular ions and fragment ions; tandem MS has clearly shown that several isobaric species are typically present at each *m/z* value in the lipid mass range (7, 9, 18).

The direct analysis of PLs from brain tissue sections could aid in the further understanding of how lipids relate to diseases such as Alzheimer's (20, 21) and Parkinson's (22) by identifying specific lipids and the location of those lipids in a brain tissue section. Imaging PLs in brain tissue may also aid in understanding how and where PLs change with age and development (19, 23). Not only determining the molecular weight of PLs (by MS) but also identifying the fatty acid tails (by MS/MS) (6) may also aid in understanding if a decrease in certain fatty acids is PL specific or global (24) and if it is localized in a certain structure of the brain or other organ. Critical to this aspect of imaging MS is the need to identify compounds directly from the tissue section using tandem MS with fatty acid regiospecificity. Coupling a MALDI source to an instrument capable of performing tandem MS provides the opportunity to identify lipids based on characteristic fragmentation patterns and to produce images based on a specific fragmentation pathway (6) rather than on a molecular ion signal, giving a much more specific image related to a single, identified lipid. This specificity is particularly critical when analyzing small molecules, as both isobaric lipids and MALDI background ions can lead to multiple interferences; the reduction in noise provided by tandem mass spectrometry offers the ability to identify compounds present at very low levels that would not be detectable in a full-scan MS experiment.

Acquiring tandem MS data for a specific parent ion across the tissue section can provide more specific images because a specific parent ion/product ion transition can be mapped. The image thus generated would be a tandem MS image. As has been shown previously (6), performing tandem MS on a ion trap provides the opportunity to separate and identify multiple isobaric PLs (as well as MALDI matrix background ions) from the isolation of parent ions of a single *m/z*. Tandem MS with MALDI thus offers the ability to completely characterize the compounds desorbed and ionized from a tissue section and differentiate them from those related to the MALDI matrix.

The methods presented here describe the ability to characterize phosphatidylcholines (PCs), sphingomyelins (SPMs), phosphatidylserines (PSs), and phosphatidylethanolamines (PEs) in the positive ion mode from rat brain tissue sections using imaging tandem MS and the creation of compound-specific images from full-scan MS and MS^2, while using MS^3 for the final identification. The use of MS^2 images for the separation of isobaric ions coupled with the ability to perform MS^3 provides the means to identify isobaric species present in the tissue section.

2. Materials

2.1. Equipment

1. Artistic airbrush (Aztek A470) from Testors (Rockford, IL, USA).
2. Thermo LTQ with vMALDI ion source (*see* **Note 1**).
3. Cryostat (Leica CM1850, Wetzlar, Germany).
4. Surfer 8.0 or newer version (Golden Software, Golden, CO, USA).

2.2. Reagents and Supplies

1. 40 mg/ml solution of 2,5-dihydroxybenzoic acid (DHB) prepared in 70/30 (v/v) HPLC-grade methanol and water. Sodium acetate is added to a concentration of 20 mM for sodium adduction, if desired.
2. Double-sided tape for attaching microscope slides to MALDI stage.
3. Glass microscope slides (*see* **Note 2**).

3. Methods

3.1. Preparation of Brain Tissue Sections

1. Whole brain tissue (from Sprague–Dawley rats in this example) should be cut with a scalpel to desired location of sectioning.
2. Set temperature of cryostat to –22°C.

3. Affix tissue to cutting target by holding tissue in one hand and pipetting a small amount of water around the tissue.
4. Set cutting thickness to 10 μm.
5. Cut tissue several times until an entire section is obtained. Discard unwanted sections.
6. When desired section is sliced, attach to a cold glass microscope slide by placing the slide over top of the section. Press down lightly on the section for it to adhere to the slide (*see* **Note 3**).
7. Remove slide with tissue from cryostat chamber. With a gloved finger, warm the underneath side of the glass below the tissue until the tissue becomes clear. This takes approximately 10 s.
8. Freeze until ready for use.

3.2. Application of MALDI Matrix with an Artistic Airbrush

1. Remove desired tissue sections from the freezer and place in a dessicator for at least 30 min.
2. Prepare a solution of DHB to a concentration of 40 mg/ml in 70/30 methanol/water with 20 mM $NaCH_3COO$ (*see* **Note 4**).
3. Set up airbrush in a ventilated hood.
4. Attach the airbrush to a ring stand using an appropriate clamp.
5. Adjust the distance from the tissue to the end of airbrush to be approximately 15 cm (*see* **Note 5**).
6. Attach a second clamp to the solvent and gas control throttle. Ensure that the trigger is clamped down and that solvent is flowing.
7. Place the blank microscope slide underneath the airbrush setup and pass underneath the spray in a back and forth motion at a steady pace.
8. After a couple of passes have been made, examine the glass slide under light to evaluate the coating evenness. A gloved finger can be used to test the adherence of the matrix by wiping the surface of the glass slide. If smearing is evident, the coating has been applied too wet, if the matrix comes off as a powder, the coating was applied without enough solvent.
9. After determining the appropriate settings, obtain the tissue sample and pass it underneath the airbrush assembly in a steady back and forth motion as before. Allow for a little bit of overlap as you move across the tissue and be sure to coat beyond the edges of the tissue. One pass is considered

a full covering of the tissue. Typically, 20 passes are needed to achieve an appropriate coating.

10. Dry the completed sample for a minimum of 30 min under a gentle stream of nitrogen.
11. Place the sample on the MALDI target plate into the predrilled wells and attach with double-sided tape.

3.3. Mass Spectrometric Analysis

1. Insert the MALDI target plate with the tissue into the instrument following the manufacturer's procedure.
2. For the instrument used here, a picture of the tissue sample needs to be obtained first. This is done by selecting the area where the tissue is located and then pressing scan to obtain the image.
3. Select the area on the tissue for analysis. Encircle the entire tissue sample using a desired geometric shape or free draw application if the whole tissue is desired for analysis (*see* **Note 6**).
4. Set the desired mass range; for PL analysis, we typically use the mass range of 180–1,000.
5. If automatic gain control (AGC) is not used, the laser power and the number of laser shots per spot must be set manually. This is done by first choosing the desired number of shots (typically 5 or 10) and then adjusting the power of the laser to achieve a signal of 1×10^5. A minimum of 10 random spots should be interrogated to determine the appropriate power. If employing AGC, the power only needs to be adjusted to obtain a similar signal as the number of laser shots are varied when AGC is employed (*see* **Note 7**).
6. Set the raster size; typically, this is set to the spot size of the laser for normal sampling (*see* **Note 8**).

3.4. Tandem MS Analysis

1. The isolation width is typically set to 1.5 amu with a collision energy of 30% (*see* **Note 9**).
2. For obtaining tandem MS images, the entire tissue section must be analyzed with the given m/z of interest. This provides for the identification of potential isobaric ions as well as the generation of MS/MS images corresponding to selected product ions.
3. If tandem MS will only be used for compound identification, a second procedure could be used. Following the instrument manufacturer's procedures, a desired list of ions can be entered. In this mode, the entire tissue will be analyzed alternating between the list of ions entered during the tissue scanning.

3.5. Image Generation

1. MS images from full-scan MS and MS/MS data are generated by extracting a desired *m/z* value with a 1.0 *m/z* wide window using a MALDI data extraction program provided by ThermoFisher. The extracted data is in text form and shows the *x*, *y* coordinates and the intensity of the ion.
2. The text version of the data is converted to a spreadsheet using a spreadsheet program (Microsoft Excel).
3. A graphical program is used to generate ion images from the spreadsheet (Surfer 8.0).

3.6. Results

3.6.1. Identifying the Lipid Signature

We have previously shown the ability to analyze phospholipids in tissue sections by intermediate-pressure MALDI (67); this method describes a comprehensive procedure for the evaluation of those phospholipids. **Figure 12.1** shows an optical image of a brain section (inset) and a mass spectrum from one point in that section. The circle and arrow in the figure indicate the area from which this mass spectrum was acquired (1 scan, 10 laser shots, scanned from *m/z* 150–1,000). The majority of the ion signal falls between *m/z* 700 and 900.

From previous results (9, 15), the ions from *m/z* 650–900 can be identified as arising from PLs; searching a database of PL

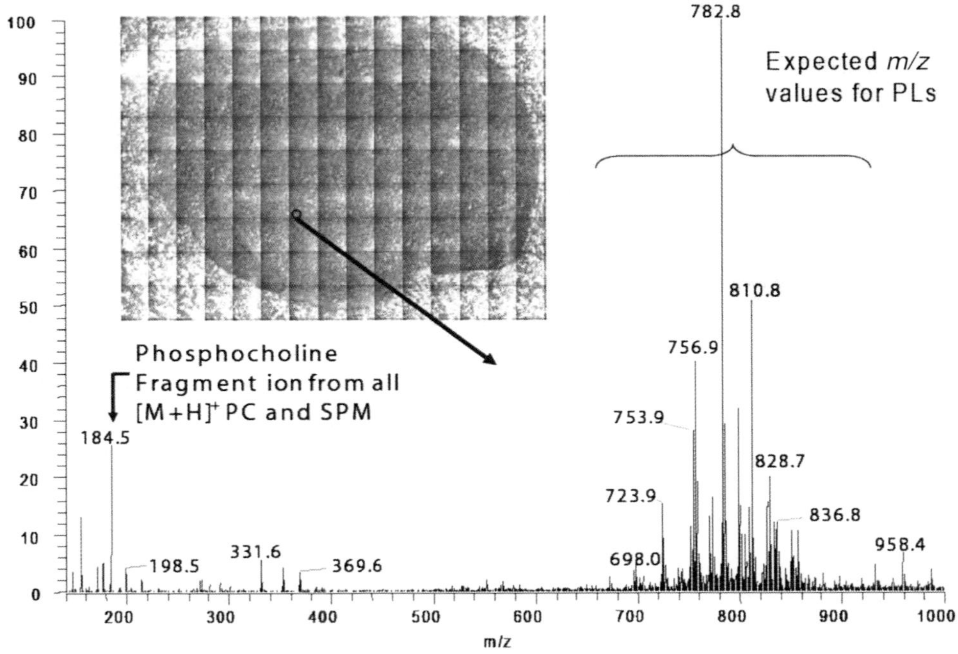

Fig. 12.1. The inset is an optical image (9 × 15 mm) generated from inside the mass spectrometer of a rat brain section coated with DHB matrix. It was acquired with 1 × 1 mm square pictures that are stitched together; this creates the lines in the picture. The mass spectrum shows the signal from the area on the tissue indicated by the *circle* and *arrow*. The *open circle* is approximately equal to the laser spot size (120 μm).The spectrum was acquired with 10 laser shots. A total of 11,156 spectra were collected across the tissue section.

ions (constructed in the lab) generated a large list of possible identities for each ion based only on the *m/z* value submitted. The list included PCs, SPMs, PEs, and PSs as either protonated, $[M+H]^+$, or sodiated, $[M+Na]^+$, ions. The ion at *m/z* 184 results from source fragmentation of only protonated PCs and SPMs; it corresponds to the phosphocholine head group from these two groups of PLs (25). The low intensity of the *m/z* 184 fragment ion is primarily due to the fact that sodium acetate was added to the matrix solution to induce the production of primarily cationized adduct ions ($[M+Na]^+$). Producing the sodiated adduct has three advantages: (1) the intensity of the fragment ion at *m/z* 184 is significantly reduced, (2) the sodiated species are a more stable ion for mass analysis (26), and (3) the fragmentation patterns for collision-induced dissociation (CID) of cationized PLs show more structural information (25, 27, 28). In addition, the intensity of the *m/z* 184 fragment ion is also reduced by the use of intermediate pressure in the MALDI source (6).

Table 12.1 shows the identification of 16 ions in the PL region from an experiment in which the tissue section was

Table 12.1
List of compounds identified from a brain tissue section using tandem MS with an imported mass list

m/z	RA (%)	PL	Ion
694.9	8.7	SPM (18:0)	$[M+Na-N(CH_3)_3]^+$
697.8	9.4	PC (16:0, 16:0)	$[M+Na-N(CH_3)_3]^+$
723.8	16.0	PC (16:0, 18:1)	$[M+Na-N(CH_3)_3]^+$
753.9	34.7	SPM (18:0)	$[M+Na]^+$
756.8	59.8	PC (16:0, 16:0)	$[M+Na]^+$
772.7	8.6	PC (16:0, 16:0)	$[M+K]^+$
782.7	100.0	PC (16:0, 18:1)	$[M+Na]^+$
798.6	33.8	PC (16:0, 18:1)	$[M+K]^+$
804.7	22.3	PC (16:0, 20:4)	$[M+Na]^+$
810.7	30.3	PC (18:0, 18:1)	$[M+Na]^+$
828.6	16.4	PC (16:0, 22:6)	$[M+Na]^+$
832.6	17.2	PC (18:0, 20:4)	$[M+Na]^+$
848.6	7.1	PC (18:0, 20:4)	$[M+K]^+$
856.7	7.0	PC (18:0, 22:6)	$[M+Na]^+$
932.4	4.4	PC (16:0, 16:0)	$[M+DHB+Na]^+$
958.4	5.8	PC (16:0, 18:1)	$[M+DHB+Na]^+$

A single compound such as PC (16:0, 18:1) can be represented by four different ions in the mass spectrum, *m/z* 723.8 ($[M+Na-N(CH_3)_3]^+$), 782.7 ($[M+Na]^+$, 798.6 ($[M+K]^+$), and 958.4 ($[M+Na+(DHB+Na-H)]^+$). The abbreviation RA stands for relative abundance scanned and MS/MS spectra were recorded for each ion in a list. In this type of experiment, a list of ions is imported into the method. When using one microscan per spot, the instrument rotates through the list while rastering across the tissue section. Thus, with a list of 16 ions (as was done here), a single ion is analyzed every 16 spots on the tissue. This experiment provided a way to collect MS/MS data for multiple ions in one tissue raster; however, MS/MS images could not be generated.

As can be seen in **Table 12.1**, only two classes of PLs were identified using this approach, SPM and PC. From these studies, it was determined that a single PC ion can have multiple ions associated with a single fatty acid arrangement. For example, PC (16:0, 18:1) has four ions that were detected, as shown in **Table 12.1**, $[M+Na–N(CH_3)_3]^+$, $[M+Na]^+$, $[M+K]^+$, and $[M+DHB+Na]^+$, as determined by tandem MS. In addition, there is also a low intensity $[M+H]^+$ ion present in the spectrum (*m/z* 760.8, ~8% RA). MS/MS was performed on this ion (in a separate experiment) and the major fragment produced was *m/z* 184.1, confirming the protonation. Tandem MS (MS/MS) allowed for the identification of these 16 ions from a single tissue scan; however, only two lipid classes were identified from this type of experiment in which selected ions were chosen for tandem MS, while scanning the tissue section. In addition, several SPMs were not characterized, SPM 24:0 (*m/z* 837.6) and SPM 24:1 (*m/z* 835.6) (both $[M+Na]^+$), by this method because they were not chosen for MS/MS analysis. These two ions were not selected because they were not very intense in a visual inspection of the spectrum. However, these two ions should only be present in the white matter of the brain according to previous research (29). The spectrum used to select ions for subsequent MS/MS experiments probably did not include as much white matter as gray matter; therefore, the intensity of these two ions would be low and would not have been selected for MS/MS. However, because a full-scan MS data set was also collected, the image for any ion can be retrieved by extracting the specific ion signals. From previous work (6 7), the most abundant ion for all the PC and SPM ions under the conditions employed here is the $[M+Na]^+$. Images generated for these two ions confirmed their greater abundance in white matter.

3.6.2. Imaging with Tandem MS Data

In order to further characterize the PLs and generate MS/MS images, another tissue section was prepared for imaging MS experiments. In several previous experiments, it was noticed that a single tissue section could be analyzed multiple times for the analysis of PLs (up to 40 times) (6); therefore, to explore the further characterization of the PL's MS/MS and MS^3 experiments were performed in which a single *m/z* value was isolated (typically with an isolation width of 1.5) and the entire tissue was analyzed by MS/MS or MS^3 for that specified ion and isolation width. **Table 12.2** shows the experiments that were performed on this tissue section and the ions isolated for tandem MS. This list represents 45 different experiments, but in total, 49 experiments were performed, including four full-scan MS experiments. Two more MS/MS experiments and two full-scan experiments were performed (after recoating the tissue section because the signal was depleted after 41 experiments).

Figure 12.2 shows the ion images from a full-scan MS experiment for the same 16 PL species identified earlier (**Table 12.1**),

Table 12.2
List of tandem MS experiments performed on a single tissue section

MS^2 (IW 1.5)	*MS^3*
742.6 → ...	*753.6 → 694.4 → ...*
753.6 → ...	756.6 → 697.3 → ...
756.6 → ...	804.6 → 745.4 → ...
782.6 → ...	808.6 → 552.3 → ...
798.6 → ...	808.6 → 749 → ...
804.6 → ...	808.6 → 765 → ...
808.6 → ...	826.6 → 767 → ...
810.6 → ...	828.6 → 741.3 → ...
820.6 → ...	828.6 → 769.4 → ...
826.6 → ...	828.6 → 785.2 → ...
828.6 → ...	832.6 → 745 → ...
832.6 → ...	832.6 → 773 → ...
835.6 → ...	835.6 → 776 → ...
835.6 (IW 1.0) → ...	835.6 → 793 → ...
836.6 (IW 1.0) → ...	836.6 → 777 → ...
837.6 → ...	837.6 → 778 → ...
848.4 → ...	837.6 → 794 → ...
856.6 → ...	856.6 → 769 → ...
	810.6 → 751 → ...
	826.6 → 761 → ...
	848.6 → 789.4 → ...
	856.6 → 797.4 → ...

Those in *italics* were performed after recoating the tissue. In addition to these experiments, two full-scan MS analyses were performed before recoating and after recoating. In total, 49 experiments were performed on the same tissue section for the analysis of phospholipids. More experiments could have been performed after recoating, but the data collected were sufficient to identify the ions of interest. The abbreviation IW stands for isolation width but from a different tissue section. A full-scan MS experiment is critical to perform for endogenous species analysis because it provides a starting point for knowing what major ions are present in the tissue section. These images were generated by extracting the intensity of each specified *m/z* with a window of 1.0 amu. These ions were all identified by tandem MS, as indicated in **Table 12.2**. As can be seen from this set of 16 ion images, the different PLs (PC or SPM) show different distributions in the rat brain. It is remarkable that the change of a single fatty acid tail, such as with *m/z* 782 (PC (16:0, 18:1)) and *m/z* 810 (PC (18:0, 18:1)), can so drastically change the distribution in the rat brain. These MS images offer a unique perspective of the rat brain, showing that the distribution of individual PL ions is varied throughout the entire brain section. Not surprisingly, *m/z* 184, the fragment ion that corresponds to the polar head group of all PC and SPM ions, shows a distribution similar to that of *m/z* 782, the most abundant PC ion detected in the positive ion mode. What is important is that imaging just the fragment ion at *m/z* 184 as has been performed in other imaging MS studies (19) does not provide enough information about the localization of other PC and SPM

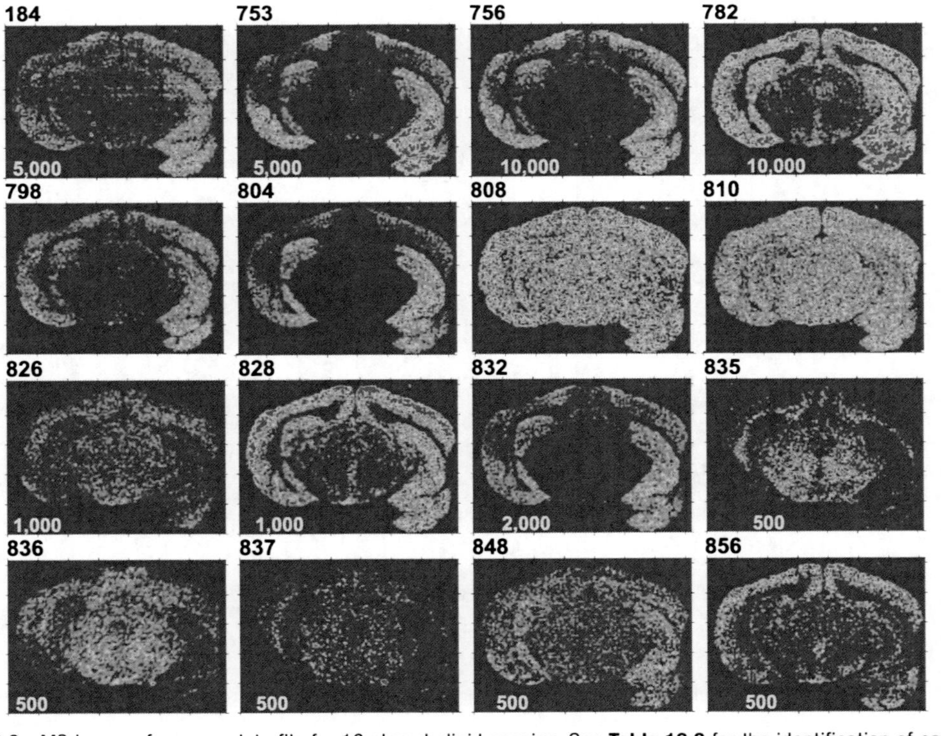

Fig. 12.2. MS images from one data file for 16 phospholipid species. See **Table 12.2** for the identification of each ion. The maximum value of the intensity scale is indicated at the bottom left of each image. All images were normalized to the total ion current.

ions and thus provides a very limited view of the chemical distribution of the rat brain.

Since MS/MS was performed across the entire tissue on 14 of the 16 ions in **Fig. 12.2**, MS/MS images were also generated from the most abundant loss produced in each MS/MS experiment. The most abundant loss for all ions isolated was a neutral loss of 59, corresponding to a loss of trimethylamine from either a PC or a SPM ion. The MS/MS ion images for 14 of these ions are shown in **Fig. 12.3**. The ion at *m/z* 798 was determined from previous results to be the $[M+K]^+$ ion for PC (16:0, 18:1), corresponding to the $[M+Na]^+$ ion of PC (16:0, 18:1) at *m/z* 782. A full tissue scan MS/MS experiment was not performed on the fragment ion at *m/z* 184 nor on the ion at *m/z* 798. The MS/MS transition for each ion is shown above each MS/MS ion image. However, performing an MS/MS experiment for *m/z* 798 might have been beneficial to determine the presence of other isobaric ions, as will be described later. For the most part, the MS/MS images

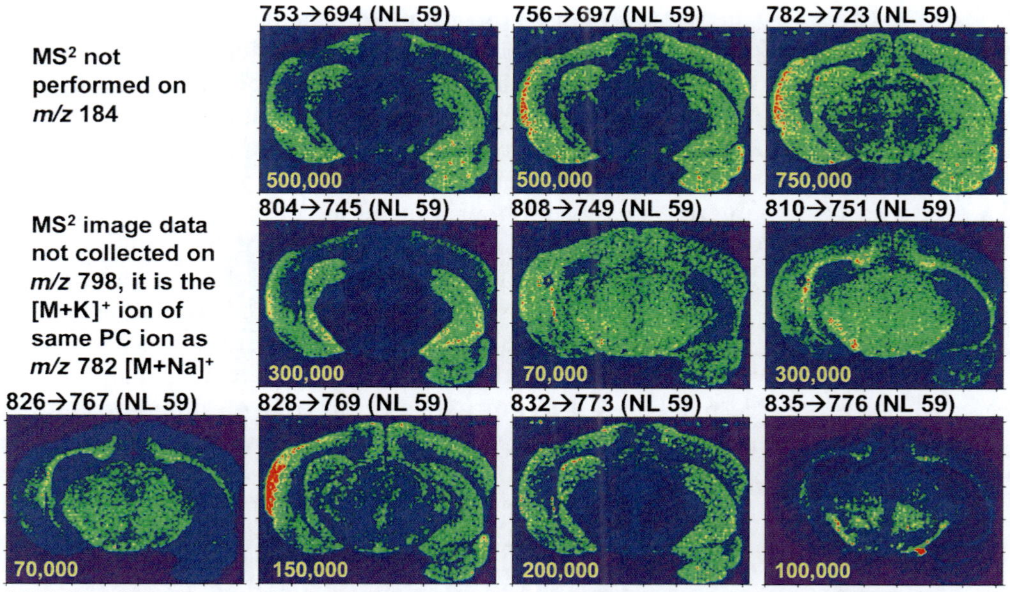

Fig. 12.3. MS/MS images for only 14 different PC and SPM species. The neutral loss of 59 for each ion is shown. See **Table 12.2** for the identification of each ion. Some ions presented represent different ions of the same compound such as 810→751 and 826→767 ($[M+Na]^+$ and $[M+K]^+$ for PC 18:0 and 18:1, respectively). These signals were not normalized to the total ion current because they are tandem MS images. Maximum scale intensity is shown in the bottom left of each figure.

are similar to the corresponding MS images. However, for some ions, this more specific image provides a different view of the ion distribution in the tissue. For example, the MS image of *m/z* 810 indicates that this ion is present in both gray and white matter of the brain and thus shows a distribution that is fairly uniform throughout the tissue. On the other hand, the MS/MS image of 810–751 shows that the distribution of the PC (18:0, 18:1) $[M+Na]^+$ ion is predominantly in the white matter of the brain. This means that the MS image of *m/z* 810 included other species at a similar intensity and of the same nominal *m/z*, but predominantly in the gray matter. For *m/z* 836, the difference between the MS and MS/MS images is striking as well. The MS image for *m/z* 836 shows a uniform localization in the white matter of the brain, while the MS/MS image for *m/z* 836 → 777 depicts localization to the substantia nigra and red nucleus, with lesser amounts in the corpus collosum of the brain. In the case of the less abundant ions (*m/z* 826, 835, 837, and 848), the MS/MS data allowed for a much clearer picture of the distribution than the MS image. This clarity is likely due to the improvement in the signal-to-noise ratio afforded by tandem MS and the selectivity of monitoring a single MS/MS daughter ion. In all cases, the MS/MS image should provide a better representation (compared to just MS alone) of an individual compound, because interfering isobaric species are removed if they fragment differently.

The ability to separate isobaric ions using tandem MS in imaging tissue has been shown previously (6), but the characterization of the many isobaric ions present has not been fully explored. For example, Cha and Yeung (18) showed the characterization of cerebrosides and PCs using graphite-assisted laser desorption and the separation of isobars including cerebrosides, but not across the entire tissue. In the current study, tandem MS data were collected across the entire tissue from over 40 different ions. The full-scan tandem mass spectra were then analyzed for possible isobaric species present; see **Table 12.2** for a list of all the tandem MS experiments. In addition, to help characterize the isobars, MS^3 was performed on selected MS/MS daughter ions to elucidate the structure of the PL species present (*see* **Table 12.2**).

3.6.3. Isobaric Ion Identification

The process for characterization of the isobars from the ion signal at a single *m/z* is illustrated in **Figs. 12.4** and **12.5**. **Figure 12.4** shows the MS^2 spectrum of *m/z* 856 averaging all 10,528 spectra across the tissue section. The most abundant product ion, *m/z* 797.4, arises from a neutral loss of 59, corresponding to a loss of trimethylamine, and thus identifies one of the isobaric ions as a PC (not an SPM because of the even *m/z* at 856). Other relatively abundant and characteristic product ions include *m/z* 701.9, 769.4, 813.0, and 838.1, arising from neutral losses of 155, 87, 43, and 18, respectively. A neutral loss of 155 is typically

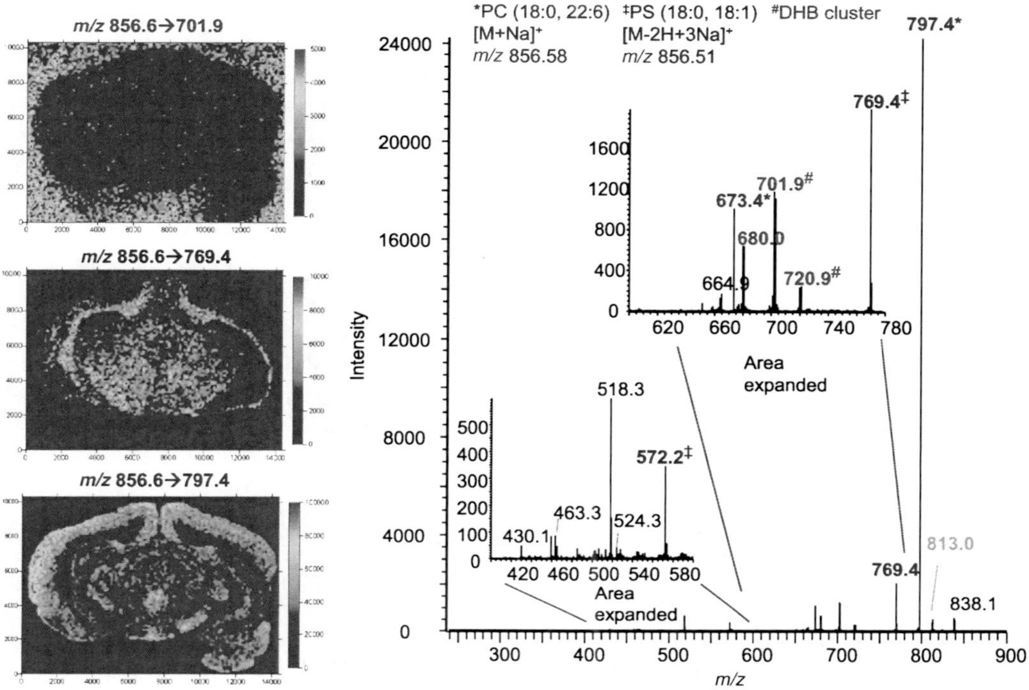

Fig. 12.4. Identification of the isobaric ions present at *m/z* 856.6 using MS/MS data. The MS/MS spectrum for *m/z* 856.6 (IW 1.5) is the average of 10,528 Spectra across the entire tissue section. Three major ions were identified at this *m/z* value and isolation width. The MS/MS images for the major fragment ion of each compound are shown in the left of the figure. The two insets in the spectrum show expanded regions as indicated. The fragment ions corresponding to each isobar are identified with *colors* and *symbols*. Axes are in micrometers.

related to a DHB cluster ion (loss of $C_7H_7O_4$), a neutral loss of 43 is related to a PE ion (loss of C_2H_5N), and a neutral loss of 87 is typically associated with a PS compound (loss of $C_3H_5NO_2$). The tandem MS images for *m/z* 856 dissociating to form *m/z* 797.2 (the PC isobar, NL of 59), 701.9 (DHB cluster, NL 155), and *m/z* 769.4 (the PS isobar NL 87) are shown at the left in **Fig. 12.4**. These data provide additional confirmation that there are multiple ionic species present at *m/z* 856.6, given that *m/z* 797.4 and *m/z* 769.4 show a near opposite distribution, as has been shown for other *m/z* values (6, 18). In addition, the MS image for the distribution of the DHB cluster shows a higher localization off the tissue, helping to confirm that this ion signal is from the matrix and not from the tissue section. The two inset spectral regions were expanded to show the importance of these less abundant ions in the identification of the isobars. Colors and symbols indicate the fragment ions that are representative of the different isobars present.

Another issue with characterizing the compounds desorbed from tissue is also determining what type of ion is detected

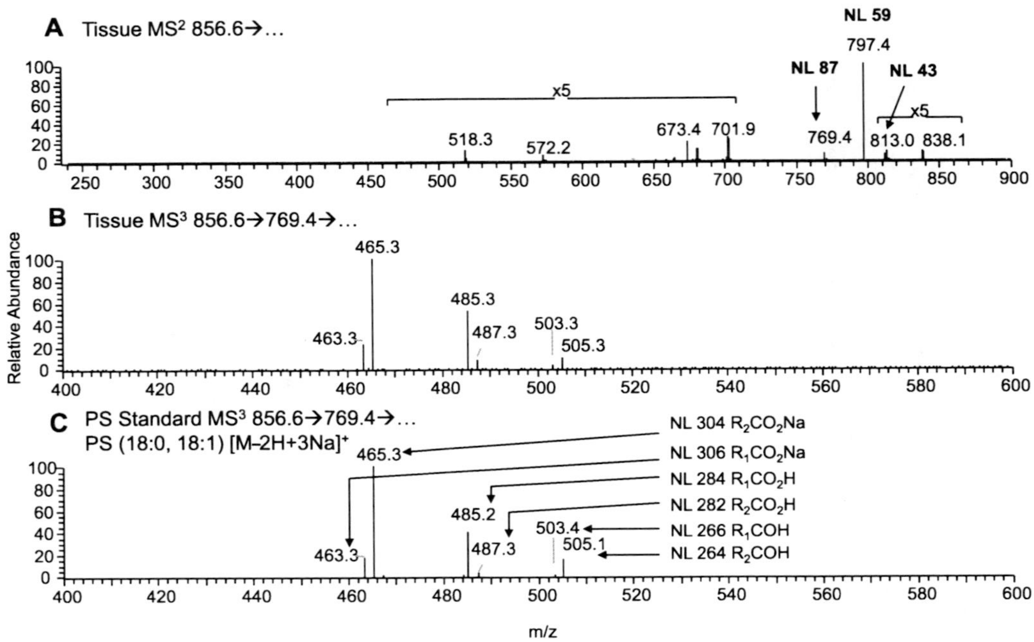

Fig. 12.5. MS/MS and MS^3 data showing the identification of one of the isobars at *m/z* 856. A is the MS/MS spectrum from tissue for *m/z* 856.6. (**a**) NL of 59 is indicative of a PC or SPM, while a NL of 87 or 43 is indicative of PS or PE, respectively. (**a**) NL of 154 or 155 corresponds to a DHB cluster ion. (**b**) The average MS^3 spectrum for 856→769→. . . . For this MS^3 experiment, the entire tissue was scanned and thus the spectrum is the average of 10,528 spectra. The fragment ion at *m/z* 769 was identified as resulting from a PS ion because of the NL of 87. MS^3 spectra from tissue (**b**) and a PS standard mixture from bovine brain (**c**). The assignment of the fatty acid chains is R_1=stearic acid (18:0) and R_2=oleic acid (18:1). The TIC values shown correspond to the signal from the standard (**a**) and from the tissue (**b**). MS^3 was required for the correct assignment of the fatty acid tails of this isobaric species and identifies this PS species as PS (18:0, 18:1) $[M–2H+3Na]^+$.

(protonated, sodiated, etc.). The further characterization of the PS ion can be used to illustrate this point. The middle spectrum (b) of **Fig. 12.5** is the MS^3 spectrum acquired from the tissue section of the ion at *m/z* 769 arising from a neutral loss of 87 (856→769→. . .), typically associated with a PS compound (loss of $C_3H_5NO_2$, serine head group) (30). As can be seen from the figure, this MS^3 fragmentation pattern shows four major product ions (*m/z* 505.3, 485.3, 465.3, and 463.3) in four very distinct patterns.

A standard mixture of PSs (bovine brain extract from Avanti lipids) was obtained to evaluate the ionization and fragmentation characteristics of this class of lipids under the conditions employed. The two most abundant fatty acid arrangements in the mixture are PS (18:0, 18:1) and PS (18:0, 22:6). MALDI was performed on this standard mixture with DHB as the matrix. For the most abundant PS compound, PS (18:0, 18:1), multiple ions of this single compound were observed in the mass spectrum, including *m/z* 790.3 ($[M+H]^+$), *m/z* 812.4 ($[M+Na]^+$), *m/z* 834.4 ($[M-H+2Na]^+$, and *m/z* 856.4 ($[M–2H+3Na]^+$).

The next most abundant PS compound observed in the standard mixture is PS (18:0, 22:6) with three ions detected at *m/z* 858.4 ([M+Na]$^+$), *m/z* 880.3 ([M−H+2Na]$^+$), and *m/z* 902.7 ([M–2H+3Na]$^+$). The four ions of the PS (18:0, 18:1) were each subjected to MS/MS and MS^3 experiments. All ions containing a sodium show the most abundant fragment ion as a neutral loss of 87 (the serine head group, $C_3H_5NO_2$) (data not shown); thus, this first stage of tandem MS can rule out that the ion is an [M+H]$^+$ ion. MS^3 was then performed on the ions arising from the NL of 87 from each of the three sodium adducts. Each MS^3 spectrum produced a different fragmentation pattern, which was used to identify the corresponding ion from the tissue tandem MS experiment. **Figure 12.5c** shows the MS^3 spectra of 856→769→. . . of the [M–2H+3Na]$^+$ ion from the standard. The identification of each product ion is shown in the top spectrum. The most abundant ion, *m/z* 465.3, corresponds to the loss of R_2CO_2Na; the related loss of the fatty acid at R1 occurs at *m/z* 463.3 (loss of R_1CO_2Na). The MS^3 spectra from the standard (C) and the tissue (B) are nearly identical, thus concluding that one of the isobaric ions at *m/z* 856.6 from the tissue is PS (18:0, 18:1) [M–2H+3Na]$^+$.

With a better idea of how to identify possible isobars at a given *m/z* value, a better characterization of the ions detected can be accomplished. The identification of the PLs for the tissue section analyzed is described in **Table 12.3**. As can be seen, there is a DHB cluster ion detected at every mass isolated (typically arising from a neutral loss of 137, 154, 155, or 176). Many of the ions isolated for tandem MS show three and even four lipid species. The three primary lipid classes detected in the positive ion mode were PC, PS, and PE. Cerebrosides were also detected, but were present at a much lower intensity due to the conditions employed (18). The primary reason for maintaining an isolation width of 1.5 for most experiments was that PLs are considered fragile ions in ion trap analysis (26, 31) and thus can be lost during isolation if the isolation width is not wide enough.

It should be noted that the isolation width was decreased from 1.5 to 1.0 for the classification of 835.6 and 836.6 because of confusion about a fragment ion observed, as detailed below. When averaging all the MS/MS data collected from across the tissue for *m/z* 835.6 with an isolation width of 1.5, the spectrum included a very intense ion at *m/z* 552.3. This corresponds to a neutral loss of 283, which would indicate an SPM (because the parent mass is odd) and the loss of the amide-linked fatty acid (18:0). However, most SPM ions do not produce an abundant neutral loss indicating the loss of an amide-linked fatty acid tail in MS/MS experiments. With an isolation width of 1.5 for a parent ion of *m/z* 835.6, this product ion could arise from ions with *m/z* 0.75 units higher or lower. The tissue was then rescanned in

Table 12.3
List of ions identified by imaging tandem MS (MS^2 full tissue scan). MS^3 was performed for further identification when needed

m/z	PC or SPM ion	Ion detected	PE ion	Ion detected	PS ion	Ion detected	Others
742	PC (31:0)	$[M+Na]^+$	PE (34:0)	$[M+Na]^+$			DHB cluster
753	SPM (18:1)	$[M+Na]^+$					DHB cluster
756	PC (16:0, 16:0)	$[M+Na]^+$					DHB cluster
782	PC (16:0, 18:1)	$[M+Na]^+$					DHB cluster
798	PC (16:0, 18:1)	$[M+K]^+$					DHB cluster
804	PC (16:0, 20:4)	$[M+Na]^+$	PE (18:1, 18:1)	$[M-H+2Na]^+$			DHB cluster
808	PC (18:1, 18:1)	$[M+Na]^+$	PE (16:0, 22:6)	$[M-H+2Na]^+$			DHB cluster
810	PC (18:0, 18:1)	$[M+Na]^+$	PE (18:1, 20:4)	$[M-H+2Na]^+$			DHB cluster
820	PC (16:0, 20:4)	$[M+K]^+$					DHB cluster
826	PC (18:0, 18:1)	$[M+K]^+$					DHB cluster
828	PC (16:0, 22:6)	$[M+Na]^+$	PE (18:0, 20:4)	$[M-H+Na+K]^+$	PS (16:0, 18:1)	$[M-2H+3Na]^+$	DHB cluster
832	PC (18:0, 20:4)	$[M+Na]^+$	PE (18:0, 22:5)	$[M+K]^+$	PS (18:1, 18:1)	$[M-H+2Na]^+$	DHB cluster
835	SPM (24:1)	$[M+Na]^+$					DHB cluster
836	PC (18:1, 20:1)	$[M+Na]^+$	PE (18:0, 22:6)	$[M-H+2Na]^+$			DHB cluster
837	SPM (24:0)	$[M+Na]^+$					DHB cluster
848	PC (18:0, 20:4)	$[M+K]^+$					DHB cluster
856	PC (18:0, 22:6)	$[M+Na]^+$			PS (18:0, 18:1)	$[M-2H+3Na]^+$	DHB cluster

MS/MS for *m/z* 835.6 with a narrower isolation width (1.0); the intensity of *m/z* 552.3 was significantly reduced. To identify the ion one *m/z* unit higher, the tissue was scanned for *m/z* 836.6 with an isolation width of 1.0. Fragments arising from the neutral loss of 43 (*m/z* 793.3), 59 (*m/z* 777.4), and 87 (*m/z* 749.4) were all detected, as well the ion at *m/z* 552.3 (neutral loss of 284). The major ion is assigned to PC (18:1, 20:1) $[M+Na]^+$, as signified by the neutral loss of 59; the PE ion was identified as PE (18:0, 22:6) $[M-H+2Na]^+$, giving rise to the neutral loss of 43. Finally, the ion related to PS (NL of 87) could be PS (40:6) $[M+H]^+$, but is not likely because a loss of 87 is less intense for protonated PS ions than for sodiated PS ions. MS^3 was not performed on *m/z* 836.6→749.4, but could be used to help identify this isobar.

3.6.4. Understanding Endogenous Interferences

In the analysis of full-scan MS data from this tissue sample, an interesting observation was made. An ion, *m/z* 616, appeared to be localized within a specific region of the rat brain section (**Fig. 12.6** inset). MS/MS was performed on this ion, and the most abundant signal was *m/z* 557.2, corresponding to a neutral loss of 59 (**Fig. 12.6a**). Coupling this result to those obtained from the analysis of all ions up to this point would suggest that

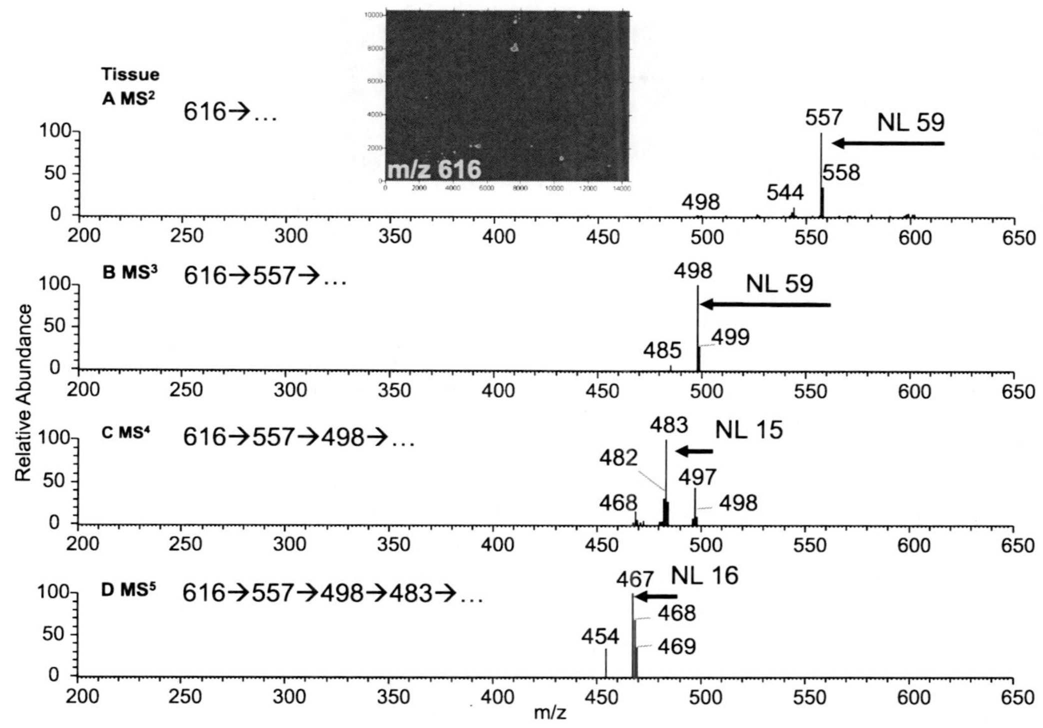

Fig. 12.6. MS/MS through MS^5 spectra of *m/z* 616 from a tissue-specific region in a brain tissue section (shown in the inset). Even after MS^5, the compound still remained unidentified.

this ion is a PC (even parent mass and neutral loss of 59). However, most PC ions are greater than *m/z* 700 when singly charged and thus one might conclude that this ion could represent a lysophosphatidylcholine (a PC ion that has lost one fatty acid tail). However, when MS^3 was performed on *m/z* 557.2, a second neutral loss of 59 was produced, which is not consistent (or possible) for a PC ion. Additional tandem MS spectra were collected up to MS^5, using the most abundant ion from each stage, to assist in the identification of the compound (**Fig. 12.6b–d**). In a personal communication, this ion was suggested to be heme (M^+). A heme standard was obtained and spotted onto a MALDI plate with α-Cyano-4-hydroxycinnamic acid (α-cyano). The most abundant ion signal detected was *m/z* 616. Tandem MS was performed up to MS^5 (**Fig. 12.7a–d**), and the tandem MS spectra show nearly identical fragmentation patterns of the ion desorbed from the brain tissue, confirming the identification of *m/z* 616 as heme.

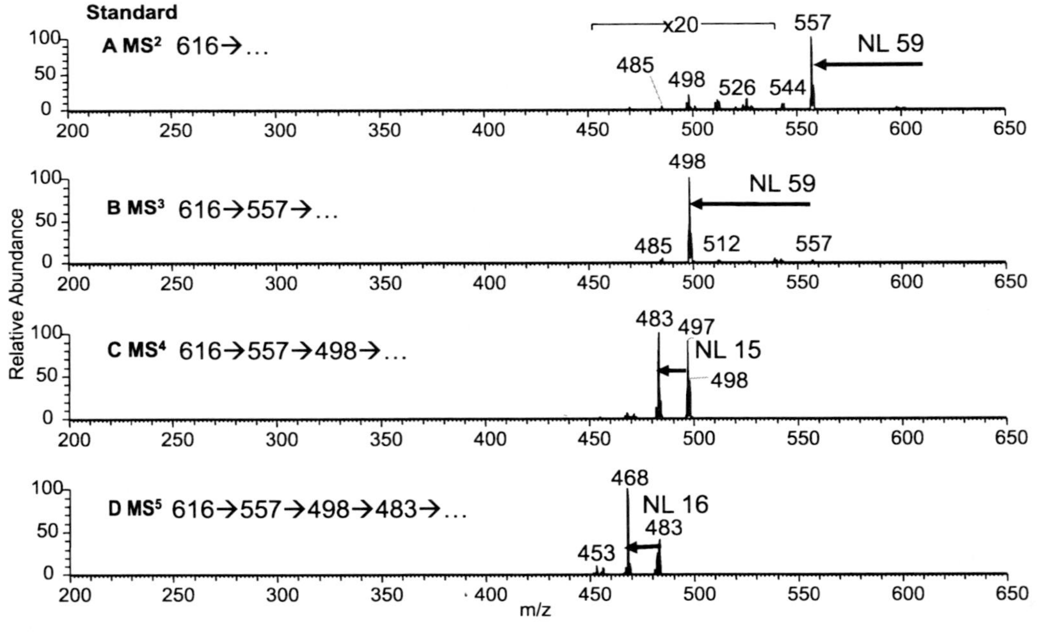

Fig. 12.7. MS/MS through MS^5 spectra from a standard of heme (*m/z* 616), showing a good match in fragmentation pathways when compared to the MS^n spectra acquired from the tissue section shown in **Fig. 12.6**.

3.7. Conclusions

The mass spectrometric images of the various PLs in rat brain showed a remarkable variation in distribution. The use of imaging tandem MS provided a means to properly identify the ions detected and to show the many isobaric species present in the lipid mass region. In addition, it was determined that there were sometimes up to five different ions found under MALDI conditions and characterized by tandem MS that represent the same compound, adding more complexity to the direct analysis of tissue for PLs. These different ions include protonated, sodiated, and potassiated species, as well as ions that are adducted with DHB matrix molecules and even ions with multiple cation adductions. Additionally, MS^5 was shown to properly identify the unexpected compound heme in the tissue. Because heme loses 59 in MS/MS, a single stage of tandem MS might identify this compound as a lysophosphatidylcholine, which would be incorrect. This finding further shows the need of tandem MS in the analysis of small molecules from tissue sections and the need for multiple stages of tandem MS to correctly identify small molecules from tissue.

Most of the images presented were generated from a single mass spectral acquisition, showing the wealth of information hidden within the data. The repeated analysis (MS, MS/MS, and MS^3) of one individual brain section was described and offers a unique opportunity to continually probe a single tissue section for further information. Images generated from an MS/MS product ion in tandem MS experiments showed better specificity and in some cases a more accurate representation of the distribution of the compound when compared to the MS ion image.

4. Notes

1. The Finnigan LTQ linear ion trap mass spectrometer equipped with the vMALDI source (ThermoFisher, San Jose, CA, USA) was used for all imaging mass spectrometry experiments. The instrument has been described in detail previously (6), but briefly, the source consists of a N_2 laser (337 nm) directed to the vacuum chamber using fiber optics. The laser spot size for these experiments was 120 μm. The spot size was evaluated by ablating a small piece of photosensitive paper and then measuring the size of the ablated area with a microscope. The source is maintained at a pressure of 0.17 Torr using N_2.
2. Even though non-conductive sample surfaces are used, the mass accuracy and peak shape are very good. Surface charging effects when using non-conductive glass microscope slides have been observed in MALDI time-of-flight imaging experiments (4), but are eliminated by the use of intermediate-pressure MALDI and a mass analyzer such as the linear ion trap that does not rely on the initial kinetic energy of the ions (6).

3. It is important for the microscope slides to be cold before the tissue is applied to them; otherwise, the tissue will thaw instantaneously and will not likely be flat. The glass slides can be placed in the chamber of the cryostat during the cutting process.
4. Addition of sodium was employed to induce cationization over protonation because cationization provided for better structural characterization of the phospholipids studied (6).
5. A small surface to lay the microscope slide upon can be used to move the microscope slide underneath the airbrush setup. We use a piece of Plexiglas that is approximately 1.3-cm thick and measures approximately 15×15 cm.
6. Although the brain tissue sections are not square, a square was chosen to select the area of interest. This allowed for the analysis of regions outside the brain tissue section as a reference to the ion signals obtained from the tissue section. It also provides for evaluation of possible analyte migration off the tissue.
7. Automatic gain control (AGC) was turned off in order to maintain the same number of laser shots at each point. To avoid deleterious space-charge effects, the number of laser shots and the power of the laser were first determined by interrogating one spot of the tissue sample. Typically, 10 laser shots with about 20–30% laser power (arbitrary units) were used for all full-scan analyses and 15–20 laser shots were used for MS/MS and MS^3 experiments.
8. The raster step size can be set to any desired value, but in the case of these experiments the step size was set to 120 μm, the same size of the laser spot size, to avoid over-sampling and to allow for repeated analysis of the same tissue section (6).
9. An isolation width of 1.5 amu is typically employed for PL analysis because the ions tend to be fragile in ion trap mass analysis (26). In some cases, a width of 1 amu is employed to remove potential background interferences.

Acknowledgments

The authors wish to acknowledge the assistance of George Stafford and Mari Prieto-Conaway at ThermoFisher Scientific (San Jose, CA, USA). We thank Dr. Nigel Calcutt of the

University of California San Diego for donation of the tissue specimens. The NIH is greatly acknowledged for support (NIH RO1 ES007355).

References

1. Monroe, E. B., Jurchen, J. C., Lee, J., Rubakhin, S. S., Sweedler, J. V. (2005) Vitamin E imaging and localization in the neuronal membrane. *J Am Chem Soc*, **127**, 12152–12153.
2. Hsieh, Y., Casale, R., Fukuda, E., Chen, J. W., Knemeyer, I., Wingate, J., Morrison, R., Korfmacher, W. (2006) Matrix-assisted laser desorption/ionization imaging mass spectrometry for direct measurement of clozapine in rat brain tissue. *Rapid Commun Mass Spectrom*, **20**, 965–972.
3. Drexler, D. M., Garrett, T. J., Cantone, J. L., Diters, R. W., Mitroka, J. G., Prieto-Conaway, M. C., Adams, S. P., Yost, R. A., Sanders, M. (2007) Utility of imaging mass spectrometry (IMS) by matrix-assisted laser desorption ionization (MALDI) on an ion trap mass spectrometer in the analysis of drugs and metabolites in biological tissues. *J Pharm Toxicol Meth*, **55**, 279–288.
4. Chaurand, P., Schwartz, S. A., Billheimer, D., Xu, B. J., Crecelius, A., Caprioli, R. M. (2004) Integrating histology and imaging mass spectrometry. *Anal Chem*, **76**, 1145–1155.
5. Maddalo, G., Petrucci, F., Iezzi, M., Pannellini, T., Del Boccio, P., Ciavardelli, D., Biroccio, A., Forli, F., Di Ilio, C., Ballone, E., Urbani, A., Federici, G. (2005) Analytical assessment of MALDI-TOF imaging mass spectrometry on this histological samples. An insight into proteome investigation. *Clin Chim Acta*, **357**, 210–218.
6. Garrett, T. J., Prieto-Conaway, M. C., Kovtoun, V., Bui, H., Izgarian, N., Stafford, G. C., Yost, R. A. (2007) Imaging of small molecules in tissue sections with a new intermediate-pressure MALDI linear ion trap mass spectrometer. *Int J Mass Spectrom*, **260**, 166–176.
7. Garrett, T. J., Yost, R. A. (2006) Analysis of intact tissue by intermediate-pressure MALDI on a linear ion trap mass spectrometer. *Anal Chem*, **78**, 2465–2469.
8. Reyzer, M. L., Hsieh, Y., Ng, K., Korfmacher, W. A., Caprioli, R. M. (2003) Direct analysis of drug candidates in tissue by matrix-assisted laser desorption/ionization mass spectrometry. *J Mass Spectrom*, **38**, 1081–1092.
9. Jackson, S. N., Wang, H.-Y. J., Woods, A. S. (2005) *In situ* structural characterization of phosphatidylcholines in brain tissue using MALDI-MS/MS. *J Am Soc Mass Spectrom*, **16**, 2052–2056.
10. Winograd, N. (2003) Prospects for imaging TOF-SIMS: from fundamentals to biotechnology. *Appl Surface Sci*, **203–204**, 13–19.
11. Caprioli, R. M., Farmer, T. B., Gile, J. (1997) Molecular imaging of biological samples: localization of peptides and proteins using MALDI-TOF MS. *Anal Chem*, **69**, 4751–4760.
12. Todd, P. J., Schaaf, T. G., Chaurand, P., Caprioli, R. M. (2001) Organic ion imaging of biological tissue with secondary ion mass spectrometry and matrix-assisted laser desorption/ionization. *J Mass Spectrom*, **36**, 355–369.
13. Wang, H.-Y. J., Jackson, S. N., McEuen, J., Woods, A. S. (2005) Localization and analysis of small drug molecules in rat brain tissue sections. *Anal Chem*, 77, 6682–6686.
14. Troendle, F. J., Reddick, C. D., Yost, R. A. (1999) Detection of pharmaceutical compounds in tissue by matrix-assisted laser desorption/ionization and laser desorption/chemical ionization tandem mass spectrometry with a quadrupole ion trap. *J Am Soc Mass Spectrom*, **10**, 1315–1321.
15. Jackson, S. N., Wang, H.-Y. J., Woods, A. S. (2005) Direct profiling of lipid distribution in brain tissue using MALDI-TOFMS. *Anal Chem*, 77, 4523–4527.
16. Sjovall, P., Lausmaa, J., Johansson, B. (2004) Mass spectrometric imaging of lipids in brain tissue. *Anal Chem*, **76**, 4271–4278.
17. Touboul, D., Kollmer, F., Niehuis, E., Laprevote, O., Brunelle, A. (2005) Improvement of biological time-of-flight-secondary ion mass spectrometry imaging with a bismuth cluster ion source. *J Am Soc Mass Spectrom*, **16**, 1608–1618.
18. Cha, S., Yeung, E. S. (2007) Colloidal graphite-assisted laser desorption/ionization mass spectrometry and MSn of small molecules. 1. Imaging of cerebrosides directly from rat brain tissue. *Anal Chem*, **79**, 2373–2385.
19. Todd, P. J., McMahon, J. M., McCandlish, C. A. (2004) Secondary ion images of the

developing rat brain. *J Am Soc Mass Spectrom*, **15**, 1116–1122.
20. Forlenza, O. V., Schaeffer, E. L., Gattaz, W. F. (2007) The role of phospholipase A(2) in neuronal homeostasis and memory formation: implications for the pathogenesis of Alzheimer's disease. *J Neural Trans*, **114**, 231–238.
21. Pettegrew, J., Panchalingam, K., Hamilton, R., McClure, R. (2001) Brain membrane phospholipid alterations in Alzheimer's disease. *Neurochem Res*, **26**, 771–782.
22. Ross, B. M., Mamalias, N., Moszczynska, A., Rajput, A. H., Kish, S. J. (2001) Elevated activity of phospholipid biosynthetic enzymes in substantia nigra of patients with Parkinson's disease. *Neuroscience*, **102**, 899–904.
23. Martinez, M., Mougan, I. (1998) Fatty acid composition of human brain phospholipids during normal development. *J Neurochem*, **71**, 2528–2533.
24. Xiao, Y., Huang, Y., Chen, Z.-Y. (2005) Distribution, depletion and recovery of docosahexanoic acid are region-specific in rat brain. *Brit J Nutr*, **94**, 544–550.
25. Han, X., Gross, R. W. (1995) Structural determination of picomole amounts of phospholipids via electrospray ionization tandem mass spectrometry. *J Am Soc Mass Spectrom*, **6**, 1201–1210.
26. Garrett, T. J. (2006) Ph.D. Dissertation, University of Florida, Department of Chemistry, Gainesville, FL.
27. Hsu, F.-F., Bohrer, A., Turk, J. (1997) Formation of lithiated adducts of glycerophosphocholine lipids facilitates their identification by electrospray ionization tandem mass spectrometry. *J Am Soc Mass Spectrom*, **9**, 516–526.
28. Hsu, F.-F., Turk, J. (2000) Structural determination of sphingomyelin by tandem mass spectrometry with electrospray ionization. *J Am Soc Mass Spectrom*, **11**, 437–449.
29. Agranoff, B. W., Benjamins, J. A., Hajra, A. A. (1998) in *Basic neurochemistry. Molecular, cellular and medical aspects* (Siegel, G. J., Agranoff, B. W., Fisher, S. K., Albers, R. W., Uhler, M. D., Eds.), 47–67, Lippincott–Raven, Philadelphia, PA.
30. Hsu, F.-F., Turk, J. (2005) Studies on phosphatidylserine by tandem quadrupole and multiple stage quadrupole ion-trap mass spectrometry with electrospray ionization: structural characterization and the fragmentation processes. *J Am Soc Mass Spectrom*, **16**, 1510–1522.
31. McClellan, J. E., Murphy, I., James P., Mulholland, J. J., Yost, R. A. (2002) Effects of fragile ions on mass resolution and on isolation for tandem mass spectrometry in the quadrupole ion trap mass spectrometer. *Anal Chem*, **74**, 402–412.

Chapter 13

Chemical Imaging with Desorption Electrospray Ionization Mass Spectrometry

Vilmos Kertesz and Gary J. Van Berkel

Abstract

Desorption electrospray ionization mass spectrometry (DESI-MS) is a relatively new ambient surface analysis method. This technique requires little or no sample preparation prior to the actual analysis. Along with other application areas, DESI-MS is used to obtain chemical images of different biomolecules in biological tissue sections. In the case of biological tissue analysis, only preparation of the tissue slices is necessary. Furthermore, the method does not require the use of an ionization matrix preventing the redistribution of analytes prior to the analysis. Chemical images are obtained by monitoring the mass spectrometric signals of compounds while moving the sample relative to the DESI sprayer.

Key words: Mass spectrometry imaging, desorption electrospray ionization, whole body, tissue section, propranolol.

1. Introduction

The first publication on desorption electrospray ionization mass spectrometry (DESI-MS) appeared in 2004 (1). That paper has been followed by a stream of publications on a variety of atmospheric pressure surface sampling/ionization techniques for mass spectrometry (2). The popularity of DESI can be owed to its ease of use. In DESI, charged solvent droplets from a pneumatically assisted electrospray ionization source impact the surface to be analyzed, desorbing and ionizing analytes. The gas-phase analyte ions are then collected and transferred into the inlet of the mass spectrometer either directly or by using a transfer tube.

S.S. Rubakhin, J.V. Sweedler (eds.), *Mass Spectrometry Imaging*, Methods in Molecular Biology 656,
DOI 10.1007/978-1-60761-746-4_13,

A potentially very useful application for DESI-MS is chemical imaging. DESI-MS has been used to image inked lettering and shapes on paper (3–6), dyes (7), and tryptic peptide digests (8) separated on TLC plates; latent fingerprints on glass (9), endogeneous lipids (3, 10, 11), and dosed clozapine and metabolites (11, 12) in rat organ thin tissue sections; and dosed propranolol in mouse whole-body thin tissue sections (13). In this chapter we describe the method to obtain DESI-MS chemical images of whole-body thin tissue sections.

2. Materials

2.1. Chemicals

1. HPLC-grade water.
2. HPLC-grade methanol.
3. Sodium chloride (anal. grade, Sigma-Aldrich, St. Louis, MO, USA).
4. D,L-Propranolol hydrochloride from Sigma-Aldrich.
5. Liquid nitrogen.
6. Solid carbon dioxide (dry ice).
7. Isoflurane (Abbott Laboratories, Abbot Park, IL, USA).
8. Hexane from Sigma-Aldrich.

2.2. Biological Tissues

1. Male CD-1 mouse (Charles River Laboratories, Shrewsbury, MA, USA).
2. Whole-body thin tissue section (*see* **Section 3.1**).
3. 2% aqueous carboxymethyl cellulose embedding media (Sigma-Aldrich, part no.: C4888, carboxymethyl cellulose sodium salt, medium viscosity, 400–800 cP).
4. 3×4 in., 1.2 mm thick glass slide (Brain Research Laboratories).
5. Photoactivated polymer adhesive (Macro-Tape-Transfer System, Instrumedics, St. Louis, MO, USA).

2.3. Equipment

1. Cryomicrotome (Model CM3600, Leica Microsystems, Inc., Bannockburn, IL, USA).
2. Vacuum desiccator (Bioworld, Dublin, OH, USA).
3. Flat-bed scanner (HP Scanjet 4370 Hewlett-Packard, Palo Alto, CA, USA).
4. Syringes (Hamilton, Reno, NV, USA).
5. Syringe pump (Model 22, Harvard Apparatus, Cambridge, MA, USA).

6. Stainless steel union (Part no. U-412, Upchurch Scientific, Inc., Oak Harbor, WA, USA).
7. Teflon tubing (approximately 40 cm of 100 μm i.d., 1/16 in. o.d., part no. 1474 from Upchurch Scientific, Inc.).
8. 5.2 cm long taper-tip fused-silica capillary (50 μm i.d., 360 μm o.d.; from New Objective, Woburn, MA, USA).
9. Microionspray II head (MDS Sciex, Concord, ON, Canada).
10. Particle discriminator interface (PDI) from MDS Sciex (14).
11. Mass spectrometer (4000 QTRAP from MDS Sciex).
12. 360° manual rotational stage (Newport, Irvine, CA, USA).
13. *XYZ* robotic platform (capable of 100 mm × 100 mm travel in the *XY* plane, Model MS2000, Applied Scientific Instrumentation, Inc., Eugene, OR, USA) (*see* **Note 1**).
14. Computer-controlled communication device with at least one digital output (DO) port enabled to send a TTL signal to trigger MS data acquisition (Model USB-1208LS USB-based data acquisition module, Measurement Computing Corp., Norton, MA, USA).
15. CCD camera and associated monitor (Protana A/S, Odense, Denmark).

3. Methods

3.1. Tissue Preparation

1. Administer propranolol intravenously to a mouse via the tail vein at 7.5 mg/kg as an aqueous solution in 0.9% NaCl (2 ml/kg of a 3.75 mg/ml propranolol solution). Euthanize the mouse 20 or 60 min after the infusion with an isoflurane overdose and immediately freeze in dry ice/hexane. Embed the frozen mouse in 2% aqueous carboxymethyl cellulose.
2. Prepare sagittal whole-body cryosections (40 μm thick) using a cryomicrotome. Transfer the frozen sections to glass slides using a tape-transfer process, which utilizes a photoactivated polymer adhesive. After transfer to the glass slides, freeze-dry the sections within the chamber of the cryomicrotome. Store all tissue sections in a desiccator at room temperature until analysis.
3. Color images of the tissue sections were acquired using a flat-bed scanner.

3.2. DESI-MS Setup

1. The DESI emitter used for these experiments was identical to that used in (7). The spray emitter was a 5.2 cm long, taper-tip, fused-silica capillary with 50 μm i.d. and 360 μm o.d. in a microionspray head with a 500 μm i.d. nebulizing gas tube providing a nebulizing gas (nitrogen) jet annulus area of approximately 1.5×10^{-7} m^2 (*see* **Note 2**).
2. Mount the DESI emitter on a 360° manual rotational stage to allow adjustment of the incident angle.
3. Set up the *XYZ* robotic platform. The travel capacity of the platform should allow analysis of the whole tissue section (*see* **Note 3**).
4. Mount the PDI with 7 cm heated chamber length (sampling capillary) on the mass spectrometer (*see* **Note 4**).
5. Mount the CCD camera and install the monitor to observe the DESI emitter and sampling capillary from an angle just above the surface (*see* **Note 5**).
6. While the instrument is in standby or off (high voltage is not applied), couple the high-voltage cable to the stainless steel body of the microionspray head and place a grounded stainless steel union in the transfer line (Teflon tubing) between the syringe pump and the DESI emitter (*see* **Note 6**).
7. Optimize mass spectrometer settings and selected reaction monitoring (SRM) transitions of the compounds you wish to monitor during chemical imaging by continuous infusion of approximately 0.1 μM (depending on the ionization efficiency) test solutions in an on-axis position about 1 cm from the entrance of the PDI. Optimized emitter voltage in our setup was 4 kV with a nebulizer gas setting of 50 (linear velocity of 282 m/s), while monitored SRM transitions were m/z 260 → 116 and m/z 260 → 183 with a collision energy of 27 eV for propranolol.
8. Optimize DESI setup geometry used during DESI imaging: spray a 0.1 μM test solution of a compound you wish to monitor onto a glass surface and monitor the SRM signal(s) of it while changing the spray tip-to-surface and spray tip-to-PDI entrance distances and the incident angle. Optimized setup for propranolol in our laboratory included positioning the emitter from approximately 2 mm from both the surface to be analyzed and from the entrance of the PDI, at an approximately 55° angle to the surface.
9. Wire a DO port of the computer-controlled communication device to the appropriate digital input (DI) port of the AUX control interface of the mass spectrometer. *Caution*: Inappropriate wiring can permanently damage the AUX interface!

3.3. DESI-MS Imaging

During an imaging experiment, the surface was moved as follows. At the beginning of an imaging experiment the spray impact plume was positioned at one corner of the area of interest and the sampling end of the PDI-heated chamber was positioned to just touch the tissue. The first lane was scanned by moving the surface parallel to the *x*-axis at a forward surface scan rate with the spray plume fixed in the optimal position in front of the sampling inlet. At the end of the first lane, the surface was lowered by 2 mm followed by moving the surface at a return speed back to the beginning of the first lane. With the surface lowered there was a 2 mm gap between the sampling capillary and the surface. As such, the desorption spray plume impacted the tissue surface outside the area of interest or at an already analyzed area preventing contamination of yet unanalyzed tissue regions. At the same time, the 2 mm gap between the sampling capillary and the surface eliminated ion signal as well. When the beginning of the first lane was reached, the surface was moved parallel to the *y*-axis with the lane spacing distance followed by raising the surface 2 mm so the sampling capillary touched the tissue again. The following lanes were scanned similar to the first lane.

1. Mount the glass slide holding the tissue section onto the robotic platform using a double-sided tape (*see* **Note 7**). Be sure that the sample is secured and completely parallel to the *XY* plane of the robotic platform.

2. Decide on the surface scan rate, the size of the area to be imaged, and the lane spacing. Set up these parameters in the stage control software and instruct the software to send a trigger signal to the mass spectrometer at the beginning of each lane scan.

3. Calculate the time necessary to scan a lane using the following formula: acquisition time per lane (ATPL) = lane length/surface scan rate. Calculate the time (TB) necessary to move the surface to the starting position of the following lane after a lane is scanned. This time is the sum of the time necessary to lower the surface by 2 mm at the end of the lane scanned, to move the surface at a return speed back to the beginning of the lane scanned, to move the surface parallel to the *y*-axis with the lane spacing distance, and to raise the surface 2 mm so the sampling capillary touches the tissue again. The acquisition time of the whole experiment can be calculated as number of lane times (ATPL + TB).

4. Fill up a glass syringe with the appropriate amount of DESI solvent. For propranolol-dosed tissues 80/20 (v/v) methanol/water solution was delivered to the emitter by a syringe pump at a flow rate of 5 μl/min (*see* **Note 8**). Make sure that the syringe contains enough DESI solvent to

acquire the chemical image of the desired size (*see* **Note 9**). This can be calculated by multiplying the flow rate and the acquisition time of the whole experiment and adding some extra solvent to allow for initial spray stabilization.

5. Create a new sample list in the mass spectrometer acquisition software. Set the number of samples in the list equal to the number of lanes to be scanned. Set the acquisition time of a sample to be equal to ATPL. Instruct the software to wait for an external trigger signal to start acquiring data before each sample.
6. Move the surface so when spraying on it the analyzed areas of interest are not contaminated (*see* **Note 10**). Start the syringe pump, turn on the nebulizer gas and high voltage, and let the spray stabilize for approximately 1 min before starting.
7. Move the surface so the spray plume interrogates the starting point. Make sure that the PDI just touches the surface (*see* **Fig. 13.1** and **Note 11**).

Fig. 13.1. Photograph of a 4000 QTRAP mass spectrometer atmospheric sampling interface region equipped with a PDI during imaging of a mouse thin tissue section showing the DESI emitter, surface, and sampling inlet. (Reproduced from ref. (13) with permission from American Chemical Society.)

8. Start data acquisition in the mass spectrometer software. The mass spectrometer should wait for a trigger signal from the stage control software.
9. Start the movement of the surface in the stage control software.
10. When all sample files are collected (*see* **Note 12**), turn off the syringe pump, wait for approximately 1 min (to stop the solvent delivery), and then turn off the nebulizer gas and the high voltage. Lower the surface so the PDI is above

the surface by approximately 2 mm. This procedure ensures that even if the spray is turned on accidentally or removal of the slide is attempted, it will not contaminate an area of interest or damage the spray tip, respectively.

11. As a first step of the data analysis, convert the data files into a format that is appropriate to the visualization software of your choice. For example, the freely available ImgConverter v3.0 software (11) is able to convert Xcalibur mass spectra files (.RAW extensions) into Analyze 7.5 format files (.img, .hdr, and .t2m) that are required by the BioMap image analysis software (http://www.maldi-msi.org, free of charge). BioMap is then able to generate 2D images of the surface versus the intensity. For Sciex platforms, we have developed an analyst processing script that converts analyst data files (.WIFF extensions) with SRM transitions into a tab-separated text file that can be easily imported into any popular data analysis and graphing software [e.g., Origin (www.originlab.com) and SigmaPlot (www.sigmaplot.com)] to plot as a 2D surface plot (intensity versus *X*, *Y* coordinates) (*see* **Note 13**). **Figures 13.2** and **13.3** show distribution of propranolol in mouse thin tissue sections obtained using surface scan rates of 0.1 and 7 mm/s, respectively.

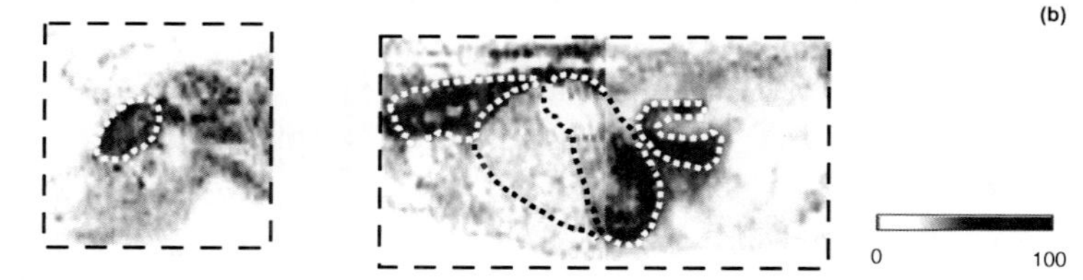

Fig. 13.2. (**a**) Scanned optical image of a 40 μm thick sagittal whole-body tissue section of a mouse dosed intravenously with 7.5 mg/kg propranolol and euthanized 20 min after dose. (**b**) Distribution of propranolol in 20×20 mm and 38×20 mm areas measured by DESI-MS/MS (SRM: *m/z* 260 → 116) using 80/20 (v/v) methanol/water as DESI solvent at a flow rate of 5 μl/min. Surface scan rate was 0.1 mm/s, dwell time was 100 ms, and the images were created from 41 lanes with 500 μm spacing. Pixel size was 84 μm (h) × 500 μm (v), and experiment times were 150 and 285 min for the 20×20 mm and 38×20 mm areas, respectively. (Reproduced from ref. (13) with permission from American Chemical Society.)

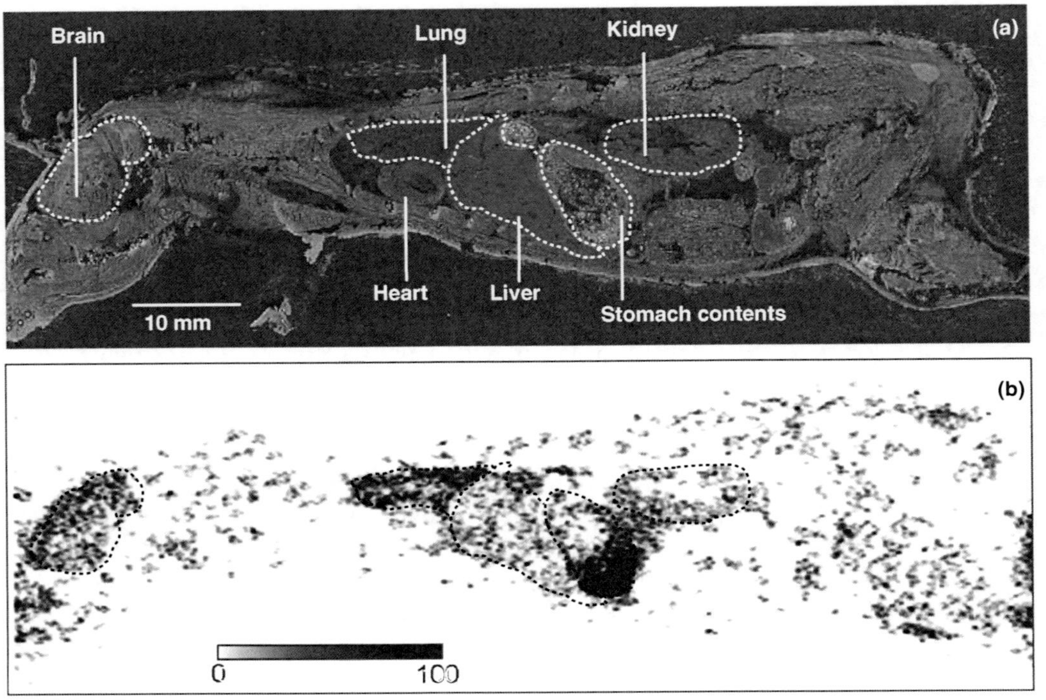

Fig. 13.3. (**a**) Scanned optical image of a 40 μm thick sagittal whole-body tissue section of a mouse dosed intravenously with 7.5 mg/kg propranolol and euthanized 60 min after dose. (**b**) Distribution of propranolol in the 94 mm × 30 mm tissue section presented in (**a**) measured by DESI-MS/MS (SRM: *m/z* 260 → 116) using 80/20 (v/v) methanol/water as DESI solvent at a flow rate of 5 μl/min. Surface scan rate was 7 mm/s, dwell time was 20 ms, and the image was created from 151 lanes with 200 μm spacing. Pixel size was 140 μm (h) × 200 μm (v), and total experiment time was 79 min. (Reproduced from ref. (13) with permission from American Chemical Society.)

4. Notes

1. The controller of the MS2000 robotic platform is also equipped with a joystick that allows manual control of the stage. This feature is extremely useful for quick manual positioning, simple testing, etc.
2. Check if the tip of the fused silica capillary is not broken or cracked. A damaged tip most likely results in a skewed spray causing irreproducible data. Replace the tip if necessary.
3. While in the most common configuration the *XY* plane is horizontal, vertical configuration of the *XY* plane can be advantageous sometimes, i.e., for easier access to the sample, monitoring purposes.
4. PDIs with different lengths (e.g., 2 cm) are available. Use of the shortest PDI to analyze the whole surface minimizes ion-transfer losses.

5. A zoomed view of the spray tip/plume/sampling capillary region is critical when setting up the ideal DESI conditions before an experiment and when keeping the sampling capillary on the surface during the chemical imaging experiment.
6. *Caution*: The DESI emitter floats at the high ES voltage, and appropriate shields and interlocks should be used to avoid accidental contact with this component.
7. Analyzing tissue samples with DESI can create harmful "bioaerosols." Appropriate safety protocols should be followed and protective equipment should be worn with respect to exposure of the operator and decontamination of equipment used in the DESI imaging studies.
8. The DESI solvent, and solvent and gas flow rates, should be optimized for the specific tissue and analyte. Under optimal conditions, only the most energetic part of the DESI spray plume should be able to extract the analyte out of the tissue, otherwise the "washing effect" (15) will result in cross contamination and signal and resolution loss. Also, if the gas and/or the solvent flow rates are high, it can extensively damage the tissue section. It is advised to sacrifice a tissue section to optimize these parameters for maximum ion signal and resolution.
9. Use of the combination of a syringe and a syringe pump that delivers a stable solvent flow is important for successful DESI imaging. Alternatively, deliver the solvent with an HPLC pump that has sufficient solvent reservoir capacity.
10. It is extremely important not to contaminate the area of interest before a DESI imaging. Be sure that there is no spray or the spray plume moves outside of the area of interest when moving the surface.
11. Using a PDI, the highest ion signal can be achieved by the PDI just touching the surface. However, with other sampling capillary configurations, e.g., using an extended ion-transfer tube available from Prosolia (Indianapolis, IN, USA) for Thermo instruments, the optimal position of the sampling capillary may be different.
12. It is advised to at least periodically check on the data in the files that are collected. For example, timely recognition of accidental clogging of the sampling capillary caused by tissue particles and indicated by DESI signal loss can save valuable research time.
13. Because of the wide range of the mass spectrometers that can be used for DESI imaging and the data formats they use, only an example of the data analysis process was given.

Most mass spectrometer manufacturers distribute a self-development kit (SDK) to access the data in the data files they support (e.g., Xcalibur SDK from Thermo, Analyst Scripting from Sciex). See the mass spectrometer manufacturer about these SDKs.

Acknowledgments

The imaging protocols described here were derived from work supported by the Division of Chemical Sciences, Geosciences, and Biosciences, Office of Basic Energy Sciences, United States Department of Energy under Contract DE-AC05-00OR22725 with ORNL, managed and operated by UT-Battelle, LLC.

References

1. Takáts, Z., Wiseman, J. M., Gologan, B., Cooks, R. G. (2004) Mass spectrometry sampling under ambient conditions with desorption electrospray ionization. *Science*, **306**, 471–473.
2. Van Berkel, G. J., Pasilis S. P., Ovchinnikova, O. (2008) Established and emerging atmospheric pressure surface sampling/ionization techniques for mass spectrometry. *J Mass Spectrom*, **43**, 1161–1180.
3. Ifa, D. R., Wiseman, J. M., Song, Q., Cooks, R. G. (2007) Development of capabilities for imaging mass spectrometry under ambient conditions with desorption electrospray ionization (DESI). *Int J Mass Spectrom*, **259**, 8–15.
4. Kertesz, V., Van Berkel, G. J. (2008) Scanning and surface alignment considerations in chemical imaging with desorption electrospray mass spectrometry. *Anal Chem*, **80**, 1027–1032.
5. Ifa, D. R., Gumaelius, L. M., Eberlin, L. S., Manicke, N. E., Cooks, R. G. (2007) Forensic analysis of inks by imaging desorption electrospray ionization (DESI) mass spectrometry. *Analyst*, **132**, 461–467.
6. Kertesz, V., Van Berkel, G. J. (2008) Improved imaging resolution in desorption electrospray ionization mass spectrometry. *Rapid Commun Mass Spectrom*, **22**, 2639–2644.
7. Van Berkel, G. J., Kertesz, V. (2006) Automated sampling and imaging of analytes separated on thin-layer chromatography plates using desorption electrospray ionization mass spectrometry. *Anal Chem*, **78**, 4938–4944.
8. Pasilis, S. P., Kertesz, V., Van Berkel, G. J., Schulz, M., Schorcht, S. (2008) HPTLC/DESI-MS imaging of tryptic protein digests separated in two dimensions. *J Mass Spectrom*, **43**, 1627–1635.
9. Ifa, D. R., Manicke, N. E., Dill, A. L., Cooks, R.G. (2008) Latent fingerprint chemical imaging by mass spectrometry. *Science*, **321**, 805.
10. Wiseman, J. M., Ifa, D. R., Song Q., Cooks, R. G. (2006) Tissue imaging at atmospheric pressure using desorption electrospray ionization (DESI) mass spectrometry. *Angew Chem Int Ed*, **45**, 7188–7192.
11. Wiseman, J. M., Ifa, D. R. Venter, A., Cooks, R. G. (2008) Ambient molecular imaging by desorption electrospray ionization mass spectrometry, *Nat Protoc*, **3**, 517–524.
12. Wiseman, J. M., Ifa, D. R., Zhu, Y., Kissinger, C. B., Manicke, N. E., Kissinger, P. T., Cooks, R. G. (2008) Desorption electrospray ionization mass spectrometry: imaging drugs and metabolites in tissues. *Proc Nat Acad Sci U S A*, **105**, 18120–18125.
13. Kertesz, V., Van Berkel, G. J., Vavrek, M., Koeplinger, K. A., Schneider, B. B., Covey, T. R. (2008) Comparison of drug distribution images from whole-body thin tissue sections

obtained using desorption electrospray ionization tandem mass spectrometry and autoradiography. *Anal Chem*, **50**, 5168–5177.

14. Leuthold, L. A., Mandscheff, J.-F., Fathi, M., Giroud, C., Augsburger, M., Varesio, E., Hopfgartner, G. (2006) Desorption electrospray ionization mass spectrometry: direct toxicological screening and analysis of illicit ecstasy tablets. *Rapid Commun Mass Spectrom*, **20**, 103–110.
15. Pasilis, S. P., Kertesz, V., Van Berkel, G. J. (2008) Surface scanning analysis of planar arrays of analytes with desorption electrospray ionization-mass spectrometry. *Anal Chem*, **79**, 5956–5962.

Chapter 14

Mass Spectrometry Imaging of Small Molecules Using Matrix-Enhanced Surface-Assisted Laser Desorption/Ionization Mass Spectrometry (ME-SALDI-MS)

Qiang Liu, Yongsheng Xiao, and Lin He

Abstract

Surface-assisted laser desorption/ionization mass spectrometry (SALDI-MS) uses inorganic particles or porous surfaces as the energy-mediating means to promote desorption and ionization of low-mass analytes of interest. With good stability during laser ablation, SALDI substrates exhibit reduced background in the low-mass region that is often crowded in conventional matrix-assisted laser desorption/ionization (MALDI) due to matrix fragmentation; a benefit renders SALDI-MS attractive in imaging low-mass species. Practical application of SALDI-MS, however, is hindered by its unsatisfied detection sensitivity for most compounds. With aims of improving MS imaging resolution and sensitivity of low-mass species, we describe an experimental protocol using a hybrid ionization method, termed matrix-enhanced SALDI (ME-SALDI), to detect crucial low-mass species with their spatial distribution in mouse brain tissue.

Key words: Surface-assisted laser desorption/ionization (SALDI), desorption ionization on porous Si (DIOS), matrix-enhanced SALDI (ME-SALDI), low-mass species, metabolite, 2D imaging.

1. Introduction

Surface-assisted laser desorption/ionization is a laser-induced ionization method developed in parallel to the more well-known matrix-assisted laser desorption/ionization (MALDI) method. Using inorganic materials as energy-mediating media, it has been exploited as a possible alternative to MALDI, especially in small molecule detection (1–4). The most successful SALDI substrate was reported by Siuzdak and coworkers, in which a silicon surface of mesoporous features was used, commonly termed as

S.S. Rubakhin, J.V. Sweedler (eds.), *Mass Spectrometry Imaging*, Methods in Molecular Biology 656,
DOI 10.1007/978-1-60761-746-4_14,

desorption/ionization on porous silicon (DIOS) (4). The use of DIOS-MS in 2D MS imaging of metabolites has been successfully demonstrated (5). Other SALDI-MS imaging of metabolites using different SALDI substrates has also been reported in the literature (6, 7).

With the aim of improving MS imaging sensitivity, SALDI and MALDI are synergistically combined to yield a hybrid ionization method, i.e., matrix-enhanced surface-assisted laser desorption ionization (ME-SALDI) (8). Comparing to conventional MALDI, this ionization source uses the porous Si surface as an effective photon absorption medium that significantly reduces laser fluence needed to bring small molecules into the gas phase; consequently less fragmentation is observed. The porous surface playing the major role in desorption also renders the desorption process less dependent on the "perfect" incorporation of matrix and analyte molecules. As a result, a much smaller amount of matrix can be applied to achieve better detection sensitivity with reduced background. In comparison to SALDI/DIOS, addition of acidic matrix molecules provides a proton-rich environment that drastically improves the ionization efficiency of the desorbed species (i.e., better sensitivity) (9) and expands the detectable mass range. Localized extraction of analytes due to the employment of matrix solutions of traditional MALDI is upheld as the extra bonus to traditional SALDI.

2. Materials

2.1. Preparation of Porous SALDI/DIOS Substrates

1. *n*-Type Sb-doped (100) single-crystalline silicon wafers at 0.005–0.02 Ω/cm (Silicon Sense Inc., Nashua, NH, USA) were stored under vacuum prior to use.
2. Hydrofluoric acid (HF) etching solution: 49% HF (Fisher Scientific) with 95% ethanol (Aaper Alcohol) (v/v 1:1).
3. Hydrofluoric acid (HF) wash solution: 49% HF (Fisher Scientific) with 95% ethanol (Aaper Alcohol) (v/v 1:10).
4. Hydrogen peroxide (H_2O_2) oxidation solution: 30% H_2O_2 (Fisher Scientific) with 95% ethanol (Aaper Alcohol) (v/v 1:1).

2.2. Tissue Sections Preparation

1. Optimal Cutting Temperature (OCT) compound (Ted Pella Inc., Redding, CA, USA).
2. Methylene blue (MP Biomedicals) staining solution: 0.15 g methylene blue powder in 100 ml of 70% ethanol. The solution was stirred overnight.

3. Leica Stainless-Steel Re-usable Microtome Knives (Fisher Scientific) were used in tissue slicing.
4. Carbon steel surgical blades (Bard-Parker) were used to pre-cut tissue samples into small portions.
5. Mouse brain tissue was received as gift.
6. Cryo-cut microtome (American Optical Corp., Buffalo, NY, USA).

2.3. Matrix Deposition

1. 2,5-Dihydroxybenzoic acid (DHB)and α-cyano-4-hydroxycinnamic acid (CHCA) (Sigma-Aldrich, St. Louis, MO, USA).
2. Silicon oil (Sigma-Aldrich) was used in the oil bath.
3. Sublimation glassware (Chemglass).
4. Edwards E2M8 high vacuum pump (Edwards High Vacuum, Burlington, ON, Canada) with a vacuum meter.
5. EG&G Princeton Potentiostat, Model 273 (EG&G Princeton Applied Research, Princeton, NJ, USA).

2.4. Mass Spectrometry

1. Applied Biosystems Voyager DE-STR MALDI-TOF mass spectrometer (Framingham, MA, USA).
2. MMSIT MALDI Imaging Tool software V2.2.0 (© 2004 by Dr. Markus Stoeckli, Novartis Institutes for BioMedical Research, Basel, Switzerland) downloadable at www.maldi-msi.org.
3. BioMap 3.7.5.4 (by Dr. Markus Stoeckli, Novartis Institutes for BioMedical Research).

2.5. Optical Microscope

1. Leica DMRX light microscope (Leica Microsystems, Wetzlar, Germany) equipped with a Donpisha XC-003P CCD (Sony, Japan).

3. Methods

3.1. Preparation of Porous SALDI Substrate

1. Silicon wafers were cut into 1-cm^2 square-shaped chips and dipped into the HF wash solution for 1 min to remove the oxidized layer. (Caution: HF can cause severe tissue damage upon contact or inhalation.) The Si chips were then washed in 95% ethanol and dried with N_2.
2. A Si chip was assembled in an anodic etching cell made of Teflon, as shown in **Fig. 14.1**. A three-electrode system was used where a Au working electrode was placed under the Si chip, a Pt counter electrode, and a Pt reference electrode

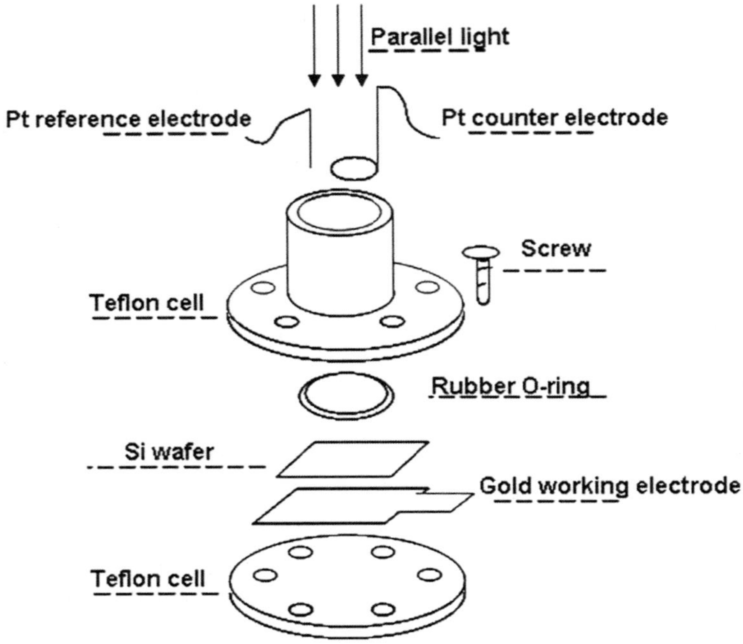

Fig. 14.1. A schematic drawing of an electrochemical etching cell for porous Si preparation.

were placed above the surface. The cell was clamped tight before it was filled with the HF etching solution. A 50-W tungsten lamp was aligned to provide uniform illumination of the Si surface. An EG&G Princeton Potentiostat, Model 273, was programmed to control etching current and time. The complete setup was placed in a chemical hood for ventilation (*see* **Note 1**).

3. The chip was electrochemically etched in the HF etching solution for 1 min at a current density of 5 mA/cm^2. HF waste was carefully transferred to a pre-labeled waste bottle in the hood dedicated to HF. The setup was washed with 95% ethanol before disassembly. The produced porous Si substrate (i.e., DIOS) was washed again with copious 95% ethanol, dried with N_2.
4. Double-etch of the DIOS substrate was carried out by first dipping the substrate into the H_2O_2 oxidation solution for 1 min, followed by washing in 95% ethanol, drying in a N_2 stream, and dipping in the HF wash solution. The substrate was then washed again in 95% ethanol and stored in the same medium till needed (*see* **Note 2**).
5. Prior to usage, the substrates were refreshed in the HF wash solution for 1 min, washed in 95% ethanol, and dried with N_2.

3.2. Tissue Section Preparation

1. Fresh mouse heart and liver tissues were pre-cut into thin slices using carbon steel surgical blades with at least one dimension at less than 2 mm. Fresh mouse brain was used directly without pre-cutting. All tissue or thin tissue slices were snap frozen in liquid N_2 immediately and stored in –80°C prior to usage (*see* **Note 3**).
2. A cryo-cut microtome equipped with Leica Stainless-Steel Re-usable Microtome Knives was used in the following description. Procedural modification may be needed when different equipment was to be used. Frozen animal tissue slices were mounted onto the microtome chunk that held the tissue in place with the OCT compound. The temperature was slowly brought up to −20°C. The frozen tissue samples were then sectioned into 10-μm-thick slices. The 10-μm-thick tissue slices were transferred onto a dry porous Si substrate and slowly brought to room temperature.
3. Meanwhile, the adjacent section was transferred onto a glass slide for conventional histological staining in which the methylene blue staining solution (0.15%, 200 μl) was deposited onto the tissue section. After 10-s staining, excess stain was washed away with 95% ethanol.
4. A Leica DMRX light microscope equipped with a Donpisha XC-003P CCD camera was used to collect the optical image of stained sections to provide visual validation of IMS data.

3.3. Matrix Deposition

1. A sublimation apparatus was set up in a chemical hood (**Fig. 14.2**). An Edwards E2M8 high vacuum pump with a vacuum meter to provide controllable vacuum in the sublimation chamber (*see* **Note 4**).
2. The tissue coated porous Si was attached to the bottom of the apparatus condenser in direct contact with the running cooling water. 0.3 g of CHCA or DHB was added to the bottom of the sublimation chamber (*see* **Note 5**).
3. The oil bath was pre-heated to 110°C for DHB or to 130°C for CHCA. The chamber pressure was maintained at ~50 torr for 2 min before immersing the apparatus into oil bath for 5 min.
4. After removing the sublimation setup from the oil bath, the vacuum was released and the DIOS coated with matrix was taken out and loaded into the MS sample chamber immediately. The vacuum pump, the cooling water, and the oil bath were turned off. Excess DHB and CHCA were discarded.

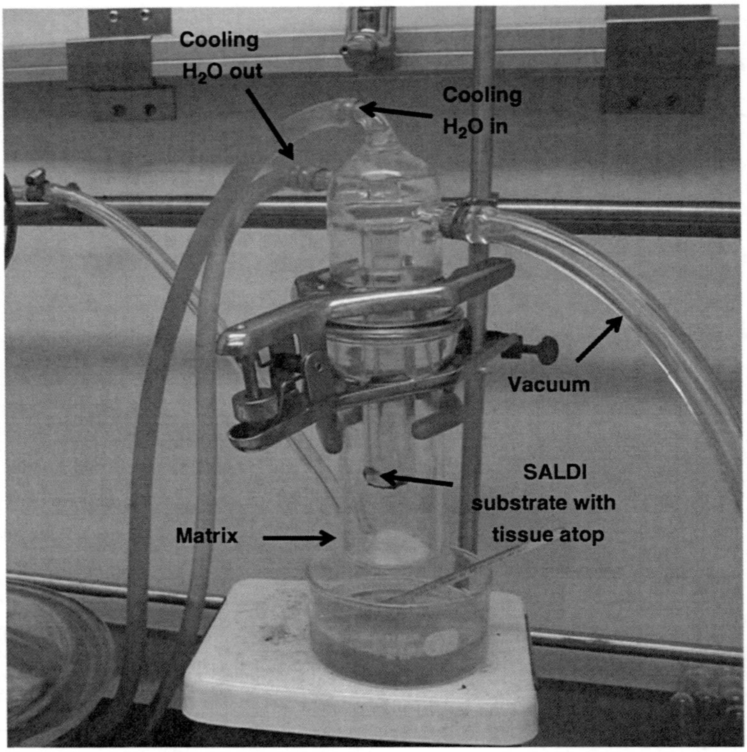

Fig. 14.2. A photo picture of in-house sublimation apparatus.

3.4. Mass Spectrometry Imaging

1. A Voyager DE-STR MALDI-TOF mass spectrometer was used for the following description. Modification in instrumental conditions may be needed to achieve optimal imaging results when different instruments are used.
2. The mass spectrometer was equipped with a 20-Hz N_2 laser. The laser irradiation energy was adjusted by a neutral density filter and the beam size was adjusted to 35 μm with an adjustable pinhole placed close to the laser entrance window. The actual laser beam size was obtained by increasing laser fluence till a burn mark was left behind to allow off-line measurements of the laser beam footprint (*see* **Note 6**).
3. The matrix-coated DIOS substrate with a piece of mouse brain tissue attached was mounted on a stainless steel MALDI plate with double-side conductive tape.
4. The instrument was operated at an accelerating voltage of 20 kV in a reflector and in a positive-ion mode. 50 Laser shots were averaged at each location to yield one accumulated spectrum for each imaging pixel. The translational stage was operated at 50-μm stepwise (*see* **Note 7**).

5. The MS instrument was controlled by MMSIT MALDI Imaging Tool software. The imaging area was manually selected along the outline of the tissue section in MMSIT (*see* **Note 8**).
6. 2D ion maps were reconstructed using BioMap 3.7.5.4. An example of results is shown in **Fig. 14.3**.

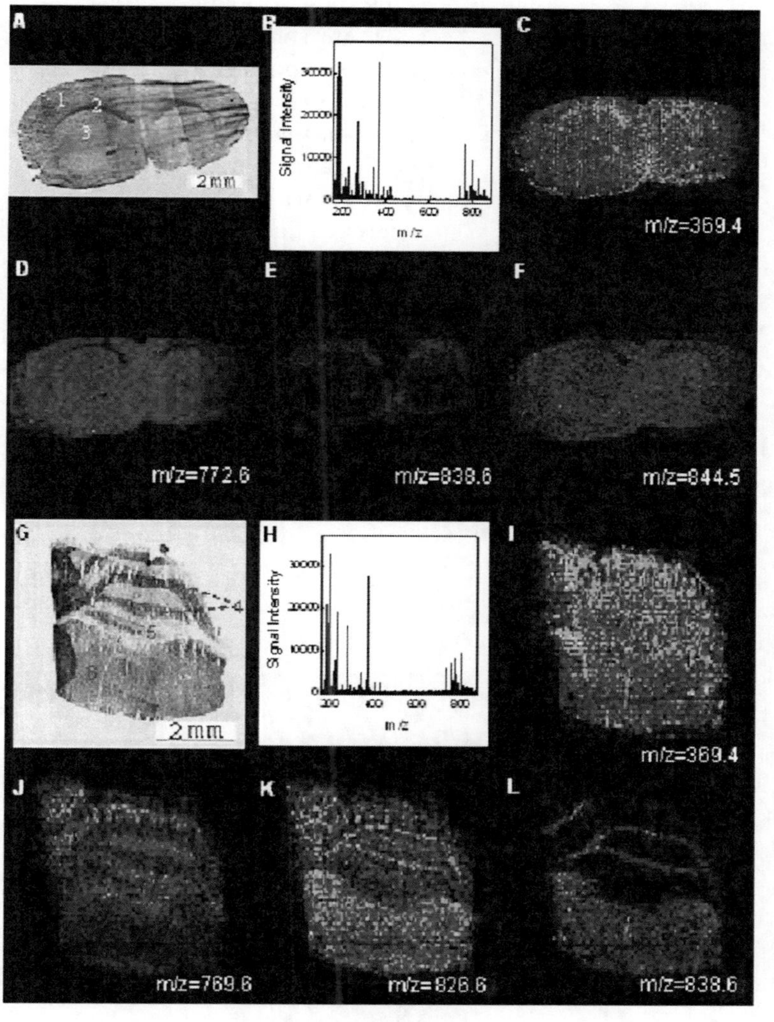

Fig. 14.3. Optical images of the coronal sections of (**a**) mouse cerebrum and (**g**) mouse cerebellum. Representative MS spectra collected from the adjacent coronal sectioning of (**b**) mouse cerebrum and (**h**) mouse cerebellum in ME-SALDI, respectively. Also shown are reconstructed 2D images for ions at (**c**) *m/z*=369.4, (**d**) *m/z* =772.6, (**e**) *m/z* = 838.6, and (**f**) *m/z* = 844.5 from the mouse cerebrum coronal section, and for ions at (**i**) *m/z* =369.4, (**j**) *m/z* =769.6, (**k**) *m/z* = 826.6, and (**l**) *m/z* = 838.6 from mouse cerebellum coronal section. Molecular identification sees the text. Different areas in the brain tissue are labeled numerically: 1, cerebral cortex; 2, corpus callosum; 3, striatum; 4, cerebellar nuclei; 5, molecular layer in cerebellum; 6, brain stem. Figure reproduced from (8) with permission.

4. Notes

1. During the etching process, gas bubbles were mildly, but continuously released. The Si chip turned from gray to bright blue first, then quickly changed to a golden color. Over time the color of the Si chip turned darker and darker, suggesting formation of a rough surface. The final DIOS substrate should exhibit a dark blue hue after dried. Formation of porous features on Si was critically related to the amount of irradiating light, the current density applied, and the enduring etching time. The color of the substrate can be used as a reference point in troubleshooting. For example, a substrate with a yellowish color suggested it was over-etched whereas a light gray hue suggested it was under-etched.
2. Severe MS performance degradation has been correlated to exposure of porous Si to air. The DIOS substrates are therefore required to be stored in ethanol until needed. For the substrates coated with tissue sample, matrix deposition should occur immediately out of the same concern.
3. Freeze and thaw of tissue samples is a critical step in preparation of tissue sections. The most common problem observed in frozen tissue sections is the ice crystal damage, which causes leaky tissues and blotched tissue surface (**Fig. 14.4**).

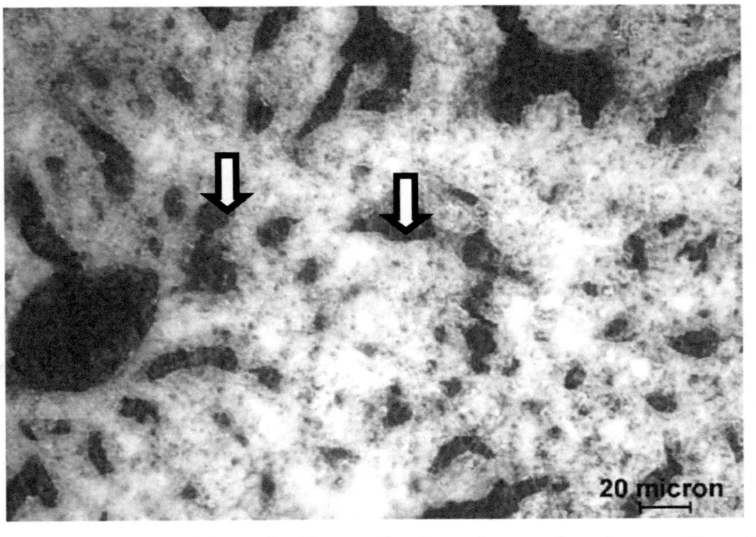

Fig. 14.4. A representative optical image of a piece of mouse liver tissue section with noticeable ice crystal damages (*arrows*).

The size of ice crystals is usually determined by the speed of the freezing process through the whole tissue. Therefore, a pre-cutting step is recommended to reduce the overall tissue size before snap freezing. Tolerance toward ice crystal damage was found at different levels for different types of tissue. For instance, the most serious ice crystal damage was observed in mouse liver sections but little in mouse brain sections.

4. Sublimation is sensitive to the vacuum pressure and the temperature (10). The conditions described here were optimized for our setup and the deposition efficiency may vary among laboratories.
5. To quantitatively control the amount and the quality of the matrix deposited, a clear glass side was coated under each sublimating condition in parallel. UV absorbance of the matrix layer on the side and the microscopic images of matrix morphology were collected to monitor the thickness and uniformity of matrix deposition. Empirically it has been found that a matrix layer with UV absorbance between 1.0 and 2.0 was, in general, suitable for ME-SALDI.
6. Irradiation energy was adjusted to achieve optimal MS performance prior to sample imaging. It was significantly lower than the energy needed for traditional MALDI.
7. Imaging resolution can be further improved through oversampling (11).
8. The overall length of an imaging experiment is mainly determined by the number of pixels (i.e., spectra, determined by the imaging spatial resolution and the tissue size), the number of laser shots per pixel, and the frequency of laser operated. For a 20-Hz laser and 50 shots/spectrum, the imaging rate is at approximately 1,440 spectra per hour, equivalent to a scanning throughput of 3.6 mm^2 in an hour. The overall size of the imaging data per experiment is determined by the total number of spectra and the mass resolution and the mass window of each spectrum.

Acknowledgments

We thank Dr. Dykstra at the Laboratory for Advanced Electron and Light Optical Methods at College of Veterinary Medicine, North Carolina State University (NCSU) for tissue preparation. We appreciate Drs. G. M. Pollack, J. Padowski, and W. Yue at

School of Pharmacy, University of North Carolina at Chapel Hill and Mrs. Welker at College of Agriculture and Life Sciences at NCSU for providing mice tissue samples.

References

1. Tanaka, K., Waki, H., Ido, Y., Akita, S., Yoshida, Y., Yoshida, T. (1988) Protein and polymer analyses up to *m/z* 100,000 by laser ionization time-of-flight mass spectrometry. *Rapid Commun Mass Spectrom*, **2**, 151–153.
2. Sunner, J., Dratz, E., Chen, Y.-C. (1995) Graphite surface-assisted laser desorption/ionization time-of-flight mass spectrometry of peptides and proteins from liquid solutions. *Anal Chem*, **67**, 4335–4342.
3. Finkel, N. H., Prevo, B. G., Velev, O. D., He, L. (2005) Ordered silicon nanocavity arrays in surface-assisted desorption/ionization mass spectrometry. *Anal Chem*, 77, 1088–1095.
4. Wei, J., Buriak, J. M., Siuzdak, G. (1999) Desorption–ionization mass spectrometry on porous silicon, *Nature*, **399**, 243–246.
5. Liu, Q., Guo, Z., He, L. (2007) Mass spectrometry imaging of small molecules using desorption/ionization on silicon. *Anal Chem*, **79**, 3535–3541.
6. Taira, S., Sugiura, Y., Moritake, S., Shimma, S., Ichiyanagi, Y., Setou, M. (2008) Nanoparticle-assisted laser desorption/ionization based mass imaging with cellular resolution. *Anal Chem*, **80**, 4761–4766.
7. Zhang, H., Cha, S., Yeung, E. S. (2007) Colloidal graphite-assisted laser desorption/ionization ms and msn of small molecules. 2. Direct profiling and MS imaging of small metabolites from fruits. *Anal Chem*, **79**, 6575–6584.
8. Liu, Q., Xiao, Y., Pagan-Miranda, C., Chiu, Y. M., He, L. (2009) Metabolite imaging using matrix-enhanced surface-assisted laser desorption ionization mass spectrometry (ME-SALDI-MS). *J Am Soc Mass Spectrom*, 20, 80–88.
9. Liu, Q., He, L. (2008) Semi-quantitative study of solvent and surface effects on analyte ionization in desorption ionization on silicon (DIOS) mass spectrometry. *J Am Soc Mass Spectrom*, **19**, 8–13.
10. Hankin, J. A., Barkley, R. M., Murphy, R. C. (2007) Sublimation as a method of matrix application for mass spectrometric imaging. *J Am Soc Mass Spectrom*, **18**, 1646–1652.
11. Jurchen, J. C., Rubakhin, S. S. and Sweedler, J. V. (2005) MALDI-MS imaging of features smaller than the size of the laser beam. *J Am Soc Mass Spectrom*, **16**, 1654–1659.

Chapter 15

Preparation of Single Cells for Imaging Mass Spectrometry

Elena S.F. Berman, Susan L. Fortson, and Kristen S. Kulp

Abstract

Characterizing the molecular contents of individual cells is critical for understanding fundamental mechanisms of biological processes. Imaging mass spectrometry (IMS) of biological systems has been steadily gaining popularity for its ability to create precise chemical images of biological samples, thereby revealing new biological insights and improving understanding of disease. In order to acquire mass spectral images from single cells that contain relevant molecular information, samples must be prepared such that cell-culture components, especially salts, are eliminated from the cell surface and that the cell contents are accessible to the mass spectrometer. We have demonstrated a cellular preparation technique for IMS that preserves the basic morphology of cultured cells, allows mass spectrometric chemical profiling of cytosol, and removes the majority of the interfering species derived from the cellular growth medium. Using this protocol, we achieve high-quality, reproducible IMS images from three diverse cell types: MCF7 human breast cancer cells, Madin-Darby canine kidney (MDCK) cells, and NIH/3T3 mouse fibroblasts. This preparation method allows rapid and routine IMS analysis of cultured cells, making possible a wide variety of experiments to further scientific understanding of molecular processes within individual cells.

Key words: ToF-SIMS, imaging mass spectrometry, single cells, molecular profiling, biological imaging.

1. Introduction

In the past few years, there has been an explosion in research utilizing imaging mass spectrometric techniques for biological applications (1–6). While most IMS research has focused on analysis of tissue samples, there is a clear need for single-cell analysis as well. Common cellular experiments utilize entire cell populations, thereby averaging the responses of all of the cells in the population and obscuring subtle differences among individual cells (7).

S.S. Rubakhin, J.V. Sweedler (eds.), *Mass Spectrometry Imaging*, Methods in Molecular Biology 656,
DOI 10.1007/978-1-60761-746-4_15, © Springer Science+Business Media, LLC 2010

By interrogating single cells, an analysis is liberated from assumptions regarding cell population homogeneity and is ensured that all cellular responses can be measured. Understanding the molecular makeup of single cells is necessary to identify small cellular changes that may underlie many biological processes including disease development.

There are several IMS techniques currently available for analyzing single cells. Although more commonly used for tissue analysis, recent advances in matrix-assisted laser desorption/ionization (MALDI) MS imaging have provided greatly enhanced lateral resolution, opening the potential for imaging individual cells (8, 9). To date, however, very few reports of single-cell MALDI imaging have been published (10). Secondary ion mass spectrometry (SIMS), in both dynamic and static mode, has been extensively applied to imaging analysis of single cells. Advances in imaging of cells by dynamic SIMS have been reviewed by Guerquin-Kern et al. (4), with many examples demonstrating superb spatial resolution and localization of elemental species within cells. Recently, a new generation of dynamic SIMS instrumentation, NanoSIMS, has been used to garner subcellular localization of a peptide vector (11), study diatom cell division (12), and perform nanoautography with stable isotope tracers (13) among others. Static SIMS has also been widely applied to cellular imaging, with recent examples including relative quantification of cholesterol in cell membranes (14), three-dimensional imaging (15–17) (among others), and distinguishing cancerous cells of differing breast cancer phenotypes (18). Cellular imaging has also been demonstrated by a variety of other mass spectrometric techniques, including desorption/ionization on silicon (19), laser post-ionization secondary neutral mass spectrometry (20), and atmospheric pressure femtosecond laser IMS (21).

We are utilizing time-of-flight secondary ion mass spectrometry (ToF-SIMS) to show the suitability of our reported cell preparation method for MS imaging. ToF-SIMS is a highly sensitive, imaging MS technique used to detect and map chemical and molecular information from sample surfaces. ToF-SIMS uses a finely focused (as small as 150 nm), pulsed primary ion beam to desorb secondary ions into a time-of-flight mass spectrometer, creating mass spectral images with excellent spatial resolution and good mass resolution. Each image consists of 256×256 pixels where each pixel is a complete mass spectrum.

As with all analytical techniques, sample preparation is critical for reproducible and meaningful IMS results. With any mass spectral technique, to obtain high-quality images of individual cells the cells must be attached to a suitable substrate, free of contamination or interfering components, and accessible to the ionization source. Traditionally for IMS analysis, cells are grown on a conductive substrate and freeze fractured prior to analysis.

The original freeze–fracture technique was reported in 1957 and has been widely used to prepare cells for membrane analysis by electron microscopy (22, 23). Working from this model, Chandra et al. developed a modified freeze–fracture method to prepare cells for imaging MS analysis (24). Currently, the majority of cellular MS imaging studies utilize some variation of this method. Improvements to Chandra's technique, most notably preservation of cryogenic temperatures throughout the fracturing and analysis (25) and freeze-etching with vapor matrix deposition (26) have resulted in improvements to cellular imaging ((14) for example).

Freeze–fracture methods, however, have several disadvantages: they require cryogenic facilities, generate low yields of suitably fractured cells, and, by design, tend to fracture cells between the leaflets of the membrane bilayer. Because the fracture plane is generally inside the membrane bilayer, the cytoplasm of the cell remains obscured by a layer of phospholipids. To circumvent some of these disadvantages, several groups have reported other cellular preparation approaches for IMS. Nygren et al. (27) freeze-dried cells in ammonium formate and then imprinted on silver foil, thus collecting only the membrane lipids for analysis. Liu et al. (19) fix cells in 70% ethanol, preserving the morphology of the cells but not the chemical contents of the cytoplasm. Parry and Winograd have reported embedding cells in a trehalose and glycerol matrix, followed by freeze-drying (28). Altelaar et al. (29) used sucrose and water washes while Fletcher et al. (15) utilized only a water wash; both groups followed these washes with freeze-drying.

This chapter provides an in-depth protocol for our recently reported cellular preparation procedure: a simple "wash and dry" procedure using an iso-osmotic ammonium acetate solution and a gentle argon drying procedure (30). This method sufficiently cleans individual cells for unobscured mass spectral analysis while at the same time preserving the molecular content of the cells (30). We have found that both the wash and dry procedures are critical to the quality of the mass spectral images produced. Components from cell-culture growth medium, most conspicuously salts, seriously interfere with spectral quality making it nearly impossible to collect meaningful molecular information. The simple "wash and dry" cellular preparation technique detailed below not only successfully removes interferences from the growth medium but also permits delicate cells to remain intact until just before the cells are completely dry in order to obtain the maximum molecular information from each cell. Furthermore, this technique is applicable to a wide variety of cell types, generates reproducible results over an extended time frame, alleviates the need for cryogenic facilities, and produces a high yield of cells suitable for mass spectrometric analysis (30). Importantly, the procedure also allows for the imaging and profiling of both membrane and cytosolic molecular contents, as evidenced by excellent localization of both phosphocholine and potassium in the mass spectral images (30). This detailed preparation technique allows routine imaging MS analysis of individual cells, making possible a wide variety of experiments to further scientific understanding of molecular processes within individual cells.

2. Materials

2.1. Silicon Imaging Substrates

1. Standard 6-inch silicon wafers (University Wafer, South Boston, MA, USA).
2. Diamond-tipped scribe.
3. 100% ethanol and low-lint wiper (Kimwipe).
4. Glass or PFA (perfluoroalkoxy) storage and transfer containers (*see* **Note 1**).

2.2. Cell Culture

1. MCF7 human breast adenocarcinoma cells, NIH/3T3 mouse fibroblasts, and MDCK canine kidney cells (American Type Culture Collection (ATCC), Manassas, VA, USA).
2. Growth medium for MCF7 Cells: Dulbecco's Modified Eagle's Medium (DMEM) supplemented with 5% fetal bovine serum (FBS), 1% non-essential amino acids, 10 μg/ml insulin, 2 mM L-glutamine, and 1% penicillin/streptomycin (*see* **Note 2**).
3. Growth medium for NIH/3T3 mouse fibroblasts: DMEM supplemented with 4 mM L-glutamine, adjusted to contain 1.5 g/l sodium bicarbonate, 4.5 g/l glucose, and 10% FBS (*see* **Note 2**).
4. Growth medium for MDCK canine kidney cells: Eagle minimum essential medium (MEM) with 2 mM L-glutamine, adjusted to contain 1.5 g/l sodium bicarbonate, 0.1 mM non-essential amino acids, 1.0 mM sodium pyruvate, and 10% FBS (*see* **Note 2**).
5. Hank's balanced salts solution (HBSS) (Sigma-Aldrich, St. Louis, MO, USA).
6. 0.25% (w/v) trypsin (Gibco, Invitrogen) diluted in Ca, Mg-free phosphate buffered saline (PBS).
7. Cell-culture flasks, 75 cm^3 canted neck (VWR Scientific, West Chester, PA, USA).

8. 60-mm glass petri dishes, autoclaved before each use (*see* **Note 1**).

2.3. Cell Proliferation

1. Standard 96-well plates for cell culture.
2. Aqueous Non-Radioactive Cell Proliferation Assay (Promega, Madison WI, USA): Briefly, this colormetric assay measures the bioreduction of a tetrazolium compound (3-(4,5-dimethylthiazol-2-yl)-5-(3-carboxymethoxyphenyl)-2-(4-sulfophenyl)-2H-tetrazolium, inner salt; MTS) to a formazan product. The conversion of MTS is directly proportional to the number of living cells in culture.
3. A standard multi-well plate reader (we use BIO-RAD Model 680 Microplate Reader) with absorbance measurement capability at 490 nm.

2.4. Washing and Drying

1. Ammonium acetate (Sigma, St. Louis, MO, USA): 150 mM in Millipore Milli-Q water (18.2 MΩ cm) with pH adjusted to 7.5 using a 1:9 solution of phosphoric acid and 1 M ammonium hydroxide and sterile filtered using a Stericup® Vacuum Filter Cup (Millipore Corporation, Billerica, MA, USA). Store at room temperature under sterile conditions (*see* **Note 3**).
2. Argon gas with attached needle valve to control argon flow rate and slit nozzle of approximately 1×2 mm (*see* **Fig. 15.1** and **Note 4**).

Fig. 15.1. Photographic illustration of the experimental setup for the presented "wash and dry" procedure which produces cells suitably cleaned for imaging mass spectrometry analysis. An ammonium acetate washing solution is quickly removed from cells on a sample substrate using a gentle argon flow.

3. Methods

3.1. Silicon Imaging Substrates

1. Standard silicon wafers are cut into approximately 1.2 cm square silicon substrates.
2. It is helpful to mark, using a diamond scribe, each substrate with a distinctive serial number or other marking on the back (non-mirror-finish) side to facilitate sample identification.
3. Each substrate is cleaned on both sides with 100% ethanol using a low-lint wiper.
4. Store substrates in glass or PFA containers to avoid contamination with silicones and other plastic components (*see* **Note 1**).
5. Just prior to cell-culture use, substrates are sterilized by UV irradiation overnight.

3.2. Cell Culture

1. MCF7 human breast adenocarcinoma cells are grown in the MCF7 DMEM growth medium at 37°C with 5% CO_2.
2. NIH/3T3 mouse fibroblasts are grown in theNIH/3T3 DMEM growth medium at 37°C with 5% CO_2.
3. MDCK canine kidney cells are grown in the MDCK MEM growth medium at 37°C with 5% CO_2.
4. For all cell types, cells are grown in cell-culture flasks and passaged when approaching 75% confluence to create new stock cultures and experimental cultures (see below).
5. Cell passaging consists of removing and discarding cell medium followed by washing with 10 ml Hank's balanced salt solution. Cells are then treated with 2.0–3.0 ml trypsin solution and incubated until the cell layer is dispersed. The trypsin is removed and discarded and cells are then split 1–4 into 25 ml of fresh growth medium for each cell type.
6. Cell cultures are not permitted to exceed 15 passages to avoid morphological changes in the cells.
7. Experimental cultures are created by plating approximately 15,000 cells in 60-mm glass petri dishes containing four sterile silicon substrates. Cells are allowed to attach overnight on the polished side of the silicon substrates at 37°C with 5% CO_2.

3.3. Cell Proliferation

1. The cellular proliferation assay is used to measure the effects of various washing solutions on cell viability (*see* **Note 5**).
2. Cells are plated at 2×10^4 cells/well in a 96-well plate and grown overnight at 37°C, 5% CO_2.

3. In each plate, one row is used per experimental condition, ensuring eight replicates per plate per experiment. Experiments are performed on at least five separate plates to provide sufficient statistical significance for observed differences in cell proliferation.
4. In each plate, at least one row is reserved for the positive control. These cells are treated by removing medium and replacing with fresh medium. This ensures that any loss of cells or viability caused simply by the change of medium is accounted for.
5. In each plate, at least one row is reserved for the negative control. These cells are treated by removing growth medium, adding 100 μl of 70% ethanol, immediately removing the ethanol, and adding fresh cell medium to the wells.
6. Growth medium is removed from cells in the experimental wells and 100 μl of a washing solution is added and immediately removed. Fresh cell medium is then returned to the wells.
7. The plate is then placed back into the incubator for 48 h to allow for cell growth.
8. Cell proliferation is quantified by using the Aqueous Non-Radioactive Cell Proliferation Assay according to the manufacturer's instructions and absorbance of the wells is measured using a standard multi-well plate reader.
9. Statistical significance of differences in cell growth is tested using a one-sample Student's *t*-test (each condition compared to control) and a two-tailed *t*-test assuming unequal variances (different washing solutions compared to each other).
10. **Figure 15.2** shows the results of comparing MCF7 proliferation in the suggested ammonium acetate washing solution with water and the 70% ethanol negative control (*see* **Note 6**).

3.4. Washing and Drying

1. The setup for the washing and drying procedure is shown in **Fig. 15.1**.
2. An individual silicon substrate is carefully removed from the petri dish in which the cells have been growing on the silicon substrate with tweezers holding one corner of the substrate.
3. The sample substrate is quickly (<15 s) dipped into a small beaker of fresh ammonium acetate solution, followed by swishing in one backward/forward motion (*see* **Note 7**).
4. The substrate is then held perpendicular to and with one edge lightly touching an absorbent towel (*see* **Fig. 15.1**).

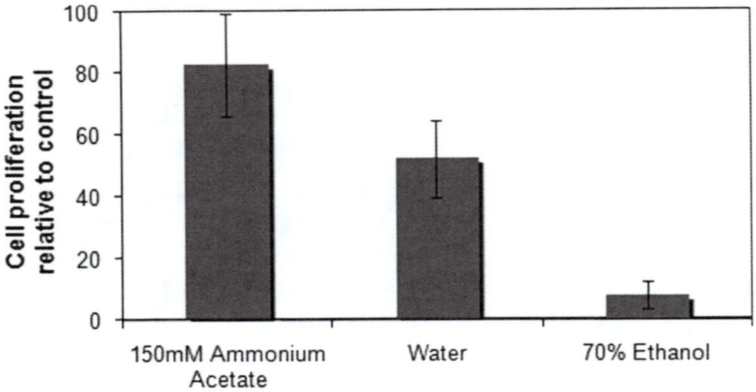

Fig. 15.2. Average effect of washing solutions on MCF7 cell proliferation. Cells were quickly washed (<30 s) with various solutions and allowed to grow in cell-culture medium for 48 h. Data are normalized to the growth of cells that were not washed (cellular medium replaced). Error bars represent the standard deviation of the mean proliferations from five separate experiments.

5. A gentle stream of argon gas is directed at the sample substrate through the slit nozzle with an argon flow rate of 12–15 cm^3/s. The nozzle is positioned so that the argon blows the liquid across the surface of the substrate and onto the absorbent towel. All of the visible liquid should be pushed off of the substrate and onto the towel, thus minimizing the collection of liquid around the cells (*see* **Note 8**). **Figure 15.3** shows an example of a suitably cleaned substrate and the resulting ToF-SIMS image.
6. During drying, care must be exercised to avoid overly vigorous or prolonged blowing on the samples, which can cause

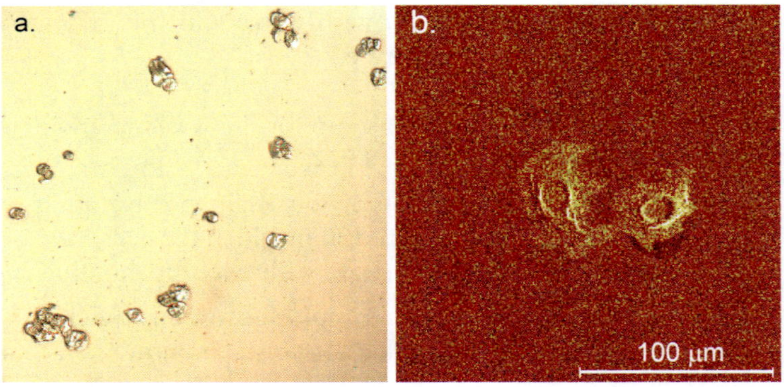

Fig. 15.3. MCF7 cells washed with iso-osmotic ammonium acetate solution and blown dry for imaging mass spectral analysis. Optical 20× (**a**) image of cells showing clean silicon between cells and ToF-SIMS total ion (**b**) image of cells showing no interference from salts or other medium or wash components.

the cells to rupture with subsequent loss of molecular information from the cytosol (*see* **Note 9**).

7. The dry substrate is placed into a glass or PFA storage container and stored at room temperature (*see* **Note 1**).

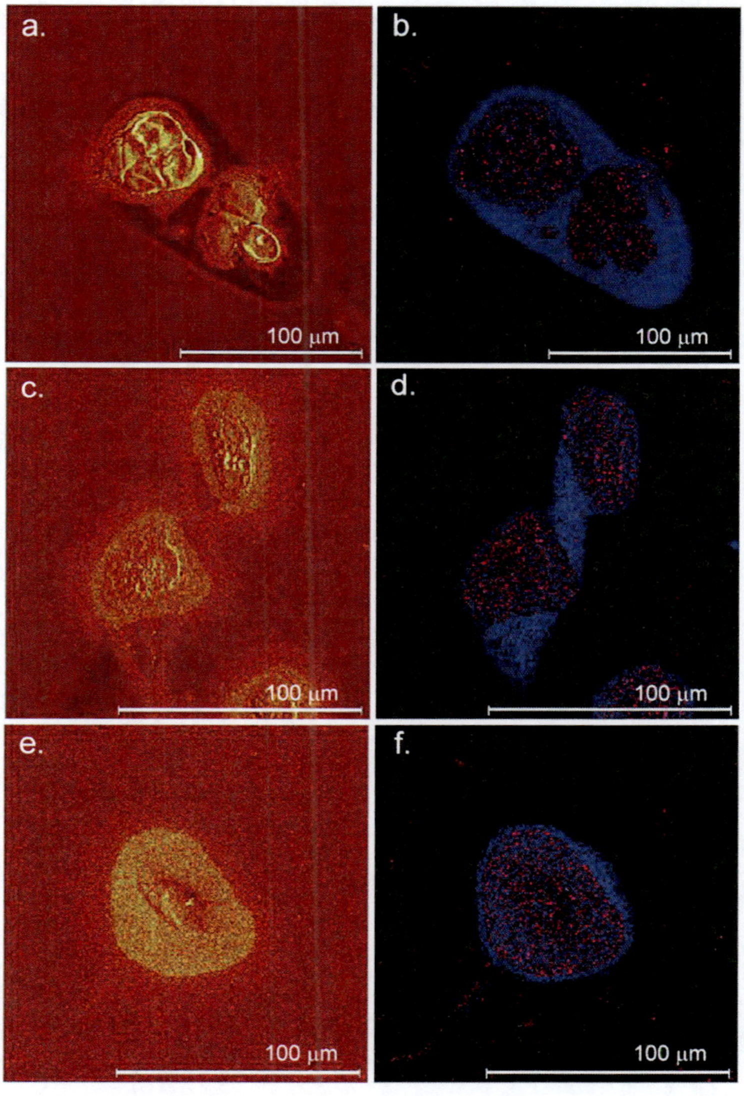

Fig. 15.4. ToF-SIMS images of three different cell types prepared with the reported optimized cellular preparation procedure. In the *left column* are total ion images, in the *right column* composite images of potassium ion (*blue*) and $m/z = 184$, a fragment of the phosphocholine head group (*red*). (**a** and **b**) MCF7 human breast cancer cells, (**c** and **d**) MDCK canine kidney cells, and (**e** and **f**) NIH/3T3 mouse fibroblast. Note the excellent localization of potassium in the area immediately surrounding the cells and the localization of phophocholine on the cellular region. Reproduced from (30) with permission from Elsevier Science.

8. Samples should be analyzed as soon as possible after preparation and in no case more than a week after washing and drying.

3.5. ToF-SIMS Analysis

1. ToF-SIMS measurements are conducted on a PHI TRIFT III instrument (Physical Electronics USA, Chanhassen, MN, USA). This protocol could easily be modified for use with other ToF-SIMS instruments. In addition, it is anticipated that the above preparation procedure will also be applicable to single-cell imaging experiments using a variety of IMS techniques, including single-cell MALDI.
2. The ToF-SIMS instrument is equipped with a gallium ($^{69}Ga^{+}$) liquid metal ion gun (LMIG) operated at 25 kV (*see* **Note 10**).
3. In order to garner the maximum molecular information, data are collected over a mass range of 1–1,850 amu.
4. Positive-ion ToF-SIMS images are generally acquired over an area of 100×100 to 200×200 μm depending on the number and size of cells being imaged. Images are acquired for a minimum of 10 min. All analyses are preformed at room temperature.
5. All ToF-SIMS spectra are calibrated to the common CH_3^+, $C_2H_3^+$, and $C_3H_5^+$ peaks before further analysis.
6. Total-ion images and images of distributions of specific ions of interest are then reconstructed from the raw ToF-SIMS spectra using the WinCadence software provided with the TRIFT III instrument. Examples of ToF-SIMS images of total ions and selected ions of interest for three different cell types are shown in **Fig. 15.4** (*see* **Note 11**).

4. Notes

1. Careful storage of both substrates and prepared samples is critical to avoid introduction of contamination. ToF-SIMS is especially sensitive to silicones and other plastics components which are easily ionizable, therefore all use of plastic containers should be fastidiously avoided. The only exception to this is that plastic cell-culture flasks are used for stock cell cultures.
2. All culture reagents and media were obtained from Invitrogen Corporation, Carlsbad, CA, USA.
3. The ammonium acetate solution is designed to be iso-osmotic to and at the same pH as cellular cytosol. The

solution can be stored under sterile conditions for up to 1 month; however, since cells are viable in the ammonium acetate solution (*see* **Fig. 15.2**), a new aliquot of solution should be used for washing each experimental group of substrates to avoid contamination.

4. We expect that any inert gas which is clean and can be well controlled would be suitable for this drying step.
5. We would recommend repeating the cell proliferation test when new types of cell culture are investigated in order to ensure that cell viability is maintained after washing.
6. Washing with ammonium acetate causes a minimal decline in the number of proliferating cells, while treatment with water shows a statistically significant ($p < 0.0001$) decrease in viable cells, as expected based on cellular physiology and our ToF-SIMS experiments. It is interesting that washing with water does not cause total arrestment of cell growth (compared to the negative control); these results may indicate why this technique has been reported as successful by others (15, 29). However, in our laboratory we have found that washing with water, even for as few as 10 s, can cause the cells to rupture (30). Based on the statistically significantly ($p = 0.013$) better viability obtained by washing with ammonium acetate and the superior imaging characteristics, we conclude that washing with ammonium acetate is preferable to washing with water for imaging MS analysis of cells (see (30) for illustration and more discussion of this point).
7. We tested a large variety of washing solutions in our laboratory, including sucrose, PBS, Tris, HEPES, magnesium acetate, sodium chloride, and ammonium acetate. All solutions were created to be iso-osmotic with cellular cytosol. Various combinations of the above solutes were also evaluated for suitability although none of the combinations were found to provide better results than a single solute. We have concluded, based on cell viability and excellent imaging characteristics, that ammonium acetate is the preferred washing solution (30).
8. Care should be taken to ensure that the argon flow rate is maintained at a constant level. It is often necessary to move the slit nozzle in a zigzag pattern across the substrate to blow all of the washing liquid off of the substrate.
9. As with any process executed by hand, some day-to-day and operator-dependent variability is expected, but with practice we have found it possible to obtain reproducibly good results with this technique.

10. Use of a gallium LMIG has the advantages of ion source stability and better than 150 nm lateral resolution. Use of a cluster ion source would allow detection of more higher molecular weight molecules and fragments, but at the expense of lateral resolution.
11. Additional images and spectra demonstrating the reproducibility of this procedure can be found in (30).

Acknowledgments

The authors gratefully acknowledge the assistance of Ligang Wu, Kyle D. Checchi, James S. Felton, Kuang Jen J. Wu, Cynthia B. Thomas, and Donald J. Sirbuly with various aspects of this research. This work was performed under the auspices of the US Department of Energy by the University of California, Lawrence Livermore National Laboratory under Contract No. W-7405-Eng-48 and supported, in part, by BCRP 10IB-0077, 11NB-0178, and LDRD 04-ERD-104 (LLNL internal funding).

References

1. Belu, A. M., Graham, D. J., Castner D. G. (2003) Time-of-flight secondary ion mass spectrometry: techniques and applications for the characterization of biomaterial surfaces. *Biomaterials*, **24**, 3635–3653.
2. Lockyer, N. P., Vickerman J. C. (2004) Progress in cellular analysis using ToF-SIMS. *Appl Surf Sci*, **231–232**, 377–384.
3. Rubakhin, S. S., Jurchen, J. C., Monroe, E. B., Sweedler, J. V. (2005) Imaging mass spectrometry: fundamentals and applications to drug discovery. *Drug Discov Today*, **10**, 823–837.
4. Guerquin-Kern, J.-L., Wu, T.-D., Quintana, C., Croisy, A. (2005) Progress in analytical imaging of the cell by dynamic secondary ion mass spectrometry (SIMS microscopy). *Biochimica et Biophysica Acta*, **1724**, 228–238.
5. Chaurand, P., Cornett, D. S., Caprioli, R. M. (2006) Molecular imaging of thin mammalian tissue sections by mass spectrometry. *Curr Opin Biotechnol*, **17**, 431–436.
6. Reyzer, M. L., Caprioli, R. M. (2007) MALDI-MS-based imaging of small molecules and proteins in tissues. *Curr Opin Chem Biol*, **11**, 29–35.
7. Elowitz, M. B., Levine, A. J., Siggia, E. D., Swain, P. S. (2002) Stochastic gene expression in a single cell. *Science*, **297**, 1183–1186.
8. Luxembourg, S. L., Mize, T. H., McDonnell, L. A., Heeren, R. M. A. (2004) High-spatial resolution mass spectrometric imaging of peptide and protein distributions on a surface. *Anal Chem*, **76**, 5339–5344.
9. Jurchen, J. C., Rubakhin, S. S., Sweedler, J. V. (2005) MALDI-MS imaging of features smaller than the size of the laser beam. *J Am Soc Mass Spectrom*, **16**, 1654–1659.
10. Rubakhin, S. S., Greenough, W. T., Sweedler, J. V. (2003) Spatial profiling with MALDI MS: distribution of neuropeptides within single neurons. *Anal Chem*, **75**, 5374–5380.
11. Romer, W., Wu, T.-D., Duchambon, P., et al. (2006) Sub-cellular localisation of a ^{15}N-labelled peptide vector using NanoSIMS imaging. *Appl Surf Sci*, **252**, 6925–6930.
12. Audinot, J.-N., Guignard, C., Migeon, H.-N., Hoffmann, L. (2006) Study of the mechanism of diatom cell division by means of ^{29}Si isotope tracing. *Appl Surfe Sci*, **252**, 6813–6815.

13. McMahon, G., Glassner, B. J., Lechene, C. P. (2006) Quantitative imaging of cells with multi-isotope imaging mass spectrometry (MIMS) – nanoautography with stable isotope tracers. *Appl Surf Sci*, **252**, 6895–6906.
14. Ostrowski, S. G., Kurczy, M. E., Roddy, T. P., Winograd, N., Ewing, A. G. (2007) Secondary ion MS imaging to relatively quantify cholesterol in the membranes of individual cells from differentially treated populations. *Anal Chem*, **79**, 3554–3560.
15. Fletcher, J. S., Lockyer, N. P., Vaidyanathan, S., Vickerman, J. C. (2007) ToF-SIMS 3D biomolecular imaging of *Xenopus laevis* oocytes using buckminsterfullerene (C_{60}) primary ions. *Anal Chem*, **79**, 2199–2206.
16. Fletcher, J. S., Rabbani, S., Henderson, A., et al. (2008) A new dynamic in mass spectral imaging of single biological cells. *Anal Chem*, **80**, 9058–9064.
17. Malmberg, P., Kriegeskotte, C., Arlinghaus, H. F., et al. (2008) Depth profiling of cells and tissues by using C_{60}^+ and SF_5^+ as sputter ions. *Appl Surf Sci*, **255**, 926.
18. Kulp, K. S., Berman, E. S. F., Knize, M. G., et al. (2006) Chemical and biological differentiation of three human breast cancer cell types using time-of-flight secondary ion mass spectrometry (ToF-SIMS). *Anal Chem*, **78**, 3651–3658.
19. Liu, Q., Guo, Z., He, L. (2007) Mass spectrometry imaging of small molecules using desorption/ionization on silicon. *Anal Chem*, **79**, 3535–3541.
20. Arlinghaus, H. F., Kriegeskotte, C., Fartmann, M., Wittig, A., Sauerwein, W., Lipinsky, D. (2006) Mass spectrometric characterization of elements and molecules in cell cultures and tissues. *Appl Surf Sci*, **252**, 6941–6948.
21. Coello, Y., Gunaratne, T. C., Dantus, M. (2009) Atmospheric pressure femtosecond laser imaging mass spectrometry. *Proceedings of* SPIE, **7182**, 71821 W.
22. Wells, W. A. (2005) The invention of freeze fracture EM and the determination of membrane structure. *J Cell Biol*, **168**, 174–175.
23. Steere, R. L. (1957) Electron microscopy of structural detail in frozen biological specimens. *J Biophys Biochem Cytol*, **3**, 45–60.
24. Chandra, S., Morrison, G. H. (1992) Sample preparation of animal tissues and cell cultures for secondary ion mass spectrometry (SIMS) microscopy. *Biol Cell*, **74**, 31–42.
25. Cannon, D. M. J., Pacholski, M. L., Winograd, N., Ewing, A. G. (2000) Molecule specific imaging of freeze-fractured, frozen-hydrated model membrane systems using mass spectrometry. *J Am Chem Soc*, **122**, 603–610.
26. Piehowski, P. D., Kurczy, M. E., Willingham, D., et al. (2008) Freeze-etching and vapor matrix deposition for ToF-SIMS imaging of single cells. *Langmuir*, **24**, 7906–7911.
27. Nygren, H., Eriksson, C., Malmberg, P., et al. (2003) A cell preparation method allowing subellular localization of cholesterol and phosphocholine with imaging ToF-SIMS. *Colloids Surf B Biointerfaces*, **30**, 87–92.
28. Parry, S., Winograd, N. (2005) High-resolution ToF-SIMS imaging of eukaryotic cells preserved in a trehalose matrix. *Anal Chem*, 77, 7950–7957.
29. Altelaar, A. F. M., Klinkert, I., Jalink, K., et al. (2006) Gold-enhanced biomolecular surface imaging of cells and tissue by SIMS and MALDI mass spectrometry. *Anal Chem*, **78**, 734–742.
30. Berman, E. S. F., Fortson, S. L., Checchi, K. D., et al. (2008) Preparation of Single cells for imaging/profiling mass spectrometry. *J Am Soc Mass Spectrom*, **19**, 1230–1236.

Chapter 16

Applying Imaging ToF-SIMS and PCA in Differentiation of Tissue Types

Ligang Wu, James S. Felton, and Kuang Jen J. Wu

Abstract

Time-of-flight secondary ion mass spectrometry (ToF-SIMS) has proven to be an extremely powerful tool for characterizing chemical distributions within biological cells and tissues. However, differentiating biological samples, e.g., cancerous cells from their normal counterparts or benign tissues from malignant tissues, presents unique challenges to ToF-SIMS. Repeatable differentiation of such samples, especially formalin-fixed paraffin-embedded (FFPE) histological specimens, could be used to improve tissue-based diagnosis and aid in prognosis decisions. In this chapter, we describe a strategy for characterizing and differentiating FFPE tissues. ToF-SIMS was used to image deparaffinized FFPE mouse embryos and differentiate tissue types. The robustness and repeatability of the method was determined by analyzing ten tissue slices from three different embryos over a period of 1 month. Using principal component analysis (PCA) to reduce the spectral data generated by ToF-SIMS, histopathologically identified tissue types of the mouse embryos can be differentiated based on the characteristic differences in their mass spectra.

Key words: ToF-SIMS, imaging mass spectrometry, mouse embryo, PCA, image PCA, FFPE.

1. Introduction

Formalin-fixed paraffin-embedded (FFPE) tumor samples are routinely used for disease diagnosis and are one of the most important and most abundant sources of clinical samples available in the tissue banks of medical centers and medical schools. Standard tumor diagnosis relies on a trained pathologist making an expert judgment based on morphological changes in the tissue section. However, using mass spectrometry to correlate specific chemical signatures to organ types, tumor types, and/or tumor

S.S. Rubakhin, J.V. Sweedler (eds.), *Mass Spectrometry Imaging*, Methods in Molecular Biology 656,
DOI 10.1007/978-1-60761-746-4_16,

stages may provide a less subjective, more reliable method for identifying and diagnosing tumors. Mass signatures can also lead to a better understanding of the carcinogenic process by defining biochemical changes in developing tumors and in tumors that metastasize to other organs.

We have utilized time-of-flight secondary ion mass spectrometry (ToF-SIMS) to image tissues by their secondary ions in deparaffinized mouse embryo sections. We have then differentiated these tissues based on differences in small molecules remaining after paraffin imbedding and fragments of the tissue proteins. These experiments serve as a preliminary study for further investigation of clinical FFPE samples from tumor and normal tissues.

ToF-SIMS has been demonstrated powerful in identifying intracellular distributions of specific biological ions such as sodium, potassium, calcium, and membrane lipid fragments (1–10). ToF-SIMS has also been successfully applied in studying the distributions of lipids in freeze-fractured human adipose tissue and human skeletal muscle (11, 12), as well as the localization of phosphocholine, galactosylceramide, and cholesterol in fresh frozen rat brain slices (13–15). These studies illustrate the ability of ToF-SIMS to sensitively analyze and localize small molecules and large molecule fragments in cells and tissue sections. We have shown the sodium and calcium distributions in the skull and rib of the deparaffinized mouse embryo section. By comparing the relative concentration of sodium and calcium within the skull and rib, we have illustrated that these two types of bones are under different development stages (16).

Differentiating biological samples presents a unique challenge for imaging mass spectrometry, since the mass spectra of biological samples are at the same time extremely complex and quite similar to one another. Therefore, data reduction and pattern recognition statistical analysis techniques must be used to differentiate similar biological materials. Principal component analysis (PCA), a standard, unsupervised multivariate statistical technique, reduces a large data matrix to a few manageable variables called principal components that represent linear combinations of the original variables and capture the greatest variation in the data set. The resulting "scores" plot illustrates the relationship between samples and the corresponding "loading" plot provides information about important mass peaks that are responsible for the separation of samples. ToF-SIMS and PCA have been utilized to successfully cluster and differentiate single proteins (17), monosaccharide isomers (18), yeast strains (19), and human breast cancer cell lines (20). By analyzing the ToF-SIMS results using PCA, we have shown a clear differentiation of the tissue types from a deparaffinized mouse embryo section based on the characteristic mass fragment patterns in their mass spectra (16). We have then demonstrated the reproducibility and robustness of the method by conducting repeated measurements of nine embryos over the period of 1 month and differentiating the tissue types of the entire data set (16).

Conventional method for interpreting ToF-SIMS images relies on univariate analysis to display selected individual variables or mass peaks. In practice, it can be difficult to investigate the many signals present in the imaging data using just a univariate analysis method. Furthermore, establishing correlations among relevant masses in a complex spectral image setting is often arduous. One approach to solving this difficulty and make full use of the large amount of information of imaging MS data is to apply multivariate analysis like image PCA to the entire image data set. Image PCA treats every pixel of a MS image as a sample and reduces the large spectral image data set to a physically realizable form that highlights the relevant chemical components of the image. The "scores" image of the calculation is displayed as a false-colored image with each color representing a principal component, while the "loading" plot of each principal component shows a group of correlated molecular information. Image PCA has been applied by Heeren et al. in studying rat brain tissue sections (21) and by Castner et al. in analyzing DNA microarrays (22). We have shown that the resulting PCA image of the skull, which has been created without a priori knowledge of the important mass peaks, is comparable to the overlay image but with more chemical information (16).

2. Materials

2.1. Animals

1. C57BL/6BAC mice are purchased from the Jackson Laboratory (Bar Harbor, ME, USA).
2. Isoflurane (Abbott Laboratories, North Chicago, IL, USA).

2.2. Mouse Embryo Tissue Slice Preparation

1. Paraformaldehyde (Fisher): Prepare a 4% (w/v) solution in PBS (4% PFA-PBS) fresh for each experiment.
2. Paraffin (Fisher).
3. Leica 819 blade (Leica Microsystems, Bannockburn, IL, USA).
4. 6 in. Si wafers (University wafer, South Boston, MA, USA) are cut into 1.2 × 1.2 cm Si wafer substrates. Number each Si substrate on the backside and clean both sides with ethanol and Kimwipes before use (*see* **Note 1**). Store Si substrates in a glass container or PFA (Perfluoroalkoxy) wafer carriers.

5. SuperFrost/Plus glass slide (Fisher).
6. Xylene substitute Clearene (Surgipath Medical Industries, Richmond, IL, USA) (*see* **Note 2**).
7. 100, 95, and 80% ethanol (Sigma, St. Louis, MO, USA).
8. Hematoxylin (Poly Scientific, Bay Shore, NY, USA).
9. Eosin (Poly Scientific, Bay Shore, NY, USA).
10. Permount (Fisher Scientific).

3. Methods

The key to the success of the method described in this chapter is to control and minimize non-biological sample to sample variations. Since ToF-SIMS is a highly surface-sensitive technique, samples should be handled with extreme care in all preparation procedures to avoid possible contaminations. All samples should be prepared using exactly the same protocols and be prepared altogether wherever possible. In the ToF-SIMS measurements, all conditions should be kept unchanged for all the tissues, including ion gun settings, instrumental settings, data acquisition settings, and vacuum conditions.

It is also essential to identify the contaminations that are unavoidable during sample preparation and storage and exclude them from the PCA calculations.

3.1. Animals

1. Female C57BL/6BAC mice are bred with male C57BL/6BAC mice to generate the embryos used in this experiment. The animals are maintained on a 12-h dark/light cycle in a temperature and humidity controlled room.
2. Animals are anesthetized with isoflurane and killed through cervical dislocation.
3. Three 16-day-old mouse embryos from different dams are harvested are designated with the numbers 1–3.

3.2. Mouse Embryo Tissue Slice Preparation

1. Embryos 1–3 are fixed in freshly made 4% paraformaldehye for 36 h at 4°C.
2. Fixed embryos 1–3 are embedded in three separate paraffin blocks. The size of the paraffin blocks should be close to the size of the Si substrates (1.2 cm).
3. These instructions assume the use of a Leica RM2165 microtome. Four micron thick saggital slice sections are cut from the embryos using a new Leica 819 blade. Slices from the

central parts of the embryo specimens are collected with desired tissues – brain, spinal cord, skull, rib, heart, and liver exposed.

4. The slice sections are transferred to the surface of a water bath with 39–42°C sterile ddH_2O and allowed to float for a full extension of the tissues.
5. The flattened slice sections are captured with clean Si substrates (*see* **Note 3**). The Si chips with tissue sections are placed upright on clean Kimwipes for 1 min and then allowed to air-dry before deparaffinization.
6. Three slice sections are cut from embryos 1–3 and are designated as sections 1–9. A fourth slice is cut from the embryo 3 and is designated as section 10.
7. For optical imaging, a fifth section is cut from embryo 3 and is designated as section 11. This section is captured with a clean SuperFrost/Plus glass slide.
8. Embryo sections 1–11 are incubated at 40°C for overnight. Sections 1–10 on Si chips are then placed between clean glass slides that are pre-loaded in a glass slide holder (*see* **Note 4**).
9. Sections 1–10 are deparaffinized by three rinses in Clearene for 5 min and then dehydrated by three rinses in 100% ethanol for 5 min (*see* **Note 5**).
10. The Si chips with tissue sections are again placed upright on clean Kimwipes for 1 min to reduce the collection of dissolved compound around the tissues. The samples are allowed to air-dry and then stored in vacuum at 1×10^{-4} torr for 24 h before ToF-SIMS analysis (*see* **Note 6**).
11. Embryo section number 11 is deparaffinized and rehydrated by rinses in Clearene, 100, 95, 80% ethanol, and DI water. It is then stained with hematoxylin and eosin (H&E) and dehydrated by rinses in 95 and 100% ethanol and Clearene. The slide with section number 11 is coverslipped using Permount and then observed under an optical microscope. An example of the optical image of an H&E stained mouse embryo section is shown in **Fig. 16.1a**. The identified tissues in this section are used as a reference for the ToF-SIMS analysis.

3.3. ToF-SIMS Imaging of Different Tissue Types

1. These instructions assume the use of a PHI TRIFT III ToF-SIMS instrument (Physical Electronics USA, Chanhassen, MN, USA) equipped with a gold liquid metal ion gun (Au LMIG). They are easily adaptable to other ToF-SIMS instruments with different ion sources.

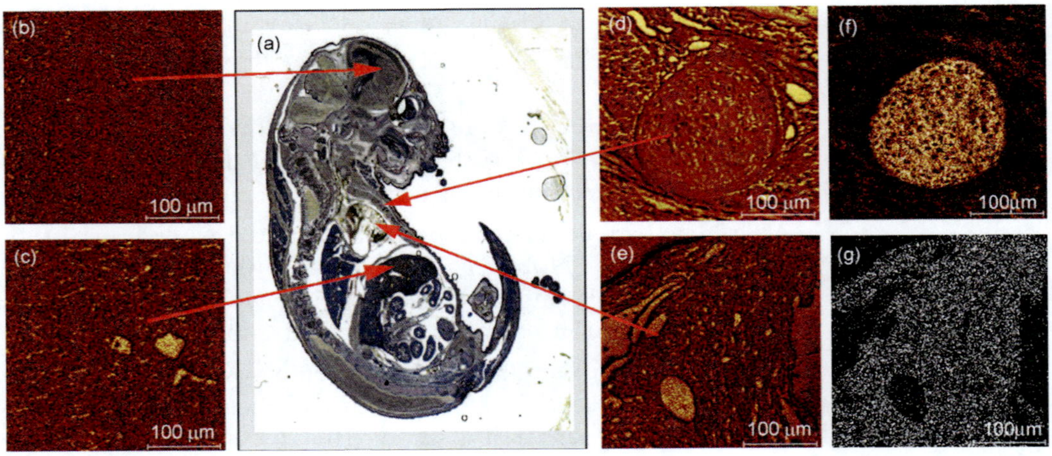

Fig. 16.1. (**a**) Optical image of a 16-day-old H&E stained mouse embryo section. (**b**) Positive total ion ToF-SIMS image of brain. (**c**) Positive total ion ToF-SIMS image of liver. (**d**) Positive total ion ToF-SIMS image of rib. (**e**) Positive total ion ToF-SIMS image of heart. Arrows point to the corresponding tissue region in the optical image. (**f**) m/z=23 (Na) distribution of (**d**). (**g**) m/z=70 (C_4H_8N) distribution of the region shown in (**e**). (Reproduced from (16) with permission from Elsevier Science.)

2. The ion gun is operated at 22 kV utilizing Au^+ ions and in an unbunched mode (high spatial resolution mode) with sub-micron spatial resolution (*see* **Note 7**). The mass resolution should be at the same time good enough to distinguish species of similar nominal mass like Ca^+ (m/z=39.963) and $C_3H_4^+$ (m/z=40.031) in order to accurately map their distributions.
3. Mouse embryo section number 10 is selected for tissue imaging analysis. Load the sample into the ToF-SIMS instrument and pump down to 5×10^{-9} torr before analysis.
4. To compensate the charge build-up of the tissue samples during the ToF-SIMS analyses, a pulsed, low-energy electron gun is utilized to provide charge neutralization. Analysis of mass range between m/z=0 and m/z=1850 is selected to capture most of the secondary ions and also allow enough time for the electron gun charge compensation to work properly.
5. Six tissue types, skull, rib, brain, spinal cord, heart, and liver, are selected for imaging. Tissues are identified based on a pathologist's designation of similar tissue regions in the H&E stained section. Positive ion ToF-SIMS measurements (RAW file) are conducted on each tissue over a 300×300 µm area for 5 min (*see* **Note 8**). All analyses are performed at room temperature.
6. The results are calibrated using common hydrocarbon fragment peaks at CH_3^+, $C_2H_3^+$, and $C_4H_7^+$. Total ion images

and specific distributions of particular masses of interest are then reconstructed by the ToF-SIMS software from different tissues. An example of the positive total ion ToF-SIMS images and images of specific ions of interest is shown in **Fig. 16.1b–g**.

3.4. Differentiating the Development of Bone with Time

The bones of skull and rib have different developmental paths, resulting in different concentrations of calcium and sodium. The flat bones of the skull developed by intramembranous ossification and in an embryo 16 days old these fetal bones have begun to ossify and concentrate calcium (23). In contrast, rib bones develop by endochondral ossification. The rib bones of the embryos under examination are still primarily cartilage (23), which consists mainly of a matrix of collagen and proteoglycans (24) with a high concentration of Na ions.

1. ToF-SIMS positive total ion images as well as the sodium, calcium, and silicon images are reconstructed from the RAW files of the skull and rib obtained in the tissue imaging analysis.
2. Sodium, calcium, and silicon images of the skull and rib are displayed in a RGB false color overlay image with the sodium and calcium signal intensities set to the same scale for direct visual comparison. An example of the total ion images and the RGB overlay images of skull and rib is illustrated in **Fig. 16.2**.
3. A further quantitative analysis is performed by calculating the ratio of the calcium to sodium signal intensity for both images. The integral ion counts of sodium and calcium peaks of both bones are read from their spectra (*see* **Note 9**). For the images shown in **Fig. 16.2**, the ratio of Ca/Na is 1.425 for the skull and 0.209 for the rib, confirming the higher concentration of Ca in the skull.

3.5. Differentiation of Tissue Types by PCA

The molecular differences of the tissues produce a different pattern of peaks in the ToF-SIMS spectra. Although the mass spectra of different tissues appear very similar, their different characteristic mass spectral patterns can be easily differentiated by PCA.

1. Mouse embryo section number 10 is further analyzed with the same ToF-SIMS instrumental and data acquisition settings as the ones in tissue imaging analysis. Ten measurements are recorded for each of the six tissue types: skull, rib, brain, spinal cord, heart, and liver, with each measurement covering a fresh, un-analyzed area.
2. Spectra for background controls are acquired by analyzing clean silicon areas around the tissues on the wafers. Contamination peaks attributed to sample handling are identified from the background control spectra. Common

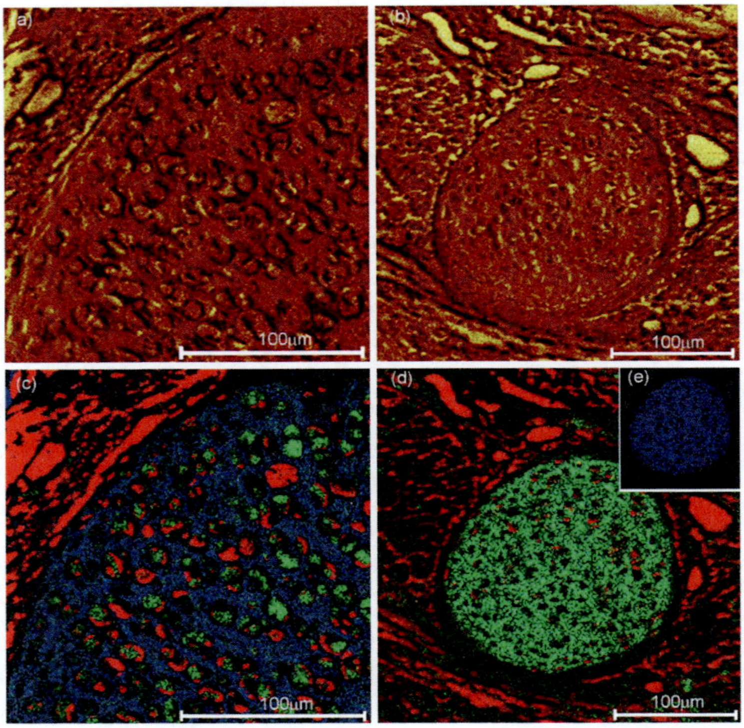

Fig. 16.2. (**a**) ToF-SIMS positive total ion image of a region of the skull. (**b**) ToF-SIMS positive total ion image of a rib. (**c**) Three-color overlay of the skull image: *red color* represents the Si substrate, *green color* represents the Na distribution, *blue color* represents the Ca distribution. (**d**) Three-color overlay of the rib image: *red color* represents the Si substrate, *green color* represents the Na, *blue color* (too low to be visible) represents the Ca distribution. Na and Ca are plotted on the same scale as (**c**). (**e**) Ca distribution with the signal intensity increased 5×. (Reproduced from (16) with permission from Elsevier Science.)

contamination peaks include fragments of polydimethylsiloxane (PDMS) *m/z*=43, 73, 147, 207, 281, 325, 355, and Irgafos peaks 647, 662, etc.

3. One average mass spectrum is reconstructed from each measurement of the tissue section. When reconstructing spectra from the measurements, regions of interest (ROI) are carefully defined to avoid substrate areas and edges of the tissues where contaminants can accumulate (*see* **Note 10**). The resulting mass spectra are calibrated using CH_3^+, $C_2H_3^+$, and $C_4H_7^+$.

4. Unit mass binning is applied to each spectrum with the exception of *m/z*=40, which contains only the calcium (Ca) peak. Masses from *m/z*=60 to 500 and *m/z*=40 (Ca) are selected for PCA analysis and are exported by the ToF-SIMS software as a table of ion counts with mass spectra as rows and masses as columns (*see* **Note 11**).

5. These instructions assume the use of MATLAB v. 7.0 (MathWorks Inc., Natick, MA, USA) along with PLS Toolbox v. 3.5 (Eigenvector Research, Manson, WA, USA) as the software platform for PCA analysis. The table of ToF-SIMS spectral data is imported into PLS Toolbox. Identified contamination peaks are excluded from the PCA data reduction.
6. For data pre-processing, the spectral data set is normalized to that spectrum's total ion counts and then mean-centered. PCA calculation is then performed and the resulting principal components (PC) are generated.
7. A scores plot and a loading plot of the first two principal component (PC1 and PC2) axes are generated by the PLS toolbox (*see* **Note 12**). The scores plot describes the relationship between samples; each point in the scores plot represents a mass spectrum acquired from one specific tissue area and spectra from different tissue types are denoted by different symbols. The corresponding loading plot illustrates important mass peaks that drive the separation in the scores plot; each point in the loading plot represents a mass peak. An example of the resulting scores and loading plots is shown in **Fig. 16.3a–b**.
8. The instructions here assume the use of the error ellipse.m code by J. Andrew Johnson of Binghamton University, acquired from the MATLAB Central File Exchange. Using the code, 90% data contours are drawn in the scores plot around the data points of each tissue based on their PC1 and PC2 coordinates.
9. Shown in **Fig. 16.3a**. The bone-derived skull and rib groups are well separated from nervous tissue (brain and spinal cord) and the heart and liver. The soft tissues, however, exhibit considerable overlap. To further separate the soft tissues, PCA analysis is conducted again with the skull and rib spectra removed. An example of the resulting scores and loading plots is shown in **Fig. 16.3c–d**.

3.6. Reproducibility and Stability of the ToF-SIMS Method

The reproducibility studies are necessary in order to identify potential sources of variation in that may occur during the analysis of paraffin-embedded clinical samples. Clinical samples are frequently stored under minimally controlled conditions for extended periods of time and may vary depending upon individual laboratory sample processing procedures. Understanding the reproducibility and stability of the method is essential for future applications of ToF-SIMS to paraffin-embedded samples.

1. Mouse embryo sections 1–10 are utilized for the reproducibility study with the same ToF-SIMS instrumental and data acquisition settings as the ones in tissue imaging analysis. Four tissue types are selected for analysis: brain, rib,

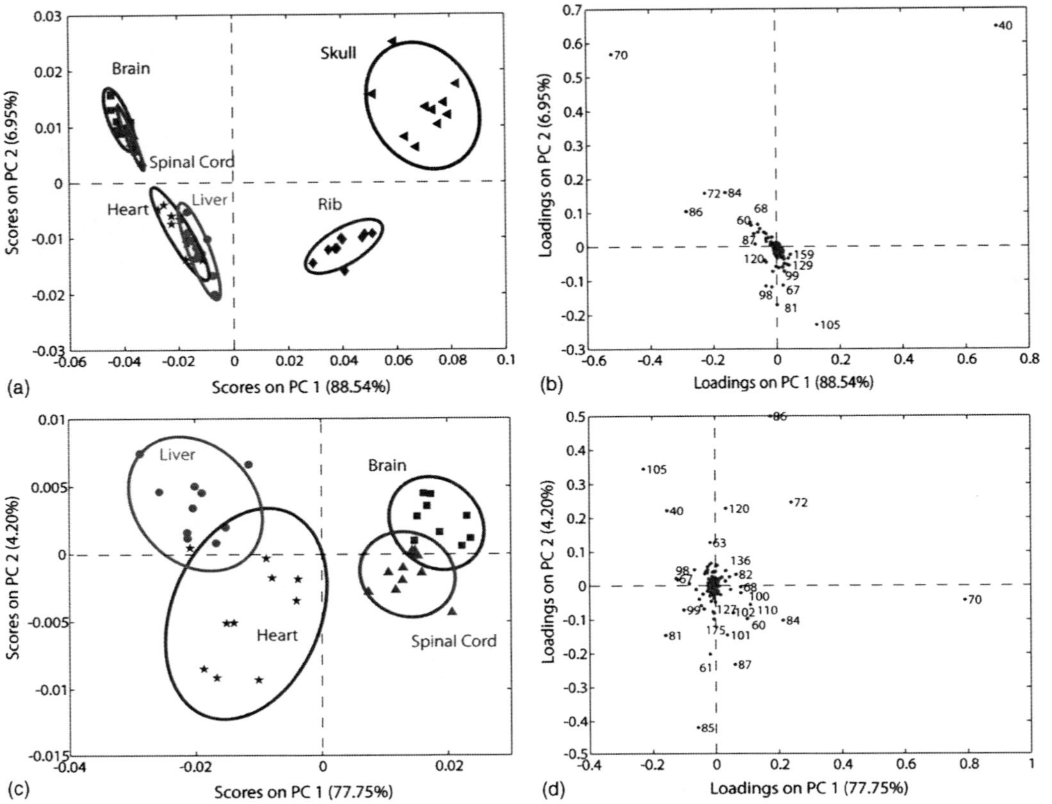

Fig. 16.3. (**a**) Scores plot of PC1 versus PC2 of spectra from six tissues with 90% data contour of each tissue. The bone-derived skull and rib groups are well separated from the other soft tissues. The soft tissues, however, exhibit considerable overlap. (**b**) Loading plot of same tissues. The separation of skull and rib is primarily driven by Ca (*m/z*=40). (**c**) Scores plot of PC1 versus PC2 of spectra from heart, liver, brain, and spinal cord. Essentially the four soft tissues are well separated. (**d**) Loading plot of same tissues. (Reproduced from (16) with permission from Elsevier Science.)

heart, and liver. Two to ten measurements are recorded for each tissue type depending on the sizes of the tissue regions available in each slice.

2. Each of the ten embryo sections is analyzed twice over a period of 1 month. The embryo sections are analyzed in a random order (*see* **Note 13**). To minimize instrument-induced variations, the instrument settings should remain unchanged throughout the whole experiment (*see* **Note 14**).
3. The resulting spectra are analyzed using the same protocol as the one described in the tissue differentiation session. The entire spectral data set are analyzed by PCA. Spectra from different tissue types are denoted by different symbols and are enclosed by 90% data contours. An example of the resulting scores and loading plots is shown in **Fig. 16.4a–b**.

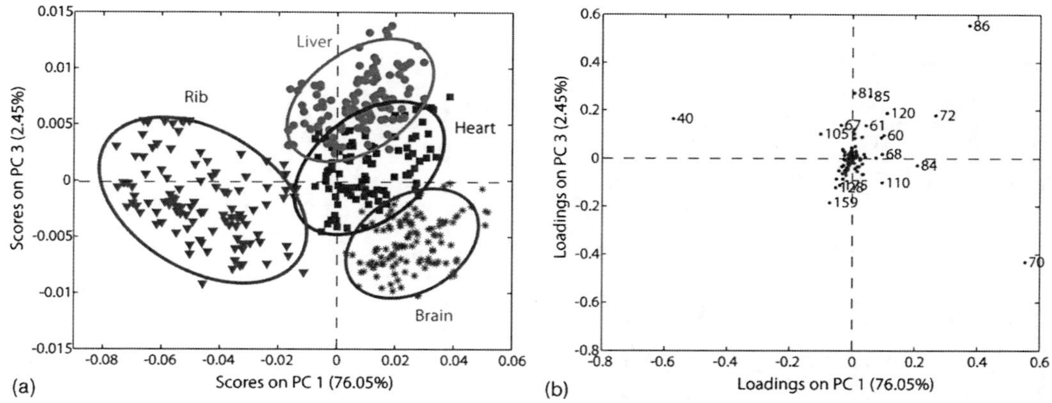

Fig. 16.4. (**a**) Scores plot PC1 versus PC3 for four tissues (brain, heart, liver, and rib) with 90% data contours drawn for each tissue. A good separation of the four tissue types is achieved (*see* **Note 15**). (**b**) Loading plot of PC1 versus PC3 for the four tissues. (Reproduced from (16) with permission from Elsevier Science.)

3.7. Image PCA Analysis of Skull

To make full utilization of the large amount of data that generated by ToF-SIMS imaging and establish correlations among relevant masses, image PCA analysis is performed on a skull image and the results are compared to the univariate false color overlay.

1. The ToF-SIMS RAW file of the skull shown in **Fig. 16.2a** is utilized for image PCA analysis. These instructions assume the use of MATLAB v. 7.0 along with PLS Toolbox v. 3.5 and MIA Toolbox 1.0 (Eigenvector Research, Manson, WA, USA) as the software platform for image PCA analysis.
2. The RAW file can be directly imported into the MIA toolbox (*see* **Note 16**). Unit mass binning is applied to the skull file and masses from $m/z=0$ to 200 are selected for image PCA analysis (*see* **Note 17**). The imported data set of the skull is a three-dimensional data set of 256 (pixels)$\times$256 (pixels)$\times$200 (masses) elements. Each element in the data set is the ion intensity of a specific mass at a specific pixel of the skull image.
3. For data pre-processing, the data set is normalized to each pixel's total ion counts and then mean-centered. Image PCA calculation is then performed by the MIA toolbox and the resulting principal components (PC) are generated. Scores image and loading plots of selected PCs are then created by the software.
4. The scores images of different PCs can be displayed in an overlay RGB false color image, with each color represents a PC. The loading plot of each PC is displayed individually. Masses with positive loading values are correlated with the scores image of the corresponding PC, while masses with

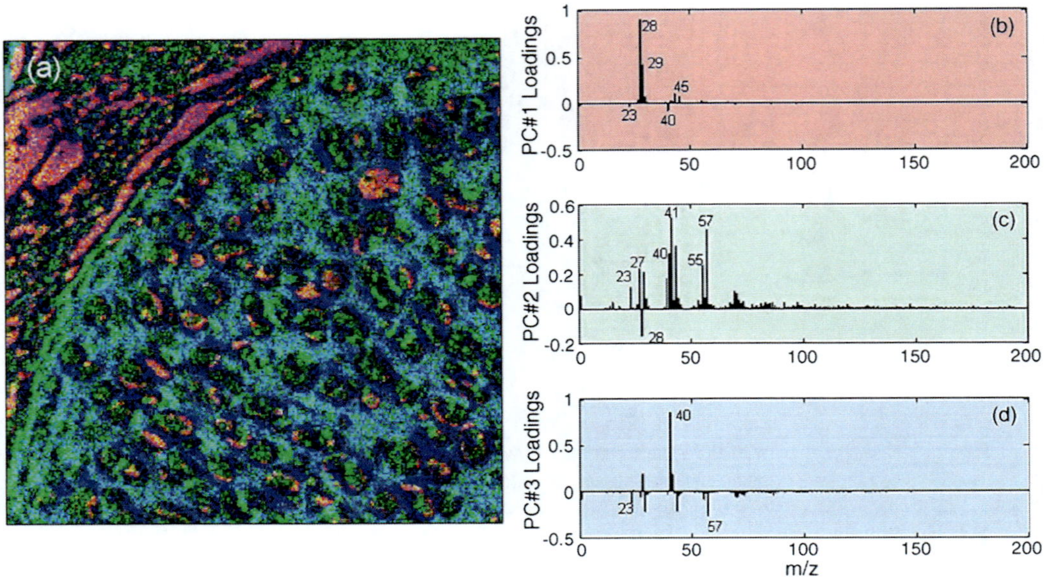

Fig. 16.5. (**a**) Scores plot PC1, PC2, and PC3 overlay of skull image. *Red color* is PC1, *green color* is PC2, *blue color* is PC3. (**b**) Loading plot of PC1. PC1 (*red area*) is correlated with the silicon substrate Si (*m/z*=28), SiH (*m/z*=29), and SiOH (*m/z*=45) (**c**) Loading plot of PC2. PC2 (*green area*) is correlated with Na (*m/z*=23) and hydrocarbon peaks (*m/z*=27, 41, 43, 55, 57, and so on), which represent the soft tissues within the bone structure. (**d**) Loading plot of PC3. PC3 (*blue color*) is mainly correlated with Ca (*m/z*=40), which shows the structure of the developing bone. (Reproduced from (16) with permission from Elsevier Science.)

negative loading values are anti-correlated with the scores image. An example of the resulting scores image and the loading plots of the first three PCs is shown in **Fig. 16.5a–d**. In this example, the PCA image (**Fig. 16.5a**), which has been created without a priori knowledge of the important mass peaks, is comparable to the overlay image (**Fig. 16.2a**) but with more chemical information.

4. Notes

1. Serial numbers scribed on the backside of the Si substrates are helpful to log the experiments and avoid sample mixing up.
2. Clearene is a safe replacement of xylene.
3. Never recycle or re-use Si substrates to avoid possible cross-contamination. The sliced embryo sections should be place at the center of the Si substrates with at least 1 mm margin

to fit in the ToF-SIMS sample holder and avoid edge effect during ToF-SIMS analysis.

4. The samples are easy to be mixed up or stick to each other during chemical filling and draining. A practical way to avoid such mistakes is to use clean glass slides to separate them in the slide holder.
5. To avoid sample to sample variation induced by the deparaffinization process, all embryo sections should be deparaffinized at a same time. Use fresh chemicals for all steps of deparaffinization and dehydration.
6. Other clean storage methods like argon or nitrogen filled tanks are also suitable. Sample variation along storage time has been observed using the vacuum storage method (*see* **Note 15**).
7. Au^+ ions instead of cluster ions are used for the analysis for better secondary ion counts and resolution. Since the tissue samples have been fixed, the benefit of cluster ions in generating high mass ions is less applicable here.
8. The time of the measurement depends on the primary current. Instructions here are based on the use of a 600 pA aperture.
9. Ca^+ (m/z=39.963) peak should be carefully defined not to include the ion counts from peak $C_3H_4^+$ (m/z=40.031).
10. It is a good habit to check the ROI using the mappings of different ions. For example, checking the ROI using the Si mapping can tell if the ROI enclose any substrate area.
11. Low mass peaks (below m/z=60, C_4, and smaller hydrocarbon clusters) are removed from the PCA calculation to exclude these hydrocarbon peaks that carry no specific chemical information. High mass peaks above m/z=500 are not included due to their low ion intensities.
12. Typically, the first two PCs contain the most variance and are enough to differentiate tissue types into groups. In some cases, the largest or second largest variance can be attributed to non-biological differences, like contamination level difference. PC3 and above should then be used.
13. A random order between 1 and 20 can be generated by MATLAB. The data acquired form section 10 for tissue differentiation can be used for the reproducibility study, so section 10 only needs to be analyzed one more time.
14. No major instrument maintenance should be scheduled during the period of the experiment. No "dirty" samples like high-outgassing materials should be analyzed during the period of the experiment to avoid potential cross-contamination.

15. The variation within the sample data set is mainly from three sources: (1) inherent biological variation within the samples; (2) environmental contamination of the sections; and (3) changes in the sample during storage, which in this case is captured by PC2. More discussions about the sample change during storage can be found in (16).
16. Ion-ToF file format can also be directly imported.
17. The maximum number masses that the MIA toolbox can handle is limited by the size of the physical memory. These instructions here assume the use of a 1 G memory.

Acknowledgments

The authors would like to thank pathologist Dr. Xiaochen Lu for his help in sample preparation and tissue identification and Dr. Kristen Kulp, Dr. Elena Berman, Dr. Erik Nelson, and Mr. Mark Knize for their advice and discussion. This work was performed under the auspices of the US Department of Energy by the University of California, Lawrence Livermore National Laboratory under contract no.W-7405-Eng-48.

References

1. Chandra, S. (2001). Studies of cell division (mitosis and cytokinesis) by dynamic secondary ion mass spectrometry ion microscopy: LLC-PK1 epithelial cells as a model for subcellular isotopic imaging. *J Microsc (Oxford)*, **204**, 150–165.
2. Chandra, S., Lorey, D. R., Smith, D. R. (2002). Quantitative subcellular secondary ion mass spectrometry (SIMS) imaging of Boron-10 and Boron-11 isotops in the same cell delivered by two combined BNCT drugs: *in vitro* studies on human glioblastoma T98G cells. *Radiat Res*, **157**, 700–710.
3. Kempson, I. M., Skinner, W. M., Kirkbride, P. K. (2003). Calcium distributions in human hair by ToF-SIMS. *Biochim Biophys Acta*, **1624**, 1–5.
4. Roddy, T. P., D.M. Cannon, J., Ostrowski, S. G., Winograd, N., Ewing, A. G. (2002). Identification of cellular sections with imaging mass spectrometry following freeze-fracture. *Anal Chem*, **74**, 4020–4026.
5. Cannon, D. M., Pacholski, M. L., Winograd, N., Ewing, A. G. (2000). Molecule specific imaging of freeze-fractured, frozen-hydrated model membrane systems using mass spectrometry. *J Am Chem Soc*, **122**, 603–610.
6. Cliff, B., Lockyer, N., Jungnickel, H., Stephens, G., Vickerman, J. C. (2003). Probing cell chemistry with time-of-flight secondary ion mass spectrometry: development and exploitation of instrumentation for studies of frozen-hydrated biological material. *Rapid Commun Mass Spectrom*, **17**, 2163–2167.
7. Cliff, B., Lockyer, N. P., Corlett, C., Vickerman, J. C. (2003). Development of instrumentation for routine ToF-SIMS imaging analysis of biological material. *Appl Surf Sci*, **203–204**, 730–733.
8. Sjovall, P., Lausmaa, J., Johansson, B. (2004). Mass spectrometric imaging of lipids in brain tissue. *Anal Chem*, **76**, 4271–4278.
9. Ostrowski, S. G., Van Bell, C. T., Winograd, N., Ewing, A. G. (2004). Mass spectrometric imaging of highly curved membranes during Tetrahymena mating. *Science*, **305**, 71–73.

10. Gazi, E., Dwyer, J., Lockyer, N., Gardner, P., Vickerman, J. C., Miyan, J., Hart, C. A., Brown, M., Shanks, J. H., Clarke, N. (2004). The combined application of FTIR microspectroscopy and ToF-SIMS imaging in the study of prostate cancer. *Faraday Discuss*, **126**, 41–59; discussion 77–92.
11. Malmberg, P., Nygren, H., Richter, K., Chen, Y., Dangardt, F., Friberg, P., Magnusson, Y. (2007). Imaging of lipids in human adipose tissue by cluster ion TOF-SIMS. *Microsc Res Tech*, **70**, 828–835.
12. Magnusson, Y. F. P., Sjövall, P, Dangardt, F, Malmberg, P, Chen, Y. (2008). Lipid imaging of human skeletal muscle using TOF-SIMS with bismuth cluster ion as a primary ion source. *Clin Physiol Funct Imaging*, **28**, 202–209.
13. Borner, K., Nygren, H., Hagenhoff, B., Malmberg, P., Tallarek, E., Mansson, J. E. (2006). Distribution of cholesterol and galactosylceramide in rat cerebellar white matter. *Biochim Biophys Acta*, **1761**, 335–344.
14. Nygren, H., Eriksson, C., Malmberg, P., Sahlin, H., Lennart, C., Kausmaa, J., Sjovall, P. (2003). *Colloids Surf*, **30**, 87.
15. Nygren, H., Borner, K., Hagenhoff, B., Malmberg, P., Mansson, J. E. (2005). Localization of cholesterol, phosphocholine and galactosylceramide in rat cerebellar cortex with imaging TOF-SIMS equipped with a bismuth cluster ion source. *Biochim Biophys Acta*, **1737**, 102–110.
16. Wu, L., Lu, X., Kulp, K. S., Knize, M. G., Berman, E. S., Nelson, E. J., Felton, J. S., Wu, K. J. (2007). Imaging and differentiation of mouse embryo tissues by ToF-SIMS. *Int J Mass Spectrom*, **260**, 137–145.
17. Wagner, M. S., Castner, D. G. (2001). Characterization of adsorbed protein films by time-of-flight secondary ion mass spectrometry with principle component analysis. *Langmuir*, **17**, 4649–4660.
18. Berman, E. S., Kulp, K. S., Knize, M. G., Wu, L., Nelson, E. J., Nelson, D. O., Wu, K. J. (2006). Distinguishing monosaccharide stereo- and structural isomers with TOF-SIMS and multivariate statistical analysis. *Anal Chem*, **78**, 6497–6503.
19. Jungnickel, H., Jones, E. A., Lockyer, N. P., Oliver, S. G., Stephens, G. M., Vickerman, J. C. (2005). Application of TOF-SIMS with chemometrics to discriminate between four different yeast strains from the species *Candida glabrata* and *Saccharomyces cerevisiae*. *Anal Chem*, **77**, 1740–1745.
20. Kulp, K. S., Berman, E. S., Knize, M. G., Shattuck, D. L., Nelson, E. J., Wu, L., Montgomery, J. L., Felton, J. S., Wu, K. J. (2006). Chemical and biological differentiation of three human breast cancer cell types using time-of-flight secondary ion mass spectrometry. *Anal Chem*, **78**, 3651–3658.
21. Altelaar, A. F., Luxembourg, S. L., McDonnell, L. A., Piersma, S. R., Heeren, R. M. (2007). Imaging mass spectrometry at cellular length scales. *Nat Protoc*, **2**, 1185–1196.
22. Lee, C. Y., Harbers, G. M., Grainger, D. W., Gamble, L. J., Castner, D. G. (2007). Fluorescence, XPS, and TOF-SIMS surface chemical state image analysis of DNA microarrays. *J Am Chem Soc*, **129**, 9429–9438.
23. Theiler, K. (1989). *The House Mouse: An Atlas of Embryonic Development*, Springer-Verlag, New York, NY.
24. Maroudas, A. (1979). *Physicochemcial Properties of Articular Cartilage*. Adult Articular Cartilage (Freeman, M. A. R., Ed.), Pitman Medical, Kent, England.

Part III

Protocols for MS Imaging of Distribution of Peptides and Proteins

Chapter 17

Direct Molecular Analysis of Whole-Body Animal Tissue Sections by MALDI Imaging Mass Spectrometry

Michelle L. Reyzer, Pierre Chaurand, Peggi M. Angel, and Richard M. Caprioli

Abstract

The determination of the localization of various compounds in a whole animal is valuable for many applications, including pharmaceutical absorption, distribution, metabolism, and excretion (ADME) studies and biomarker discovery. Imaging mass spectrometry is a powerful tool for localizing compounds of biological interest with molecular specificity and relatively high resolution. Utilizing imaging mass spectrometry for whole-body animal sections offers considerable analytical advantages compared to traditional methods, such as whole-body autoradiography, but the experiment is not straightforward. This chapter addresses the advantages and unique challenges that the application of imaging mass spectrometry to whole-body animal sections entails, including discussions of sample preparation, matrix application, signal normalization, and image generation. Lipid and protein images obtained from whole-body tissue sections of mouse pups are presented along with detailed protocols for the experiments.

Key words: Imaging mass spectrometry, whole-body sections, drugs, proteins, lipids.

1. Introduction

MALDI imaging mass spectrometry (MS) is a powerful tool for acquiring molecularly specific profiles directly from tissue sections. Many analyte classes present in biological sections, including peptides, proteins, lipids, and endogenous metabolites, may be analyzed using this technology. Thus, one can extend this technology to the localization of very different compounds throughout a whole-body section. Whole-body imaging can therefore illuminate a broad picture of the biological state of an animal at

S.S. Rubakhin, J.V. Sweedler (eds.), *Mass Spectrometry Imaging*, Methods in Molecular Biology 656,
DOI 10.1007/978-1-60761-746-4_17,

a given point in time in both health and disease. One can identify what analytes co-localize together, where analytes are present or absent, and what analytes may be expressed at the same time but in different places. This type of information will be invaluable in expanding our understanding of basic biological processes, the origin and progression of disease as well as the response to drug treatments.

The pharmaceutical industry routinely uses quantitative whole-body autoradiography (QWBA) in order to map the distribution of drug compounds in animals with time. QWBA requires the use of a radiolabeled drug compound, and indeed only the radiolabel is followed in the experiment. Nonetheless, the results give a global picture of the localization of the drug in the body and may alert investigators to unforeseen potential toxicological issues. Performing the QWBA experiment with imaging MS provides much more information because the compounds of interest are detected with high molecular specificity, with the dosed drug differentiated from metabolites that differ in mass. Protein targets of the drug may also be analyzed from the same or a serial section, providing additional information.

Imaging MS can be directly applied to larger, whole-body tissue specimens for small molecule, lipid, or protein analyses. The general theory, instrumentation, and advantages and limitations to the experiment are the same. The main differences will be discussed in the following section.

1.1. Unique Aspects of the Whole-Body MS Imaging Experiment

Due to the large size and multiple tissue types typically present in a whole-body section, there are specific considerations that must be taken into account when preparing the tissue for analysis, as well as for acquiring and processing the data.

1.1.1. Specimen Size and Sectioning

Acquiring sections from whole animals requires special considerations due to the increased size of the specimens. The bodies of adult rats can be up to approximately 30-cm long, while adult mice can approach 12-cm long, and obtaining sections from these whole animals requires a cryostat designed for larger specimens. As a result, the sample preparation for whole animals is more complex than the simple flash-freezing approach used for most single organs (1).

After sacrifice, the animal must be frozen, preferably after exsanguination to minimize blood pooling and artifacts that arise from compounds present in the blood. Freezing must occur while the animal is in a satisfactory position for obtaining sections in the desired plane (sagittal, coronal, or axial). This is typically accomplished by resting the animal in a foil or plastic boat, on their back or on their side, while soaking in a dry ice/hexane bath. Prior to sectioning, the frozen animal must be encased in a block of ice or other embedding media in order to stabilize the whole body and minimize tearing of the sections. As with smaller sections, care must be taken to ensure the embedding media (typically a small percentage of carboxymethylcellulose (CMC)) does not interfere with the MALDI signal. Pure water usually stabilizes the specimen adequately without adversely affecting the mass spectrum (2). This is illustrated in **Fig. 17.1**, which shows a 3-day-old mouse pup embedded in a block of ice (A) and a 1-day-old mouse pup mounted on a cryostat chuck with OCT embedding compound (B), after sagittal sections have been cut.

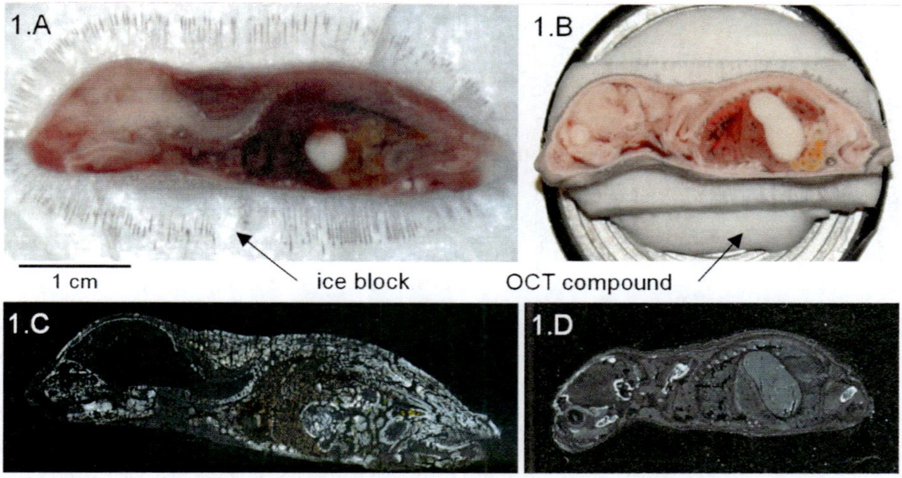

Fig. 17.1. Generation of whole-body tissue sections from fresh frozen mouse pups. Photomicrographs of (**a, c**) a 3-day-old pup frozen in a block of ice from which 15-μm thick sections were cut using a whole-body cryostat and mounted on conductive glass slides using the CryoJane system; (**b, d**) a 1-day-old pup held into position using OCT from which 12-μm thick sections were cut in a biological specimen cryostat and directly thaw-mounted on conductive glass slides. See text for details.

Due to the presence of multiple organs in a given section, additional stabilization is required to keep the organs together and maintain their spatial integrity once removed from the animal. Typically this is accomplished by acquiring the section onto a piece of tape. For whole-body autoradiography, this is acceptable, as it does not interfere with the subsequent radioactivity analysis. However, for MALDI, this is problematic for several reasons. First, the tapes used (acetate film tape) are non-conductive and may induce charging effects and mass shifts, depending on the instrument used. Second, they must be affixed to standard MALDI sample plates for insertion into the instruments. This is accomplished using with conductive double-sided tape. Care must be taken in any event to ensure even transfer of the section to the MALDI plate (avoiding air bubbles, etc.). Third, certain compounds may preferentially interact with the tape and may not be easily extracted into the matrix. Alternatively, signals originating from the tape itself may cause ionization suppression or otherwise adversely affect the MALDI signal. Overall, this whole-body section mounting approach has been found useful for the imaging MS of drugs and their metabolites on MALDI QqTOF and ion trap systems. In these instruments, where the ionization source is decoupled from the mass analyzer, the non-conductive behavior of the sample is not a factor.

An alternative to tape collection is to collect whole-body sections on rice paper which is affixed to the acetate tape. When the sections are cut they adhere to the rice paper instead of the acetate tape. These are then are transferred (thaw-mounted) to the MALDI target plate. In this case, the sections are directly mounted on the target plate without the use of tape. This is beneficial because the adverse affects of the tape are avoided; however, this approach generally results in poor section quality because transfer efficiency is often unequal for different organs and across the section.

Another protocol may be used that minimizes the conductivity and ion suppression issues present with acetate tape. The tape transfer methodology (CryoJane tape transfer system, Instrumedics, Inc., St. Louis, MO, USA) (3) utilizes a proprietary tape instead of the acetate tape. Once the section has been obtained, it is placed on a glass slide pre-coated with a UV-sensitive adhesive. Irradiation of the section with a flash of UV light releases the tape and binds the tissue to the activated adhesive present on the slide. Whole-body sections obtained using the CryoJane tape transfer system have been found to be of much higher quality than those obtained using the rice paper approach. In order for the slide to be used as a MALDI sample plate, it must be in a shape and size that can fit in the instrument holder and must be conductive (ITO coated) when used in time-of-flight mass spectrometers. Conductive glass slides and plates are available in larger formats and have been successfully used for MALDI analysis (4) and in particular with this tape transfer system. For example, **Fig. 17.1c** shows a 20-μm thick whole-body section cut from the ice block containing the 3-day-old mouse pup (shown in **Fig. 17.1a**) using a full-body cryostat and mounted on a conductive glass slide. Although the CryoJane transfer approach removes the tape from the sample that will be analyzed, it does introduce a new variable in the adhesive that is applied to the glass slide. The effects of the adhesive on the MALDI experiment still need to be fully examined.

Finally, due to the large size of some specimens, the entire section may be too big to fit onto a standard MALDI plate. For whole-body sections, the larger microtiter format MALDI sample plates can hold a whole-mouse section or approximately half of an adult rat section. The largest commercially available conductive glass slides (and matching target holder) are 50×75 mm and will hold at best a (small) whole-mouse section. Thus whole sections must be divided, often into several smaller sections, and each section affixed to its own MALDI plate. This can induce experimental variations, as data will be acquired from each sub-section independently.

Size may not be a limiting factor as some specimens are small enough in size to be cut using standard biomedical cryostats. This is demonstrated in **Fig. 17.1d** which shows a 12-μm thick section cut from the 1-day-old mouse pup in **Fig. 17.1b**. This pup has a length of only ~3 cm; sections can then be directly thaw-mounted on MALDI target plates, in this case a conductive glass slide. When using rice paper, the CryoJane transfer approach or direct thaw-mounting, sample conductivity is not an issue since the sections are in direct contact with the MALDI target plates. These approaches are therefore preferred for imaging MS of lipids, peptides, and proteins using MALDI TOF mass spectrometers.

1.1.2. Matrix Application

Proper application of matrix is critical in obtaining a meaningful image in MALDI mass spectrometry (1). Application of matrix to whole-body tissue sections requires additional considerations. Choice of matrix to be used depends on the analyte of interest, as with single organs. Commonly, sinapinic acid is used for proteins, CHCA is used for peptides, and DHB is used for lipids and most pharmaceutical compounds. To date, matrix deposition for imaging MS of whole-body sections has been performed by manual spray deposition. The spray system usually consists of a small pneumatic nebulizer (typically used for thin layer chromatography) coupled to a high-pressure nitrogen gas outlet via a regulator. Matrix is prepared in the appropriate solvent system and manually sprayed on the sections. For whole-body sections mounted on tape, the solvent that the matrix is dissolved in must be compatible with the tape used. Dissolution of the tape, loss of adhesion of the sample to the plate, and/or poor MALDI signal may result from a matrix solution too high in organic composition. A homogeneous matrix coating is progressively built by alternating spraying and drying cycles. Matrix deposition by spray on large surfaces such as from whole-body sections is fast (typically ~1 h); however, the risk of delocalizing analytes during the spray process is high. (For more information on analyte extraction and migration, *see* **Chapter 8**.) Further, the quality of the mass spectra in terms of signal-to-noise and the overall number of detected mass peaks is significantly poorer compared to the spectral quality obtained from discrete matrix spots. This is particularly true for peptides and proteins (5).

Matrix may crystallize differently on different organs (2), and the signal response for a single compound may be affected by the unique microenvironments present in different organs (6). This is obviously an issue for whole-body analyses, where multiple organs are present in each section. Addition of an internal standard to the matrix solution may allow normalization to account for these differences, although this is not straightforward either. However, such a normalization approach is of value for imaging MS of single compounds such as administered pharmaceuticals. In this case, it is preferable to use a drug compound with a similar structure but a different molecular weight (ideally, a stable isotope-labeled analog). For lipid or protein imaging MS, it is difficult to find one internal standard that would be of use for all compounds of interest in the tissue. In these cases, normalization to the total ion current (TIC) may be a useful approach (see **Section 1.1.5**).

1.1.3. Imaging MS Data Acquisition

Imaging MS data acquisition from whole-body sections is performed in the same manner as from any other tissue section. (For more information on imaging of biological tissue sections, *see* **Chapter 15**.) Phospholipids, peptides, and proteins are directly acquired in the MS mode generally using MALDI TOF instruments in the linear mode (proteins) or reflex mode (peptides and phospholipids). Other MALDI mass spectrometers may also be used. To minimize acquisition time per spot (or pixel), the minimal number of laser shots necessary to acquire quality data needs to be optimized. Although imaging MS of administered drugs (and their metabolites) can be acquired in the MS mode, MS/MS mode (or selected reaction monitoring, SRM) is often used to increase sensitivity and specificity by following a unique precursor/fragment transition (7, 8). In this case, MALDI QqTOF or ion trap mass spectrometers are generally used. (For more information on imaging MS of pharmaceuticals, *see* **Chapter 5**.)

1.1.4. Time

The increased sample size has effects on the time required for the experiment. As with all imaging experiments, the time required is a function of resolution (essentially the number of pixels) and acquisition time per pixel. These times are further influenced by instrumental parameters, such as laser frequency and time required for motor movements, as well as the duty cycle of the instrument used. The most obvious difference for whole-body images is an increase in the total number of pixels. Even for a moderate resolution image (~500 μm), the number of pixels it takes to image a whole-rat section (assume ~6×17 cm) is ~40,000. This assumes a rectangular shape; if the instrument can be programmed to only scan the tissue, this could reduce the total number by several thousand to ten thousand pixels. Nonetheless, assuming a relatively fast scan speed of 1 s/pixel, mass spectral acquisition would still require approximately 10–12 h.

1.1.5. Data Processing

The processing of imaging MS data files has proven to be useful in removing matrix deposition artifacts resulting in sharper recovered ion images (9). Processing imaging MS data typically consists of a series of pre-processing steps including background/noise reduction and smoothing and may include steps of signal realignments and data normalization. Signal realignment is typically based on a series of stable signals (~10) selected throughout the acquired mass range. For whole-body imaging MS data in particular, realignment can be performed only if enough mass signals are found to be stable throughout the section. Considering the huge variations in tissue types and molecular compositions encountered through whole-body sections, this is rarely the case. However, spectral realignment is only necessary if obvious mass shifts are observed during the course of data acquisition potentially due to sample plate charging or progressive source contamination from the ablated matrix. Starting with a clean ion source usually permits imaging MS data acquisition without observation of significant peak shifting.

Normalization of imaging MS data by total ion current (TIC) within a predetermined m/z range has proven to be useful in improving the quality of the recovered ion images (9). For whole-body imaging MS, normalization by TIC carries the risk of skewing the data resulting in incorrect ion signal images and thus needs to be carefully evaluated. MS data normalization by TIC assumes that the total ion signal for each pixel is somewhat constant or at least does not significantly deviate from an average value. Considering the very different tissue types encountered during the course of imaging MS of whole-body sections, this is rarely the case. For example, MS signals recovered from bone tissue typically produce a very low TIC in comparison with brain or liver tissues which produce very high TIC. Further, some tissues contain in high abundance one or several proteins that will overwhelm the spectra and produce obvious ion suppression effects. This, for example, is the case for lung tissue from which intense hemoglobin signals are detected. This is again a situation where the validity of MS data normalization by TIC needs to be carefully evaluated. For whole-body imaging MS, other normalization alternatives still need to be explored.

1.1.6. Ongoing Improvements

As outlined above, challenges still exist in all of the aspects of whole-body imaging from tissue sectioning and handling, matrix application, and imaging MS data acquisition, visualization, and analysis. Because of the potentially large size of the whole-body tissue sections, significant improvements within the analytical workflow have the potential to reduce the overall time of analysis. As seen above, alternative tape transfer methodologies (3) to more efficiently transfer and stabilize whole-body sections onto (conductive) glass slides are emerging in the field of whole-body imaging MS. Background signals coming from binding polymer have, however, been observed and may interfere with the expression of tissue signals. Most manufacturers of MALDI mass spectrometers have adopted the Society for Biomolecular Screening microtiter sample plate format which allows the mounting of larger tissue sections. Special sample plate holders able to accommodate for larger dimension glass slides have also been designed by some manufacturers. This allows imaging MS of whole (mouse)-body sections in a single experiment avoiding analysis of multiple independent tissue pieces, from which the results have to be electronically reassembled.

Matrix deposition can also be a time-limiting step. Usually, matrix deposition for whole-body imaging is performed by manual spray coating. Although rapid to perform, significant section-to-section coating variability can be observed, even in experienced hands. Beyond analyte delocalization issues, in most instances, MALDI MS signal quality, especially for peptides and proteins, can be significantly degraded with respect to the quality recovered from individual matrix spots (5). Current matrix deposition alternatives involve automated matrix deposition using acoustic (see below) or Piezo-based matrix printing systems (10, 11). These offer very reproducible matrix deposition conditions without analyte delocalization as well as high-quality MALDI MS signals. Other alternatives include systems for automated spray deposition. These allow a precise control of the spray conditions including spray height, volume, temperature, and movement over the sections (9). For phospholipid imaging MS, matrix deposition via sublimation (12) or dry coating (13) is also a rapid and attractive alternative (see below).

The acquisition speed of mass spectrometers has also greatly improved within the past decade. Whereas early imaging MS experiments were performed with MALDI TOF instruments operating at laser repetition rates of a few hertz, modern systems routinely acquire MS data at repetition rates of 200–1000 Hz. The next generation of (TOF) mass spectrometers is expected to work at repetition rates of several kilohertz with continuous sample stage movement. This will further streamline data acquisition by reducing the overhead data writing time to the computer. Such improvements should significantly reduce imaging MS data acquisition times, for whole-body imaging in particular.

1.2. Examples from the Literature

To date, several reports have been published using MALDI imaging MS on whole-body (rat and mouse) tissue sections. Images of peptides, proteins, drugs, and metabolites have been presented. All of the published reports have used spray coating for application of matrix, but otherwise, they have differed in application and methodology. The earliest detailed report describes the acquisition of protein images as well as drug and metabolite images from rat whole-body sections (2). MS data were acquired on a MALDI TOF instrument for the protein analysis, while tandem mass spectra were acquired via a MALDI QqTOF instrument for drug and metabolite imaging. Interestingly, for the protein analysis, a mix of matrices (sinapinic acid and 2,5-dihydroxybenzoic acid (DHB), 4:1) was found to provide the best conditions in terms of tissue coverage and signal intensity, while DHB alone was utilized for the small molecule experiments. This report was the first to show imaging MS of a dosed drug (olanzapine, *m/z* 313) and two of its phase-one metabolites from one experiment as shown in **Fig. 17.2**. The MS/MS imag-

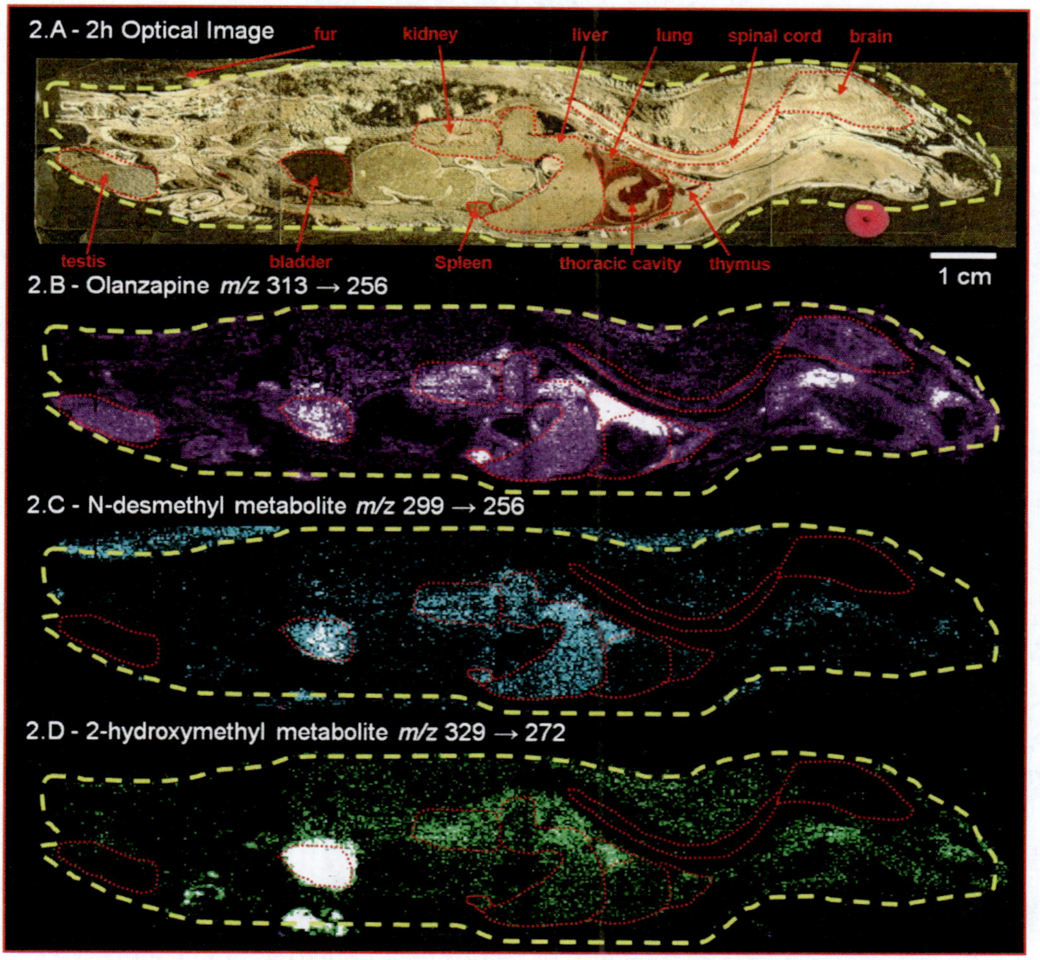

Fig. 17.2. Olanzapine (oral administration, 8 mg/kg) and metabolite distribution at 2 h post-dose in a whole-rat sagittal tissue section monitored by imaging MS. Organs are outlined in *red* on the photomicrograph of the section (**a**). MS/MS ion images of olanzapine (**b**) and its *N*-desmethyl (**c**) and 2-hydroxymethyl (**d**) metabolites are displayed. (Adapted from Ref. (2) with permission.)

ing results for olanzapine correlated quite well for location with autoradiographic results and quantity with LC–MS/MS results from individual organ extracts.

Another approach to imaging dosed compounds involves using MS scans alone (14, 15). This was demonstrated for both an injected β-peptide (15) and a small molecule with its metabolites (14) in whole-body rats and mice. The advantage to this approach is mostly speed of acquisition, as all signals obtained in the full mass range can be imaged in one experiment. However, it lacks the unambiguous identification of compounds that is typically obtained from an MS/MS experiment.

One recent application has incorporated ion mobility separation into the whole-body imaging experiment (16). In this case, the anticancer drug, vinblastine, was detected from whole-body rat sections. Comparisons of images obtained via conventional MS imaging to imaging with ion-mobility separation show that the latter significantly reduces chemical noise by eliminating interferences from isobaric ions, likely lipids. For example, the product ion of vinblastine at *m/z* 355 was shown overlaid onto a background of an endogenous lipid in the brain, with separation by ion mobility clearly showing vinblastine localized only to the ventricle.

Whole-body imaging MS has also started to be used to explore the proteome of other (non-vertebrate) animals. For example, some efforts have been focused on identifying protein signals associated with the nervous system of the medicinal leech (17, 18). Examples of peptide and protein whole-body imaging from zebrafish sections have also been reported by the Bruker Daltonics Corporation at the 2005 HUPO meeting.

2. Materials

2.1. Sample Preparation

1. Embedding polymer, such as optimal cutting temperature (OCT) polymer used to affix the specimen to the cryostat chuck (OCT, –50°C frozen section medium, Richard-Allan Scientific).
2. MALDI-compatible indium-tin oxide-coated glass slides (Delta Technologies Ltd., Stillwater, MN, USA).
3. Cryostat (CM 3050 S, Leica Microsystems, GmbH, Wetzlar, Germany).
4. 70 and 95% v/v isopropanol in water for section fixation and washing prior to protein imaging MS (HPLC grade).

2.2. Histological Staining

1. Microscope slides (ColorFrost microscope slides with clipped corners, Fisher Scientific).
2. Hematoxylin solution: 1 g hematoxylin, 25 g aluminum potassium sulfate, 400 ml water, 100 ml glycerol, 0.1 g sodium iodate.
   Dissolve hematoxylin in glycerol. Dissolve aluminum potassium sulfate in 375 ml of water. Dissolve sodium iodate in 25 ml of water. Add aluminum potassium sulfate solution to hematoxylin solution, mix well. Add sodium iodate solution, mix well. Filter complete solution before use. The remaining solution should be wrapped in foil to protect from the light.
3. Eosin and pholoxine solution: 57 ml 1% pholoxine, 1 g pholoxine B, 100 ml water (Milli-Q), 50 ml eosin.
4. Rinsing solutions: ethanol (200 proof), water (Milli-Q), xylene (ACS grade, Acros).
5. Cytoseal XYL mounting medium (Richard-Allen Scientific).
6. Slide coverslips (Fisher Scientific).

2.3. Matrix Application

1. For protein imaging: 20 mg/ml solution of sinapinic acid (Sigma, St. Louis, MO, USA) in 50:50 v/v acetonitrile:water with 0.2% trifluoroacetic acid.
2. Robotic spotter (Portrait 630, Labcyte, Sunnyvale, CA, USA).
3. For lipid imaging: 2,5-dihydroxybenzoic acid, finely ground (Sigma, St. Louis, MO, USA).

2.4. Data Acquisition and Processing

1. For protein imaging: Autoflex II MALDI TOF mass spectrometer (Bruker Daltonics, Billerica, MA, USA).
2. For lipid imaging: Ultraflex II MALDI TOF/TOF mass spectrometer (Bruker Daltonics, Billerica, MA, USA).
3. For image acquisition and visualization: FlexImaging 2.0 software (Bruker Daltonics, Billerica, MA, USA).

3. Methods

We present below two examples of protein and lipid whole-body imaging acquired by MALDI MS from sagittal sections obtained from a 1-day-old mouse pup.

3.1. Cryostat Sectioning

1. The pup was sacrificed by isoflurane gas.
2. Since the animal was small, about 1 × 3 cm, it was directly mounted on a cryostat chuck (**Fig. 17.1b**).

3. Thin 10-μm sections were cut at–21°C using a cryostat.
4. The sections to be imaged were thaw-mounted on MALDI compatible (indium-tin oxide-coated) glass slides (4).
5. Serial sections were also cut, thaw-mounted on regular microscope slides, and stained by hematoxylin and eosin (H&E). This allows alignment of the imaging MS data with mouse histology (**Figs. 17.3a** and **17.4a**).

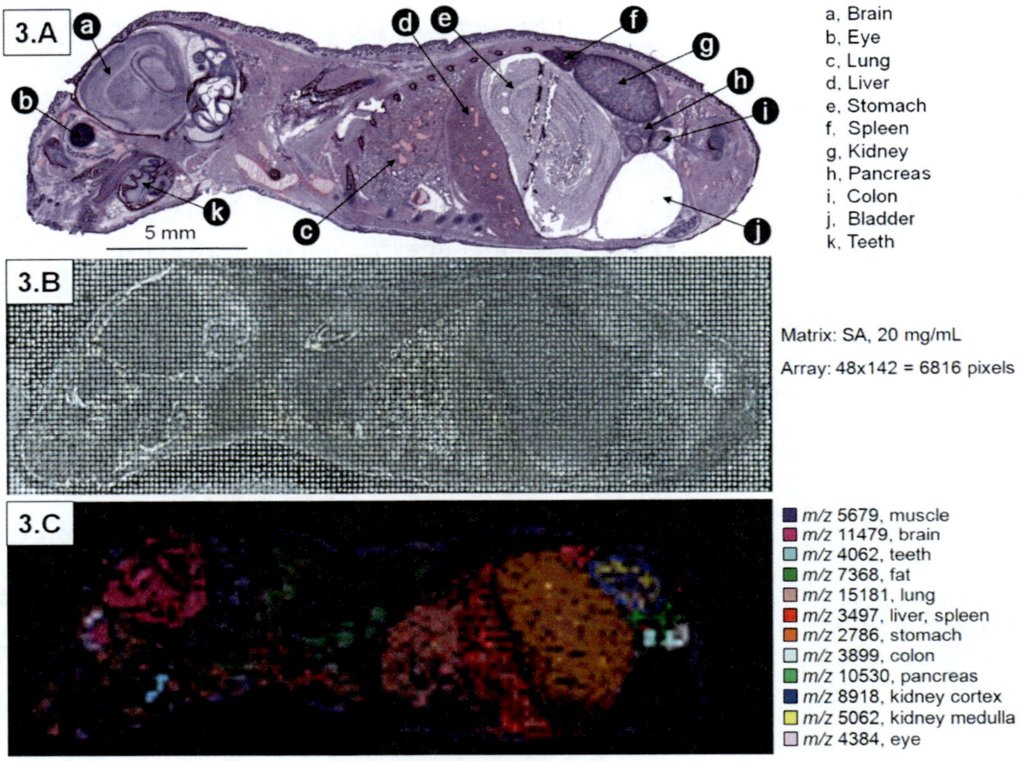

Fig. 17.3. Whole-body imaging MS of proteins from a 1-day-old mouse pup. (**a**) H&E stained section from which numerous organs are clearly visible. (**b**) Serial section used for imaging MS on which matrix (sinapinic acid) has been automatically deposited in an array manner with a final center-to-center spacing of 200 μm. (**c**) Overlay of 12 individual organ or tissue-specific protein images, each presented with a different color. See text for details.

3.2. H&E Staining

1. Hematoxylin and eosin staining for thin tissue sections mounted on glass slides was performed in a truncated manner from that typically performed for high-resolution histological approaches. The quality of the stain is sufficient for alignment of the MS ion images to the histological image. The truncated H&E staining protocol used is as follows: Serially immerse the slide in 95% ethanol (30 s), 70% ethanol (30 s), water (30 s), hematoxylin solution (2 min), water

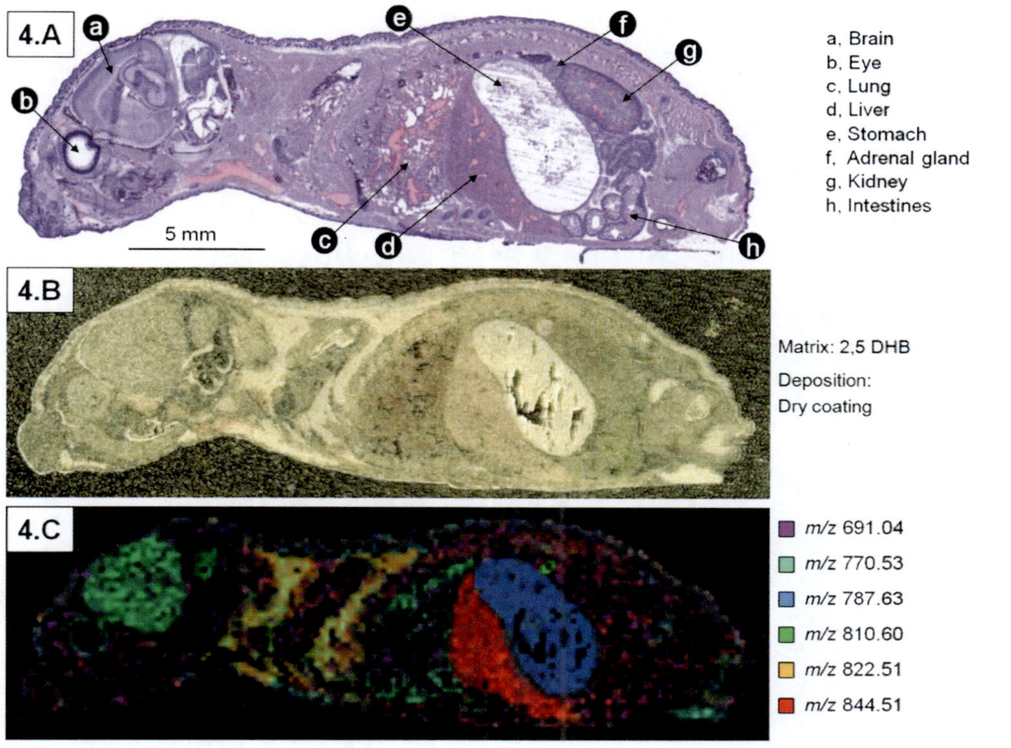

Fig. 17.4. Whole-body imaging MS of lipids from a 1-day-old mouse pup. (**a**) H&E stained section from which numerous organs are clearly visible. (**b**) Serial section used for imaging MS on which matrix (DHB) has been homogeneously applied using the dry deposition approach. (**c**) Overlay of six individual organ or tissue-specific lipid images, each presented with a different color. See text for details.

(20 s), 70% ethanol (30 s), 95% ethanol (30 s), eosin solution (1 min), 95% ethanol (30 s), 100% ethanol (30 s), and xylene (2–2.5 min).

2. Remove excess xylene by gentle wiping.
3. Deposit a small drop of cytoseal solution on section and gently deposit the coverslip slide on top of the cytoseal drop. Make sure the coverslip covers the entire tissue section. For whole-body sections, several coverslip slides may be necessary.

3.3. Protein Image Protocols

1. The section to be analyzed for proteins was rinsed with 70 and 95% isopropanol solutions according to published protocols (19) to improve overall signal detection from all parts of the body.
2. Sample coating for imaging MS was achieved by printing arrays of small matrix droplets using a Portrait 630 reagent multispotter (10). The volume for each droplet was

estimated to be ~170 pL. Sinapinic acid was printed in arrays with a center-to-center distance of 400 μm. By interlacing four print patterns, a final center-to-center spacing between spots of 200 μm was achieved.

3. Each print pattern was repeated for 30 cycles ejecting one drop of matrix per cycle. The final array consisted of a rectangular $48 \times 142 = 6,816$ spot pattern spaced by 200 μm center-to-center (**Fig. 17.3b**) (*see* **Note 1**).
4. The section was analyzed by MALDI TOF MS using an Autoflex II mass spectrometer operated in positive linear mode geometry under delayed extraction conditions time focused at ~*m/z* 15,000. In this case, signal resolutions ($M/\Delta M$ measured at full width at half maximum) close to ~1,000 were attainable for ions at or around *m/z* 15,000, whereas acceptable resolutions in the range of 600 were obtainable for *m/z* values in the range of 5,000–25,000.
5. One mass spectrum was acquired per matrix spot. Each spot was analyzed in the same manner by averaging signals from 250 laser shots acquired in five series of 50 shots with each series acquired at a different location within the spot using a random walk pattern.
6. The collected mass spectra were baseline corrected before image assembly.
7. Imaging MS data was acquired, assembled, and visualized using FlexImaging 2.0 software. Data acquisition of the 5,325 matrix spots on tissue took 6 h 47 min generating 3.85 GB of data.
8. From each matrix spot (pixel), between 200 and 300 protein signals were typically detected with some of these being very organ or tissue specific. **Figure 17.3c** presents an overlay of 12 ion images from protein signals with some specificity for different tissues or organs. For example, the signal at *m/z* 4,062 was found to have a very strong expression in dental tissue whereas the signal at *m/z* 10,530 was found in higher abundance in the pancreas. Other signals such as *m/z* 5,679 specific for muscle and *m/z* 7,368 consistent with fat deposits were found throughout the section.

3.4. Lipid Image Protocols

1. The section to be analyzed by imaging MS for lipids was left unrinsed.
2. Finely ground 2,5-dihydroxybenzoic acid (DHB) used as matrix was homogeneously deposited onto the section using a dry-coating technique, in which solid matrix particles were

filtered directly onto the tissue through a 20-μm stainless steel sieve as previously described (13) (*see* **Note 2**).

3. Imaging MS data acquisition was performed using an Ultraflex II TOF/TOF mass spectrometer operated in positive reflex mode geometry under optimized delayed extraction conditions time focused at ~*m/z* 800. In this case, signal resolutions ($M/\Delta M$) above 10,000 were routinely attainable for the phospholipid mass range from *m/z* 600–1,000.
4. Data acquisition was performed over the entire tissue section at a spatial resolution of 300 μm accumulating signals from 300 successive laser shots per position.
5. The collected mass spectra were baseline corrected before image assembly.
6. Imaging MS data was acquired, assembled, and visualized using FlexImaging 2.0 software. Data acquisition of the 5,444 spots on tissue took 5 h 23 min generating 1.0 GB of data.
7. From each position, intense phospholipid signals were observed with some of these being very organ or tissue specific. **Figure 17.4c** presents an overlay of six ion images from phospholipid signals with some specificity for different tissues or organs. For example, the signal at *m/z* 844.51 was found to have a very strong expression in liver, whereas the signal at *m/z* 810.60 was uniquely found in the adrenal gland. Other signals such as *m/z* 691.04 abundant in muscle and *m/z* 822.51 consistent with fat deposits were found throughout the section.

3.5. Conclusions

Clearly, imaging MS has tremendous analytical potential for interrogating whole-body tissue sections. Many analyte classes present in biological specimens – including lipids, peptides, proteins, endogenous metabolites, and administered drug compounds – can be analyzed and imaged via MALDI mass spectrometry. As with so many advanced technologies, “the devil is in the details”. Application of imaging MS to large, very heterogeneous whole-body sections requires special considerations in terms of experimental design, sample handling, data acquisition, data processing, and image generation in order to obtain meaningful data. While not trivial, these extra challenges can be overcome, and the tremendous stores of biological information present in a whole animal can begin to be understood. The methods and protocols presented in this chapter should help interested researchers successfully apply imaging MS to their own applications.

4. Notes

1. For this particular example, because of the relatively small dimensions of the section, matrix deposition using an automated printing approach was possible within a time frame of about 4 h. Matrix has also been successfully deposited on larger sections from rats and mice using the same approach but with larger center-to-center distances in order to keep printing times within a two working day period. Compared to manual matrix spray deposition, the spatial resolution with printed arrays is limited by the center-to-center distances between spots. Since imaging MS from whole-body sections is typically done at lower resolutions (200–500 μm) to keep acquisition times manageable, this is typically not an issue.
2. The resulting crystalline film of matrix is fairly homogeneous (**Fig. 17.4b**) and allows imaging by MALDI MS with spatial resolutions as small as 30 μm (13). Further, because of its simplicity, this matrix coating technique is easily applicable to larger whole-body tissue sections. Since matrix is applied dried on the section, this eliminates any risks of analyte delocalization.

References

1. Schwartz, S. A., Reyzer, M. L., Caprioli, R. M. (2003) Direct tissue analysis using matrix-assisted laser desorption/ionization mass spectrometry: practical aspects of sample preparation. *J Mass Spectrom*, **38**, 699–708
2. Khatib-Shahidi, S., Andersson, M., Herman, J. L., Gillespie, T. A., and Caprioli, R. M. (2006) Direct molecular analysis of whole-body animal tissue sections by imaging MALDI mass spectrometry. *Anal Chem*, **78**, 6448–6456
3. Nissanov, J., Bertrand, L., Tretiak, O. (2001) Cryosectioning distortion reduction using tape support. *Microsc Res Tech*, **53**, 239–240
4. Chaurand, P., Schwartz, S. A., Billheimer, D., Xu, B. J., Crecelius, A., Caprioli, R. M. (2004) Integrating histology and imaging mass spectrometry. *Anal Chem*, **76**, 1145–1155
5. Chaurand, P., Caprioli, R. M. (2002) Direct profiling and imaging of peptides and proteins from mammalian cells and tissue sections by mass spectrometry. *Electrophoresis*, **23**, 3125–3135
6. Crossman, L., McHugh, N. A., Hsieh, Y., Korfmacher, W. A., Chen, J. (2006) Investigation of the profiling depth in matrix-assisted laser desorption/ionization imaging mass spectrometry. *J Mass Spectrom*, **20**, 284–290
7. Reyzer, M. L., Caprioli, R. M. (2007) MALDI-MS-based imaging of small molecules and proteins in tissues. *Curr Opin Chem Biol*, **11**, 29–35
8. Reyzer, M. L., Hsieh, Y., Ng, K., Korfmacher, W. A., Caprioli, R. M. (2003) Direct analysis of drug candidates in tissue by matrix-assisted laser desorption/ionization mass spectrometry. *J Mass Spectrom*, **38**, 1081–1092
9. Norris, J. L., Cornett, D. S., Mobley, J. A., Andersson, M., Seeley, E. H., Chaurand, P., Caprioli, R. M. (2007) Processing MALDI mass spectra to improve mass spectral direct tissue analysis. *Int J Mass Spectrom*, **260**, 212–221

10. Aerni, H. R., Cornett, D. S., Caprioli, R. M. (2006) Automated acoustic matrix deposition for MALDI sample preparation. *Anal Chem*, **78,** 827–834
11. Sugiura, Y., Shimma, S., Setou, M. (2006) Two-step matrix application technique to improve ionization efficiency for matrix-assisted laser desorption/ionization in imaging mass spectrometry. *Anal Chem*, **78**, 8227–8235
12. Hankin, J. A., Barkley, R. M., Murphy, R. C. (2007) Sublimation as a method of matrix application for mass spectrometric imaging. *J Am Soc Mass Spectrom*, **18**, 1646–1652
13. Puolitaival, S. M., Burnum, K. E., Cornett, D. S., Caprioli, R. M. (2008) Solvent-free matrix dry-coating for MALDI imaging of phospholipids. *J Am Soc Mass Spectrom*, **19**, 882–886
14. Stoeckli, M., Staab, D., Schweitzer, A. (2007) Compound and metabolite distribution measured by MALDI mass spectrometric imaging in whole-body tissue sections. *Int J Mass Spectrom*, **260**, 195–202
15. Stoeckli, M., Staab, D., Schweitzer, A., Gardiner, J., Seebach, D. (2007) Imaging of a [beta]-peptide distribution in whole-body mice sections by MALDI mass spectrometry. *J Am Soc Mass Spectrom*, **18**, 1921–1924
16. Trim, P. J., Henson, C. M., Avery, J. L., McEwen, A., Snel, M. F., Claude, E., Marshall, P. S., West, A., Princivalle, A. P., Clench, M. R. (2008) Matrix-assisted laser desorption/ionization-ion mobility separation-mass spectrometry imaging of vinblastine in whole body tissue sections. *Anal Chem*, **80**, 8628–8634
17. Jardin-Mathé, O., Bonnel, D., Franck, J., Wisztorski, M., Macagno, E., Fournier, I., Salzet, M. (2008) MITICS (MALDI imaging team imaging computing system): a new open source mass spectrometry imaging software. *J. Proteomics*, **71**, 332–345
18. Wisztorski, M., Croix, D., Macagno, E., Fournier, I., Salzet, M. (2008) Molecular MALDI imaging: an emerging technology for neuroscience studies. *Dev Neurobiol*, **68**, 845–858
19. Seeley, E. H., Oppenheimer, S. R., Mi, D., Chaurand, P., Caprioli, R. M. (2008) Enhancement of protein sensitivity for MALDI imaging mass spectrometry after chemical treatment of tissue sections. *J Am Soc Mass Spectrom*, **19**, 1069–1077

Chapter 18

MALDI Direct Analysis and Imaging of Frozen Versus FFPE Tissues: What Strategy for Which Sample?

Maxence Wisztorski, Julien Franck, Michel Salzet, and Isabelle Fournier

Abstract

Significant advances have been made in the past decade in the field of mass spectrometry imaging with MALDI ion sources (MALDI-MSI). While MALDI-MSI has high potential in the field of biology and in the clinic, a challenge for MALDI-MSI has been to adapt itself to a greater range of sample types. In particular, much of the biological archived materials for pathology studies are tissue biopsies fixed with paraformaldehyde and embedded in paraffin (FFPE tissues) because of the high stability of such samples. Thus, there has been a need to develop strategies for analyzing FFPE samples as this would allow retrospective studies of past clinical cases on large cohorts of existing samples. Obviously, PAF fixation, by inducing protein cross-linking, causes problems for molecular analysis by MS. We developed on tissue digestion strategies for overcoming these difficulties and allowing molecular data to be retrieved from FFPE samples no matter how long they have been stored. These digestion strategies preserve localization from digested proteins making MALDI-MSI of proteins possible by monitoring the resulting peptides. We present methods and protocols for FFPE samples. These strategies have proven to be valuable for all tested FFPE samples and have opened archived tissues from hospital banks to MALDI-MSI.

Key words: Matrix-assisted laser desorption/ionization, time-of-fight, mass spectrometry, mass spectrometry imaging, formalin fixed and paraffin embedded, frozen tissue, on tissue digestion, proteins identification.

1. Introduction

After 10 years of development, MALDI-MSI has shown a great potential for applications to a variety of fields. In particular, applications are found in pharmacology for drug evaluations (localization of drugs and their metabolites) (1, 2) or in clinics for

S.S. Rubakhin, J.V. Sweedler (eds.), *Mass Spectrometry Imaging*, Methods in Molecular Biology 656,
DOI 10.1007/978-1-60761-746-4_18, © Springer Science+Business Media, LLC 2010

biomarkers discovery and study or molecular classification of pathologies (3–6). MALDI imaging was successfully carried out and demonstrated on frozen tissue sections after snap freezing procedures without fixation (7, 8). But the possibility of simultaneously detecting and localizing hundreds of different molecules in a sample without prior labeling become a tremendous challenge: how does one identify them? In fact, clinical applications require the identification and validation of target molecules (9).

Importantly, a vast number of samples contained in hospital tissue banks have proven to be incompatible with MS analysis and are thus had not been usable for MALDI-MSI. The majority of archived samples have been stored after fixation and paraffin embedding. The fixation step is necessary to increase sample stability and preserves tissue fine structures for observations after tissue sectioning. The most commonly used fixative is paraformaldehyde (PAF) or formalin. When PAF fixation is used prior to paraffin embedding, the resulting samples are called formalin fixed and paraffin embedded or FFPE. Because FFPE samples can be stored for decades, many samples are available from hospital tissue banks allowing creating larger cohorts even for rare pathologies or subpathologies (8, 10–12). In addition, the clinical outcomes of the donors and the pathologist diagnosis are known, making FFPE tissues a mine of information, especially if the imaged compounds can be identified. While PAF fixation is ideal for conservation and structure preservation, it causes difficulties for molecular analysis. In fact, PAF fixation induces molecular cross-linking, especially of peptides and proteins.

Chemical reaction processes involved in PAF fixation are extremely complex. Protocols and outcomes of PFA fixations have been extensively studied, but only few literature reports directly address the issues of the specific molecular chemical reactions underlying the fixation process (13–15). In a simplified way, we can state that PAF reacts principally with free amine groups by the formation of methylene bridges. The primary reaction of the aldehyde to the protein has fast kinetics (13–15). On the contrary, secondary reactions lead to the formation of the methylene bridge as a much slower process taking place over days (13–15). Thus, reactions still proceed even after paraffin embedding. Therefore, the proteins become further imprisoned over time as methylene bridges slowly form.

As proteins are all cross-linked together, MS analysis of such tissue samples after tissue sectioning and paraffin removal generates few peaks. As reactions still proceed inside the paraffin blocks, MS spectra generated from these samples are dependent on storage time. For recent FFPE blocks (~0–6 months), ion signals are retrieved (8). These signals are mainly observed in the peptide mass range and detailed study of the corresponding mass spectra shows that peaks have broadened profiles with +12 u.

mass adduct (16). Such phenomena were previously observed and described in other MS studies and are due to sub-reactions from PAF fixation. These sub-reactions also present slow kinetics and proceed after paraffin embedding. The intensity of +12 u. adducts peaks and their number increases with storage time as is observed in the MS spectra. More peptide signals are observed when using specific MALDI matrices such as the reactive matrix 2,4-dinitrophenylhydrazine (2,4-DNPH) (8). Surprisingly, for such a matrix, MS spectra do not contain the +12 u. adduct ions. This can be explained by assuming that 2,4-DNPH has a benefic effect by reacting with any free aldehyde groups (PAF) remaining in the tissues. For longer stored samples (>1 year), MS analyses do not allow the detection of exploitable signals even using reactive matrices such as 2,4-DNPH. Proteins cross-linking has also shown to be problematic for immunohistochemistry (IHC) experiments, by hampering antibody access to the epitope of the antigen. To overcome such problems, pathologists have extensively studied epitope unmasking procedures. Different antigen retrieval (AR) protocols compatible with IHC are now well known and described (17–23). One popular procedure is heating the sample at high pressure (17–23). But, if AR allows epitope unmasking, such procedures do not reverse protein cross-linking. AR can be used before MS analysis of tissues and can improve analysis, although this has not proved not sufficient as the cross-linked proteins render these samples difficult to use for direct analysis by MALDI and MALDI-MSI.

New strategies are then needed to analyze FFPE tissue sections in an efficient manner, especially an approach that is independent of storage time. We have developed an approach based on "tissue enzymatic digestion" allowing one to obtain peaks from pieces of proteins (**Fig. 18.1**). In this strategy, the protein is enzymatically digested and the resulting peptides are subsequently analyzed on the tissue. MALDI-MSI after on tissue enzymatic digestion is possible if the localization of the generated peptides is maintained, i.e., if enzyme can be precisely deposited and localized onto the tissue. This is achievable by dropping small droplets of enzyme in a control manner. For the highest reproducibility, this can be done via an automatic device. If digestion peptides locations are precisely maintained, then MALDI-MSI of proteins can be deduced from the images generated from the peptide ion signals. In addition, these peptides are also useful for protein identification. In fact, because MALDI produces ions with only a few charges, MALDI is not well suited for direct identification of proteins by the means of "top-down" approaches; thus, "bottom-up" strategies are more common. Here we digest the proteins and use MALDI-MSI bottom-up approaches by performing MS/MS experiments on the peptides from the tryptic digestion from the FFPE preserved proteins. Such strategies are

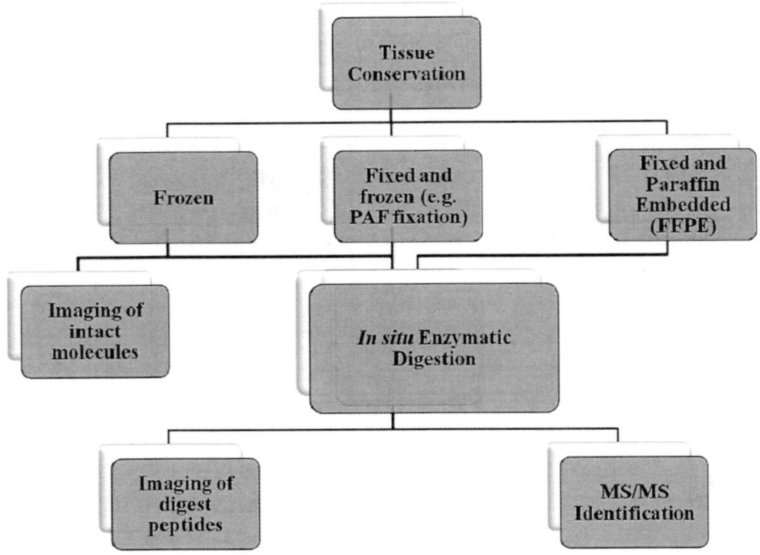

Fig. 18.1. Strategies for direct analysis and imaging mass spectrometry of FFPE, PAF, and frozen tissue section.

work with frozen samples when the proteins need to be identified (**Fig. 18.1**). In comparison with FFPE samples, frozen samples are simpler to analyze and image as they do not systematically require an on tissue digestion strategy. While frozen samples are easier for MALDI-MSI, they do not present the same stability. Degradation of the proteome can be observed for samples stored over long periods at –80°C.

In this chapter, we describe methods and protocols for direct analysis of frozen and FFPE tissue samples, including on tissue digestion for both frozen and FFPE samples for protein identification.

2. Materials

2.1. Preparation of Frozen Tissue Sections

2.1.1. Snap Frozen Tissues

1. Isopentane cooled at –45°C with dry ice. Vapors may cause drowsiness and dizziness, so work in a hood.

2.1.2. Tissue Cryosection and Thaw Mounted

1. Optimal cutting temperature polymer, OCT.
2. Indium tin oxide (ITO) coated glass slides or other holder compatible with mass spectrometry analysis.

3. A cryomicrotome, Leica CM150S (Leica Microsystems, Nanterre, France).

2.1.3. Pre-analysis Treatment – Tissue Fixation

1. Ethanol 75% (–20°C): 75 ml of absolute ethanol (≥99.8%) and water (HPLC grade) to 100 ml. Prepare fresh. Store at –20°C.
2. Ethanol 95% (–20°C): 95 ml of absolute ethanol (≥99.8%) and water (HPLC grade) to 100 ml. Prepare fresh. Store at –20°C.

2.1.4. Pre-analysis Treatment – Removal of Lipids

1. Chloroform (–20°C): 100 ml of chloroform (≥99.9%). Store at –20°C. Chloroform is harmful by inhalation, so work in the hood.

2.2. Preparation of FFPE Tissue Section

2.2.1. FFPE Tissue Section

1. Indium tin oxide (ITO) coated glass slides or other holder compatible with mass spectrometry.
2. Water: 100 ml of water (HPLC grade). Prepare fresh.
3. A microtome and a hotplate warm at 50°C.

2.2.2. FFPE Tissue Dewaxing

1. Xylene: 100 ml of xylene (≥99.9%). Xylene is harmful by inhalation, so work in the hood.
2. Ethanol 100%. Prepare fresh.
3. Ethanol 95%: 95 ml of absolute ethanol (99.8%) and water to 100 ml. Prepare fresh.
4. Ethanol 75%: 75 ml of absolute ethanol (≥99.8%) and water to 100 ml. Prepare fresh.
5. Ethanol 30%: 30 ml of absolute ethanol (≥99.8%) and water to 100 ml. Prepare fresh.
6. Water: 100 ml of water (HPLC grade). Prepare fresh.

2.3. On Tissue Digestion

2.3.1. Using a Microspotter

1. Trypsin, sequencing grade modified (Promega). Suspend in 20 mM NH_4HCO_3 buffer at 20 μg/ml (*see* **Note 1**).
2. Methanol 50%: 50 ml of absolute methanol and water to 100 ml. Prepare fresh. Methanol is toxic, so work in the hood.
3. Chemical Inkjet Printer CHIP-1000 (Shimadzu Biotech, Kyoto, Japan).

2.3.2. Using an Automatic Sprayer

1. Trypsin, sequencing grade modified (Promega). Resuspend in 20 mM NH_4HCO_3 buffer at 40 μg/ml (*see* **Note 1**).

2. Methanol 50%: 50 ml of absolute methanol and water to 100 ml. Prepare fresh. Methanol is toxic, so work in the hood.
3. ImagePrep (Bruker Daltonics, Bremen, Germany).

2.4. Matrix Deposition

2.4.1. For Protein Analysis Using a Microspotter

1. SA/ANI solution: 1 equivalent of aniline (ANI) was added to a solution containing 40 mg/ml of sinapinic acid (SA) in acetonitrile/aqueous TFA 0.1% (6:4, v/v) (*see* **Note 2**). Aniline and TFA are toxic, so work in the hood.
2. Chemical Inkjet Printer CHIP-1000 (Shimadzu Biotech, Kyoto, Japan).

2.4.2. For Protein Analysis Using an Automatic Sprayer

1. SA/ANI solution: 1 equivalent of aniline (ANI) was added to a solution containing 40 mg/ml of sinapinic acid (SA) in acetonitrile/aqueous TFA 0.1% (6:4, v/v) (*see* **Note 2**). Aniline and TFA are toxic, so work in the hood.
2. ImagePrep (Bruker Daltonics, Bremen, Germany).

2.4.3. For Peptide Analysis Using a Microspotter

1. HCCA/ANI solution: 1.5 equivalent of aniline (ANI) was added to a solution containing 10 mg/ml of α-Cyano-4-hydroxycinnamic acid (HCCA) in acetonitrile/aqueous TFA 0.1% (6:4, v/v) (*see* **Note 3**). Aniline and TFA are toxic, so work in the hood.
2. Chemical Inkjet Printer CHIP-1000 (Shimadzu Biotech, Kyoto, Japan).

2.4.4. For Peptide Analysis Using an Automatic Sprayer

1. HCCA/ANI solution: 1.5 equivalent of aniline (ANI) was added to a solution containing 10 mg/ml of α-Cyano-4-hydroxycinnamic acid (HCCA) in acetonitrile/aqueous TFA 0.1% (6:4, v/v) (*see* **Note 3**). Aniline and TFA are toxic, so work in the hood.
2. ImagePrep (Bruker Daltonics, Bremen, Germany).

2.5. Mass Spectrometry Analysis

2.5.1. MALDI-MSI Experiment

1. Peptide calibration standard II (Bruker Daltonics, Wissenbourg, France): angiotensin II, angiotensin I, substance P, bombesin, ACTH clip 1–17, ACTH clip 18–39, somatostatin 28, bradykinin fragment 1–7, renin substrate tetradecapeptide porcine. Covered mass range: ~700–3,200 Da. Store at –20°C.

2. Protein Calibration Standard I (Bruker Daltonics, Wissenbourg, France): insulin, ubiquitin I, cytochrome C, myoglobin. Covered mass range: ~5,000–17,500 Da. Store at –20°C.
3. An Ultraflex II TOF–TOF equipped with a smartbeam laser and all the Flex software suite (FlexControl, FlexAnalysis, and FlexImaging) (Bruker Daltonics, Bremen, Germany).

2.5.2. MS/MS Analysis

1. An Ultraflex II TOF–TOF equipped with a smartbeam laser and all the Flex software suite (FlexControl, FlexAnalysis, and FlexImaging) (Bruker Daltonics, Bremen, Germany).
2. Biotools (Bruker Daltonics, Bremen, Germany).

3. Methods

In what follows, the methods are presented according to the MALDI-MSI workflow. We primarily describe the steps leading to tissue section preparation prior to MALDI-MSI which is an extremely important part of the method. In particular, snap frozen tissues require specific tissue treatment (washing procedures) prior to other steps. One of the procedures is fixation of tissues using ethanol (24). The second step removes small organic compounds (namely lipids) that hampers detection of peptides in the low *m/z* range by use of organic solvents. These steps are independent of each other, but are strongly recommended when performing in situ digestion and identification of proteins. We also describe how to prepare FFPE tissue sections for MALDI-MSI analysis. Because in situ digestion and/or matrix deposition must be performed with precision and reproducibility to prevent molecule delocalization, it is recommended that one use an automated deposition device. Two types of instruments are described (**Fig. 18.2**):

- An automatic microspotter delivering picoliter volumes of solution in defined coordinates allow the coverage of the entire surface of the tissue with an array of spot spaced by ~200 μm.
- An automatic sprayer can spray small droplets (~25 μm size in average) onto the top of the tissue section. In this case the tissue section is fully covered by the spray.

The main advantages of these devices are to limit delocalization of molecules. This depends on the solvents used, but they still must allow optimal extraction of the proteins and peptides, allowing enzyme activity and the incorporation of analytes into matrix crystals during matrix crystallization. These devices currently

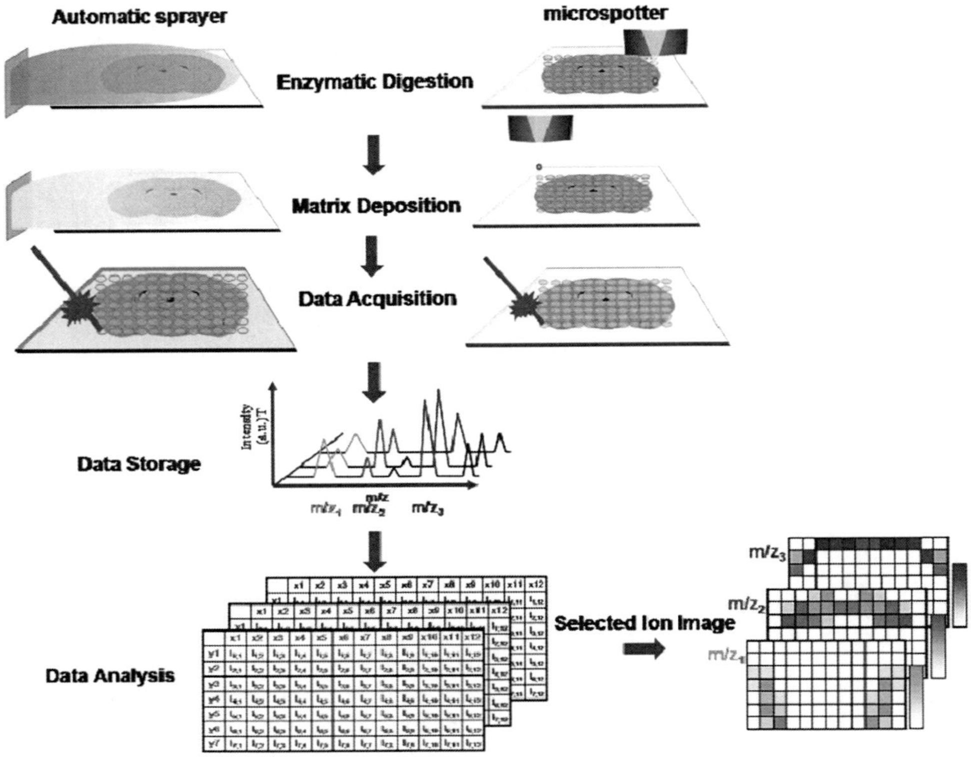

Fig. 18.2. Schematic representation of the strategies for in situ digestion using an automatic sprayer or microspotter devices.

represent the best compromise of minimizing lateral diffusion of analytes while maintaining enzymatic digestion and analyte incorporation efficiency (**Fig. 18.3**).

For matrix deposition using these automatic devices, it is recommended to use solid ionic matrices (25). In fact, solid ionic matrices have proven to be more efficient for MALDI-MSI. They have several advantages including higher salt tolerances, higher stabilities under vacuum conditions (lower sublimation rates because of higher sublimation temperatures), lower ablation volumes and thus produce higher spectral quality (increased number of detected ions, higher signal intensity). Moreover, these matrices are compatible with the automatic deposition devices. Classical matrices often lead to clogging problems of the systems when used at optimal analytical concentrations. If concentration of matrix solutions is decreased to reduce clogging, this dramatically decreases the resulting MS spectra quality (less detected signals with a lower average intensity). Overall, the classical matrices used with automated spotting devices do not currently offer a robust system for high-quality analyses of tissues. On the contrary, solid ionic matrices have demonstrated that they are well suited with automatic devices. In addition, solid ionic matrices present

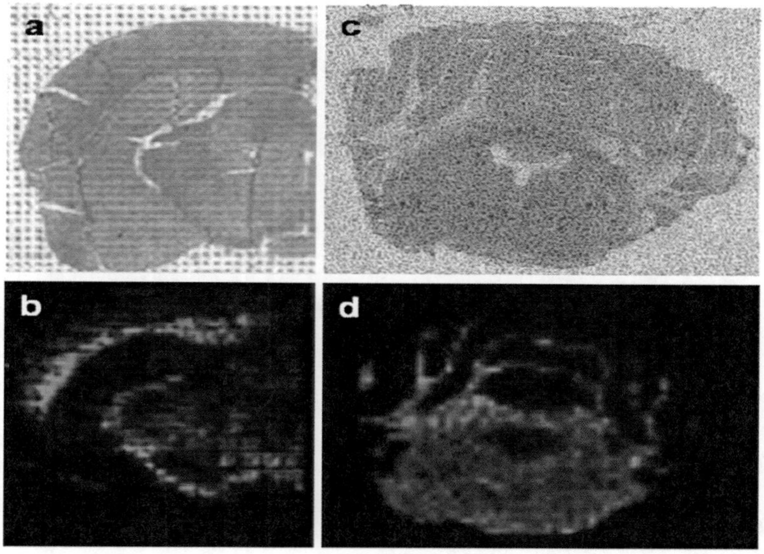

Fig. 18.3. Optical picture and MALDI image using microspotting (respectively **a** and **b**) or using an automatic sprayer (respectively **c** and **d**) for enzymatic digestion and matrix deposition.

different physicochemical properties than classical matrices. And in particular their surface tension is lower. Thus, it is possible to avoid clogging problems using solid ionic matrices while working at the optimum analytical concentration. By using solid ionic matrices with automated devices, we have a robust system that can be used in an automate fashion without any control giving higher printing (more defined spots) and analytical quality.

In the last part of methods, mass spectrometry analyses of tissue sections (FFPE or Frozen) after in situ digestion and matrix deposition are described. Typically on tissue digestion produces complex MS spectra depending on the local composition of the tissue (**Fig. 18.4**). MS/MS experiments can be performed by selecting a parent ion to achieve structural information. Identification is generally obtained by database interrogation using search algorithm such as MASCOT. Because of spectral complexity, the approach used is generally to first identify one the fragment peptides of a protein and then, based on the protein sequence, search for the presence of the other digestion fragments of the putative protein. If the additional fragments are present, they are submitted to MS/MS experiments to confirm that their fragmentation pattern fits with the expected peptide sequence. In parallel, localization of the different digestion peptides expected to be part of the same protein is generated in order to confirm identification; all digestion fragments of a single protein should present the same spatial distribution in the tissue (**Fig. 18.5**).

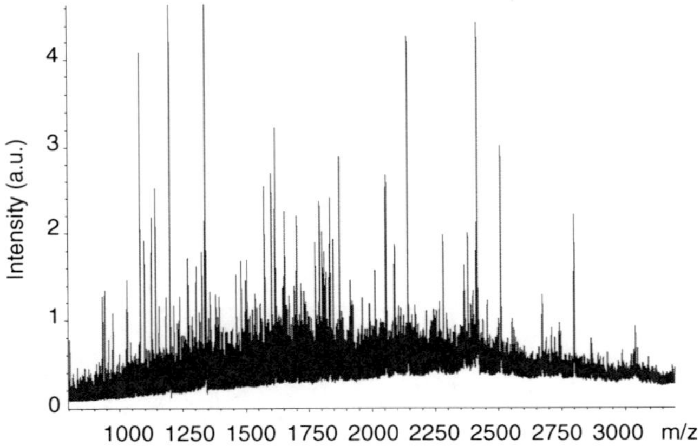

Fig. 18.4. Example of a typical mass spectrum recorded on a rat brain tissue section after in situ enzymatic digestion of a FFPE or frozen tissue.

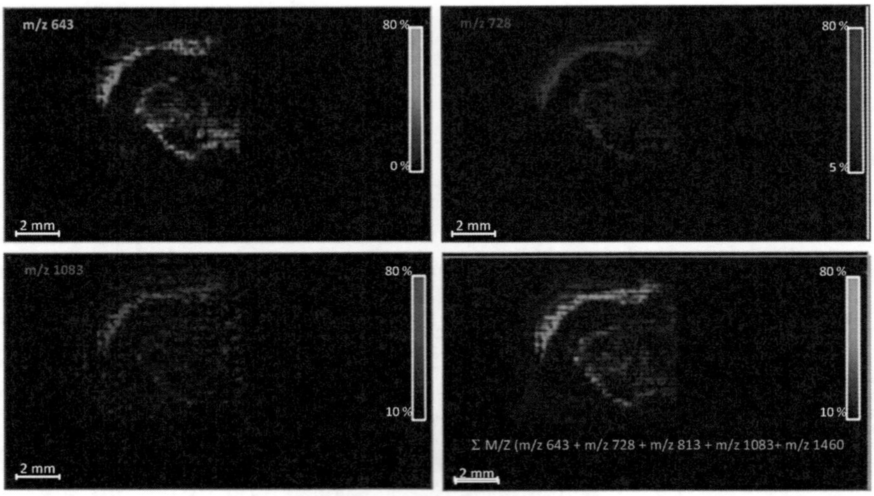

Fig. 18.5. Example of MALDI images of digest fragments of the same protein (here CRMP2) recorded from FFPE tissue section stored for 10 year after in situ enzymatic digestion of a 6-OHDA injected rat brain. Reprinted from (10).

3.1. Preparation of Frozen Tissue Section

3.1.1. Tissue Snap Frozen

1. The organ is dissected and rinsed with a saline solution suitable for the tissue. This removes blood and other tissue fragments from the surface. Alternative: prior to killing, the animal can be perfused with the saline solution to remove blood inside the organ.
2. Morphology of the organ needs to be carefully maintained. Thus, the tissue should not be placed in a tube or wrapped

in an aluminum foil to avoid deformation of the organ (otherwise, it adapts to the outlines of the container).

3. Snap freezing procedures are applied for tissue preservation to maintain tissue morphology and to prevent ice crystal formation and cell explosion. In fact, different cooling rates for some parts of the organ or direct dipping of the organ into liquid nitrogen lead to the formation of cracks and fragmentation of the tissue. Therefore, the use of isopentane cooled at –45°C with dry ice is recommended. Freezing time is dependent on the size of the organ. It is preferable not to use an embedding media. For very small organs or surgical pieces, cutting without embedding material increases deformations and damages of the tissue sections. In such a case, a solution containing nonpolymeric compounds such as 10% gelatin solutions help to obtain high-quality tissue section. Tissue is embedded in 10% gelatin directly after dissection and frozen as previously described.
4. After snap freezing, tissue is removed from isopentane and stored at –80°C. We heartily recommend not surpassing a storage period of 6 months. After 6 months of storage, variations in the molecular profiles are observed if no sample stabilization procedure is performed. Preferentially, tissues should be analyzed a few days or weeks after snap freezing.

3.1.2. Tissue Cryosection and Thaw Mounted

1. The use of cryopreservative solutions containing organic polymers such as optimal cutting temperature (OCT) polymer should be restricted to the attachment of the tissue to the sample holder and not used for complete embedding of the tissue. Moreover, all parts of the cryostat in contact with the tissue must be cleaned to prevent any contamination between two different samples or with the polymer containing solution. In the case of contact between the tissue and cryopreservative solutions containing polymers, MS spectra will be dominated by polymer signals such as PEG.
2. Tissue is placed in the cryostat during sufficient time before sectioning for slow warming of the sample to the cryostat temperature. If the tissue is too cold, poor quality sections are obtained.
3. 10 μm thickness tissue sections are cut using cryomicrotome at –20°C. Different tissue types may need other temperature settings. 10 μm thickness is optimal (*see* **Note 4**).
4. Collect the tissue sections onto ITO glass slides pre-cooled at –20°C. Transfer is performed by applying the cooled ITO slide onto the section. The cuts thus stick on the cold slide (*see* **Note 5**). Adhesion of the frozen sections to the glass slides is obtained by heating, by putting fingers under the

slide, or by placing the slide at room temperature. This transfer procedure, contrary to classical thaw mounting, prevents formation of ice crystals at the surface of the cryostat microtome cutting plate.

5. Mounted sections are stored in a sealed container at –80°C until their use.

3.1.3. Pre-analysis Treatment – Tissue Fixation

1. A closed container store at –80°C is warmed at room temperature in a vacuum desiccator to prevent water condensation at the surface of the frozen slide.
2. After complete drying, the ITO slide is washed. Washing steps are optional and dependent on the molecules to be analyzed. Careful washing is crucial for conserving spatial localization of molecules.
3. For analysis of small molecules like lipids or drugs, no washing steps are used. For macromolecules such as peptides or proteins, washing procedures are generally used. Washing is performed by immersing the glass slide softly in ice-cold 75% ethanol during 30 s. No agitation or shake is needed. This step washes out salts, cells fragments, or residual fluids.
4. Take the slide out and remove the excess of liquid around the section. A stream of nitrogen over the surface could help to remove excess of ethanol.
5. The ITO glass slide is then placed in a vacuum desiccator to complete dry of the tissue. The time of drying is dependent on the size of the section.
6. *Optional:* A second bath of fresh ice-cold 75% ethanol during 30 s followed by a complete drying under vacuum desiccator can be achieved.
7. After complete drying, the sample is dipped into cold 95% ethanol during 30 s. No agitation or shake is needed. This step prevents degradation of proteome by dehydration and fixation of the tissue.
8. The slide is completely dried as in Steps 4 and 5.

3.1.4. Pre-analysis Treatment – Lipid Removal

1. After complete drying, immerse the glass slide softly in ice-cold chloroform during 30 s (*see* **Note 6**). No agitation or shake is needed. This step removes lipids (especially phospholipids) present in high concentration in the tissue (components of cell membranes) and may cause signal suppression in MS spectra (26).
2. Take the slide out and place it in the vacuum desiccator for complete drying of the tissue.

3.2. Preparation of FFPE Tissue Section

3.2.1. FFPE Tissue Section

1. 10 μm thickness FFPE tissue sections are cut using a microtome at room temperature. Paraffin block can be cooled down –20°C prior sectioning to facilitate tissue sectioning.
2. Sections are transferred onto a conductive ITO glass slide on top of a water droplet.
3. Glass slide is warmed up on a hotplate to leave the cuts unfolds.
4. Excess of water is removed and glass slide is stored in an incubator at 30°C during 20 min for good adherence. Subsequently obtained glass slides with FFPE tissue sections can be stored over months at room temperature.

3.2.2. FFPE Tissue Dewaxing

1. After complete drying, the glass slide is gently dipped into a bath of xylene during 5 min. This procedure is repeated two times. No agitation or shaking is needed.
2. The slide is then washed in stepwise immersion, 5 min duration each, into 100% ethanol twice, 95% ethanol, 75% ethanol, and 30% ethanol for rehydration of tissue sections.
3. The ITO glass slide is placed in the vacuum desiccator for complete drying of the sections.

3.3. On Tissue Digestion

3.3.1. Using a Microspotter

1. Microspotter like the chemical Inkjet Printer, CHIP-1000 (Shimadzu Biotech, Kyoto, Japan) can be used. CHIP-1000 is a piezoelectric solvent delivery system able to deliver picoliter volumes of reagents to define locations of various surfaces. It is equipped with an on board scanning device, which allows imaging the area of interest and easily selecting the right print location. An array of microspots that cover a specific area or the entire tissue section can be defined. Spots of 200 μm in diameter were spaced by 250 μm.
2. An ITO slide after washing step for frozen tissue or dewaxing for FFPE tissue is used.
3. On each defined spot, 40 nl of trypsin solution is applied. Five droplets of 100 pl are deposited at each spot per cycle, then 80 iterations are necessary to obtain the total volume.
4. The ITO glass slide is then incubated at 37°C for 2 h in a humidity chamber containing 50% methanol in water.
5. The ITO glass slide is placed in the vacuum desiccator to completely dry the tissue prior to matrix deposition.

3.3.2. Using an Automatic Sprayer

1. Automatic and control sprayer like the ImagePrep (Bruker Daltonics, Bremen, Germany) can be used for on tissue digestion. The ImagePrep system is equipped with an optical sensor that monitors light scattering from the solution deposited at the surface of the slide to control all relevant preparation parameters in real-time, namely deposition periods and intervals, wetness, drying rate.
2. A method with different step of spraying, incubation, and drying phase is used. For trypsin deposition, the spray time needs to be defined depending on the surface of the tissue section. Spray time is generally less than 2 s.
3. A particular attention is required to set correctly the drying time to achieve a complete drying of the section before a new spray cycle. If the time is too short, the section will be too wet and a delocalization of molecules may be observed. Normally, 40 cycles of spray are sufficient for efficient digestion of a rat brain section.
4. For FFPE tissue, a larger amount of trypsin is required to achieve a better digestion. In this case the number of cycle is increased.
5. The ITO glass slide is then incubated at 37°C for 2 h in a humidity chamber containing 50% methanol in water.
6. The ITO glass slide is placed in the vacuum desiccator for complete dry of the tissue prior to matrix deposition.

3.4. Matrix Deposition

3.4.1. For Protein Analysis Using a Microspotter

1. An ITO slide after washing step for frozen tissue is used.
2. On each defined spot, 20 nl of SA/ANI solution is applied. Five droplets of 100 pl are deposited at each spot per cycle, then 40 iterations are necessary to obtain the total volume.
3. Check matrix coverage using an optical microscope (*see* **Note 7**).
4. A rapid MS analysis on one spot is recommended to verify that a sufficient amount of matrix is deposited. Increase of iterations number may improve MSI when signal intensity appears to low.

3.4.2. For Protein Analysis Using an Automatic Sprayer

1. An ITO slide after washing step for frozen tissue is used.
2. A method with different steps of spraying, incubation, and drying phase is required. The ImagePrep method for SA/ANI deposition is based on the normal SA method included in the ImagePrep. Optimization is necessary for

each different type of tissue. Briefly, the spray time is around 2 s (depending on the surface of the tissue section). An incubation time of 30 s (except for initialization phase: 10 s) allows an efficient extraction of proteins. A particular attention is required to set correctly the drying time to achieve complete crystallization on the tissue section. If the time is too short, the section will be too wet and a delocalization of molecules will be observed. The minimum drying time is around 45 s.

3. Check matrix coverage using an optical microscope (*see* **Note 7**).
4. A rapid MS analysis at one position can be performed to check out that a sufficient amount of matrix has been deposited. If not, some cycles of the last phase of deposition can be done again and may improve MSI when signal intensity is too low.

3.4.3. For Peptide Analysis Using a Microspotter

1. An ITO slide after washing step for frozen tissues or digestion for FFPE or frozen tissues is used.
2. On each defined spot, 20 nl of HCCA/ANI solution is applied. Five droplets of 100 pl are deposited at each spot per cycle, then 40 iterations are necessary to obtain the total volume. For slides after digestion, the matrix is deposited with the same array than the one used for trypsin deposition. In this case matrix is deposited exactly at the same position than the trypsin.
3. Check matrix coverage using an optical microscope (*see* **Note 7**).
4. A rapid MS analysis on one spot is recommended to verify that a sufficient amount of matrix is deposited. Increasing the number of iterations may improve MSI when signal intensity appears to low.

3.4.4. For Peptides Analysis Using an Automatic Sprayer

1. An ITO slide after washing step for frozen tissue or digestion for FFPE or frozen tissues is used.
2. A method with different step of spraying, incubation, and drying phase is needed. The ImagePrep method for HCCA/ANI deposition is based on the normal HCCA method included in the ImagePrep. Optimization is required for each type of tissue. Briefly, the spray time is around 2 s (depending on the surface of tissue section). An incubation time of 20 s (except for initialization phase: 10 s) allow an effective extraction of proteins. A particular attention is drawn to correctly set the drying time for complete crystallization on the tissue section. If the time is too short,

the section will be too wet and a delocalization of molecules will be observed. The minimum drying time is around 120 s.

3. Check matrix coverage using an optical microscope (*see* **Note 7**).
4. A rapid MS analysis at one position can be performed to check out that a sufficient amount of matrix has been deposited. If not, some cycles of the last phase of deposition can be done again and may improve MSI when signal intensity is too low.

3.5. Mass Spectrometry Analysis

3.5.1. Mass Spectrometry Analysis for Protein MALDI-MSI (For Frozen Tissue Analysis Exclusively)

1. 0.5 μl of protein calibration solution is deposited near to the tissue section and mix with 0.5 μl of SA/ANI solution.
2. The mass spectrometer is calibrated with the calibration spot.
3. Using FlexImaging an area of interest is selected on the tissue after definition of the teaching points.
4. The distance between each measurement point is set. Distance between measurement points is dependent on the method used for matrix deposition:

   With Chip-1000 deposition, the spots are generally spaced by 250 μm center to center. It is possible to define the same raster than the one defined during matrix deposition. Due to the size of the spot it is possible to accumulate spectra at different position in the same spot. This increase statistics and reduce spot-to-spot variability.

   With ImagePrep deposition, distance between two measurements can be chosen by the user. Generally the resolution is around 100 μm.
5. In FlexControl, the adequate methods for proteins analysis is set in positive linear mode and a total of 500 spectra are acquired at each position at a laser frequency of 100 Hz.
6. The images are saved and reconstructed using FlexImaging 2.1.

3.5.2. Mass Spectrometry Analysis for Peptide MALDI-MSI (For Nondigest Frozen Tissue and Digest Frozen or FFPE Tissue)

1. 0.5 μl of peptide calibration solution is deposited near to the tissue section and mix with 0.5 μl of ANI solution.
2. The mass spectrometer is calibrated with the calibration spot.
3. Using FlexImaging an area of interest is selected on the tissue after definition of the teaching points.

4. The distance between each measurement point is set. Distance between measurement points depends on the method used for matrix deposition.
   4.1. With Chip-1000, deposition spots are generally spaced by 250 μm center to center. It is possible to define the same raster than for matrix deposition. Due to the size of the spots spectra can be accumulated at different positions in the same spot.
   4.2. With ImagePrep deposition, distance between two measurements is chosen by users. Generally the resolution is around 100 μm.
5. In FlexControl, the adequate methods for peptides analysis are set in positive reflector mode and a total of 500 spectra are acquired at each position at a laser frequency of 100 Hz. Although, negative reflectron mode can also be used for specific class of peptides.
6. The images are saved and reconstructed using FlexImaging 2.1.

3.5.3. MS/MS Analysis

1. Ultraflex II TOF–TOF is equipped with a LIFT III cell. For each MS/MS spectrum, 5,000 total shots are averaged including 1,000 for parent ions and 4,000 for fragments.
2. Peptides are identified by searching MS/MS spectra against an appropriate database using Biotools software for MASCOT (Matrix Science) interrogation. For MALDI data, peptide mass tolerance is set at 2 Da and MS/MS tolerance at 1 Da. Oxidation of methionine is selected as variable modification.
3. When a peptide is identified as a digest fragment of a protein, the protein sequence is used for in silico digestion. Expected digestion fragments of this protein are then search in the total MS spectrum. For the observed digestion fragments of the protein, MS/MS is performed to confirm that the peptide match to the expected sequence.
4. Using FlexImaging, molecular images of predicted digest fragments for one protein find in the data are realized to check out that they do present the same localization.

4. Notes

1. In some cases, the trypsin can be suspended in various solutions. For example, trypsin in water can be used for frozen sections for which the pH at the tissue surface is close to

the optimal pH value required for enzyme efficiency. Mix of water:methanol (1:1 v/v) can also be used with trypsin to achieve better extraction and permit a better access of cleavage sites to the enzyme. Use of water or water/methanol is recommended when using CHIP-1000 device for easier stabilization of droplets ejection.

2. Prior to deposition, 2 μl of matrix solution could be deposited with a micropipette on a classical MALDI sample plate to check the crystallization. Crystallization is expected close to that of SA, with homogenous smalls and fines white crystals.
3. Prior to deposition, 2 μl of matrix solution could be deposited with a micropipette on a classical MALDI sample plate to check the crystallization. Crystallization is expected to be uniform, with longs and fines white crystals as shown in table 2a of (25).
4. Smaller sections have not enough molecules for extraction and thicker sections may cause problems of conductivity (due to the insulating nature of tissues) and charge effects by charge accumulation at the sample surface during MALDI analysis. Charge effects will decrease spectral quality in axial TOF configuration instruments resulting in a progressive peak shifting toward the high m/z ratio.
5. Care must be taken of air bubbles formation at the surface of the tissue section that may leads to artifacts during MS analysis.
6. Other organic solvents could be used to perform this step. More information could be obtained in (26, 27). You must be careful not to increase the number of washing steps to avoid molecules delocalization.
7. To check matrix coverage, you need to ensure that the crystallization is dense, uniform and composed of small crystals.

Acknowledgments

This study was supported by grants from the Centre National de la Recherche Scientifique (DPI), Ministère de L'Enseignement Supérieur et de la Recherche, the Agence Nationale de la Recherche PCV (To IF).

References

1. Reyzer, M. L., Hsieh, Y., Ng, K., Korfmacher, W. A., Caprioli, R. M. (2003) Direct analysis of drug candidates in tissue by matrix-assisted laser desorption/ionization mass spectrometry. *J Mass Spectrom*, **38**, 1081–1092.
2. Rubakhin, S. S., Jurchen, J. C., Monroe, E. B., Sweedler, J. V. (2005) Imaging mass spectrometry: fundamentals and applications to drug discovery. *Drug Discov Today*, **10**, 823–837.
3. Deininger, S. O., Ebert, M. P., Futterer, A., Gerhard, M., Rocken, C. (2008) MALDI imaging combined with hierarchical clustering as a new tool for the interpretation of complex human cancers. *J Proteome Res*, 7, 5230–5236
4. Fournier, I., Wisztorski, M., Salzet, M. (2008) Tissue imaging using MALDI-MS: a new frontier of histopathology proteomics. *Expert Rev Proteomics*, **5**, 413–424.
5. Franck, J., Arafah, K., Elayed, M., Bonnel, D., Vergara, D., Jacquet, A., Vinatier, D., Wisztorski, M., Day, R., Fournier, I., Salzet, M. (2009) MALDI IMAGING: state of the art technology in clinical proteomics. *Mol Cell Proteomics*, **8**, 2023-2033.
6. Walch, A., Rauser, S., Deininger, S. O., Hofler, H. (2008) MALDI imaging mass spectrometry for direct tissue analysis: a new frontier for molecular histology. *Histochem Cell Biol*, **130**, 421–434.
7. Groseclose, M. R., Andersson, M., Hardesty, W. M., Caprioli, R. M. (2007) Identification of proteins directly from tissue: *in situ* tryptic digestions coupled with imaging mass spectrometry. *J Mass Spectrom*, **42**, 254–262.
8. Lemaire, R., Desmons, A., Tabet, J. C., Day, R., Salzet, M., Fournier, I. (2007) Direct analysis and MALDI imaging of formalin-fixed, paraffin-embedded tissue sections. *J Proteome Res*, **6**, 1295–1305.
9. Franck, J., Wisztorski, m., El ayed, M., Bonnel, D., Barnes, A., Fournier, I., Salzet, M. (2008) Automatic spotting solution for MALDI Imaging: process optimization and new developments. *56th Annual Conference on Mass Spectrometry and Allied Topics*, Denver, CO, June 1–5, 2008.
10. Stauber, J., Lemaire, R., Franck, J., Bonnel, D., Croix, D., Day, R., Wisztorski, M., Fournier, I., Salzet, M. (2008) MALDI imaging of formalin-fixed paraffin-embedded tissues: application to model animals of Parkinson disease for biomarker hunting. *J Proteome Res*, 7, 969–978.
11. Groseclose, M. R., Massion, P. P., Chaurand, P., Caprioli, R. M. (2008) High-throughput proteomic analysis of formalin-fixed paraffin-embedded tissue microarrays using MALDI imaging mass spectrometry. *Proteomics*, **8**, 3715–3724.
12. Ronci, M., Bonanno, E., Colantoni, A., Pieroni, L., Di Ilio, C., Spagnoli, L. G., Federici, G., Urbani, A. (2008) Protein unlocking procedures of formalin-fixed paraffin-embedded tissues: application to MALDI-TOF imaging MS investigations. *Proteomics*, **8**, 3702–3714.
13. Plenat, F., Antunes, L., Haller, T., Piet-Ounnoughene, M., Klein-Monhoven, N., Champigneulle, J., Chenal, P., Bland, V., Garcia-Pimenta, F., Labouyrie, E. (2001) Formaldehyde fixation in the third millennium. *Ann Pathol*, **21**, 29–47.
14. Plenat, F., Montagne, K., Weinbreck, N., Corby, S., Champigneulle, J., Antunes, L., Bonnet, C., Maire, C., Monhoven, N. (2006) Molecular consequences of fixation and tissue processing: the examples of nucleic acids and proteins. *Ann Pathol*, **26**, 8–21.
15. Kieman, J. A. (2000) Formaldehyde, formalin, paraformaldehyde and glutaraldehyde: what they are and what they do. *Micros Today*, **8**, 8–14.
16. Redeker, V., Toullec, J. Y., Vinh, J., Rossier, J., Soyez, D. (1998) Combination of peptide profiling by matrix-assisted laser desorption/ionization time-of-flight mass spectrometry and immunodetection on single glands or cells. *Anal Chem*, **70**, 1805–1811.
17. Shi, S. R., Cote, R. J., Taylor, C. R. (1997) Antigen retrieval immunohistochemistry: past, present, future. *J Histochem Cytochem*, **45**, 327–343.
18. Shi, S. R., Cote, R. J., Taylor, C. R. (2001) Antigen retrieval techniques: current perspectives. *J Histochem Cytochem*, **49**, 931–937.
19. Shi, S. R., Cote, R. J., Taylor, C. R. (2001) Antigen retrieval immunohistochemistry and molecular morphology in the year 2001. *Appl Immunohistochem Mol Morphol*, **9**, 107–116.
20. Shi, S. R., Liu, C., Young, L., Taylor, C. (2007) Development of an optimal antigen retrieval protocol for immunohistochemistry of retinoblastoma protein (pRB) in formalin fixed, paraffin sections based on comparison of different methods. *Biotech Histochem*, **82**, 301–309.
21. Shi, S. R., Shi, Y., Taylor, C. R. (2007) Updates on antigen retrieval techniques for

immunohistochemistry. Chin J Pathol, **36,** 7–10.
22. Taylor, C. R., Shi, S. R. (2000) Antigen retrieval: call for a return to first principles. *Appl Immunohistochem Mol Morphol*, **8,** 173–174.
23. Xu, H., Yang, L., Wang, W., Shi, S. R., Liu, C., Liu, Y., Fang, X., Taylor, C. R., Lee, C. S., Balgley, B. M. (2008) Antigen retrieval for proteomic characterization of formalin-fixed and paraffin-embedded tissues. *J Proteome Res*, **7,** 1098–1108.
24. Chaurand, P., Latham, J. C., Lane, K. B., Mobley, J. A., Polosukhin, V. V., Wirth, P. S., Nanney, L. B., Caprioli, R. M. (2008) Imaging mass spectrometry of intact proteins from alcohol-preserved tissue specimens: bypassing formalin fixation. *J Proteome Res*, **7,** 3543–3555.
25. Lemaire, R., Tabet, J. C., Ducoroy, P., Hendra, J. B., Salzet, M., Fournier, I. (2006) Solid ionic matrixes for direct tissue analysis and MALDI imaging. *Anal Chem,* **78,** 809–819.
26. Lemaire, R., Wisztorski, M., Desmons, A., Tabet, J. C., Day, R., Salzet, M., Fournier, I. (2006) MALDI-MS direct tissue analysis of proteins: improving signal sensitivity using organic treatments. *Anal Chem,* **78,** 7145–7153.
27. Seeley, E. H., Oppenheimer, S. R., Mi, D., Chaurand, P., Caprioli, R. M. (2008) Enhancement of protein sensitivity for MALDI imaging mass spectrometry after chemical treatment of tissue sections. *J Am Soc Mass Spectrom*, **19,** 1069–1077.

Chapter 19

On Tissue Protein Identification Improvement by N-Terminal Peptide Derivatization

Julien Franck, Mohamed El Ayed, Maxence Wisztorski, Michel Salzet, and Isabelle Fournier

Abstract

Identification of potential markers of a physiological stage (e.g., pathology) discovered using MALDI-MSI is an important step in the understanding of signaling pathways or for providing sets of diagnosis and prognosis markers for clinical applications. Classically, identification can be achieved by extraction from a piece of tissue and proteomics strategies. However, this induces loss of information especially for low-abundance proteins or proteins localized to a specific region of the tissue. In this respect, identification directly at the tissue level is an attractive alternative. Because the molecular charge states in MALDI are low, on tissue identification is possible using bottom-up MALDI-MSI strategies. Enzymatic digestion using an enzyme such as trypsin can be performed at the micro-scale level to generate peptide collections while avoiding these peptides to be delocalized. It is, therefore, possible to image proteins through the molecular images of their digested peptides. These peptides can also be used to retrieve information on protein sequences by performing MS/MS, although databank interrogation or de novo sequencing using MS/MS spectra does not always lead to a successful or confident identification because on tissue complexities render PMF data problematic. Identification can be improved by increasing MS/MS spectra quality and simplifying their interpretation. This can be achieved by derivatization of peptides. In fact, derivatization of peptides leads to increases in fragmentation yields and orients fragmentations toward a specific series of fragment ions. In this respect, N-terminal chemical derivatization has proven to be particularly efficient. N-terminal chemical derivatization of tryptic peptides has been developed to be performed at the tissue level after on tissue digestion. Specific focus is given to 4-sulfophenyl isothiocyanate (4-SPITC), 3-sulfobenzoic acid NHS ester (3-SBASE), and (*N*-succinimidyloxycarbonylmethyl)tris(2,4,6-trimethoxyphenyl)phosphonium bromide (TMPP) derivatizations. This provides a complete strategy for protein identification in a bottom-up MALDI-MSI approach and opens the way for novel biomarker identification.

Key words: Mass spectrometry imaging, de novo sequencing, derivatization, protein identification.

S.S. Rubakhin, J.V. Sweedler (eds.), *Mass Spectrometry Imaging*, Methods in Molecular Biology 656, DOI 10.1007/978-1-60761-746-4_19, © Springer Science+Business Media, LLC 2010

1. Introduction

MALDI-MSI has shown potential in the field of proteomics to study the distribution of proteins within tissue sections. Specifically, MALDI-MSI can be used to follow protein markers along physiological stages in order to provide information on signaling pathways. In some cases, MALDI-MSI is used to study the distribution of known proteins at different physiological stages. But often, MALDI-MSI leads to the discovery of molecular weight protein markers. In such cases, protein identification remains to be performed. Protein identification can be performed by classical proteomic strategies after extraction from the remaining tissue piece. However, the most straightforward strategy is to perform protein identification directly from the tissue section itself. As for classical proteomics, two approaches may be considered, namely "top-down" and "bottom-up" approaches. Top-down at the level of the tissue section is clearly the most direct way to protein identification since identification is performed directly by fragmentation into the gas phase of the intact molecule ions. Unfortunately, MALDI generates low charge stages, which make it not well-suited for most of the analyzers or ion activation methods. Therefore, MALDI-MSI "bottom-up" strategies were needed. Such strategies are less direct and require the optimization of enzymatic digestion at the tissue level (**Fig. 19.1**). On tissue digestions can be performed while maintaining localization of digested proteins using micro-spotting of the enzyme. A collection of digestion peptides are generated and protein localization can therefore be obtained indirectly by imaging digestion peptides. $[M+H]^+$ peptide ions are then available for structural elucidation via MS/MS using classical activation methods such as collision-induced dissociation (CID) or infrared multiphoton dissociation (IRMP) or metastable decay. On tissue MS/MS performed after on tissue digestion is accessible although it must be noted that fragmentation yields appear lower when working on tissue sections. This could be explained by a lower internal energy of the parent ion by a higher relaxation of energy during the desorption process itself. This is especially true for MALDI-TOF systems that often lead to weak fragmentation especially in the higher m/z range of the MS/MS spectrum and also leads to different series of fragments creating incomplete sets of series only giving access to small sequence tags. Peptide sequencing can be difficult to establish owing to the multitude of fragment ions generated during MS/MS experiments, like immonium ions (1), internal fragments, N-terminal ions (2, 3), or C-terminal ions (4).

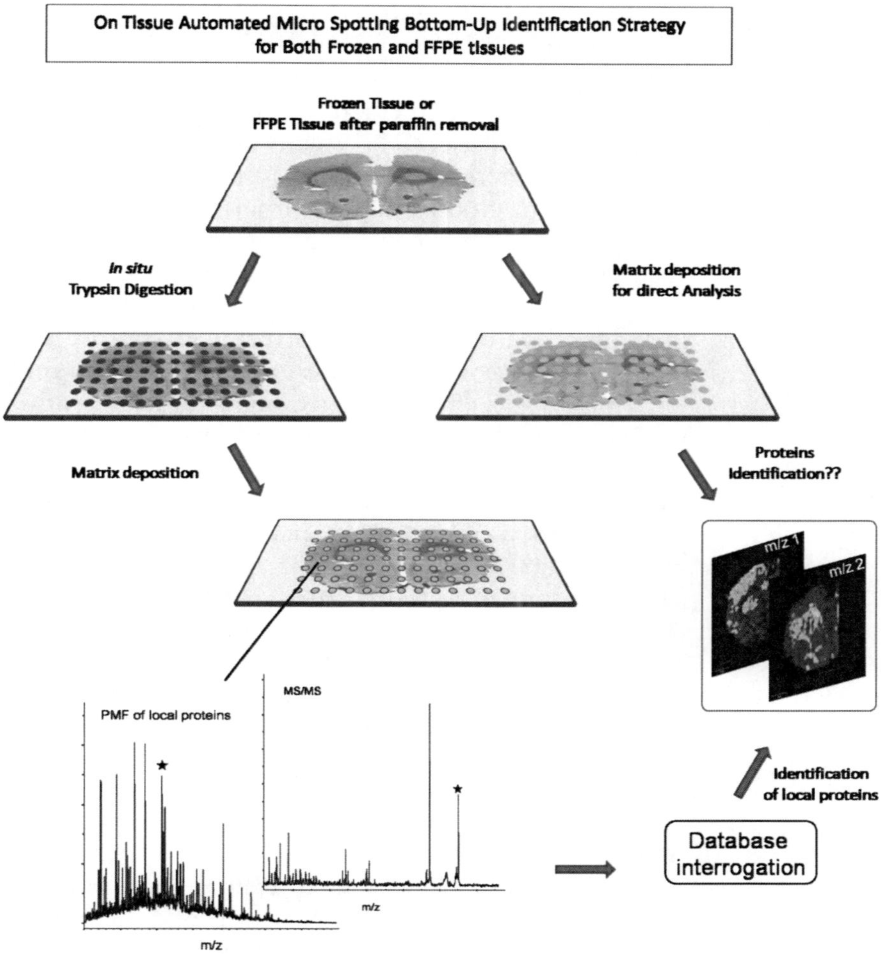

Fig. 19.1. Schematic workflow of MALDI-MSI "bottom-up" strategy used to perform protein identification.

Under classical identification strategies efforts were given to simplify data interpretation by increasing fragmentation ion yield and orienting fragmentations toward a specific series of fragment ions. In this respect, derivatization at the C- or N-terminal part of peptides by addition of positive or negative charges has proved to be an efficient strategy (5), especially for post-source decay (PSD) spectra. In general N-terminal modifications are easier to achieve because of the primary amine reactivity and have proved to be more efficient for fragmentations. Sulfonation at the N-terminal part of peptides seems especially advantageous when working on tryptic digestion peptides that are naturally positive at their C-termini. Keough and coworkers have shown a possible orientation of fragmentations toward the yi^+ series of fragment ions by addition of chlorosulfonylactetyl chloride (6, 7) or 2-sulfobenzoic acid cyclic anhydride (8) at the N-terminal side of peptides. Fragmentation orientation greatly eases MS/MS spectra interpretation allowing for de novo sequencing to be performed (9). This method was later on improved by using 3-sulfopropionic acid NHS ester as derivatization agents (10) to allow the reaction to be performed in aqueous phase. Alternatively, derivatization for liquid chromatography (11) and quantification with isotope coded (12) were proposed using derivative agents such as 4-sulfophenyl isothiocyanate (13–15). Such derivatizations are fast and easy, showing good reaction yields, although they often lead (according to the peptide sequence) to the loss of the derivative group as one of the fragmentation pathway lowering the benefits of derivatization. More recently, another water-compatible reagent, the 3-sulfobenzoic acid NHS ester, was introduced (16). This sulfonation agent has the advantage not to lead to the loss of the derivative group consecutively resulting in the observation of a complete and unique y ion series. Alternatively, addition of a positively charged group at the N-terminal part of peptides using (*N*-succinimidyloxycarbonylmethyl)tris (2,4,6-trimethoxyphenyl)phosphonium bromide (TMPP) was successfully used for fragmentation orientation (17). This derivatization has the advantage of being independent of the presence and/or position of basic amino acids (mainly arginine) in the peptide chain and leads to an orientation toward a_i^+ fragment ion series with the presence of a few b_i^+, c_i^+, and d_i^+ ions. TMPP is useful for peptide identification in bottom-up strategies.

Because MS/MS data interpretation from tissue sections is particularly complex, development of peptide derivatization is a good alternative to help protein identification. In fact, on tissue digestion leads to numerous digestion peptides that are not all belonging to the same protein and not separated. In this context, peptide mass fingerprint data interrogation is useless and only MS/MS data interrogation is used. This can render the identification difficult if MS/MS spectra do not contain enough information. Thereof, providing higher quality MS/MS spectra is important for easier and more confident identification of proteins from the tissue sections. Because derivatization of peptides increases fragmentation yields and above all channel fragmentation to specific pathways, protein identification is improved with MS/MS from on tissue experiments.

We describe here on tissue N-terminal chemical derivatization of peptides performed after on tissue enzymatic digestion. Since N-terminal derivatization has proven to be more efficient than the corresponding C-terminal reactions in solution, focus has been given to N-terminal chemical derivatizations. More specifically, among the tested derivatization approaches, three derivatization protocols appear more compatible and efficient when performed at the tissue level, namely 4-sulfophenyl isothiocyanate (4-SPITC), 3-sulfobenzoic acid NHS ester (3-SBASE), and (*N*-Ssccinimidyloxycarbonylmethyl)tris(2,4,6-trimethoxyphenyl)phosphonium bromide (TMPP) (18). These derivatives promote efficient charge-site-initiated cleavage of backbone amide bonds enabling selective detection of only a single series of fragment ions that contain either the original C-terminus of the peptide (y_i^+ ions) for 4-SPITC and 3-SBASE or the N-terminus of the peptide (TMPP). All three derivatization reagents have been shown to ease identification by databank interrogation (increase in the identification scores) and with de novo sequencing. Often peptide sequences can also be manually obtained if desired. We also notice that TMPP derivatization is a bit easier to perform at a micro-scale level using a robot than 4-SPITC and 3-SBASE mainly because the room temperature reactions for TMPP enable molecular images to be performed after the derivatization step. However, molecular images after on tissue derivatization are rarely required, which makes all three derivatization approaches available for identification. We describe here procedures for on tissue digestion using trypsin as well as N-terminal chemical derivatization using 4-SPITC, 3-SBASE, and TMPP. For 4-SPITC and 3-SBASE global on tissue derivatization protocols are given whereas micro-spotted derivatization protocol is provided for TMPP.

2. Materials

2.1. Preparation of Frozen Tissue Section

1. Optimal cutting temperature polymer, OCT.
2. Indium tin oxide (ITO)-coated glass slides or other holders compatible with mass spectrometry.
3. A cryomicrotome, Leica CM150S (Leica Microsystems, Nanterre, France).
4. Ethanol 75% (−20°C): 75 ml of absolute ethanol (≥99.8%) and HPLC-grade water to 100 ml. Prepare fresh and store at −20°C.
5. Ethanol 95% (−20°C): 95 ml of absolute ethanol (≥99.8%) and HPLC-grade water to 100 ml. Prepare fresh and store at −20°C.
6. Chloroform (−20°C): 100 ml of chloroform (≥99.9%). Store at −20°C. Chloroform is harmful by inhalation, so work in the hood.

2.2. In Situ Enzymatic Digestion

2.2.1. Using Micropipette

1. Trypsin, sequencing grade modified (Promega, Charbonnieres, France). Suspend in 20 mM NH_4HCO_3 buffer at 40 μg/ml (*see* **Note 1**).
2. Methanol 50%: 50 ml of absolute methanol completed with water up to 100 ml. Prepare fresh. Methanol is toxic, so work in the hood.

2.2.2. Using a Microspotter

1. Trypsin, sequencing grade modified (Promega, Charbonnieres, France). Suspend in 20 mM NH_4HCO_3 buffer at 40 μg/ml (*see* **Note 1**).
2. Methanol 50%: 50 ml of absolute methanol completed with water up to 100 ml. Prepare fresh. Methanol is toxic, so work in the hood.
3. Chemical Inkjet Printer CHIP-1000 (Shimadzu Biotech, Kyoto, Japan).

2.3. In Situ N-Terminal Derivatization

2.3.1. Synthesis of 3-Sulfobenzoic Acid Succinimidyl Ester (3-SBASE)

1. DMSO: 10 ml of dimethylsulfoxide.
2. 3-SBA: 1 g of 3-sulfobenzoic acid, purity: 98%. 3-SBA is irritating to the respiratory system, so work in the hood.
3. DCC: 1.38 g of *N,N′*-dicyclohexylcarbodiimide, purity 99%. DCC is toxic, so work in the hood.
4. NHS: 770 mg of *N*-hydroxysuccinimide, purity 98%.
5. Cold acetone: 60 ml of acetone (≥99.9%), store at 4°C. Vapors may cause drowsiness and dizziness, so work in the hood.

2.3.2. On Tissue N-Terminal Derivatization with 4-SPITC

1. 20 mg/ml 4-SPITC in pure water.
2. 50% methanol: 50 ml of absolute methanol completed with water up to 100 ml. Prepare fresh. Methanol is toxic, so work in the hood.

2.3.3. On Tissue N-Terminal Derivatization with 3-SBASE

1. 20 mg/ml 3-SBASE in pure water.
2. 50% methanol: 50 ml of absolute methanol completed with water up to 100 ml. Prepare fresh. Methanol is toxic, so work in the hood.

2.3.4. On Tissue N-Terminal Derivatization with TMPP

1. TMPP: 1 mg/ml (*N*-succinimidyloxycarbonylmethyl)tris (2,4,6-trimethoxyphenyl)phosphonium bromide, puriss. p.a., for protein sequence analysis (by MALDI-MS), ≥98.5% in acetonitrile/H_2O (3:7, v/v).
2. Solution of acetonitrile/H_2O/TEA (triethylamine) (30:69:1, v/v/v). Prepare fresh (*see* **Note 2**).
3. Chemical Inkjet Printer CHIP-1000 (Shimadzu Biotech, Kyoto, Japan).

2.4. Matrix Deposition for Peptide Analysis

2.4.1. Using Micropipette

HCCA/ANI solution: 1.5 equivalents of aniline (ANI) were added to a solution containing 20 mg/ml of α-cyano-4-hydroxycinnamic acid (HCCA) in acetonitrile/aqueous 0.1% TFA (6:4, v/v) (*see* **Note 3**). Aniline and TFA are toxic, so work in the hood.

2.4.2. Using a Microspotter

1. HCCA/ANI solution: 1.5 equivalents of aniline (ANI) were added to a solution containing 20 mg/ml of α-cyano-4-hydroxycinnamic acid (HCCA) in acetonitrile/aqueous 0.1% TFA (6:4, v/v) (*see* **Note 3**). Aniline and TFA are toxic, so work in the hood.
2. Chemical Inkjet Printer CHIP-1000 (Shimadzu Biotech, Kyoto, Japan).

2.5. Mass Spectrometry Analysis

2.5.1. MALDI-MSI Experiment

1. Peptide calibration standard II (Bruker Daltonics, Wissembourg, France): angiotensin II, angiotensin I, substance P, bombesin, ACTH clip 1–17, ACTH clip 18–39, somatostatin 28, bradykinin fragment 1–7, renin substrate tetradecapeptide porcine. Covered mass range: ~700–3,200 Da. Store at –20°C.
2. Ultraflex II TOF–TOF equipped with a Smartbeam laser and all the Flex software suite (FlexControl, FlexAnalysis, and FlexImaging) (Bruker Daltonics, Bremen, Germany).

2.5.2. MS/MS Analysis

1. Ultraflex II TOF–TOF equipped with a Smartbeam laser and all the Flex software suite (FlexControl, FlexAnalysis, and FlexImaging) (Bruker Daltonics, Bremen, Germany).
2. Biotools (Bruker Daltonics, Bremen, Germany).

3. Methods

In situ enzymatic digestion creates a library of peptides from the proteins present at that spot. These peptides can be fragmented using MS/MS; using this peptide fragmentation data and the appropriate database, the peptide can be identified. Using this information, the corresponding protein from which it came can be identified at the tissue level. This strategy can be performed at the scale of an entire tissue section, manually using micropipette deposition providing a fast and sensitive detection of peptides generated from protein digestion. However, this strategy does not provide information on the spatial distribution of the proteins in the tissue. Derivatization can also be performed at the microscale level and can be used to prevent migration/delocalization of compounds within the tissue, via an automatic device system such as automatic micro-spotting. Here, a protocol is described for the automatic piezoelectric micro-spotting using the CHIP-1000 system. The micro-spotting device allows the deposition of the different reagents including trypsin and matrix solutions at the surface of the whole tissue sections or at a predefined location. Using this strategy of defined deposition, protein and peptides are not delocalized, thus providing the opportunity for imaging peptides after the derivatization step.

Figure 19.1 presents the global workflow for on tissue protein identification in a bottom-up strategy including chemical derivatization at a micro-scale level. The peptides that are characterized can be submitted to databank interrogation after MS/MS experiments to retrieve the protein identification. However, databank identification is not always successful nor does it provide highly confident assignments. To improve protein identification, N-terminal derivatization is used to facilitate peptide fragmentations toward a specific series of fragment ions. It is thereof possible to increase the information content of the resulting MS/MS spectra. 4-SPITC and 3-SBASE derivatizations use sulfonic acid reagents and provide complete y_i^+ series of ions for a peptide, ideally without missing cleavages. We notice that for 4-SPITC, loss of the 4-SPITC group is often observed as one fragmentation pathway in the MS/MS spectra, and this can slightly reduce the resulting MS/MS spectral quality. In addition, spots (i.e., missing cleavages) in the y_i^+ series will be observed if a basic amino acid is present in the peptide sequence. For these reasons, care must be taken with on tissue enzymatic digestion. Because of the higher optimum reaction temperatures (35–60°C), 4-SPTIC and 3-SBASE are a bit more difficult to use than TMPP for micro-spotting.

Figure 19.2 shows an example of 3-SBASE derivatization obtained from a rat brain tissue section. After interrogation of the database with a search algorithm like MASCOT, the score achieved is at least twofold better than the score obtained for the unmodified peptides as shown in **Table 19.1**. **Figure 19.3** presents TMPP derivatizations used in this micro-spotting strategy. Using MS/MS, the derivatized peptides provide an abundant series of a^+ ions. As shown in **Fig. 19.3**, this reagent is easily compatible with MALDI imaging.

3.1. Preparation of Frozen Tissue Section

1. 10-μm thick tissue sections are obtained from snap-frozen tissue using a cryomicrotome at –20°C. The tissue is placed in the cryomicrotome ~15 min prior to sectioning. If the tissue temperature is too low, tissue sections can pleat and crack. If the temperature is too high, tissue sections will roll up.
2. The tissue sections are then applied onto ITO-coated conductive glass slides and placed in a desiccator under vacuum for a minimum of 30 min to dry out the tissue sections.

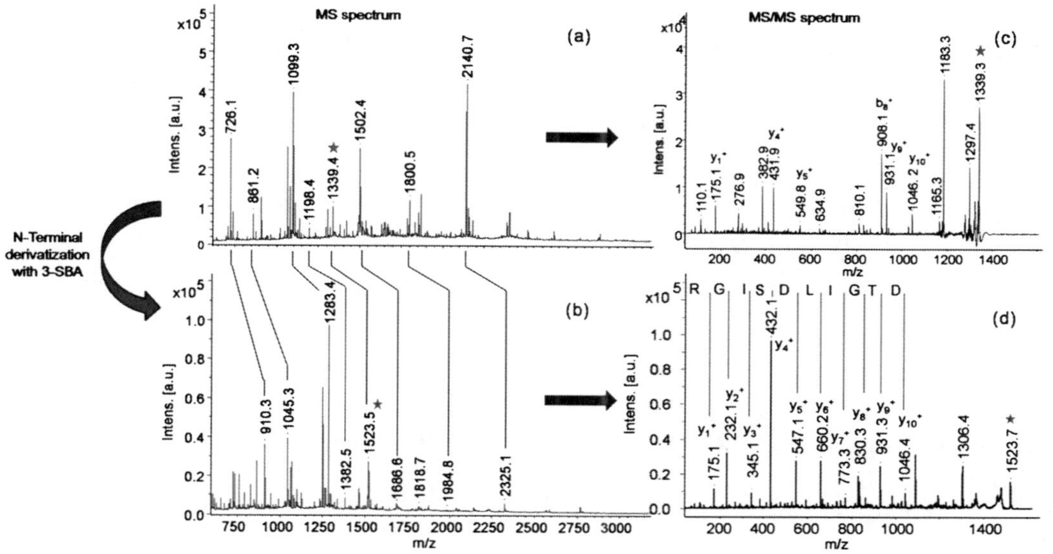

Fig. 19.2. (**a–b**) MALDI-MS spectra recorded on a rat brain tissue section after on tissue trypsin digestion (**a**) and on a tissue with trypsin digestion followed by derivatization with 3-SBA (**b**). (**c–d**) MALDI-MS/MS spectra of one of the peptides generated from digest (*m/z* 1,339.3) (**c**) and the same peptide after derivatization with 3-SBA (*m/z* 1,523.7) (**d**).

3. The tissue is fixed. The slide is immersed softly in an ice-cold 75% ethanol bath for 30 s and dried in the vacuum desiccator. After complete drying, the sample is dipped into cold 95% ethanol for 30 s and then dried completely with the vacuum desiccator. This important step removes salts contained in the tissue and prevents sample degradation by dehydration.
4. The lipids are removed. The sample is washed by immersion of the tissue section in a chloroform bath for 30 s to remove abundant phospholipids; these lipids lead to highly abundant signals in the 500–1,500 mass range hampering the observation of peptide signals, generating ionization suppression effects and poor matrix crystallization efficiency (*see* **Note 4**).

3.2. In Situ Enzymatic Digestion

3.2.1. Using a Micropipette

1. Use a tissue section after washing step (step 4 of 3.1). Whole tissue section is covered by the solution of trypsin using a micropipette. 10 μl/cm^2 is used (according to the tissue section size).
2. Tissue sections are then incubated for 2 h at 37°C in an atmosphere saturated with MeOH/H_2O (1:1, v/v).
3. The slide is placed in the vacuum desiccator to completely dry out the tissue prior to matrix deposition.

3.2.2. Using a Microspotter

1. Microspotter like the chemical Inkjet Printer, CHIP-1000, can be used. CHIP-1000 is a piezoelectric solvent delivery

Table 19.1
Comparison of identification scores as found from MASCOT interrogation using Swiss-Prot database and *R. norvegicus* taxonomy for four peptides before and after automated derivatization with 3-SBASE

	m/z observed	*m/z* expected	Ion score	Rank	Delta	Expect	Sequence
Peptide	643.35	642.32	NI	NI	NI	NI	RPSQR myelin basic protein
Peptide + derivatization (3-SBASE)	827.9		46	1	−0.0314	0.0064	
Peptide	726.23	725.22	28	2	−0.1765	0.044	HGFLPR myelin basic protein
Peptide + derivatization (3-SBASE)	910.1		30	1	−0.3046	0.029	
Peptide	1,339.24	1,339.71	42	1	−0.4706	0.0015	HRDTGILDSIGR myelin basic protein
Peptide + derivatization (3-SBASE)	1,523.7		93	1	−0.277	9.6×10^{-19}	
Peptide	1,701.92	1,701.91	57		0.0162	6.2×10^{-5}	AVFVDLEPTVIDEVR tubulin alpha-1A chain
Peptide + derivatization (3-SBASE)			142		−0.0058	1.1×10^{-12}	

system, able to deliver picoliter volumes of reagents at defined locations of a surface. The instrument is equipped with an onboard scanning device for imaging the area of interest and thus precisely define and select the print location. An array of micro-spots covering a specific area or the entire tissue section can be defined. Spots of 150 μm in diameter are generated. Micro-spots are spaced by 250 μm.

2. The slide is then loaded in the microspotter and scanned to precisely define the array to be printed with trypsin enzyme. Selection of the whole tissue section, one location, or several locations is possible.

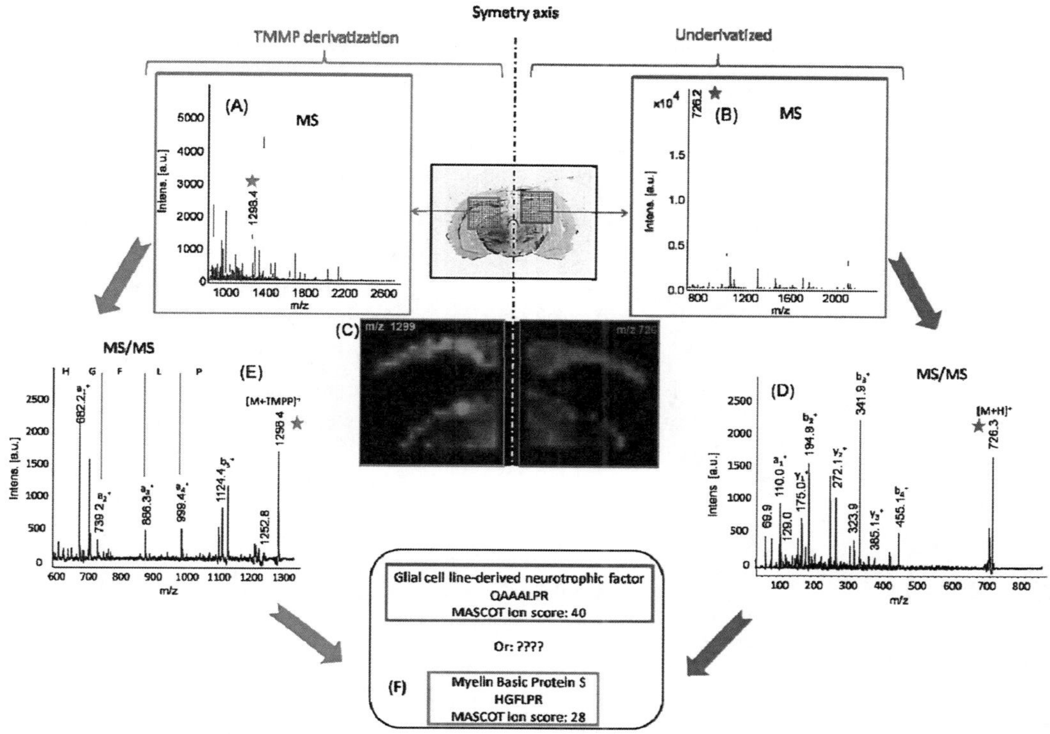

Fig. 19.3. (**a–b**) MALDI-MS spectra recorded after (**a**) enzymatic digestion followed by TMPP derivatization and (**b**) enzymatic digestion. (**c**) Images of the digestion peptide at *m/z* 726 and its corresponding TMPP derivative at *m/z* 1,299 showing an identical distribution for both ions. (**d–e**) MALDI-MS/MS spectra recorded from a rat brain tissue section in the same area after (**d**) on tissue trypsin digestion (**e**) and on tissue trypsin digestion followed by TMPP derivatization. (**f**) Identification results from the databank using *Rattus norvegicus* taxonomy.

3. A total of 20 nl of trypsin solution is applied on each spot. This is obtained by dropping off five droplets of approximately 100 pl of solution at each spot per cycle. 40 iterations are thus necessary to obtain the final volume.
4. Tissue sections are then incubated at 37°C for 2 h in a MeOH/H_2O (1:1, v/v) saturated atmosphere.
5. After incubation, the slide is placed in the vacuum desiccator for complete drying of the tissue section prior to matrix deposition.

3.3. In Situ N-Terminal Derivatization

3.3.1. On Tissue N-Terminal Derivatization with 4-SPITC

1. Whole tissue trypsin digestions are performed by applying 10 μl/cm^2 (according to the tissue section size) of trypsin solution using a micropipette at the surface of the section.
2. Tissue sections are then incubated for 2 h at 37°C in MeOH/H_2O (1:1, v/v) saturated atmosphere.
3. The slide is placed in the vacuum desiccator to let the tissue dry completely prior to 4-SPITC deposition.

4. Manual derivatization is performed by applying the solution of 4-SPITC in pure water. In fact, no buffer is required since NH_4HCO_3 is still present on the tissue after trypsin deposition (*see* step 1). Volume of solution is adjusted according to the size of the tissue section or the size of the area to be studied within a tissue section (~10 μl/cm^2).
5. Tissue sections are then incubated for 1 h at 55°C in $MeOH/H_2O$ (1:1, v/v) saturated atmosphere.
6. *Optional step*: The tissue can be washed with cold 95% EtOH for 20 s to remove salts from buffer solutions to avoid signal loss due to poor ionization of derivate peptides. This procedure must be performed carefully to avoid peptide losses. Thus, washing steps in cold 95% EtOH must not exceed 20 s.
7. Place the slide in the vacuum desiccator for complete drying prior to matrix deposition.

3.3.2. Synthesis of 3-Sulfobenzoic Acid Succinimidyl Ester (3-SBASE)

1. 1.0 g of 3-SBA is completely dissolved in 10 ml of DMSO.
2. 1.5 molar excess of NHS is then added to the solution up to complete dissolution.
3. 1.5 molar excess of DCC is then added to the solution up to complete dissolution.
4. The mixture is stirred overnight at room temperature.
5. The precipitated dicyclohexylurea (DCU), a side product of the reaction, is filtered and discarded.
6. After filtration, 60 ml of cold acetone (4°C) is slowly added to precipitate 3-SBASE.
7. The product is then washed four times with cold acetone and dried under vacuum.

3.3.3. On Tissue N-Terminal Derivatization with 3-SBASE

1. Whole tissue trypsin digestions are performed by applying 10 μl/cm^2 (according to the tissue section size) of trypsin solution using a micropipette at the surface of the section.
2. Tissue sections are then incubated for 2 h at 37°C in $MeOH/H_2O$ (1:1, v/v) saturated atmosphere.
3. The slide is placed in the vacuum desiccator to let the tissue dry completely prior to 3-SBASE deposition.
4. Manual derivatization is performed by applying the solution of 3-SBASE in pure water. In fact, no buffer is required since NH_4HCO_3 is still present on the tissue after trypsin deposition (*see* step 1). Volume of solution is adjusted according to the size of the tissue section or the size of the area to be studied within a tissue section (~10 μl/cm^2).
5. Tissue sections are then incubated for 1 h at 37°C in $MeOH/H_2O$ (1:1, v/v) saturated atmosphere.

6. *Optional step*: The tissue can be washed with cold 95% EtOH for 20 s to remove salts from buffer solutions to avoid signal loss due to poor ionization of derivate peptides. This procedure must be performed carefully to avoid peptide losses. Thus, washing steps in cold 95% EtOH must not exceed 20 s.
7. Place the slide in the vacuum desiccator for complete drying prior to matrix deposition.

3.3.4. On Tissue N-Terminal Derivatization with TMPP

1. TMPP is deposited by micro-spotting using the piezoelectric device (CHIP-1000). The derivatization must follow the same array as the trypsin one. For trypsin digestion follow the procedure . The automated micro-spotting is performed either following the global scheme of the array or by dividing the total array into smaller areas of 10 × 10 spots which will each be printed one after the other to increase the yield of the derivatization reaction.
2. 10 nl of a solution of TMPP is applied at each spot. Five droplets of approximately 100 pl are deposited at each spot per cycle. 20 iterations are thus necessary to obtain the final volume on each spot.
3. 20 nl of a solution of ACN/H_2O/TEA is then spotted allowing reaction of derivatization and preventing the use of buffers like NH_4HCO_3 which could induce poor ionization of derivate peptides (*see* **Note 2**). Five droplets of approximately 100 pl are deposited at each spot per cycle. 40 iterations are thus necessary to obtain the final volume on each spot.
4. The slide is then placed in the vacuum desiccator for complete drying of the tissue section prior to matrix deposition.

3.4. Matrix Deposition for Peptide Analyses

3.4.1. Using a Micropipette

1. Whole tissue matrix deposition is performed by applying a solution of HCCA/ANI using a micropipette (~10–20 μl according to the surface to cover). Careful attention must be drawn on this step to avoid touching the tissue with the micropipette tip and not induce tissue damages.
2. Check matrix coverage using an optical microscope (*see* **Note 5**).
3. Perform a rapid MS analysis of one spot to check that a sufficient amount of matrix is deposited.

3.4.2. Using a Microspotter

1. On each defined spot, 20 nl of HCCA/ANI solution is applied. Five droplets of 100 pl are deposited at each spot per cycle, then 20 iterations are necessary to obtain the total volume.
2. Check matrix coverage using an optical microscope (*see* **Note 5**).

3. A rapid MS analysis at one spot is recommended to check through signal intensity that a sufficient amount of matrix has been deposited. Increasing the number of iterations may improve MSI when signal intensity seems too low but too much matrix may decrease *S/N* ratio by increasing the background noise and decreasing ionization.

3.5. Mass Spectrometry Analysis

3.5.1. MALDI-MSI Experiments

1. 0.5 μl of calibration solution is applied near the tissue section and mixed with 0.5 μl of HCCA/ANI matrix solution.
2. The mass spectrometer is calibrated with the calibration spot.
3. Using FlexImaging and after defining teach points, an area of interest is selected on the tissue section.
4. The distance between each measurement point is set. Distance between measurement points is dependent on the method used for matrix deposition. For piezoelectric deposition, spots are generally spaced by 250 μm center to center. If required, the same array used for matrix deposition can be used for acquisition. Due to the size of the spot it is also possible to accumulate spectra at different positions within one spot.
5. In FlexControl, the method for peptide analysis involves using positive reflectron mode and a total of 500 spectra acquired at each position at a laser frequency of 100 Hz.
6. Images are saved and reconstructed using FlexImaging 2.1.

3.5.2. MS/MS Analysis

1. Ultraflex II TOF–TOF is equipped with LIFT III cell. For each MS/MS spectrum, 5,000 total shots are averaged including 1,000 shots for the parent ion and 4,000 for fragments.
2. Peptides are identified by searching MS/MS spectra against an appropriate database using Biotools software for MASCOT (Matrix Science) interrogation. For MALDI data, peptide mass tolerance is set at 0.5 Da and MS/MS tolerance at 1 Da. Oxidation of methionine is selected as variable modification.
3. When a peptide is identified as a digest fragment of a protein, the total sequence of this protein is used for in silico digestion and other digest fragments of this protein are searched in the total spectrum. MS/MS fragmentation is performed to confirm the identification of the proteins based on these fragments.
4. Using FlexImaging, digestion peptide distribution within the tissue is generated to check that all fragments of the same protein really provide the same localization.

4. Notes

1. In some cases, the trypsin can be suspended in various solutions. For example, trypsin in water can be used for frozen sections for which the pH at the tissue surface is close to the optimal pH value required for enzyme efficiency. Mix of water:methanol (1:1, v/v) can also be used with trypsin to achieve better extraction and permit a better access of cleavage sites to the enzyme. Use of water or water/methanol is recommended when using CHIP-1000 device for easier stabilization of droplets ejection.
2. For TMPP derivatization, instead of using a solution containing ACN/H_2O/TEA (30/69/1), a solution of ACN/NH_4HCO_3 (pH=10) (3:7, v/v) can be used, but poor ionization of derivatized peptides can be observed due to the presence of salts.
3. Prior to deposition, 2 μl of matrix solution could be deposited with a micropipette on a classical MALDI sample plate to check the crystallization. Crystallization is expected to be uniform, with long and fine white crystals as shown by Lemaire et al. (*see* Table 2a in ref. (19)).
4. Other organic solvents could be used to perform this step. More information can be obtained in ref. (20, 21). You must be careful not to increase the number of washing steps to avoid delocalization of the peptides.
5. To check matrix coverage, you need to ensure that the crystallization is dense, uniform, and composed of small crystals.

References

1. Ambihapathy, K., Yalcin, T., Leung, H. W., Harrison, A. G. (1997) Pathways to immonium ions in the fragmentation of protonated peptides. *J Mass Spectrom*, **32**, 209–215.
2. Kaufmann, R., Kirsch, D., Spengler, B. (1994) Sequencing of Peptides in a time-of-flight mass spectrometer – evaluation of postsource decay following matrix-assisted laser-desorption ionization (MALDI). *Int J Mass Spectrom Ion Processes*, **131**, 355–385.
3. Yalcin, T., Csizmadia, I. G., Peterson, M. R., Harrison, A. G. (1996) The structure and fragmentation of *B–n* (*n*>=3) ions in peptide spectra. *J Am Soc Mass Spectrom*, **7**, 233–242.
4. Biemann, K. (1990) Sequencing of peptides by tandem mass spectrometry and high-energy collision-induced dissociation. *Methods Enzymol*, **193**, 455–479.
5. Roth, K. D., Huang, Z. H., Sadagopan, N., Watson, J. T. (1998) Charge derivatization of peptides for analysis by mass spectrometry. *Mass Spectrom Rev*, **17**, 255–274.
6. Keough, T., Youngquist, R. S., Lacey, M. P. (1999) A method for high-sensitivity peptide sequencing using postsource decay matrix-assisted laser desorption ionization mass spectrometry. *Proc Natl Acad Sci U S A*, **96**, 7131–7136.
7. Keough, T., Lacey, M. P., Youngquist, R. S. (2000) Derivatization procedures to facilitate de novo sequencing of lysine-terminated tryptic peptides using postsource decay matrix-assisted laser desorption/ionization

mass spectrometry. *Rapid Commun Mass Spectrom*, **14,** 2348–2356.
8. Samyn, B., Debyser, G., Sergeant, K., Devreese, B., Van Beeumen, J. (2004) A case study of de novo sequence analysis of N-sulfonated peptides by MALDI TOF/TOF mass spectrometry. *J Am Soc Mass Spectrom*, **15,** 1838–1852.
9. Keough, T., Youngquist, R. S., Lacey, M. P. (2003) Sulfonic acid derivatives for peptide sequencing by MALDI MS. *Anal Chem*, **75,** 156A–165A.
10. Keough, T., Lacey, M. P., Strife, R. J. (2001) Atmospheric pressure matrix-assisted laser desorption/ionization ion trap mass spectrometry of sulfonic acid derivatized tryptic peptides. *Rapid Commun Mass Spectrom*, **15,** 2227–2239.
11. Lee, Y. H., Kim, M. S., Choie, W. S., Min, H. K., Lee, S. W. (2004) Highly informative proteome analysis by combining improved N-terminal sulfonation for de novo peptide sequencing and online capillary reverse-phase liquid chromatography/tandem mass spectrometry. *Proteomics*, **4,** 1684–1694.
12. Lee, Y. H., Han, H., Chang, S. B., Lee, S. W. (2004) Isotope-coded N-terminal sulfonation of peptides allows quantitative proteomic analysis with increased de novo peptide sequencing capability. *Rapid Commun Mass Spectrom*, **18,** 3019–3027.
13. Gevaert, K., Demol, H., Martens, L., Hoorelbeke, B., Puype, M., Goethals, M., Van Damme, J., De Boeck, S., Vandekerckhove, J. (2001) Protein identification based on matrix assisted laser desorption/ionization-post source decay-mass spectrometry. *Electrophoresis*, **22,** 1645–1651.
14. Marekov, L. N., Steinert, P. M. (2003) Charge derivatization by 4-sulfophenyl isothiocyanate enhances peptide sequencing by post-source decay matrix-assisted laser desorption/ionization time-of-flight mass spectrometry. *J Mass Spectrom*, **38,** 373–377.
15. Wang, D., Kalb, S. R., Cotter, R. J. (2004) Improved procedures for N-terminal sulfonation of peptides for matrix-assisted laser desorption/ionization post-source decay peptide sequencing. *Rapid Commun Mass Spectrom*, **18,** 96–102.
16. Alley, W. R., Jr., Mechref, Y., Klouckova, I., Novotny, M. V. (2007) Improved collision-induced dissociation analysis of peptides by matrix-assisted laser desorption/ionization tandem time-of-flight mass spectrometry through 3-sulfobenzoic acid succinimidyl ester labeling. *J Proteome Res*, **6,** 124–132.
17. Huang, Z. H., Wu, J., Roth, K. D., Yang, Y., Gage, D. A., Watson, J. T. (1997) A picomole-scale method for charge derivatization of peptides for sequence analysis by mass spectrometry. *Anal Chem*, **69,** 137–144.
18. Franck, J., El Ayed, M., Wisztorski, M., Salzet, M., Fournier, I. (2009) On tissue N-terminal peptide derivatizations for enhancing proteins identification in bottom-up Imaging strategies. *Anal Chem*, **81,** 8305–8317.
19. Lemaire, R., Tabet, J. C., Ducoroy, P., Hendra, J. B., Salzet, M., Fournier, I. (2006) Solid ionic matrixes for direct tissue analysis and MALDI imaging. *Anal Chem*, **78,** 809–819.
20. Lemaire, R., Wisztorski, M., Desmons, A., Tabet, J. C., Day, R., Salzet, M., Fournier, I. (2006) MALDI-MS direct tissue analysis of proteins: improving signal sensitivity using organic treatments. *Anal Chem*, **78,** 7145–7153.
21. Seeley, E. H., Oppenheimer, S. R., Mi, D., Chaurand, P., Caprioli, R. M. (2008) Enhancement of protein sensitivity for MALDI imaging mass spectrometry after chemical treatment of tissue sections. *J Am Soc Mass Spectrom*, **19,** 1069–1077.

Chapter 20

Specific MALDI-MSI: TAG-MASS

Jonathan Stauber, Mohamed El Ayed, Maxence Wisztorski, Michel Salzet, and Isabelle Fournier

Abstract

MALDI imaging as a molecular mass spectrometry imaging technique (MSI) can provide accurate information about molecular composition on a surface. The last decade of MSI development has brought the technology to clinical and biomedical applications as a complementary technique of MRI and other molecular imaging. Then, this IMS technique is used for endogenous and exogenous molecule detection in pharmaceutical and biomedical fields. However, some limitations still exist due to physical and chemical aspects, and sensitivity of certain compounds is very low. Thus, we developed a multiplex technique for fast detection of different compound natures. The multiplex MALDI imaging technique uses a photocleavable group that can be detect easily by MALDI instrument. These techniques of targeted imaging using Tag-Mass molecules allow the multiplex detection of compounds like antibodies or oligonucleotides. Here, we describe how we used this technique to detect huge proteins and mRNA by MALDI imaging in rat brain and in a model for regeneration; the leech.

Key words: Matrix-assisted laser desorption/ionization, time-of-fight, mass spectrometry, mass spectrometry imaging, mRNA, antigens, photocleavage.

1. Introduction

As an innovative technique, MALDI-IMS is a powerful tool for direct detection and localization of endogenous and exogenous molecules within biological samples (1, 2). The last developments have led to use this technique to obtain the distribution of various compounds such as lipids, drugs, peptides, and proteins within tissue sections (1–4). Non-targeted aspect of MALDI-IMS is one of the big advantages of the technology compared to other imaging techniques as well as strength of MS for structural elucidation.

S.S. Rubakhin, J.V. Sweedler (eds.), *Mass Spectrometry Imaging*, Methods in Molecular Biology 656,
DOI 10.1007/978-1-60761-746-4_20, © Springer Science+Business Media, LLC 2010

Many successful applications of this technique have been undertaken recently. In particular, MALDI-IMS of lipids, peptides, and proteins for clinical applications by studying pathologies has shown to be a very promising application by providing information on the variations of abundance and localization of markers (3–14). Moreover, biological processes bring into play many different signaling pathways involving various classes of molecules ranging from oligonucleotides, to proteins, peptides, and lipids. In particular, the correlation of mRNA expression with their corresponding protein regulation, or more generally the correlation of transcriptome to proteome, is of special interest for better understanding of biological mechanisms. This is especially an essential aspect when studying pathologies for earlier diagnosis.

However, some specific classes of biomolecules such as oligonucleotides or sugars are still non- or hardly accessible to the direct analysis by MALDI, as are also, very hydrophobic proteins, membrane proteins, high mass proteins (>30 kDa), or lower abundance ones. Ideally, oligonucleotides should be directly detected from tissues, although their large size and low abundance in cells added to analytical difficulties in mass spectrometry (salts adducts and gas phase instability) render their analysis difficult (15). In the same time, multiplex techniques are necessary for diagnosis and prognosis. More and more tissue micro arrays (TMAs) are used today to analyze large number of disease tissues and new, fast, and reproducible multiplex techniques are necessary (9).

We, thus, have proposed a new concept of possible multiplex and specific detection and tracking of biomolecules with a special focus on mRNA and proteins for transcriptome/proteome correlations. This concept relies on affinity detection by using a specific designed probe, called Tag-Mass which can be detected by mass spectrometry (16). Tag-Mass offers more selectivity to MALDI-MSI for selectively and specifically tracking known markers of physiological stages in cohorts of samples with a high sensitivity (3, 16, 17).

1.1. The Tag-Mass Concept: Selective Multiplexed Imaging of Biomolecules from Tissue Sections

The "Tag-Mass" strategy is an affinity-based strategy where a probe is directed against a specific target, using a probe that can be imaged by MALDI-MSI (3, 16, 17). The Tag-Mass is a modified probe bearing a reporter group where the reporter group is used in MALDI-MSI to indirectly obtain the image of the probe. The reporter is designed to be a molecule of known molecular mass that is easily detectable under MALDI conditions taking care to use a molecule that is not corresponding to an endogenous compound. To image a probe indirectly via the direct image of the reporter, the reporter must be linked to the probe and released in the final step just before or during the MALDI sequence.

In the Tag-Mass, the release of the reporter group is obtained by photodissociation under the MALDI laser irradiation using a photocleavable moiety that binds the reporter to the probe (**Fig. 20.1a**). Thus, the reporter is detached from the probe during the MALDI-MSI acquisition. Many different reporters can be used for this purpose, but most of the times peptides were used. The photocleavable linker is chosen to present a specific absorption band in the UV at a wavelength (340 nm) very closed to that of MALDI lasers (i.e., 337–355 nm). Thus, after hybridization of the modified probe to its target, a classical MALDI-MSI sequence is performed. At a specific location of the acquisition, the presence of a probe will be signed by the presence of the reporter released under the MALDI laser irradiation, which traduced by the observation of a peak at the m/z expected for the reporter (**Fig. 20.1b**). Reconstruction of the reporter molecular image gives, then, the image of the probe, i.e., those of the targeted molecule (3, 16, 17). Such a concept is compatible with all types of probes including mRNA probes, antibody probes, or even lectins or aptamers owing to image with high selectivity, respectively, mRNA, antigens, oligosaccharides (including glycosylated proteins), and drugs. Tag-mass workflow, MALDI-MSI, is combined with hybridization techniques including in situ hybridization (ISH) and immunohistochemistry (IHC) (3, 16, 17). Specific MALDI-MSI or Tag-Mass MSI by using reporter moieties that can be distinguished by their change in m/z is a technique that can be used in multiplex conditions. Theoretically, there are no limits in the multiplexing conditions except the hybridization step itself because of kinetic competition during the affinity reactions or steric obstruction problems. Tag-Mass can also be use for semi-quantification in multiplex conditions by using a reporter presenting the same physico-chemicals properties, i.e., same analytical behavior using, for example, isotopically labeled reporter such as differentially deuterated peptides.

This concept can be extended by looking for alternative ways of releasing the reporter moiety, e.g., chemical released or even released by prompt fragmentation pathways (i.e., before the end of the delay time period). The reporter can also be designed to be observable in LDI conditions avoiding, thus, the use of the MALDI matrix (18, 19). Although this latest solution could be less sensitive than using MALDI conditions but would increase spatial resolution of images. Extension to other ion production sources can also be searched. For secondary ion mass spectrometry (20, 21) (SIMS) or laser ablation inductively coupled plasma mass spectrometry (22–24) (LA-ICP), probes bearing directly a monoatomic element easily detectable by these techniques should be used, if the element as a good sensitivity of analysis and is not present naturally in the surface to study. Such techniques induce quite important fragmentation yields and the reporter element

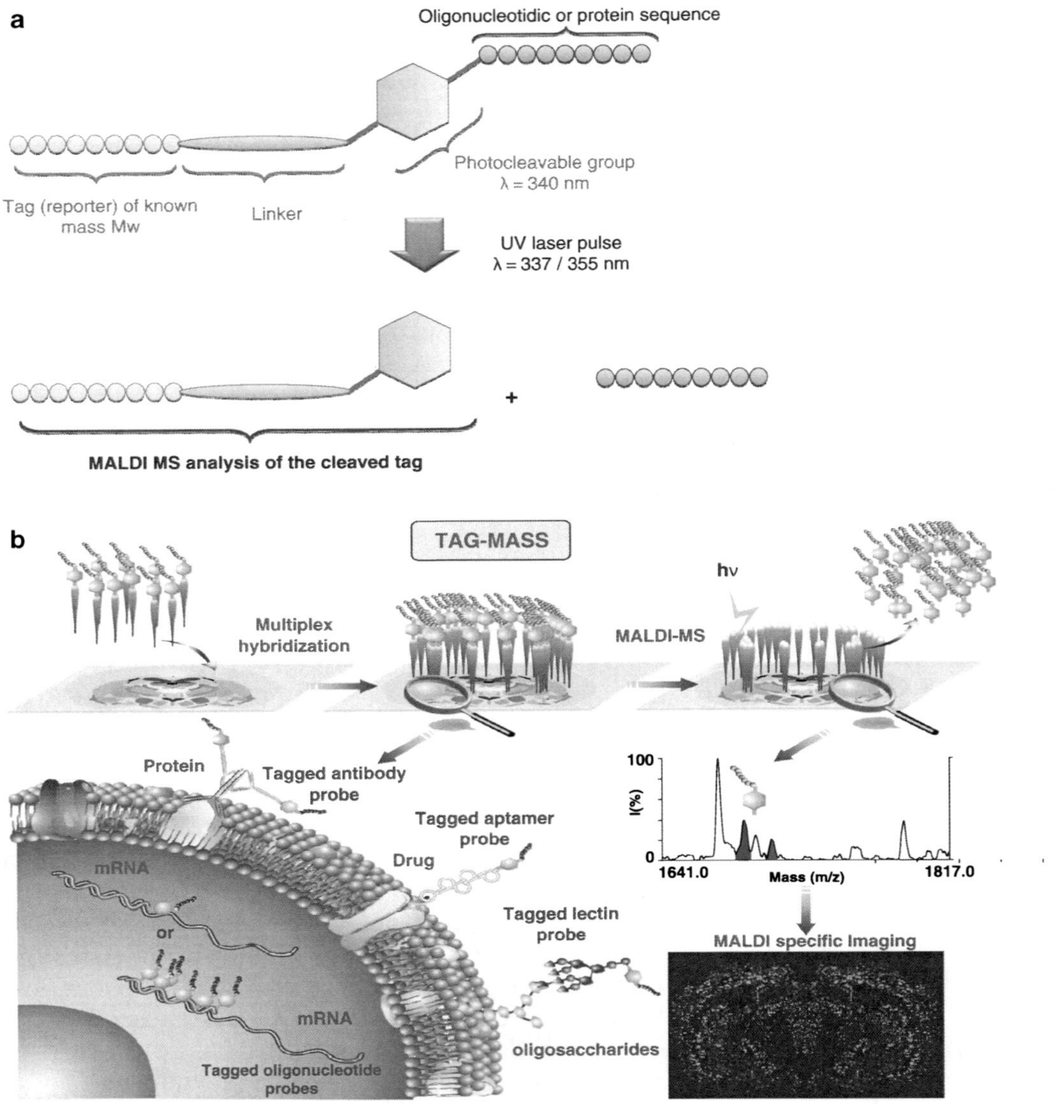

Fig. 20.1. (**a**) Schematic representation of the reporter release under photodissociation by the MALDI laser using a photocleavable-reporter system coupled to the probe. (**b**) Workflow of multiplex specific MALDI-MSI (Tag-Mass).

will appear as a fragmentation product. For example, gold-labeled secondary antibodies are a good solution for imaging antigens in LA-ICP MS at a spatial resolution below 10 μm.

We present, here, the workflows for Tag-Mass of antigens and mRNA using a photocleavable probe bearing a peptide as reporter moiety. For antibodies, preference was given to use indirect IHC with a primary–secondary antibody system. Indeed, indirect IHC is known to present better performances by decreasing steric obstruction problems and increasing detection level, since secondary antibodies will recognize consensus epitope present in the primary antibody sequence allowing attachment of several secondary antibodies. Moreover, secondary antibodies are easier to produce since they require much less specificity. For mRNA, modified uracile nucleotides were used. This requires a specific synthesis in order to add the photocleavable group and reporter moiety directly on the nucleotide basis for keeping both 3′ and 5′ termini free. In fact, the modified nucleotide is to be used for the probe amplification. In former experiments, modified primers (by the addition of a photocleavable-reporter system) were used. This approach had revealed several disadvantages including lake of sensitivity (only one reporter per probe), high cost (specific synthesis required for each mRNA to be localized), and impossibility to amplify the probe by in vitro translation (only one terminus of the primer free). Development of modified uracile nucleotides was a great advance in this respect. Modified nucleotides are available for all probes construction, the sensitivity is increased by the incorporation of several reporters in the probe sequence (amplification of the signal) and probes can be obtained by in vitro translation. Only tagged Uracile strategy will be presented here. Specific MALDI-MSI can also be performed in multiplexing conditions. An example of duplex imaging of two antigens (Cystatin B/Cathepsin D) from a FPE tissue section of the leech *T. tessulatum* are presented here as an example for multiplexing.

2. Materials

2.1. Preparation of Frozen Tissue Sections

2.1.1. Snap-Frozen Tissues

1. Isopentane cooled at –45°C with dry ice. Vapors may cause drowsiness and dizziness, so work in a hood.

2.1.2. Tissue Cryosection and Thaw Mounted

1. Optimal cutting temperature polymer, OCT.
2. Indium tin oxide (ITO)-coated glass slides or other holder compatible with mass spectrometry analysis.
3. A cryomicrotome, Leica CM150S (Leica Microsystems, Nanterre, France).

2.1.3. Pre-analysis Treatment: Tissue Fixation

1. Ethanol 75% (–20°C): 75 ml of absolute ethanol (≥99.8%) and water (HPLC grade) to 100 ml. Prepare fresh. Store at –20°C.
2. Ethanol 95% (−20°C): 95 ml of absolute ethanol (≥99.8%) and water (HPLC grade) to 100 ml. Prepare fresh. Store at −20°C.

2.1.4. Pre-analysis Treatment: Lipids Removal

1. Chloroform (−20°C): 100 ml of chloroform (≥99.9%). Store at −20°C. Chloroform is harmful by inhalation, so work in the hood.

2.2. Preparation of FFPE Tissue Section

2.2.1. FFPE Tissue Section

1. Indium tin oxide (ITO)-coated glass slides or other holder compatible with mass spectrometry.
2. Water: 100 ml of water (HPLC grade). Prepare fresh.
3. A microtome and an hotplate warm at 50°C.

2.2.2. FFPE Tissue Dewaxing

1. Xylene: 100 ml of xylene (≥99.9%). Xylene is harmful by inhalation, so work in the hood.
2. Ethanol 100%. Prepare fresh.
3. Ethanol 95%: 95 ml of absolute ethanol (≥99.8%) and water to 100 ml. Prepare fresh.
4. Ethanol 75%: 75 ml of absolute ethanol (≥99.8%) and water to 100 ml. Prepare fresh.
5. Ethanol 30%: 30 ml of absolute ethanol (≥99.8%) and water to 100 ml. Prepare fresh.
6. Water: 100 ml of water (HPLC grade). Prepare fresh.

2.3. Hybridization Buffers for In Situ Hybridization (ISH)

1. Buffer solution glycine 0.1 M/Tris HCl buffer (pH 7.4).
2. RNAse inhibitoring activity solution: Proteinase K (1 μg/μl in 1 M/Tris HCl and 0.5 M EDTA, pH 8).
3. Post-fixation buffer: 4% paraformaldehyde (0.1 M Phosphate/ 5 mM MgCl2 buffer, pH 7.4), 15 min then triethanolamine (0.1 M, pH 8), 10 min.
4. Washing solution: 20× SSC buffer: standard sodium citrate solution : for a 20× SSC solution dissolve 701.28 g NaCl and 352.92 g NaCitrate in a recipe to make 4 l (check to have pH to 7.0) and bring final volume to 4 l , then Autoclave the solution in order to be RNAse free).
5. Dehydration by ethanol (30°, 70°, 96°).
6. Probes denaturation (100°C, 10 min).
7. Hybridization buffer: 0.01 M dextran sulfate, 0.2 M formamide, 20× SSC 20%, 100 × Denhardt's 10%). 16 h, 55°C.
8. Non-hybridized probe degradation buffer: RNase (10 μg/ml), 37°C, 30 min.

9. Rinsed steps:
   a. 20 and 10 mM 2-mercaptoethanol solutions, 10 min.
   b. 0.5× and 0.1× SSC.
   c. Ultrapure water.

2.4. Hybridization Buffers for Immunocytochemistry (IHC)

1. Incubation buffer: 0.1 M PBS/1% BSA/1% normal goat serum/0.05% Triton X100.
2. Primary antibody incubation, overnight, 4°C, on rocking.
3. Washing solution: phosphate-buffered saline (PBS). Prepare 10× stock with 1.37 M NaCl, 27 mM KCl, 100 mM Na2HPO4, 18 mM KH2PO4 (adjust to pH 7.4 with HCl if necessary) and autoclave before storage at room temperature. Prepare working solution by dilution of one part with nine parts water.
4. Secondary antibody solution: antibody diluted in incubation buffer at room temperature, on rocking. Antibody is either anti-rabbit IgG 1/100 developed in goat (Jackson Immunoresearch, Inc., Europe LTD); FITC-conjugated secondary antibody anti-rabbit IgG 1/100 developed in goat (Jackson Immunoresearch, Inc., Europe LTD) or photocleaved tagged antibody 1/100 (Imabiotech, France).
5. Revelation
   a. Photocleavable tagged antibody, precleavage under UV 5 min Staining substrate for peroxidase antibody in chloronapthol with 0.05% H_2O_2 for detection.
   b. For FITC ICC, slices were prepared using phenylenediamine in glycerol.

2.5. Matrix Deposition for Proteins Analysis

2.5.1. Using a Microspotter

1. SA/ANI solution: 1.5 equivalent of aniline (ANI) were added to a solution containing 40 mg/ ml of sinapinic acid (SA) in acetonitrile/aqueous TFA 0.1% (6:4, v/v). Aniline and TFA are toxic, so work in the hood.
2. Chemical Inkjet Printer CHIP-1000 (Shimadzu Biotech, Kyoto, Japan).

2.5.2. Using an Automatic Sprayer

1. SA/ANI solution: 1.5 equivalent of aniline (ANI) were added to a solution containing 40 mg/ ml of sinapinic acid (SA) in acetonitrile/aqueous TFA 0.1% (6:4, v/v). Aniline and TFA are toxic, so work in the hood.
2. ImagePrep (Bruker Daltonics, Bremen, Germany).

2.6. Matrix Deposition for Peptides Analysis

2.6.1. Using a Microspotter

1. ANI solution: 1.5 equivalent of aniline (ANI) were added to a solution containing 10 mg/ml of α-cyano-4-hydroxycinnamic acid (HCCA) in acetonitrile/aqueous TFA 0.1% (6:4, v/v). Aniline and TFA are toxic, so work in the hood.
2. Chemical Inkjet Printer CHIP-1000 (Shimadzu Biotech, Kyoto, Japan).

2.6.2. Using an Automatic Sprayer

1. ANI solution: 1.5 equivalent of aniline (ANI) were added to a solution containing 10 mg/ml of α-cyano-4-hydroxycinnamic acid (HCCA) in acetonitrile/aqueous TFA 0.1% (6:4, v/v). Aniline and TFA are toxic, so work in the hood.
2. ImagePrep (Bruker Daltonics, Bremen, Germany).

2.7. Mass Spectrometry Analysis

2.7.1. MALDI-MSI Experiment

1. Peptide calibration standard II (Bruker Daltonics, Wissenbourg, France): Angiotensin II, angiotensin I, substance P, bombesin, ACTH clip 1–17, ACTH clip 18–39, Somatostatin 28, Bradykinin Fragment 1–7, Renin Substrate Tetradecapeptide porcine. Covered mass range: ~700–3,200 Da. Store at −20°C.
2. Protein Calibration Standard I (Bruker Daltonics, Wissenbourg, France): Insulin, ubiquitin I, cytochrome C, myoglobin. Covered mass range: ~5,000–17,500 Da. Store at −20°C.
3. An Ultraflex II TOF–TOF equipped with a Smartbeam laser and all the Flex software suite (FlexControl, FlexAnalysis, and FlexImaging) (Bruker Daltonics, Bremen, Germany).

2.7.2. MS/MS Analysis

1. An Ultraflex II TOF–TOF equipped with a Smartbeam laser and all the Flex software suite (FlexControl, FlexAnalysis, and FlexImaging) (Bruker Daltonics, Bremen, Germany).
2. Biotools (Bruker Daltonics, Bremen, Germany).

2.8. MALDI Imaging Analysis

1. MALDI matrix used for antibody detection: α-Cyano-4-hydroxycinnamic acid (HCCA) and 3-hydroxypicolinic acid (3-HPA) (Sigma-Aldrich).

2. Calibrant solutions of angiotensin II, Des-Arg-bradykinin, substance P, ACTH 18–39, ACTH 7–38, and bovine insulin (Sigma-Aldrich).
3. Trifluoroacetic acid (TFA) (Applied Biosystems). Acetonitrile and methanol (J.T. Baker).
4. Sprayer of matrix (ImagePrep, Bruker Daltonics).

3. Methods

The methods are described according to their sequential used in experiments. One part is dedicated to the protocols or hybridization themselves, i.e., ISH and IHC using the photocleavable modified probe for both cryostat sections (from frozen samples or fixed and frozen samples) and microtome sections (for fixed and paraffin-embedded [FPE] samples). It must be noticed that the protocols are tissue and target dependent. This is usual in hybridization. Thus, proposed protocols are given for rat brain tissue sections and shall be slightly modified and optimized for other applications according to the tissue and the probe (specificity, selectivity). **Figures 20.2** and **20.3** give examples of MS spectra and molecular images obtained, respectively, for Proenkephalin mRNA imaging and Carboxypeptidase D (CPD) protein imaging, respectively, using the modified dU for the oligonucléotides probe construction and a rabbit secondary modified antibody for CPD in conjunction with a primary antibody directed against rat CPD and raise in rabbits as antibody. The images are reconstructed based on the reporter signal with a peptide close the to bradykinin sequence. **Figure 20.2** gives the molecular distribution of Proenkephalin mRNA as obtained by specific MALDI-MSI compared to the localization obtained by classical ISH. A good correlation is observed in between ISH and MALDI-MSI images. **Figure 20.3** compared the specific MALDI-MSI localization of CPD with its distribution obtained for classical colorimetric reaction or fluorescence. Specific MALDI-MSI presents higher sensitivity than classical revelation and is not so far from fluorescence detection. Specific MALDI-MSI enables multiplex detection of different epitopes. An example of duplex experiments for proteins is given **Fig. 20.4** where Cystatin B/Cathepsin D system is studied in FPE tissue sections of the leech *Theromyzon tessulatum*. Experiments were performed using two photocleavable secondary antibodies bearing different reporter peptides represented by different m/z. In such an experiment, one primary antibody is raised in rabbit

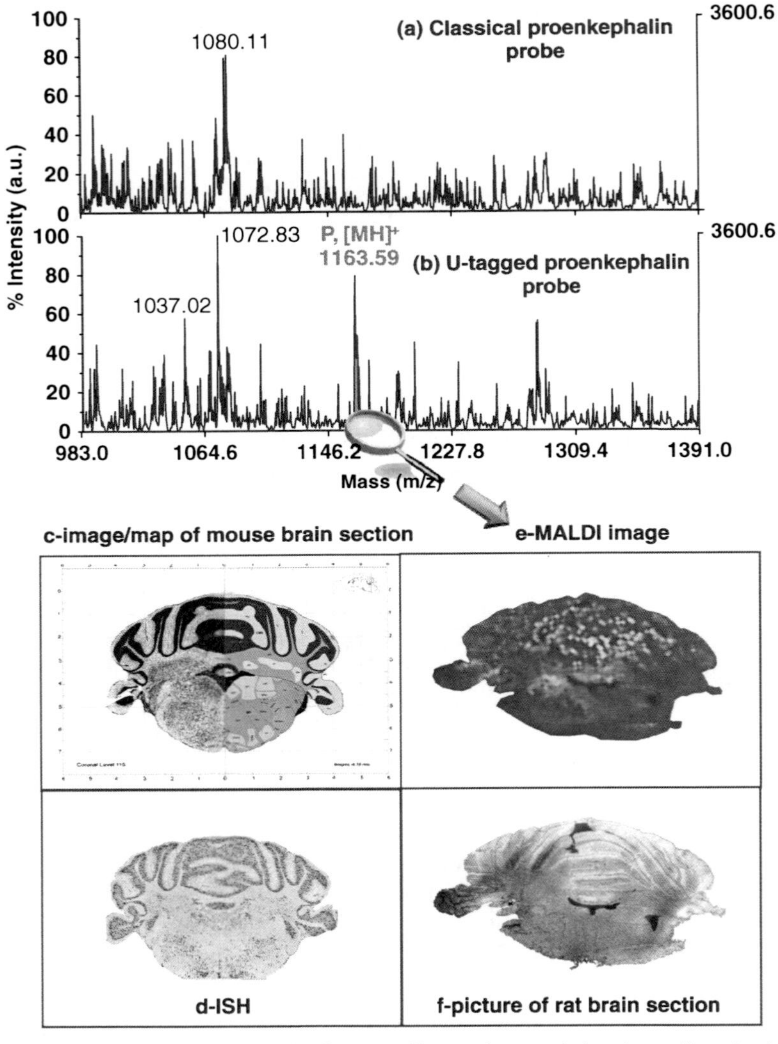

Fig. 20.2. Compared MALDI mass spectra in the linear positive mode recorded on two adjacent rat brain sections in the same region of the brain after ISH of double strand oligonucleotide cDNA probe corresponding to proenkephalin for classical untagged proenkephalin probe (**a**) and the U-tagged proenkephalin probe (**b**). (**f**) Corresponding reconstructed MALDI image on the basis of the tag signal obtained by scanning the tissue section after ISH experiment (7,000 spots separated each of 100 μm) compared (**e**) to proenkephalin mRNA localization in 8-weeks old male C57BL/6 J mouse brain using digoxigenin ISH technique by the Allen Institute (http://www.brainatlas.org/aba). For this experiment, colorimetric detection of bound probe is generated by the alkaline phosphatase substrates nitroblue tetrazolium (NBT) and 5-bromo-4-chloro-3-indolyl phosphate (BCIP) that produce a vivid blue/purple particulate reaction product. Figures (**a**) presents the map/picture representation of the mouse brain and figure (**f**) the picture of the rat brain section prior to ISH –MALDI imaging experiment. Reprinted from (14) with permission.

with an anti-rabbit photocleavable secondary antibody, whereas the second one is raised in mouse with an anti-mouse photocleavable secondary antibody. The experiment-exemplified multiplex conditions for proteins, although multiplexing, can also be achieved for mRNA.

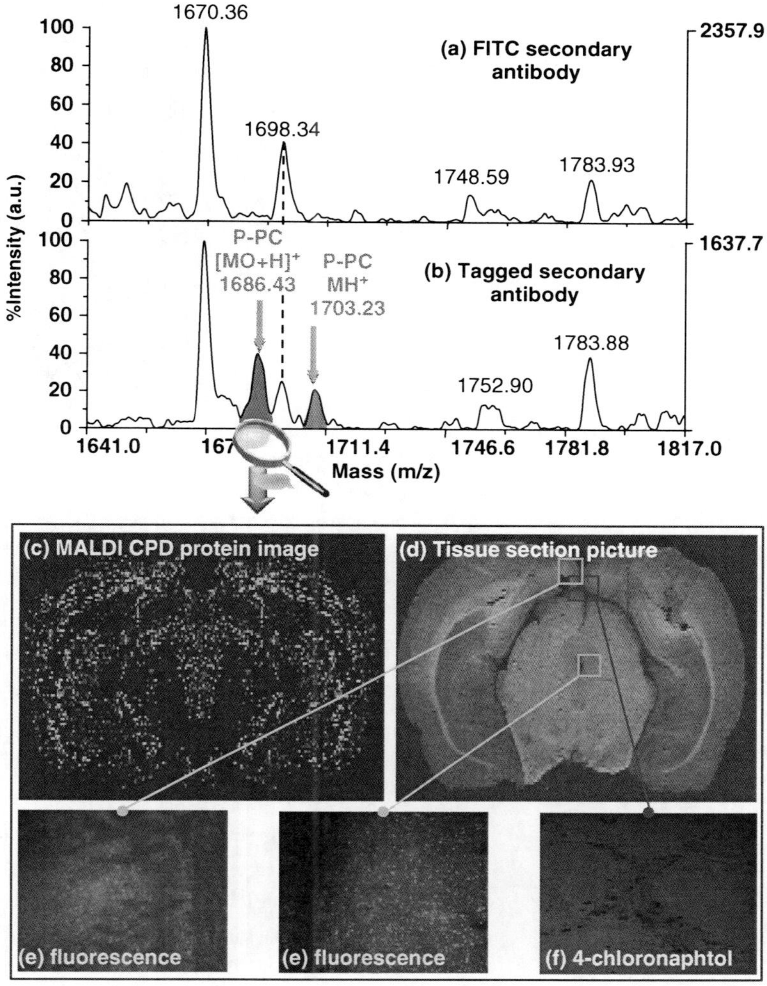

Fig. 20.3. Specific MALDI-MSI of CPD protein from a rat brain tissue section performed using a rabbit anti-rat primary antibody directed against CPD and the modified photo-cleavable goat anti-rabbit secondary antibody. (**a–b**) MALDI-MS spectra recorder for serial rat brain tissue sections after the IHC experiments for the tagged secondary antibody (**b**) compared to a FITC secondary antibody (**a**) for the same IHC conditions. (**c**) Molecular MALDI images reconstructed using the signal at *m/z* 1,686.43 of the reporter moiety (**d**) tissue section image before IHC experiments. (**e** and **f**) Comparison with fluorescence and 4-cholonaphaol detection using, respectively, a FITC or preroxydase tagged secondary antibody. Reprinted from (14) with permission.

3.1. Tissue Treatment

3.1.1. Tissue Snap-Frozen

1. The organ is dissected and rinsed with a saline solution suited for the considered tissue to remove blood and other tissue fragment of the surface. Alternative: prior to sacrifice, the animal can be perfused with the saline solution to remove blood inside the organ.

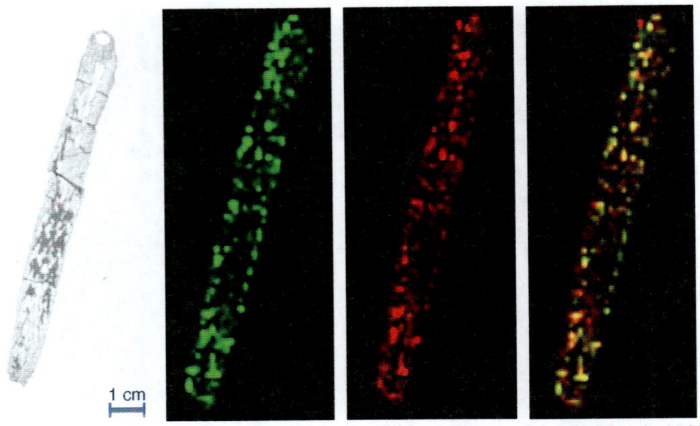

Fig. 20.4. Duplex specific MALDI-MSI of Cystatin B/Cathepsin D from a FPE tissue section of the leech *T. tessulatum* performed using rabbit anti-leech Cathepsin D primary antibody/photocleavable goat anti-rabbit secondary antibody and mouse anti-leech Cystatin B primary antibody/photocleavable goat anti-mouse secondary antibody. Molecular images have been reconstructed on the signals of the two reporter peptides, i.e., *m/z* 1,309 for Cathepsin D and *m/z* 15.

2. Morphology of the organ needs to be carefully maintained. Thus, the tissue should not be placed in a tube or wrap in an aluminum foil to avoid deformation of the organ (adaptation to the outlines of the container).
3. Snap-freezing procedure is applied for tissue conservation to maintain tissue morphology and to prevent ice crystals formation and cell explosion. In fact, different rate cooling of parts of the organ or direct dipping of the organ into liquid nitrogen leads to the formation of cracks and fragmentation of the tissue. Therefore the use of isopentane cooled at −45°C with dry ice is recommended. Freezing time is dependent on the size of the organ. It is preferable not to use an embedding media. For very small organs or surgical pieces, cutting without embedding material increases deformations and damages of the tissue sections. In such a case, a solution containing non-polymeric compounds, namely 10% gelatin solutions, helps to obtain high-quality tissue section. Tissue is embedded in 10% gelatin directly after dissection and frozen as previously described.
4. After snap freezing, tissue is removed from isopentane and stored at −80°C. We heartily recommend not overpassing a storage period of 6 months. Over 6 months storage, variation in the molecular profiles could be observed if no sample stabilization procedure is performed. Preferentially tissues should be analyzed a few days or weeks after snap freezing.

3.1.2. Tissue Cryosection and Thaw Mounted

1. The use of cryopreservative solutions containing organic polymers such as optimal cutting temperature (OCT) polymer should be restricted to the attachment of the tissue to the sample holder and not used for complete embedding of the tissue. Moreover, all parts of the cryostat in contact with the tissue must be cleaned to prevent any contamination between two different samples or with the polymer containing solution. In the case of contact between the tissue and cryopreservative solutions containing polymers, MS spectra will be dominated by polymer signals such as PEG signal distribution.
2. Tissue is placed in the cryostat during sufficient time before sectioning for slow warming of the sample to the cryostat temperature. If the tissue is too cold, poor-quality sections are obtained.
3. 10 μm thickness tissue sections are cut using cryomicrotome at −20°C. Different tissue types may need other temperature settings. 10 μm thickness is optimal. Smaller sections have not enough molecules for extraction and thicker sections may cause problems of conductivity (due to the insulating nature of tissues) and charge effects by charge accumulation at the sample surface during MALDI analysis. Charge effects will decrease spectral quality in axial TOF configuration instruments resulting in a progressive peak shifting toward the high m/z ratio.
4. Collect the tissue sections onto ITO glass slides pre-cooled at −20°C. Transfer is performed by applying the cooled ITO slide onto the section. The cuts are thus stick on the cold slide. Adhesion of the frozen sections to the glass slides is obtained by heating putting fingers under the slide or by placing the slide at room temperature. This transfer procedure, contrarily to classical thaw mounting, prevents formation of ice crystals at the surface of the cryostat microtome cutting plate.
5. Care must be taken of air bubbles formation at the surface of the tissue section that may leads to artifacts during MS analysis.
6. Mounted sections are stored in a sealed container at −80°C until their use.

3.1.3. Pre-analysis Treatment: Tissue Fixation

1. A closed container store at −80°C is warmed at room temperature in a vacuum desiccator to prevent water condensation at the surface of the frozen slide.
2. After complete drying, the ITO slide is washed. Washing steps are optional and dependent on the molecules to be

analyzed. Careful washing is crucial for conserving spatial localization of molecules.

3. For analysis of small molecules like lipids or drugs, no washing steps are used. For macromolecules analysis like peptides or proteins washing procedures are generally used. Washing is performed by immersing the glass slide softly in ice-cold 75% ethanol during 30 s. No agitation or shake is needed. This step washes out salts, cells fragments, or residual fluids.
4. Take the slide out and remove the excess of liquid around the section. A stream of nitrogen over the surface could help to remove excess of ethanol.
5. The ITO glass slide is then placed in a vacuum desiccator to completely dry of the tissue. The time of drying is dependent to the size of the section.
6. Optional: a second bath of fresh ice-cold 75% ethanol during 30 s followed by a complete drying under vacuum desiccator can be achieved.
7. After complete drying, the sample is dipped into cold 95% ethanol during 30 s. No agitation or shake is needed. This step prevents degradation of proteome by dehydration and fixation of the tissue.
8. The slide is completely dried like in steps 4 and 5.

3.1.4. Pre-analysis Treatment: Lipids Removal

1. After complete drying, immerse the glass slide softly in ice-cold chloroform (30 s). No agitation or shake is needed. This step removes lipids (especially phospoholipids) present in high concentration in the tissue (components of cell membranes) and may cause signal suppression in MS spectra.
2. Take the slide out and place it in the vacuum desiccator for complete drying of the tissue.

3.2. Preparation of FFPE Tissue Section

3.2.1. FFPE Tissue Section

1. 10 μm thickness FFPE tissue sections are cut using a microtome at room temperature. Paraffin block can be cooled down −20°C prior sectioning to facilitate tissue sectioning.
2. Sections are transferred onto a conductive ITO glass slide on top of a water droplet.
3. Glass slide is warmed up on a hotplate to leave the cuts unfolds.
4. Excess of water is removed and glass slide is stored in an incubator at 30°C during 20 min for good adherence.

Subsequently obtained glass slides with FFPE tissue sections can be stored during over months at room temperature.

3.2.2. FFPE Tissue Dewaxing

1. After complete drying, the glass slide is gently dipped into a bath of xylene during 5 min. This procedure is repeated to fold. No agitation or shake is needed.
2. The slide is then washed in stepwise immersion, 5 min duration each, into 100% ethanol twice, 95% ethanol, 75% ethanol, and 30% ethanol for rehydration of tissue sections.
3. The ITO glass slide is placed in the vacuum desiccator for complete drying of the sections.

3.3. Tissue Preparation for In Situ Hybridization (ISH)

1. 10 μm thickness FFPE tissue sections obtained as previously described are used for ISH.
2. Paraffin is removed by using xylene baths (two times, 15 min), and then tissue is hydrated during 5 min in three steps of different mixed ethanol/water baths (96°, 70°, 30°).
3. Sections were prepared according to classical ISH protocols. Tissues were incubated in glycine buffer, and then treated for 15 min with proteinase K for protein digestion.
4. After post-fixation with 4% paraformaldehyde for 15 min, then a bath with triethanolamine (0.1 M, pH 8) was carried out for 10 min.
5. Sections were washed with 2× SSC, then ultrapure water for 5 min. Probes were denaturized at 100°C for 10 min, and after a 3 step tissue dehydration (30°, 70°, 96°), hybridization was done for 16 h at 55°C dissolving cDNA probes in hybridization buffer (Dextran sulfate 10%, formamide 50%, 20× SSC 20%, 100× Denhardt's 10%).
6. Tissues were incubated with RNase, then rinsed 10 min with successive SSC solutions and twice 0.5×SSC solutions at 55°C for 30 min. After rinsing slices with 0.1×SSC for 5 min at room temperature, one bath of ultrapure water was carried out to remove the excess of polymers. Tissues were kept drying at room temperature before MALDI matrix application.

3.4. Tissue Preparation for Immunocytochemistry (ICC)

1. Frozen sections of rat obtained as previously described are used for ICC.
2. They were incubated at room temperature with 500 μl of incubation buffer for 30 min. The same buffer was used to dilute anti-rat Carboxypeptidase D (CPD) antibody (1:400), and incubation was performed overnight at 4°C.

3. For the leech, Cystatin B and Cathepsin D primary antibodies are used at different concentrations (1/500 Cystatin B and 1/400 Cathepsin D).
4. After washing three times in PBS, sections were incubated with peroxidase conjugated secondary antibody or FITC-conjugated secondary antibody or using photocleavable tagged antibody for 80 min at room temperature.
5. After another three washing steps in PBS buffer, the sections for peroxydase ICC were incubated in chloronapthol with 0.05% H_2O_2 for detection. Reaction was stopped with several PBS and ultrapure water baths. For FITC ICC, slices were prepared using phenylenediamine in glycerol. For photocleavable tagged antibody, tissues were rinsed three times for 5 min with ultrapure water to remove salts, and sections were kept drying at room temperature in dark before matrix application. Tissues were then compared using microscopy.

3.5. Peptide Reporter Analysis

3.5.1. Peptide Reporter Analysis Using Dry Droplet

For classical analysis, 1 μl of sample solution and 1 μl of matrix solution (HCCA/ANI) were mixed on the MALDI plate using the dried-droplet technique as a standard control for the different Tag-Mass molecules before imaging.

3.5.2. Peptide Reporter Analysis Using Microspotter

1. An ITO slide after washing step for frozen tissues or digestion for FFPE or frozen tissues is used.
2. On each defined spot, 20 nl of HCCA/ANI solution is applied. 5 droplets of 100 pl are deposited at each spot per cycle, then 40 iterations are necessary to obtain the total volume. For slides after digestion, the matrix is deposited with the same array than the one used for trypsin deposition. In this case matrix is deposited exactly at the same position than the trypsin.
3. Check matrix coverage using an optical microscope.
4. A rapid MS analysis on one spot is recommended to verify that a sufficient amount of matrix is deposited. Increase of iteration number may improve MSI when signal intensity appears to be low.

3.5.3. Peptide Reporter Analysis Using an Automatic Sprayer

1. An ITO slide after washing step for frozen tissue or digestion for FFPE or frozen tissues is used.

2. A method with different step of spraying, incubation, and drying phase is needed. The ImagePrep method for HCCA/ANI deposition is based on the normal HCCA method included in the ImagePrep. Optimization is required for each type of tissue. Briefly, the spray time is around 2 s (depending the surface of tissue section). An incubation time of 20 s (except for initialization phase: 10 s) allows an effective extraction of proteins. A particular attention is drawn to correctly set the drying time for complete crystallization on the tissue section. If the time is too short, the section will be too wet and a delocalization of molecules will be observed. The minimum drying time is around 120 s.
3. Check matrix coverage using an optical microscope.
4. A rapid MS analysis at one position can be performed to check out that a sufficient amount of matrix has been deposited. If not, some cycles of the last phase of deposition can be done again and may improve MSI when signals intensity is too low.

3.6. Mass Spectrometry Analysis

3.6.1. MALDI-MSI Experiment: In Linear Mode

Acquisition parameters were set to acceleration voltage, 20 kV; first grid voltage, 94%; guide-wire voltage, 0.05%; extraction delay time, 200 ns. Each spectrum was an average of 500 laser shots at 100 Hz.

3.6.2. Mass Spectrometry Analysis for Proteins MSI (For Frozen Tissue Analysis Exclusively)

1. 0.5 μl of protein calibration solution is deposited near to the tissue section and mix with 0.5 μl of HCCA /ANI solution.
2. The mass spectrometer is calibrated with the calibration spot.
3. Using FlexImaging an area of interest is selected on the tissue after definition of the teaching points.
4. The distance between each measurement point is set. Distance between measurement points is, depending on the method, used for matrix deposition:
   With Chip-1000 deposition, the spots are generally spaced by 250 μm center to center. It is possible to define the same raster than the one defined during matrix deposition. Due to the size of the spot it is possible to accumulate spectra at different position in the same spot. This increase statistics and reduce spot-to-spot variability.

With ImagePrep deposition, distance between two measurements can be chosen by the user. Generally the resolution is around 100 μm.

5. In FlexControl, the adequate methods for proteins analysis is set in positive linear mode and a total of 500 spectra are acquired at each position at a laser frequency of 100 Hz.
6. The images are saved and reconstructed using FlexImaging 2.1.

3.6.3. Mass Spectrometry Analysis for Peptides MSI

1. 0.5 μl of peptide calibration solution is deposited near to the tissue section and mixed with 0.5 μl of ANI solution.
2. The mass spectrometer is calibrated with the calibration spot.
3. Using FlexImaging an area of interest is selected on the tissue after definition of the teaching points.
4. The distance between each measurement point is set. Distance between measurement points is dependent of the method used for matrix deposition.
   - 4.1. With Chip-1000, deposition spots are generally spaced by 250 μm center to center. It is possible to define the same raster than for matrix deposition. Due to the size of the spots spectra can be accumulated at different positions in the same spot.
   - 4.2 With ImagePrep deposition, distance between two measurements is chosen by users. Generally the resolution is around 100 μm.
5. In FlexControl, the adequate methods for peptides analysis is set in positive reflector mode and a total of 500 spectra are acquired at each position at a laser frequency of 100 Hz. Although negative reflectron mode can also be used for specific class of peptides.
6. The images are saved and reconstructed using FlexImaging 2.1.

4. Notes

4.1. Photocleavable Tagged Oligonucleotide

The peptide is synthesized on Symphony (Protein Technologies Inc.) and purified on a Delta-Pak C18 15 μm 100A column (Waters). The oligonucleotide is synthesized from 3′ to 5′ on Expedite (Applied BioSystems). The amine function with photocleavable linker is added in 5′ before cleavage and deprotection. These steps are performed using a NH_4OH 28% solution during 24 h in the dark. The amino oligonucleotide is then purified on a Delta-Pak C18 15 μm 300A column (Waters). The amino function of the oligonucleotide is coupled with a heterobifunctional reagent comprising a maleimide function. The maleimido oligonucleotide is solubilized in water and added to a 1.2 equivalent of peptide in solution. The mixture is stirred for 16 h.

The oligo-peptide conjugate is then purified on a Delta-Pak C18 15 μm 300A column (Waters) and characterized by MALDI-MS (Voyager STR, Applied BioSystems).

4.2. Photocleavable Tagged Antibody

Peptides were custom made by Eurogentec S.A. using solid phase peptide synthesis (SPPS) on a 0.25 mmol (millimole) scale using Fmoc (9-fluorenylmethyloxycarbonyl amino-terminus protection) standard synthesis protocols (4 equivalents of Fmoc-AA) with double-coupling reactions (twice 40 min) using TBTU/NMM as activator on a Symphony (Rainin Instrument Co, Woburn, MA, USA) synthesizer. The photocleavable linker (4 equivalents) was introduced manually using DIPCDI/DIPEA (2 h) as activator. Purifications were performed by RP-HPLC on a Waters (Milford, MA, USA) Delta-Pak C18 (15 μm–100A–25×100 mm) column using a Waters liquid chromatography system consisting of Model 600 solvent delivery pump, a Rheodine injector, and a automated gradient controller (solvent A: H_2O-0.125% TFA; solvent B: CH3CN-0.1% TFA, gradients: 5–15 to 30–60% B in 20 min). Detection was carried out using Model M2487 variable wavelength UV detector connected to the Waters Millennium software control unit. The Quality Control was performed by analytical RP-HPLC on a Waters Delta-Pak C18 (5 μm–100A–150×3.9 mm) column (solvent A: H_2O-0.125% TFA; solvent B: CH3CN-0.1% TFA, gradient: 100% A–60% B in 20 min) using a Waters Alliance 2690 Separation Module equipped with a Waters 996 Photodiode Array Detector and by MALDI-TOF MS (Voyager STR, Applied BioSystems).

The Functionalization with the photolinker derivatized peptide A was done as follow: a solution of 0.5 mg of MBS in 300 μl of DMF is added to a solution of 4 mg of goat anti-rabbit IgG in 2 ml of PBS and mixed for 30 min. The solution is then desalted on a PD 10 column using 50 mM phosphate buffer at pH =6. To this desalted activated IgG, a solution of 1 mg of the photocleavable derivatized peptide in 300 μl of DMF and 1 ml of PBS is added and stirred for 3 h at room temperature. Afterward, the reaction mixture is dialyzed overnight against PBS (membrane cut-off 12–14,000).

In order to prepare this triphosphate, a Fmoc-protected CPG resin was required. The succinylate was prepared from GT115A (100 mg) (Scheme **20.1**). The sample was relatively pure but contained a small amount (by TLC) of a higher running non-tritylated compound (originates from the Sonogashira reaction

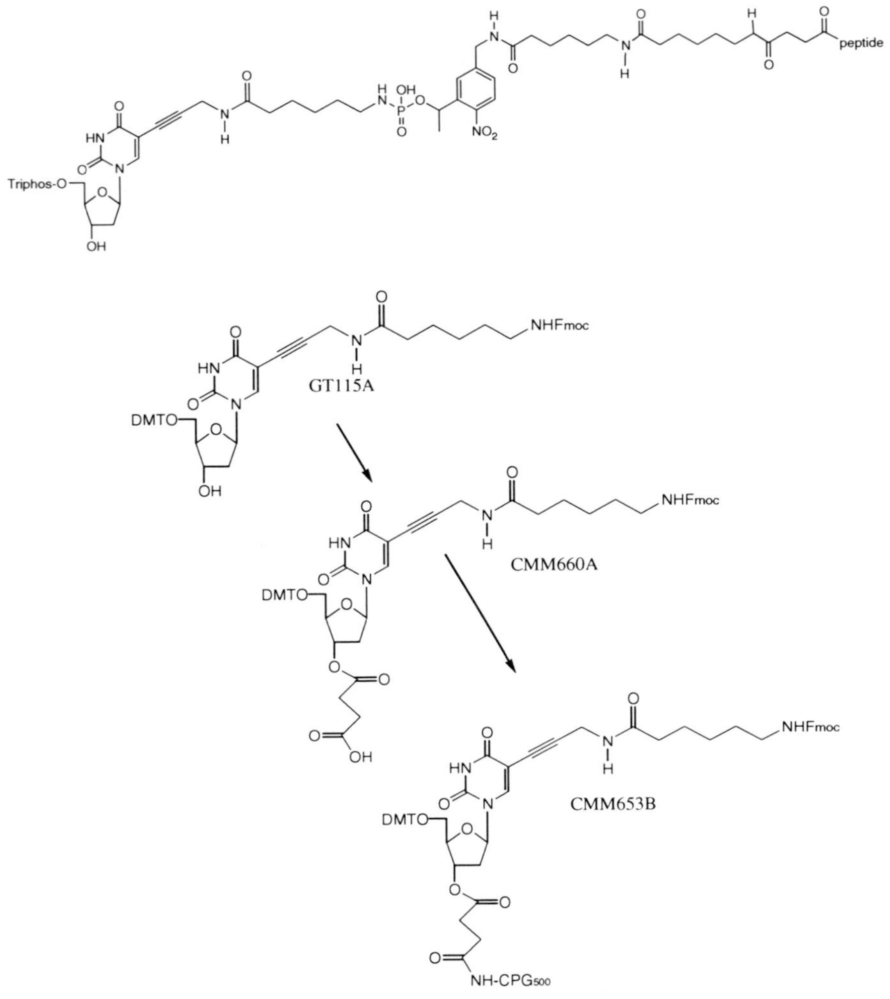

Scheme 20.1. Synthesis of a dUTP-peptide conjugates with a photocleavable linker (see text for details).

and does not interfere with subsequent reactions and was not visible in the NMR spectra of the sample). Since it was not possible to purify the succinate, the reaction was modified slightly. It is normal to add two equivalents of succinic anhydride to the reaction to get quantitative yield but if this is not removed completely, the amino residues of the cpg resin can become blocked during functionalization. Therefore, 1.5 equivalents were used since the exact purity of the product is undetermined. The reaction did not go to completion (from TLC this was more than 50%) by comparing the intensity of the components on the TLC by UV (254 nm) and the intensity of the DMT cation on treatment with HCl fumes. Since the non-succinylated product will not react, the resin was functionalized using this mixture. The resin was prepared but the loading is very low, 5.4 μ mol g^{-1} (180 mg).

The resin was detritylated using 2% TCA/DCM washed with DCM and the process repeated until no orange color due to the DMT cation was observed. This was then dried (suction under argon) and the resin soaked in pyr/DMF 1:3 (0.4 ml) for 5 min before a solution of 0.1 M Eckstein's reagent in dioxane was added (0.1 ml). The reaction was allowed to stand for 15 min after which time the resin was washed (dioxane, MeCN) and dried (suction under argon). The resin was then soaked in a solution on 0.5 M *bis*-(tributylammonium) pyrophosphate in anhydrous DMF and tri-*n*-butylamine for 20 min and the resin washed (DMF, MeCN) and dried (suction under argon). The product was oxidized (iodine/water/pyridine/THF for 30 min), washed (MeCN), and dried (suction under argon). The Fmoc protecting group was removed (20% piperidine/DMF, 0.5 ml, 20 min) and the resin washed thoroughly (DMF, MeCN) and dried (suction under argon). This was then washed with DCI and a solution of DCI/photolabile amino linker CEP (1:1, 0.5 ml) was added and the reaction was allowed to stand for 20 min. The solution was removed and the resin washed (MeCN) and dried (suction under argon). A mixture of cap A/cap B (1:1, 0.5 ml) was added and the resin soaked for 5 min before removing the capping reagents and washing and drying the resin as before. The product was oxidized (I_2/THF/pyr/H_2O, 5 min) and the resin washed and dried as before. This was cleaved from the resin with cNH_4OH at room temperature for 30 min, then purified by anion exchange HPLC on a Dionex NucleoPac100 HPLC column using the following solvent system Buffer A:0.1 M NH_4Cl with 10% acetonitrile; Buffer B: 1 M NH_4Cl with 10% acetonitrile; flow rate 2.5 ml /min. using 6Triphos.mth. This gave three fractions (A:–7 min, B:–7.9 min, and C:–10.3 min). All three fractions were lyophilized overnight before being desalted by reverse phase HPLC Buffer A: Water; Buffer B: acetonitrile; flow rate 4 ml/min. The three fractions were again lyophilized overnight before being suspended in 200 μl of water. MS showed that CMM661A pk 1 was definitely not the triphosphate but it could be either CMM661pk 2 or 3 (very similar MS profiles). (CMM662A was formed from CMM661A pk 2 and CMM663A was formed from CMM661A pk 3). Both samples were then used in the subsequent reaction. Bicarbonate buffer (10 μl) and the maleimide NHS ester (50 μl) were added to each sample and the reactions agitated overnight. The samples were diluted with milliQ water (500 μl) and filtered. The samples were purified by RP-HPLC, buffer A: 0.1 M TEAA, buffer B: MeCN, flow rate 4 ml /min. using MeCN50.mth and the coupling of the peptide was carried out on these fractions.

The use of a cryopreservative solution containing polymer compounds such as a solution with an optimal cutting temperature (OCT) polymer should be restricted to the attachment of the tissue to the sample holder and not for wholly embedded the tissue. Moreover, all parts of the cryostat in contact with the tissue need to be cleaned to prevent any contamination between two different samples or with a polymer contain solution. In the case of contact between the tissue and a polymer containing cryopreservative solution, MS spectra will be dominated by polymer signals.

Acknowledgments

This study was supported by grants from the Centre National de la Recherche Scientifique (DPI), Ministère de L'Enseignement Supérieur et de la Recherche, the Agence Nationale de la Recherche PCV (To IF)

References

1. Caprioli, R. M. (2008) Perspectives on imaging mass spectrometry in biology and medicine. *Proteomics*, **8**, 3679–3680.
2. Cornett, D. S., Reyzer, M. L., Chaurand, P., Caprioli, R. M. (2007) MALDI imaging mass spectrometry: molecular snapshots of biochemical systems. Nat Methods **4**, 828–833.
3. Franck, J., Arafah, K., Elayed, M., Bonnel, D., Vergara, D., Jacquet, A., Vinatier, D., Wisztorski, M., Day, R., Fournier, I., Salzet, M. (2009) MALDI imaging: state of the art technology in clinical proteomics. *Mol Cell Proteomics*, **8**, 2023–2033.
4. Murphy, R. C., Hankin, J. A., Barkley, R. M. (2009) Imaging of lipid species by MALDI mass spectrometry. *J Lipid Res*, **50 Suppl**, S317–S322.
5. Grey, A. C., Chaurand, P., Caprioli, R. M., Schey, K. L. (2009) MALDI imaging mass spectrometry of integral membrane proteins from ocular lens and retinal tissue (dagger). *J Proteome Res*, **8**, 3278–3283.
6. Dekker, L. J., van Kampen, J. J., Reedijk, M. L., Burgers, P. C., Gruters, R. A., Osterhaus, A. D., Luider, T. M. (2009) A mass spectrometry based imaging method developed for the intracellular detection of HIV protease inhibitors. *Rapid Commun Mass Spectrom*, **23**, 1183–1188.
7. Wolthuis, R., Travo, A., Nicolet, C., Neuville, A., Gaub, M. P., Guenot, D., Ly, E., Manfait, M., Jeannesson, P., Piot, O. (2008) IR spectral imaging for histopathological characterization of xenografted human colon carcinomas. *Anal Chem*, **80**, 8461–8469.
8. Wisztorski, M., Croix, D., Macagno, E., Fournier, I., Salzet, M. (2008) Molecular MALDI imaging: an emerging technology for neuroscience studies. *Dev Neurobiol*, **68**, 845–858.
9. Walch, A., Rauser, S., Deininger, S. O., Hofler, H. (2008) MALDI imaging mass spectrometry for direct tissue analysis: a new frontier for molecular histology. *Histochem Cell Biol*, **130**, 421–434.
10. Stauber, J., Lemaire, R., Franck, J., Bonnel, D., Croix, D., Day, R., Wisztorski, M., Fournier, I., Salzet, M. (2008) MALDI imaging of formalin-fixed paraffin-embedded tissues: application to model animals of Parkinson disease for biomarker hunting. *J Proteome Res*, **7**, 969–978.
11. Fournier, I., Wisztorski, M., Salzet, M. (2008) Tissue imaging using MALDI-MS: a new frontier of histopathology proteomics. *Expert Rev Proteomics*, **5**, 413–424.
12. Chaurand, P., Rahman, M. A., Hunt, T., Mobley, J. A., Gu, G., Latham, J. C., Caprioli, R. M., Kasper, S. (2008) Monitoring mouse prostate development by profiling and imaging mass spectrometry. *Mol Cell Proteomics* **7**, 411–423.
13. Burnum, K. E., Tranguch, S., Mi, D., Daikoku, T., Dey, S. K., Caprioli, R. M. (2008) Imaging mass spectrometry reveals unique protein profiles during

embryo implantation. Endocrinology, **149**, 3274–3278.

14. Lemaire, R., Menguellet, S. A., Stauber, J., Marchaudon, V., Lucot, J. P., Collinet, P., Farine, M. O., Vinatier, D., Day, R., Ducoroy, P., Salzet, M., Fournier, I. (2007) Specific MALDI imaging and profiling for biomarker hunting and validation: fragment of the 11S proteasome activator complex, Reg alpha fragment, is a new potential ovary cancer biomarker. *J Proteome Res*, **6**, 4127–4134.
15. Nordhoff, E., Kirpekar, F., Karas, M., Cramer, R., Hahner, S., Hillenkamp, F., Kristiansen, K., Roepstroff, P., Lezius, A. (1994) Comparison of IR- and UV-matrix-assisted laser desorption/ionization mass spectrometry of oligodeoxynucleotides. *Nucleic Acids Res*, **22**, 2460–2465.
16. Lemaire, R., Stauber, J., Wisztorski, M., Van Camp, C., Desmons, A., Deschamps, M., Proess, G., Rudlof, I., Woods, A. S., Day, R., Salzet, M., Fournier, I. (2007) Tag-mass: specific molecular imaging of transcriptome and proteome by mass spectrometry based on photocleavable tag. *J Proteome Res*, **6**, 2057–2067.
17. Stauber, J., Lemaire, R., Wisztorski, M., Ait-Menguellet, S., Lucot, J.P., Vinatier, D., Desmons, A., Deschamps, M., Proess, G., Rudolf, I., Salzet, M., Fournier, I. (2006) New developments in MALDI imaging mass spectrometry for pathological proteomic studies; introduction to a novel concept, the specific MALDI imaging. *Mol Cell Proteomics*, **5**, S247–S49.
18. Thiery, G., Anselmi, E., Audebourg, A., Darii, E., Abarbri, M., Terris, B., Tabet, J. C., Gut, I. G. (2008) Improvements of TArgeted multiplex mass spectrometry imaging. Proteomics, **8**, 3725–3734.
19. Thiery, G., Shchepinov, M. S., Southern, E. M., Audebourg, A., Audard, V., Terris, B., Gut, I. G. (2007) Multiplex target protein imaging in tissue sections by mass spectrometry–TAMSIM. *Rapid Commun Mass Spectrom*, **21**, 823–829.
20. Brunelle, A., Laprevote, O. (2009) Lipid imaging with cluster time-of-flight secondary ion mass spectrometry. *Anal Bioanal Chem*, **393**, 31–35.
21. Touboul, D., Halgand, F., Brunelle, A., Kersting, R., Tallarek, E., Hagenhoff, B., Laprevote, O. (2004) Tissue molecular ion imaging by gold cluster ion bombardment. *Anal Chem*, **76**, 1550–1559.
22. Becker, J. S., Zoriy, M., Matusch, A., Wu, B., Salber, D., Palm, C. (2009) Bioimaging of metals by laser ablation inductively coupled plasma mass spectrometry (LA-ICP-MS). *Mass Spectrom Rev*, **29**, 156–175
23. Becker, J. S., Dobrowolska, J., Zoriy, M., Matusch, A. (2008) Imaging of uranium on rat brain sections using laser ablation inductively coupled plasma mass spectrometry: a new tool for the study of critical substructures affined to heavy metals in tissues. *Rapid Commun Mass Spectrom*, **22**, 2768–2772.
24. Becker, J. S., Zoriy, M. V., Dobrowolska, J., Matucsh, A. (2007) Imaging mass spectrometry in biological tissues by laser ablation inductively coupled plasma mass spectrometry. *Eur J Mass Spectrom (Chichester, Eng)*, **13**, 1–6.

Chapter 21

Structurally Selective Imaging Mass Spectrometry by Imaging Ion Mobility-Mass Spectrometry

John A. McLean, Larissa S. Fenn, and Jeffrey R. Enders

Abstract

This chapter describes the utility of structurally based separations combined with imaging mass spectrometry (MS) by ion mobility-MS (IM-MS) approaches. The unique capabilities of combining rapid (μs-ms) IM separations with imaging MS are detailed for an audience ranging from new to potential practitioners in IM-MS technology. Importantly, imaging IM-MS provides the ability to rapidly separate and elucidate various types of endogenous and exogenous biomolecules (e.g., nucleotides, carbohydrates, peptides, and lipids), including isobaric species. Drift tube and traveling wave IM-MS instrumentation are described and specific protocols are presented for calculating ion–neutral collision cross sections (i.e., apparent ion surface area or structure) from experimentally obtained IM-MS data. Special emphasis is placed on the use of imaging IM-MS for the analysis of samples in life sciences research (e.g., thin tissue sections), including selective imaging for peptide/protein and lipid distributions. Future directions for rapid and multiplexed imaging IM-MS/MS are detailed.

Key words: Ion mobility, ion mobility-mass spectrometry, IM-MS, imaging mass spectrometry, IMS, MSI, imaging ion mobility-mass spectrometry, structural separations, MALDI, IM-MS/MS.

1. Introduction

One of the recent advances in mass spectrometry (MS) instrumentation is the incorporation of post-ionization separations on the basis of ion mobility (IM) combined with subsequent MS analysis (IM-MS). Importantly, IM-MS adds an additional dimension of separations on the basis of analyte structure to facilitate interpretation of MS spectra directly from complex biological samples. Typically separations in the IM dimension are completed in 100 s of microseconds to milliseconds, thus imaging

S.S. Rubakhin, J.V. Sweedler (eds.), *Mass Spectrometry Imaging*, Methods in Molecular Biology 656,
DOI 10.1007/978-1-60761-746-4_21,

matrix-assisted laser desorption/ionization (MALDI)-IM-MS is performed on the same timescale as contemporary imaging MALDI-MS experiments. In practice, the combination of imaging IM-MS can be thought of as rapid gas-phase electrophoresis at each spatial dimension yielding a 5D data set (i.e., mobility, *m/z*, relative abundance, and *x*, *y* spatial coordinates). This information can be supplemented by performing fragmentation studies at each spatial position in imaging IM-MS/MS experiments (**Section 1.4.1**). Furthermore, IM gas-phase separations yield direct structural information (ion–neutral collision cross sections, or apparent ion surface area (Å^2)) for analytes that can be interpreted by complementary molecular simulations using either ab initio and/or molecular dynamic techniques.

The main focus of this chapter is to illustrate the advantages of merging IM-MS with imaging MALDI-MS, which is aimed primarily at new or potential users of imaging MALDI-IM-MS. The ability to separate molecules based on collision cross section allows for the simplification and validation of complex spectra, such as those commonly encountered in biological imaging MS. The recent commercial availability of IM-MS instruments should ultimately result in this technology becoming a staple in many structural MS and biological laboratories. In order to avoid repetition with other works in this edition, this chapter focuses mainly on the theory of IM-MS separations and the benefits of using this technique for biological imaging experiments. The selected examples described and illustrated herein are intended to underscore the advantages and limitations of imaging MALDI-IM-MS rather than being a comprehensive review of IM-MS research. The reader is directed to several recent reviews for a more detailed description of IM-MS (1–5).

1.1. Ion Mobility Applications to the Life Sciences

Although gas-phase IM separations have existed for well over a century (6) and coupling IM–MS has existed since the early 1960s (7, 8), the utility of IM-MS for biomolecular separations was not fully realized until combined with soft ionization techniques, such as electrospray ionization (ESI) and MALDI (9, 10). The first applications of IM-MS to determine peptide and protein structures were performed in the late 1990s (11–13). Subsequent to these pioneering studies, research over the past decade has extended IM-MS techniques to the study of complex biological samples, such as whole cell lysates (14), plasma (15–18), homogenized tissue (14, 19, 20), non-covalent complexes (21–23), or directly from thin tissue sections (24, 25). However, until very recently, IM-MS was essentially available in only a limited number of laboratories where custom instruments were constructed. The recent introduction of commercially available IM-MS instrumentation, in several forms, has further fueled the integration of IM-MS techniques into life sciences research programs. The following sections describe the theory of

IM separations (**Section 1.2**), an overview of IM-MS instrumentation (**Section 1.3**), and data interpretation in IM-MS conformation space (**Section 1.4**). Materials, methods, and protocols for performing imaging IM-MS of complex biological samples are then detailed (**Sections 2** and **3**).

1.2. Overview and Theory of Ion Mobility Separations

Ion mobility-mass spectrometers are composed of an ion source, a mobility separation cell, a mass analyzer, and a detector as depicted in **Fig. 21.1a**. There are many variations to the general

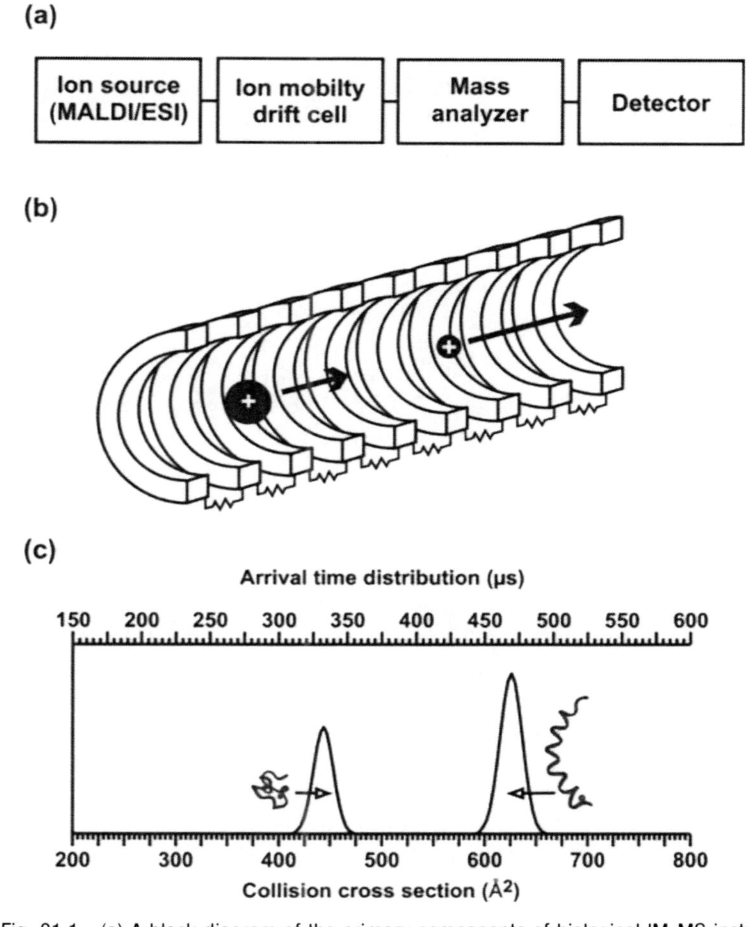

Fig. 21.1. (**a**) A block diagram of the primary components of biological IM-MS instrumentation. (**b**) A conceptual depiction of an IM drift cell. A stack of ring electrodes are connected via resistors in series to form a voltage divider, which is typically designed to generate a relatively uniform electrostatic field along the axis of ion propagation. Ions of larger apparent surface area experience more collisions with the neutral drift gas and therefore elute slower than ions of smaller apparent surface area. (**c**) A hypothetical IM separation for peptide ions exhibiting two distinct structural sub-populations corresponding to globular (*left*) and to helical (*right*) conformations. The arrival time distribution data (*top axis*), or what is measured, can be transformed to a collision cross-section profile (*bottom axis*) via equation [4] and described in **Section 3.1**. Adapted with kind permission from Springer Science+Business Media (1) Fig. 1.

design such as different ion sources (i.e. ESI, MALDI) and types of ion mobility separation cells used (i.e., whether the ions are dispersed in space or time). For imaging IM-MS applications typically time-of-flight (TOF) mass analyzers are used for timescale considerations as described below (*see* **Section 1.3**). This chapter focuses on temporal ion dispersion through the use of drift tube or traveling wave ion mobility (DTIM and TWIM, respectively). In contrast with high-energy ion–neutral gas-phase collisions used in collision-induced dissociation (CID), both DTIM and TWIM separations utilize low-energy gas-phase collisions to separate ions on the basis of predominantly molecular surface areas. Briefly, ions are injected into a drift tube filled with a neutral drift gas and migrate under the influence of a weak electrostatic field gradient (**Fig. 21.1b**). Larger ions have a lower mobility than smaller ions which result in longer drift times versus shorter drift times, respectively. This field is electrostatic for drift tube and electrodynamic for traveling wave separations, respectively. The migration of these ions is impeded by collisions with the neutral drift gas to a degree that is proportional to apparent surface area or collision cross section. Although the experimental parameter obtained from IM separations is the ion arrival time distribution (t_{atd}) or the time between ion injection and ion detection, it can be converted to collision cross section or apparent surface area as illustrated in **Fig. 21.1c**. The following description details how this conversion is performed based on the kinetic theory of gases for drift tube separations. For a derivation of ion–neutral collision cross-section theory, the reader is directed to several excellent texts and reviews (26–28). Procedures to estimate collision cross section using traveling wave IM are described elsewhere (29, 30). This section summarizes several of the key equations and practical considerations for determining ion–neutral collision cross sections in uniform electrostatic field IM instrumentation. Further details for experimental implementation will follow in the methods section.

1.2.1. Transforming Drift Time to Collision Cross Section

The separation of ions in a weak electrostatic field (E) is measured as the ion drift velocity (v_{d}) and is related by the proportionality constant, K, which is the mobility of the ion in a particular neutral gas:

$$v_{\text{d}} = KE \qquad [1]$$

The drift cell is of a fixed length (L), and the velocity of the ion packet is determined by measuring the drift time (t_{d}) of the packet across the drift cell. In evaluating K, the drift velocity of the ion packet depends not only on the electrostatic field strength but also on the pressure (p, torr) of the neutral drift gas and the temperature (T, kelvin) of separation. Therefore, it is conventional practice to report K as the standard or reduced mobility

(K_0), which normalizes the results to standard temperature and pressure conditions (i.e., 0°C and 760 torr):

$$K_0 = K \frac{p}{760} \frac{273}{T} \quad [2]$$

For applications where IM is used to obtain structural information about the ion, such as those in structural proteomics and biophysics, the IM separations are performed using weak electrostatic fields (*ca.* 20–30 V cm^{-1} $torr^{-1}$). Provided the field strength is sufficiently weak, or under so-called low-field conditions, a closed equation for the ion–neutral collision cross section can be expressed from the kinetic theory of gases (*see* **Note 1**). When the IM separations are performed in low-field conditions, i.e., constant K, the mobility is related to the collision cross section of the ion–neutral pair:

$$K_0 = \frac{(18\pi)^{1/2}}{16} \frac{ze}{(k_B T)^{1/2}} \left[\frac{1}{m_i} + \frac{1}{m_n}\right]^{1/2} \frac{760}{p} \frac{T}{273} \frac{1}{N_0} \frac{1}{\Omega} \quad [3]$$

where these parameters include the charge of the ion (ze), the number density of the drift gas at STP (N_0, 2.69×10^{19} cm^{-3}), the reduced mass of the ion–neutral collision pair (ion and neutral masses of m_i and m_n, respectively), Boltzmann's constant (k_B), and the ion–neutral collision cross section (Ω). Inspection of equation [3] shows that the mobility of an ion is inversely related to its collision cross section or apparent surface area. Substituting for K_0 in equation [3] and rearranging to solve for the collision cross section yields:

$$\Omega = \frac{(18\pi)^{1/2}}{16} \frac{ze}{(k_B T)^{1/2}} \left[\frac{1}{m_i} + \frac{1}{m_n}\right]^{1/2} \frac{t_d E}{L} \frac{760}{p} \frac{T}{273} \frac{1}{N_0} \quad [4]$$

which is the typical functional form of the equation used to solve for collision cross sections from IM data (*see* **Section 3**). Conceptually, the ion–neutral collision cross section can be thought of as the radius of the orientationally averaged projection of the ion in combination with the drift gas, i.e., $\Omega = \pi (r_i + r_{He})^2$, as depicted in **Fig. 21.2** (27).

In both equations [3] and [4], the collisions of ions with neutrals are considered to be a completely elastic process. Thus, the collision cross section obtained is termed the hard-sphere collision cross section. When compared to molecular simulations, these collision cross section measurements can provide detailed structural information about the analyte (31–34).

1.3. Instrumentation Overview

The two time-dispersive methods of IM separation are DTIM and TWIM. Drift tube IM facilitates absolute collision cross section

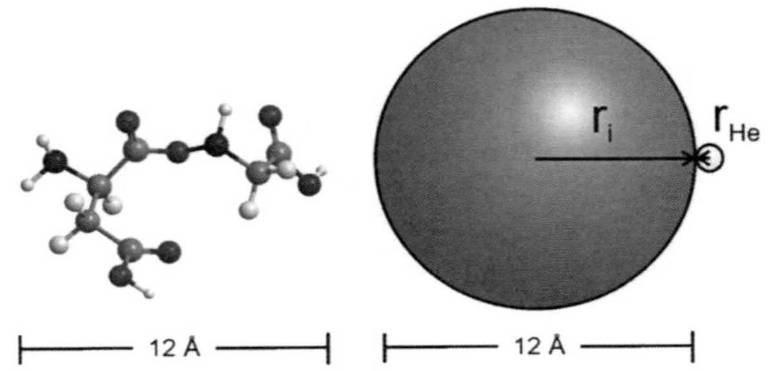

Fig. 21.2. Visual representation of the ion–neutral collision cross section measured using DTIM strategies. The radii of the ion (r_i) and the drift gas helium atom (r_{He}) can be used to approximate the collision cross section (Ω) using the simplified equation $\Omega = \pi(r_i + r_{He})^2$ (27).

calculations (35–38). This data can then be compared to molecular simulation results to interpret analyte structural and conformational details. Traveling wave IM utilizes electrodynamic fields, which only provides estimated collision cross sections when measurements are compared to internal standards with previously measured DTIM absolute collision cross sections (29, 30). This is because gas-phase theory is insufficiently developed for the fundamental physical processes in TWIM separations, although recent efforts in this regard have been reported (39). Nevertheless, both DTIM and TWIM instrumentations are increasingly used for imaging IM-MS applications.

1.3.1. Drift Tube Ion Mobility

The first ion mobility instruments were developed on a drift tube design (40). As described in **Section 1.2**, DTIM-MS instruments are conceptually analogous to typical imaging MALDI-MS instruments with the exception of inserting an IM drift cell between the MALDI source and mass analyzer. Specifically for imaging applications it is important to consider the timescales of separation, which are largely affected by drift cell pressure and length. The fundamental limit to throughput in imaging applications is dependent on the slowest component, which in this case is either the duration of IM separation or the pulse-to-pulse repetition rate of the MALDI laser. Although atmospheric pressure drift cells and reduced pressure (1–10 torr) drift cells have been constructed for IM-MS (41) the latter are much better suited for imaging applications. For example, atmospheric pressure drift cells typically separate ions on a timescale of 10 s of ms (42), whereas reduced pressure drift cells can provide separations in <1 ms for relatively short drift cells (*ca.* 10 cm (2)). This is significant because it dictates the fastest rate at which the MALDI laser can be operated without losing time correlation in signals arising from different MALDI events, i.e., 10 Hz for 100 ms versus 1,000 Hz for 1 ms drift times, respectively (43).Thus by having short drift times one can perform imaging IM-MS without compromising imaging rates in traditional imaging MALDI-MS experiments. MALDI repetition rates using DTIM in imaging IM-MS have ranged from 300 to 1,000 Hz (24, 25, 44). Even with fast separation times, IM resolution typically ranges from 30 to 50 ($r=t/\Delta t$ at FWHM), whereas longer, cryogenically cooled, or higher pressure drift tubes have been reported with resolutions exceeding 100 (45–47). Furthermore, drift tube instruments can also be equipped for imaging IM-MS/MS experiments as described in **Section 1.4.1**.

1.3.2. Traveling Wave Ion Mobility

The recent commercial availability of traveling wave ion mobility (TWIM) instrumentation (Waters Corp.) has made imaging IM-MS accessible to a large number of users. Similar to drift tube instruments, TWIM separates ions by time dispersion through collisions with a background buffer gas, but in contrast, it uses electrodynamic fields rather than electrostatic fields (1, 48). This is accomplished by transmitting voltage pulses sequentially across a stack of ring electrodes (similar to **Fig. 21.1b**), which creates the so-called traveling wave (49). Conceptually, TWIM separations are performed based on the susceptibility of different ions to the influence of the specific wave characteristics and have been described as the ability of ions to "surf" on waves (48). Adjustable wave parameters include traveling wave pulse height, wave velocity, and ramping either of these variables. The commercial platform (Synapt HDMS) is comprised of a MALDI (200-Hz pulse repetition rate) or ESI source, a mass-resolving quadrupole, a trapping region for injecting pulses of ions into the TWIM, the TWIM drift cell, an ion transfer region, and an orthogonal TOFMS ($r=m/\Delta m$ at FWHM of >17,500). As described in **Section 1.4.1**, CID can be performed in the regions before and after the TWIM drift cell (*see* **Note 2**). Generally resolution in the TWIM is <15, but this is sufficient for the separation of many molecular classes of interest (e.g., isobaric peptides from lipids) in the imaging community (*see* **Section 1.4**). For example TWIM has been used to separate biomolecular signals from complex samples (50) and to study the structure of peptides following CID in the trapping region (51). Although protocols have been proposed to approximate collision cross-section values using TWIM experimental data, the calculations still rely on absolute values obtained using drift tube instruments (29, 30).

1.4. Data Interpretation in Conformation Space

Typical data for a 2D IM-MS separation are presented in **Fig. 21.3** for the separation of lipids and oligonucleotides. Conformation space data (**Fig. 21.3a**) are termed such because they represent biomolecular structure, or conformation, as a function of m/z (*see* **Note 3**). An integrated mass spectrum over all arrival time distributions is shown in **Fig. 21.3b**, which is what would be observed in the absence of IM. An integrated IM arrival time

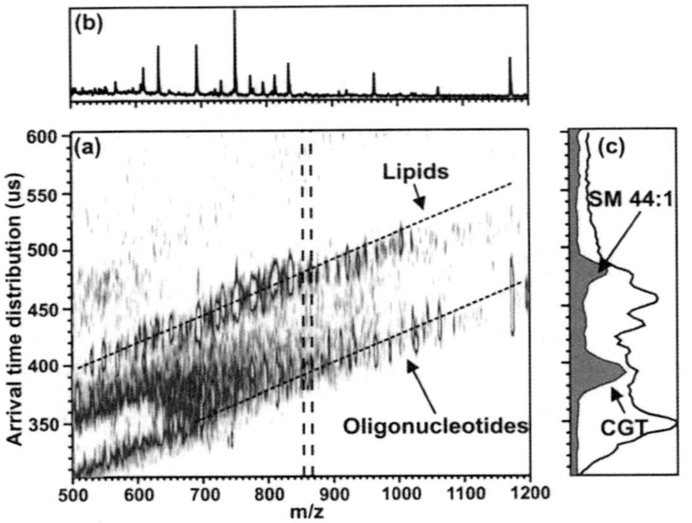

Fig. 21.3. (**a**) A 2D IM-MS conformation space plot for several lipid and oligonucleotide standards. This data illustrate the variation of gas-phase packing efficiencies for different types of biomolecules. (**b**) An integrated mass spectrum over all arrival time distributions. (**c**) The *white curve* represents the integrated arrival time distribution over the mass range of 500–1,200 *m/z*, whereas the gray curve represents an extracted arrival time distribution over the mass range of 860–870 *m/z*. The latter illustrates baseline resolution of two peaks with similar masses but different mobilities (lipid-[sphingomyelin 44:1+Na]$^+$, $m/z = 866.7$ Da; oligonucleotide-[CGT+H]$^+$, $m/z = 860.2$ Da).

distribution is illustrated by the white curve of **Fig. 21.3c** which would be obtained by placing the detector directly after the IM drift cell. By plotting the data in 2D conformation space two distinct correlations are observed, one for lipids and one for oligonucleotides, respectively. Note that either extracted mass spectra or arrival time distributions can be derived from conformation space data. For example, an extracted arrival time distribution over the *m/z* range of 860–870 is represented by the gray curve in **Fig. 21.3c**. The latter illustrates baseline resolution for a sodium-coordinated lipid (sphingomyelin 44:1, $m/z = 866.7$ Da) and a protonated oligonucleotide (CGT, $m/z = 860.2$ Da) of nearly the same *m/z*.

One of the main challenges in the analysis of complex biological samples, such as those encountered in biomolecular imaging MS, is the large diversity of molecular species with the high probability for isobaric molecules at a particular *m/z*. Structural separations on the basis of ion mobility allow isobaric species belonging to different biomolecular classes to be easily resolved (**Fig. 21.4**), resulting in more confident peak assignments. Although biomolecules are generally composed of a limited combination of elements (e.g., C, H, O, N, P, and S), different biomolecular classes preferentially adopt structures at a given *m/z* correspondent to the prevailing intermolecular

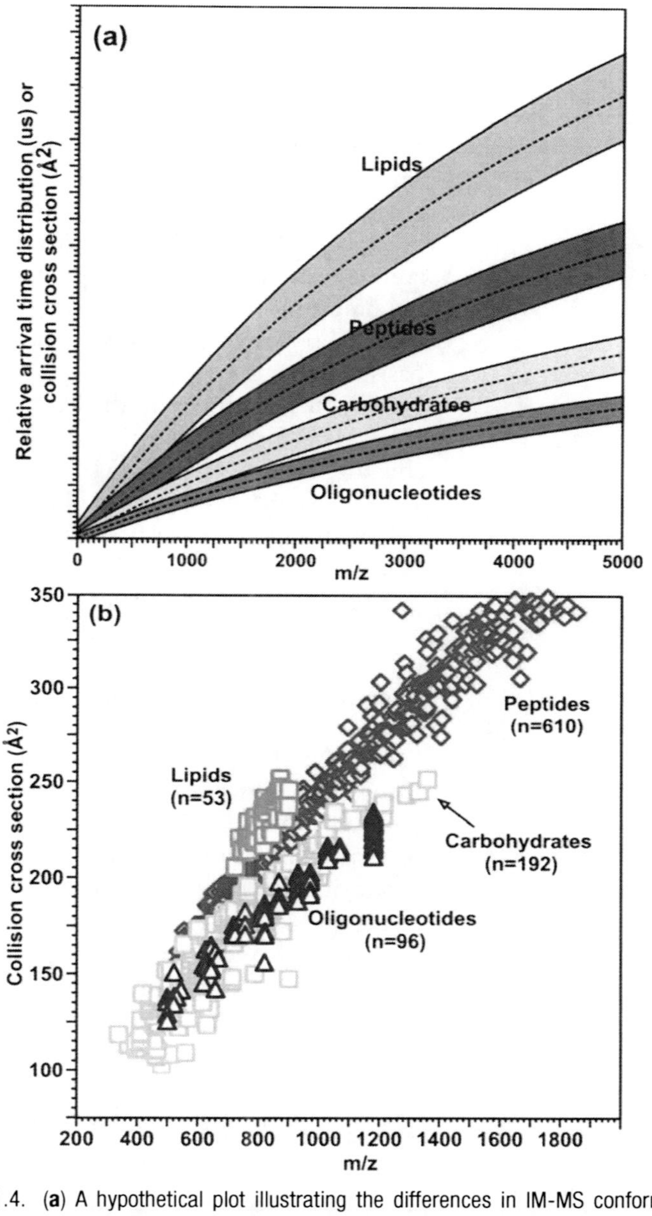

Fig. 21.4. (**a**) A hypothetical plot illustrating the differences in IM-MS conformation space for different molecular classes based on different gas-phase packing efficiencies. (**b**) A plot of collision cross section as a function of *m/z* for different biologically relevant molecular classes, including oligonucleotides ($n = 96$), carbohydrates ($n = 192$), peptides ($n = 610$), and lipids (53). All species correspond to singly charged ions generated by using MALDI, where error $\pm 1\sigma$ is generally within the data point. Values for peptides species are from (35). (**a**) is adapted with kind permission from Springer Science+Business Media: (1) 2a. (**b**) is adapted with kind permission from Springer Science+Business Media: (59), 1a.

folding forces for that class. A hypothetical plot delineating regions of conformation space for which different biomolecular classes (e.g., nucleotides, carbohydrates, peptides, lipids) are predicted to occur is presented in **Fig. 21.4a**. This occurs as a result of different average densities for different molecular classes (nucleotides>carbohydrates>peptides>lipids (1)). Although more pronounced for an *m/z* range over 2,000, at lower ranges as that depicted in **Fig. 21.4b** separations are still feasible but the correlations begin to overlap. Nevertheless, imaging IM-MS allows for obtaining more informative images on the basis of structure and *m/z* whereby isobaric chemical noise is selectively rejected on the basis of structure.

The first examples in the literature of combining imaging IM-MS was performed using DTIM and demonstrated two important advantages over imaging MS strategies, namely (i) the ability to separate isobaric species on the basis of structure and *m/z* and (ii) enhanced signal-to-noise ratios by the separation of chemical noise (24, 25). The former is demonstrated through the selective differentiation of nominally isobaric peptide and lipid species as illustrated in **Fig. 21.5a**. In proof-of-concept experiments, both the lipid and the peptide were spotted onto a thin tissue section of mouse liver (12 μm) using a reagent spotter in either a "\" direction for the peptide or a "/" direction for the lipid (**Fig. 21.5a, top left**). In the 2D IM-MS plot (**Fig. 21.5a, top right**), the signal for the protonated forms of the lipid, phosphotidylcholine 34:2, and the peptide, RPPGFSP, overlaps in the *m/z* spectrum, but is baseline resolved in the IM arrival time distribution with the peptide and lipid centered at 449 and 504 μs, respectively. In **Fig. 21.5a bottom**, the × represents what would be obtained using conventional imaging MS in the absence of IM, which is a convolution of both the peptide and lipid signals. The right two images are for the same 1 Da mass range (759–760 *m/z*), but selectively for structures corresponding to putative peptides "\" and lipids "/," respectively (25). Thus, structurally selective imaging on the basis of molecular class results in more

Fig. 21.5. (**a**) Imaging DTIM-MS of a nominally isobaric peptide (RPPGFSP) and lipid (PC 34:2) deposited onto a mouse liver thin tissue section (12 μm) in the pattern of an "×". The "\" line is RPPGFSP, while the "/" line is a phosphotidylcholine extract, respectively (*top left*). An optical image of the patterned matrix/analyte spots deposited on the tissue section (*top right*). A zoomed view in the region of PC 34:2 and RPPGFSP for a representative 2D IM-MS conformation space plot of a mixture of the two analytes. IM-MS signal intensity is indicated by false coloring, where *purple* and *yellow* corresponds to the least and most intense signals, respectively. (bottom left) An extracted ion intensity map over the mass range of 759–760 Da representing what would be obtained using conventional imaging MALDI-MS. (*bottom*

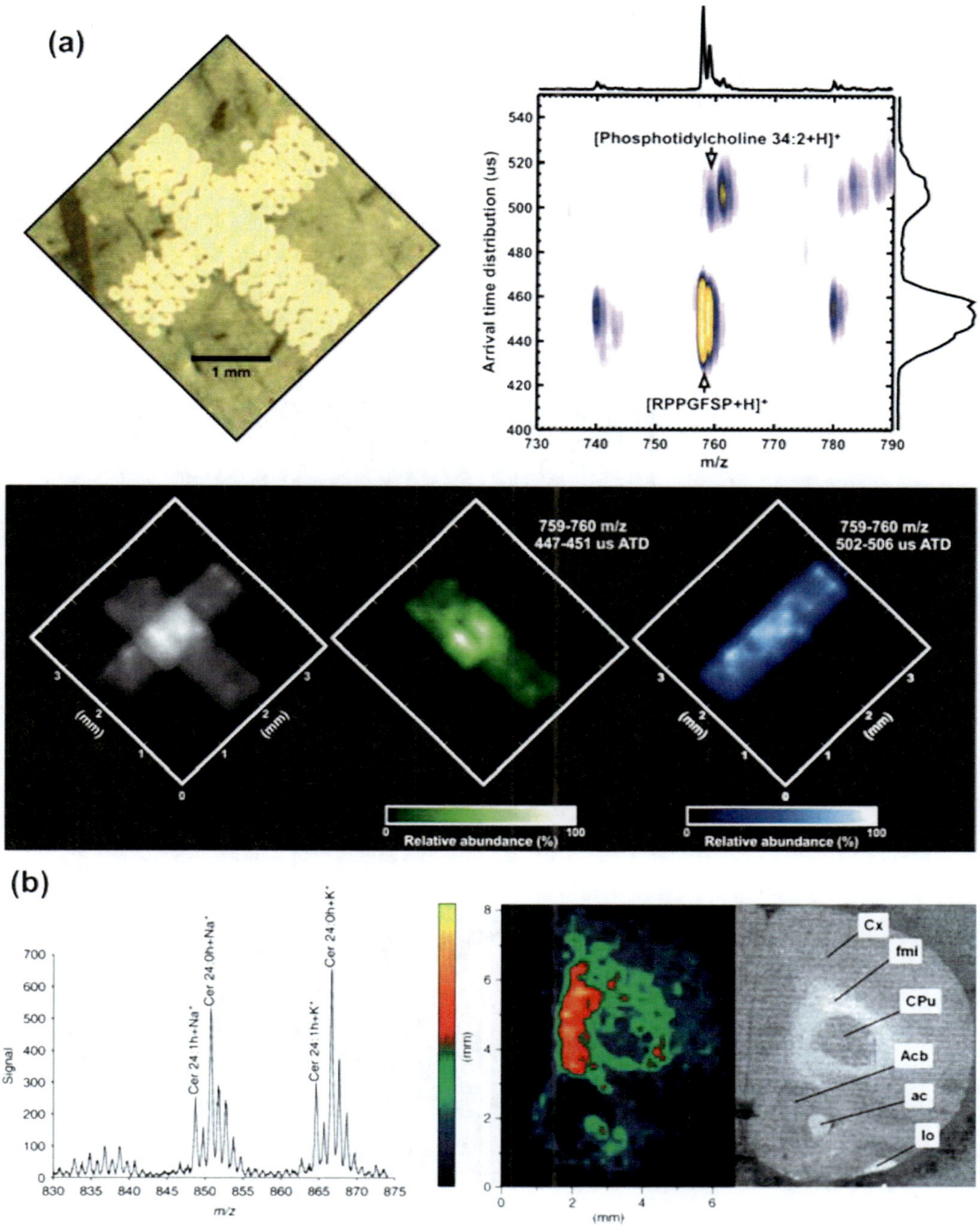

Fig. 21.5. (continued) *middle* and *right*) Extracted ion intensity maps for imaging DTIM-MS of the peptide and lipid over the mass range of 759–760 Da and DTIM drift times of 447–451 μs and 502–506 μs, respectively. (**b**, *left*) An integrated mass spectrum of cerebrosides directly from rat brain tissue. (**b**, *right*) Imaging DTIM-MS of the sodium-coordinated cerebroside 24:0 OH ($m/z = 850.7$). (**b**, *far right*) An optical image of an adjacent rat brain section. Histological abbreviations are Cx – cortex; fmi – forceps minor of the corpus callosum; CPu – caudate putamen (striatum); Acb – nucleus accumbens; ac – anterior commissure; and lo – lateral olfactory tract. Figures (**a**) and (**b**) are adapted from (25) and (24), respectively, with permission. Copyright© 2007 Wiley-Liss, Inc.

accurate images in contrast with conventional imaging MS alone. In **Fig. 21.5b**, imaging IM-MS is demonstrated for a coronal rat brain section (16 μm) where the image to the right corresponds structurally to lipids and specifically to the sodium-coordinated cerebroside 24:0 OH ($m/z = 850.7$ Da). This results in enhanced signal to noise for species of interest through the separation of chemical noise and contaminants with IM (24).

More recently, imaging IM-MS using a TWIM drift cell has been demonstrated (**Fig. 21.6**). Although the IM resolution is more limited using TWIM, it is sufficient for the separation of lipids from peptides as illustrated for a coronal thin tissue rat brain

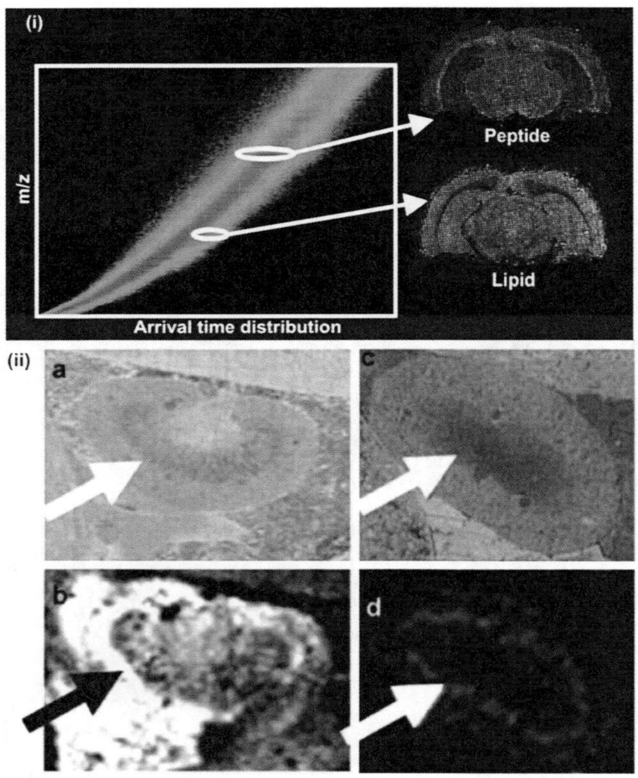

Fig. 21.6. (i) Imaging TWIM-MS data of a rat brain thin tissue section illustrating selective imaging of peptides and lipids on the basis of structure. (ii) Imaging TWIM-MS data obtained in the analysis of a small drug molecule, common name vinblastine, in thin tissue kidney sections of vinblastine-dosed rats. (**a**) An optical image of the kidney from the whole body section dosed at 6 mg kg^{-1} IV vinblastine before matrix application. (**b**) The same tissue section as shown in (**a**) but imaged by TWIM-MS showing the distribution of vinblastine within the kidney, with the highest intensity (*white*) showing a broken ring of intensity between the cortex and the medulla. (**c**) Optical image of the kidney within the whole body section dosed with ^{3}H vinblastine. (**d**) Whole body autoradiography of the tissue section shown in (**c**) indicated is the broken ring of slightly higher intensity (*white*) between the cortex and the medulla. Reproduced with permission from (52). Copyright 2008 American Chemical Society.

section (**Fig. 21.6i**, *see* **Note 4**). The utility of imaging TWIM-MS for mapping the distribution of a drug, vinblastine, in a kidney from a whole body section is shown in Fig. **21.6ii, b**. The accuracy of the obtained image is increased through the addition of TWIM due to the removal of isobaric interferences common in highly complex biological samples, such as tissue (52). Since commercial TWIM instrumentation equipped for imaging applications was released in 2008, the present number of reports is rather limited but expected to increase substantially in the near future.

1.4.1. Imaging IM-MS/MS Measurements

When performing imaging experiments, additional confirmation of an unidentified peak is often required. A common practice for increasing confidence in peak assignments is to image in a selected reaction monitoring mode for a fragmentation channel characteristic of the analyte of interest (e.g., using CID (53)). Like traditional MS/MS, the coupled arrangement of IM and MS yields the ability to obtain structure information in the MS dimension by performing IM-MS/MS. In IM-MS/MS mode, MS^1 can be accomplished in two ways: (i) time dispersion in the drift tube can perform parent ion selection (2) or (ii) a mass analyzer can be used (49). By placing an ion activation region between the drift tube and mass analyzer ions may be selected for fragmentation according to drift time. Performing the ion selection in this manner provides a multiplex advantage in that all fragment ions will possess the same drift time as the parent ion. This is significant in imaging applications because the sample is limited to the spatial coordinates of a particular pixel. In imaging IM-MS/MS experiments, multiple parent ions can be fragmented whereby fragment ions are correlated to their respective parent ions by drift time. A demonstration of the potential utility of IM-MS/MS is illustrated in **Fig. 21.7** for a carbohydrate, lacto-*N*-fucopentaose 1 (LNFP1). In-source decay fragmentation for this carbohydrate is illustrated in **Fig. 21.7a** where fragment ions occur at different times in the IM separation. Correlated fragmentation spectra can also be obtained as illustrated in **Fig. 21.7b** by using both in-source fragmentation and post-IM CID. The CID fragmentation results in ions correlated to the drift time of the parent. Importantly, this results in redundancy of the fragment ions that are observed for higher confidence that particular fragment ions arise from the parent ion of interest. For example, the integrated mass spectra for in-source and CID fragmentation (i.e., pre- and post-IM separation) are illustrated in **Fig. 21.7c, d**, respectively. When applied to multiple ions, this operation allows for multiple reaction monitoring (MRM) in a single scan. This application is a highly promising yet virtually untapped resource for biomolecular imaging MS, where limited sample exists at each pixel location.

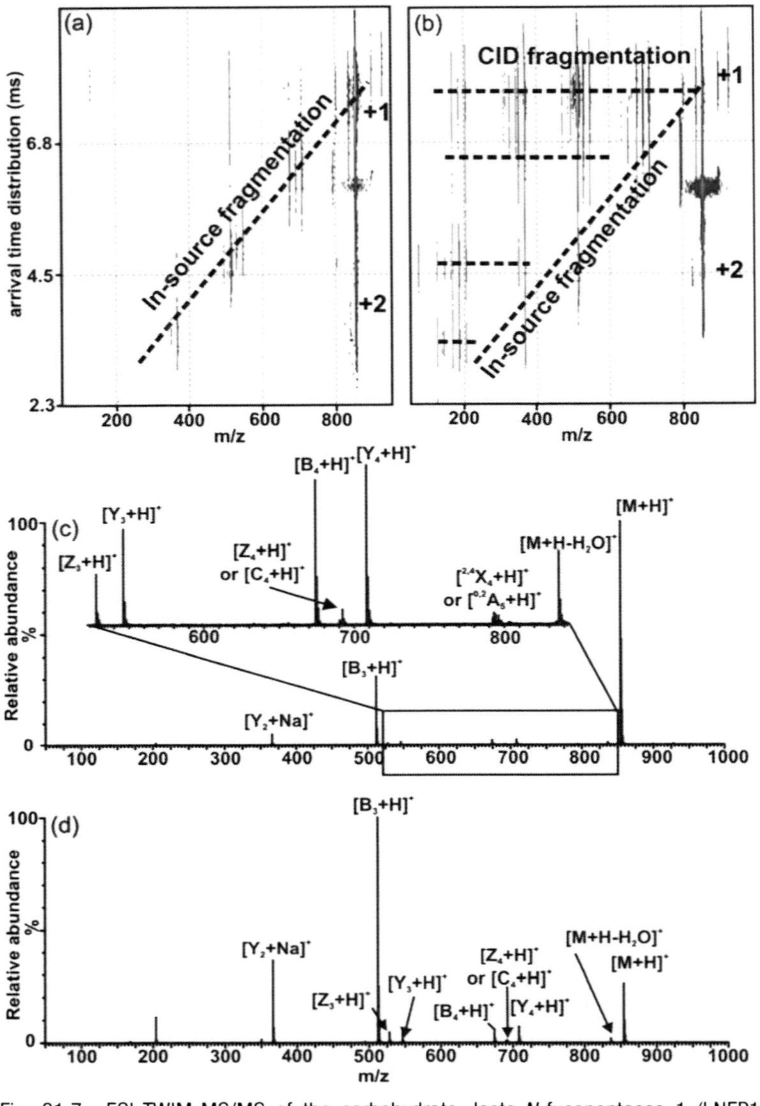

Fig. 21.7. ESI-TWIM-MS/MS of the carbohydrate, lacto-*N*-fucopentaose 1 (LNFP1), illustrating two modes of IM-MS/MS. (**a**) In-source decay fragmentation of LNFP1 followed by TWIM analysis of the fragment ions. (**b**) In-source decay prior to TWIM separation and collision-induced dissociation following TWIM separation for LNFP1. The latter results in fragment ions to be observed at the same drift time as the parent leading to the possibility for simultaneous CID for various ions at the same time. In both CID and ISD, cross-ring cleavages were seen, but glycosidic bond cleavages were the most abundant type of fragmentation. (**c**) In-source decay fragmentation spectrum that was extracted from the IM-MS plot above. Along with a zoom in view of the region from ~500 to 850 Da. Nomenclature for the fragmentation pattern of carbohydrates was first used by Domon and Costello (54). All in-source fragmentation and collision-induced dissociation peaks are labeled utilizing this nomenclature. (**d**) CID spectrum of the *top dotted line* for LNFP1 extracted from (**b**). (**c**) and (**d**) can be compared to examine the difference between the two different means for fragmenting carbohydrates. *Dotted lines* are for illustration purposes of the fragmentation peaks.

2. Materials

1. Sample prepared for MALDI analysis (thin tissue section washed, fixed to MALDI plate, and MALDI matrix applied, *see* **Chapters 4**, **7**, **11**, **16**, **20**, and **21** for detailed methodologies).
2. Mass and drift tube IM standards/calibrants. Mass standards correspond to peptides and proteins bracketing the mass range of interest. Ion mobility structural standards for DTIM are typically C_{60} and C_{70} fullerenes, because they exist in one structural form. These can be used for evaluating DTIM resolution and for day-to-day evaluation of instrument performance. Additionally, fullerenes can be used as mass standards as they are structurally separated from biomolecules in conformation space (*see* **Fig. 21.4**) and provide a wide range of gas-phase reaction products resulting in peaks spanning a large mass range in increments of 24 Da. To validate gas pressure in DTIM, typically the peptide bradykinin (RPPGFSPFR) is used to compare collision cross-section measurement with the accepted value of 242 ± 2 Å^2 (12). Bradykinin can be mixed with matrix of choice or a 1 mg/ml standard solution in H_2O can be combined 1:1 v/v with 20 mg/ml α-cyano-4-hydroxycinnamic acid in 50% methanol. Both calibrants can be applied to MALDI plate using the dried droplet method (55).
3. Traveling wave IM standards/calibrants. As discussed in **Section 1.3**, estimated collision cross sections obtained by TWIM require internal standards with corresponding absolute collision cross-section values obtained using DTIM. Published absolute collision cross sections can be obtained from several published databases, including (i) peptide collision cross sections determined by ESI (56, 57), (ii) intact protein collision cross sections determined by ESI (58), (iii) peptide collision cross sections determined by MALDI (36), and (iv) biologically relevant carbohydrate, lipid, and oligonucleotide collision cross sections determined by MALDI (59).

3. Methods

3.1. Performing Collision Cross-Section Measurements Using DTIM

1. In order to take measurements, the samples for imaging (tissue, etc.) should be prepared the same as for conventional imaging MALDI-MS (*see* **Note 5**).
2. Following insertion of the sample target into the instrument, mass and ion mobility standard/calibrants are measured. In

particular, to MALDI-IM-MS methods the laser pulse serves as the start signal (t_0) for measuring the IM arrival time distribution (t_{atd}). These time distinctions are necessary for the calculations in Step 4.

3. Following separation in the IM drift cell filled with an inert gas (1–10 torr, *see* **Note 6**), ions are directed through a skimming and differential pumping region where the pressure is reduced from 1–10 to $\sim 10^{-8}$ torr for mass analysis in the orthogonal TOFMS. The stop time for t_{atd} corresponds to the ion injection time for the TOFMS measurement.
4. To perform the calculations as described in **Section 1.2.1** (e.g., equation [4]) the arrival time distribution must be corrected for time spent in regions outside of the drift cell (i.e., time spent traversing from the MALDI plate into the drift cell, in skimming and differential pumping regions, and ion optic regions prior to the source of the TOFMS). This will result in the drift time (t_d) of the ions within the IM drift cell used in the calculation of collision cross section:

$$t_d = t_{atd} - t_{dtc}$$

5. To determine the value of t_{dtc}, IM separations are performed by varying the voltage across the drift cell while maintaining all other experimental parameters constant. The arrival time distribution measured at each drift voltage is then plotted versus the inverse of drift voltage ($1/V$). Provided the range of voltages used maintains ion separations under low-field conditions, this plot will result in a linear correlation. If non-linearity is observed, a calculation of the low-field limit should be performed (*see* **Note 1**), because curvature in this plot indicates that mobility is not constant over the voltage range used. A linear regression of this data results in a *y*-intercept corresponding to t_{dtc} (*see* **Note 7**). Preferably at least five voltages should be used to define this line although for high-precision measurements as many voltages as is practical should be used (*see* **Note 8**).
6. After the t_d has been determined from the t_{atd}, it can now be used to calculate the collision cross section, Ω, of the ion of interest through the equation [5] (*see* **Notes 9** and **10** (26)).
7. After the collision cross section has been calculated, this can be further related to the structure using molecular dynamic simulations. More information about these computational methods can be found in more detail in other resources (31–34). An excellent overview and tutorial of these strategies can be found elsewhere (60).

8. For calculating relative collision cross sections using traveling wave ion mobility-MS, there are two main procedures used which can be found in the literature (29, 30).

4. Notes

1. For electrostatic fields higher than the low-field limit, the ion velocity distribution depends less strongly on the temperature of the separation and the mean ion energy increases as it traverses the drift region. Consequently, *K* is no longer constant and depends on the specific ratio of the electrostatic field to the gas number density (*E/N*) (*see* (26) for a derivation of calculating the low-field limit for a particular analyte).
2. In principle, the Synapt has sufficient activation/ dissociation regions to perform up to MS^5, although typically up to MS^3 is practically feasible.
3. Presentation of 3D conformation space data (IM arrival time distribution, *m/z*, signal intensity) is typically projected with false coloring or gray scale representing signal intensity to project 3D data in a 2D plot.
4. There is presently no consensus on the reporting of IM-MS conformation space data, i.e., TWIM-MS data are generated with arrival time distribution on the abscissa and *m/z* on the ordinate axes, but it is either reported using this convention or where the axes are inverted. This reporting of conformation space data parallels historical preferences in the reporting of DTIM-MS data (e.g., *see* **Figs. 21.3**, **21.4**, and **21.5**).
5. Typically ionization is performed at the pressure of the DTIM (e.g., 1–10 torr), which results in moderate pressure MALDI. Thus some collisional cooling typically takes place after ionization and can result in matrix-adducted and cluster species. These can be dissociated prior to DTIM by performing injected ion experiments. Furthermore, matrix optimization may be required. One effect of moderate pressure MALDI that we have observed is that higher ion currents can be achieved at slightly lower matrix-to-analyte ratios (i.e., 1,000–100:1) than those used in high-vacuum MALDI (i.e., 10,000–1,000:1).
6. As developed in more detail in the theory section, typically He is used because of its low mass and low polarizability relative to other inert gases. However, other drift gases or

drift gas additives can be used to promote long-range interactions between the ion and drift gas. This is analogous to tuning selectivity in HPLC by changing the mobile or stationary phase that is used.

7. The *y*-intercept of this plot corresponds to t_{dtc} because it represents the limit of $t_d \rightarrow 0$ at infinite drift cell voltage. Also note that the accuracy with which this correction should be made is more important for shorter drift times and its significance is less important at longer drift times. For example, in fast separations as described for imaging IM-MS experiments the values of t_{dtc} can approach the relative magnitude of t_d.
8. For the most accurate results, the drift time correction should be evaluated for each component in the IM profile. The motivation for evaluating individual drift time corrections arises from additional ion–neutral collisions in the differential pumping regions at the entrance and/or exit of the IM drift cell. In these regions the gas dynamics typically transition from viscous to molecular flow, e.g., at the exit aperture of the drift cell at 1–10 torr to the high vacuum ($\sim 10^{-8}$ torr) of the mass spectrometer, respectively.
9. Note that the equation for calculating collision cross section is derived from classical electrodynamics, and as such, great care should be exercised in the dimensionality of the units used. Specifically, the units for E should be expressed in CGS Gaussian units, i.e., statvolts cm^{-1}, where 1 statvolt equals 299.79 V. Note that statvolts cm^{-1} is equivalent to statcoulombs cm^{-2} and that elementary charge, e, is 4.80×10^{-10} statcoulombs.
10. By comparing empirically determined cross sections with theoretical results, it has been shown that the hard-sphere approximation is best suited for analytes larger than *ca.* 1,000 Da, which is typically the size range in which many biological measurements are made. However, as the size of the analyte approaches the size scale of the drift gases used for separation, long-range interaction potential between the ion and neutral must be considered for accurate results (33, 61, 62).

Acknowledgments

We thank Whitney B. Ridenour and Richard M. Caprioli (Vanderbilt University) for assistance and use of the Synapt HDMS (data shown in **Figs. 21.6** and **21.7**), which is supported by the Vanderbilt University Mass Spectrometry Research Core. Financial support for this work was provided by the National Institutes of Health-NIDA (#HHSN271200700012C), Vanderbilt University College of Arts and Sciences, Vanderbilt Institute of Chemical Biology, Vanderbilt Institute for Integrated Biosystems Research and Education, the American Society for Mass Spectrometry (Research award to J.A.M), the Spectroscopy Society of Pittsburgh, Waters Corp., and Ionwerks Inc.

References

1. Fenn, L. S., McLean, J. A. (2008) Biomolecular structural separations by ion mobility-mass spectrometry. *Anal Bioanal Chem*, **391**, 905–909.
2. McLean, J. A., Ruotolo, B. T., Gillig, K. J., Russell, D. H. (2005) Ion mobility-mass spectrometry: a new paradigm for proteomics. *Int J Mass Spectrom*, **240**, 301–315.
3. Wyttenbach, T., Bowers, M. T. (2003) Gas-phase conformations: the ion mobility/ion chromatography method. *Modern Mass Spectrom*, **225**, 207–232.
4. Jarrold, M. F. (2000) Peptides and proteins in the vapor phase, *Annu Rev Phys Chem*, **51**, 179–207.
5. Clemmer, D. E. Jarrold, M. Fd. (1997) Ion mobility measurements and their applications to clusters and biomolecules. *J Mass Spectrom*, **32**, 577–592.
6. Eiceman, E. A., Karpas, Z. (2004) *Ion Mobility Spectrometry*, 2nd Ed., CRC Press, Boca Raton, FL, Chapter 1.
7. McAfee, K. B., Jr., Edelson, D. (1963) Identification and mobility of ions in a townsend discharge by time-resolved mass spectrometry, *Proc Phys Soc Lond*, **81**, 382–384.
8. Barnes, W. S., Martin, D. W., McDaniel, E. W. (1961) Mass spectrographic identification of the ion observed in hydrogen mobility experiments, *Phys Rev Lett*, **6**, 110–111.
9. Gieniec, J., Mack, L. L., Nakamae, K., Gupta, C., Kumar, V., Dole, M. (1984) Electrospray mass spectroscopy of macromolecules: application of an ion-drift spectrometer. *Biomed Mass Spectrom*, **11**, 259–268.
10. Von Helden, G., Wyttenbach, T., Bowers, M. T. (1995) Inclusion of a MALDI ion source in the ion chromatography technique: conformation information on polymer and biomolecular ions. *Int J Mass Spectrom Ion Process*, **146/147**, 349–364.
11. Shelimov, K. B., Clemmer, D. E., Hudgins, R. R., Jarrold, M. F. (1997) Protein structure *in vacuo*: the gas phase conformations of BPTI and cytochrome *c*, *J Am Chem Soc*, **119**, 2240–2248.
12. Wyttenbach, T., Von Helden, G., Bowers, M. T. (1996) Gas-phase conformation of biological molecules: bradykinin. *J Am Chem Soc*, **118**, 8355–8364.
13. Von Helden, G., Wyttenbach, T., Bowers, M. T. (1995) Conformation of macromolecules in the gas phase: use of matrix-assisted laser desorption methods in ion chromatography. *Science*, **267**, 1483–1485.
14. Myung, S., Lee, Y. J., Moon, M. H., Taraszka, J., Sowell, R., Koeniger, S., Hilderbrand, A. E., Valentine, S. J., Cherbas, L., Cherbas, P., Kaufmann, T. C., Miller, D. F., Mechref, Y., Novotny, M. V., Ewing, M. A., Sporleder C. R., Clemmer, D. E. (2003) Development of high-sensitivity ion trap ion mobility spectrometry time-of-flight techniques: a high-throughput nano-LC-IMS-TOF separation of peptides arising from a *Drosophila* protein extract. *Anal Chem*, **75**, 5137–5145.
15. Isailovic, D., Kurulugama, R. T., Plasencia, M. D., Stokes, S. T., Kyselova, Z., Goldman, R., Mechref, Y., Novotny, M. V. and Clemmer, D. E. (2008) Profiling of human serum glycans associated with liver cancer and cirrhosis by IMS-MS. *J Proteome Res*, 7, 1109–1117.
16. Liu, X., Valentine, S. J., Plasencia, M. D., Trimpin, S., Naylor, S., Clemmer, D. E. (2007) Mapping the human plasma proteome by SCX-LC-IMS-MS. *J Am Soc Mass Spectrom*, **18**, 1249–1264.
17. Valentine, S. J., Plasencia, M. D., Liu, X., Krishnan, M., Naylor, S., Udseth, H. R., Smith, R. D., Clemmer, D. E. (2006) Toward plasma proteome profiling with ion mobility-mass spectrometry. *J Proteome Res*, **5**, 2977–2984.
18. Liu, X., Plasencia, M., Ragg, S., Valentine, S. J., Clemmer, D. E. (2004) Development of high throughput dispersive LC–ion

mobility–TOFMS techniques for analysing the human plasma proteome. *Brief Funct Genomic Proteomic*, **3**, 177–186.
19. Liu, X., Miller, B. R., Rebec, G. V., Clemmer, D. E. (2007) Protein expression in the striatum and cortex regions of the brain for a mouse model of Huntington's disease. *J Proteome Res*, **6**, 3134–3142.
20. Taraszka, J. A., Kurulugama, R., Sowell, R. A., Valentine, S. J., Koeniger, S. L., Arnold, R. J., Miller, D. F., Kaufman, T. C., Clemmer, D. E. (2005) Mapping the proteome of *Drosophila melanogaster*: analysis of embryos and adult heads by LC-IMS-MS methods. *J Proteome Res*, **4**, 1223–1237.
21. Benesch, J. L. P., Ruotolo, B. T., Simmons, D. A., Robinson, C. V. (2007) Protein complexes in the gas phase: technology for the structural genomics and proteomics. *Chem Rev*, **107**, 3544–3567.
22. Ruotolo, B. T., Hyung, S.-J., Robinson, P. M., Giles, K., Bateman, R. H., Robinson, C. V. (2007) Ion mobility-mass spectrometry reveals long-lived, unfolded intermediates in the dissociation of protein complexes. *Angew Chem Int Ed*, **46**, 8001–8004.
23. Ruotolo, B. T., Giles, K., Campuzano, I., Sandercock, A. M., Bateman, R. H., Robinson, C. V. (2005) Evidence for macromolecular rings in the absence of bulk water. *Science*, **310**, 1658–1661.
24. Jackson, S. N., Ugarov, M., Egan, T., Post, J. D., Langlais, D., Schultz, J. A., Woods, A. S. (2007) MALDI-ion mobility-TOFMS imaging of lipids in rat brain tissue. *J Mass Spectrom*, **42**, 1093–1098.
25. McLean, J. A., Ridenour, W. B., Caprioli, R. M. (2007) Profiling and imaging of tissues by imaging ion mobility-mass spectrometry. *J Mass Spectrom*, **42**, 1099–1105.
26. Mason, E. A., McDaniel, E. W. (1988) *Transport Properties of Ions in Gases*. John Wiley & Sons, New York, NY.
27. Mason, E. A. (1984) Ion mobility: its role in plasma chromatography, in *Plasma Chromatography*, (Carr, T. W. ed.), Plenum Press, New York, NY, 43–93.
28. Revercomb, H. E. and Mason, E. A. (1975) Theory of plasma chromatography/gaseous electrophoresis – a review. *Anal Chem*, **47**, 970–983.
29. Ruotolo, B. T., Benesch, J. L. P., Sandercock, A. M., Hyung, S.-J. Robinson, C. V. (2008) Ion mobility-mass spectrometry analysis of large protein complexes. *Nat Protoc*, **3**, 1139–1152.
30. Williams, J. P., Scrivens, J. H. (2008) Coupling desorption electrospray ionization and neutral desorption/extractive electrospray ionization with a travelling-wave based ion mobility mass spectrometer for the analysis of drugs. *Rapid Commun Mass Spectrom*, **22**, 187–196.
31. Gidden, J., Bowers, M. T. (2003) Gas-phase conformations of deprotonated and protonated mononucleotides determined by ion mobility and theoretical modeling. *J Phys Chem B*, **107**, 12829–12837.
32. Gidden, J., Bowers, M. T. (2003) Gas-phase conformations of deprotonated trinucleotides (dGTT(−), dTGT(−), and dTTG (−)): the question of zwitterion formation. *J Am Soc Mass Spectrom*, **14**, 161–170.
33. Wyttenbach, T., Witt, M., Bowers, M. T. (2000) On the stability of amino acid zwitterions in the gas phase: the influence of derivatization, proton affinity, and alkali ion addition. *J Am Chem Soc*, **122**, 3458–3464.
34. Shvartsburg, A. A., Jarrold, M. F. (1996) An exact hard-spheres scattering model for the mobilities of polyatomic ions. *Chem Phys Lett*, **261**, 86–91.
35. Dwivedi, P., Wu, P., Klopsch, S. J., Puzon, G. J., Xun, L., Hill, H. H. (2008) Metabolic profiling by ion mobility mass spectrometry (IMMS). *Metabolomics*, **4**, 63–80.
36. Tao, L., McLean, J. R., McLean, J. A., Russell, D. H. (2007) A collision cross-section database of singly-charged peptide ions. *J Am Soc Mass Spectrom*, **18**, 1232–1238.
37. Ruotolo, B. T., Verbeck, G. F., Thomson, L. M., Woods, A. S., Gillig, K. J., Russell, D. H. (2002) Distinguishing between phosphorylated and nonphosphorylated peptides with ion mobility-mass spectrometry. *J Proteome Res*, **1**, 303–306.
38. Furche, F., Ahlrichs, R., Weis, P., Jacob, C., Gilb, S., Bierweiler, T., Kappes, M. M. (2002) The structures of small gold cluster anions as determined by a combination of ion mobility measurements and density functional calculations. *J Chem Phys*, **117**, 6982–6990.
39. Shvartsburg, A. A., Smith, R. D. (2008) Fundamentals of traveling wave ion mobility spectrometry. *Anal Chem*, **80**, 9689–9699.
40. Mason, E. A., McDaniel, E. W. (1988) Measurement of drift velocities and longitudinal diffusion coefficients, in *Transport Properties of Ions in Gases*, John Wiley & Sons, New York, NY, 31–102.
41. Kanu, A. B., Dwivedi, P., Tam, M., Matz, L., Hill, H. H., Jr. (2008) Ion mobility-mass spectrometry, *J Mass Spectrom*, **43**, 1–22.
42. Steiner, W. E., Clowers, B. H., English, W. A., Hill, H. H., Jr. (2004) Atmospheric

pressure matrix-assisted laser desorption/ionization with analysis by ion mobility-mass spectrometry. *Rapid Commun Mass Spectrom*, **18**, 882–888.
43. McLean, J. A., Russell, D. H. (2003) Subfemtomole peptide detection in ion-mobility-time-of-flight mass spectrometry measurements. *J Proteome Res*, **2**, 427–430.
44. McLean, J. A., Ridenour, W. B., Caprioli, R. M. (2007) Imaging ion mobility mass spectrometry: advantages, challenges, and future prospects. *Proceedings of the 55th Annual Meeting of the American Society for Mass Spectrometry*, Indianapolis, IN.
45. Merenbloom, S. I., Glaskin, R. S., Clemmer, D. E. (2009) High resolution ion cyclotron mobility spectrometry. *Anal Chem*, **81**, 1482–1487.
46. Wyttenbach, T., Kemper, P. R., Bowers, M. T. (2001) Design of a new electrospray ion mobility mass spectrometer. *Int J Mass Spectrom*, **212**, 13–23.
47. Dugourd, Ph., Hudgins, R. R., Clemmer, D. E., Jarrold, M. F. (1997) High-resolution ion mobility measurements. *Rev Sci Instrum*, **68**, 1122–1129.
48. Giles, K., Pringle, S. D., Worthington, D. L., Wildgoose, J. L., Bateman, R. H. (2004) Applications of travelling wave-based radio-frequency-only stacked ring ion guide. *Rapid Commun Mass Spectrom*, **18**, 2401–2414.
49. Pringle, S. D., Giles, K., Wildgoose, J. L., Williams, J. P., Slade, S. E., Thalassinos, K., Bateman, R. H., Bowers, M. T., Scrivens, J. H. (2007) An investigation of the mobility separation of some peptide and protein ions using a new hybrid quadrupole/travelling wave IMS/oa-ToF instrument. *Int J Mass Spectrom*, **261**, 1–12.
50. Vakhrushev, S. Y., Langridge, J., Campuzano, I., Hughes, C., Peter-Katalinic, J. (2008) Ion mobility mass spectrometry analysis of human glycourinome. *Anal Chem*, **80**, 2506–2513.
51. Riba-Garcia, I., Giles, K., Bateman, R.H., Gaskell, S.J. (2008) Evidence for structural variants of a- and b- type peptide fragment ions using combined ion mobility/mass spectrometry. *J Am Soc Mass Spectrom*, **19**, 609–613.
52. Trim, P. J., Henson, C. M., Avery, J. L., McEwen, A., Snel, M. F., Claude, E., Marshall, P. S., West, A., Princivalle, A. P., Clench, M. R. (2008) Matrix-assisted laser desorption/ionization-ion mobility separation-mass spectrometry imaging of vinblastine in whole body tissue sections. *Anal Chem*, **80**, 8628–8634.
53. Reyzer, M. L., Hsieh, Y., Ng, K., Korfmacher, W .A., Caprioli, R. M. (2003) Direct analysis of drug candidates in tissue by matrix-assisted laser desorption/ionization mass spectrometry. *J Mass Spectrom*, **38**, 1081–1092.
54. Domon, B., Costello, C. E. (1988) A systematic nomenclature for carbohydrate fragmentations in FAB-MS/MS spectra of glycoconjugates. *Glycoconjugate J*, **5**, 397–409.
55. Karas, M., Hillenkamp, F. (1988) Laser desorption ionization of proteins with molecular masses exceeding 10,000 daltons. *Anal Chem*, **60**, 2299–2301.
56. Henderson, S. C., Valentine, S. J., Counterman, A. E., Clemmer, D. E. (1999) ESI/ion trap/ion mobility/time-of-flight mass spectrometry for rapid and sensitive analysis of biomolecular mixtures. *Anal Chem*, **71**, 291–301.
57. Hoaglund, C. S., Valentine, S. J., Sporleder, C. R., Reilly, J. P., Clemmer, D. E. (1998) Three-dimensional ion mobility/TOFMS analysis of electrosprayed biomolecules. *Anal Chem*, **70**, 2236–2242.
58. Clemmer cross section database (2006) (Assessed January 28th, 2009 at http://www.indiana.edu/~clemmer/Research/cross section database/cs database.htm)
59. Fenn, L. S., Kliman, M., Mahsutt, A., Zhao, S. R., McLean, J. A. (2009) Characterizing ion mobility-mass spectrometry conformation space for the analysis of complex biological samples. *Anal Bioanal Chem*, **394**, 235–244.
60. Theories and analysis of IM-MS data (Assessed January 28th, 2009 at http://bowers.chem.ucsb.edu/theory_analysis/index.shtml).
61. Wyttenbach, T., von Helden, G., Batka, J. J., Jr., Carlat, D., Bowers, M. T. (1997) Effect of the long-range interaction potential on ion mobility measurements. *J Am Soc Mass Spectrom*, **8**, 275–282.
62. Mesleh, M. F., Hunter, J. M., Shvartsburg, A. A., Schatz, G. C., Jarrold, M. F. (1996) Structural information from ion mobility measurements: effects of the long-range potential. *J Phys Chem*, **100**, 16082–16086.

Chapter 22

Tutorial: Multivariate Statistical Treatment of Imaging Data for Clinical Biomarker Discovery

Sören-Oliver Deininger, Michael Becker, and Detlev Suckau

Abstract

Cancer research is one of the most promising application areas for the new technology of MALDI tissue imaging. Cancerous tissue can easily be distinguished from healthy tissue by dramatically changed metabolism, growth, and apoptotic processes. Of even higher interest is the fact that MALDI imaging allows to unveil molecular differentiation undetectable by classical histological techniques. Thus, MALDI imaging has tremendous potential as a tool to characterize the therapeutic susceptibility of tumors in biopsies as well as to predict tumor progression in endpoint studies. However, some aspects are important to consider for a successful MALDI imaging-based cancer research. Cancer sections are usually very heterogeneous – different biochemical pathways can be active in individual tumor clones, at different development stages or in various tumor microenvironments. Understanding tissue at this level is only possible for experienced histopathologists working on high-resolution optical images. Therefore, the largest benefit from the use of MALDI imaging results in histopathology will arise if molecular images are related to classical high-resolution histological images in a simple way without the need to interpret mass spectra directly. Each MALDI imaging data set effectively provides information on hundreds of molecules and permits the generation of molecular images displaying the relative abundance of these molecules across the tissue. The interpretation of these in the histological context is a major challenge in terms of expert analysis time. This is true especially for clinical work with hundreds of tissue specimens to be analyzed by MALDI, interpreted, and compared. Therefore, a MALDI imaging workflow is described here that enables fast and unambiguous interpretation of the MALDI imaging data in the histological context. Preprocessing of the image data using statistical tools allows efficient and straightforward interpretation by the histopathologist. In this chapter, we explain the use of principal component analysis (PCA) and hierarchical clustering (HC) for the efficient interpretation of MALDI imaging data. We also outline how these methods can be used to compare specific disease states between patients in the search for biomarkers.

Key words: MALDI imaging, molecular histology, tumor, principal component analysis, hierarchical clustering, biomarker.

S.S. Rubakhin, J.V. Sweedler (eds.), *Mass Spectrometry Imaging*, Methods in Molecular Biology 656,
DOI 10.1007/978-1-60761-746-4_22, © Springer Science+Business Media, LLC 2010

1. Introduction

Cancer research is one of the most promising application areas for the new technology of MALDI tissue imaging (for review, see, e.g., (1)). Cancerous tissue can easily be distinguished from healthy tissue due to dramatically changed metabolism, growth, and apoptotic processes. In addition, tumors themselves are highly heterogeneous and can consist of areas of invasive tumor, the invasion front, tumor cells of different development stages, lymphocyte infiltrations, inflamed tissue, and others. Different metabolic states inside a tumor, such as hypoxic areas, or differentially regulated cellular proteases further complicate the overall picture. Non-tumorous tissue in the vicinity of the tumor cells, the stroma, may influence the tumor as well.

MALDI imaging is able to unveil molecular differentiation undetectable by classical histological techniques and can therefore serve as a tool to characterize tumor tissue from biopsies. In differential studies, therapeutic susceptibility of tumors as well as tumor progression in endpoint studies can be assessed (2).

Understanding MALDI imaging results at this level of heterogeneity is a task for experienced pathologists and usually requires the parallel evaluation of histologically stained tissue at microscopic resolution. A major challenge for MALDI imaging in clinical research is the correlation of histology and the molecular information, i.e., the precise matching of two image dimensions.

A further challenge is the efficient interpretation of MALDI imaging data in the histological context, which is of particular importance in clinical studies where large numbers of specimen are analyzed. Peak-by-peak interpretation of MALDI imaging data is time-consuming and tedious. In fact, the detailed analysis of numerous tissue sections by a pathologist, as necessary in clinical studies, is not feasible due to time constraints.

One way to eliminate this limitation in inspection time has been suggested by (3). In this workflow, the pathologist defines specific tissue regions based on conventional histology prior to the MALDI imaging analysis. Using these assignments, it is then possible to compare patient samples to look for disease (stage) markers. However, a major advantage of MALDI imaging is the ability to detect molecular differentiation not apparent from histology alone. Although of particular interest, this information is not fully exploited when the interpretation of the molecular images is left in the hands of a mass spectrometrist.

It is mandatory that both the histological and the molecular information in MALDI imaging experiments are evaluated by the pathologists themselves. Such evaluation on a reasonable timescale requires the use of statistical methods that reduce the molecular data in a comprehensive way. One commonly used way to achieve this is the reconstruction of images based on principal component analysis (PCA). This approach has long been used in secondary ion mass spectrometry (4–6) and is gaining popularity in MALDI imaging (7, 8). Using PCA, the information of a MALDI imaging data set is condensed to a small number of molecular images allowing automated feature extraction. It is a very robust unsupervised multivariate approach to report the results of MALDI imaging experiments if no further information on the tissue is available. However, PCA does not increase the detail level of the analysis or enable interactive work with the data set. These limitations are overcome by hierarchical clustering combined with a user interface that allows the reconstruction of images based on spectra similarity (1, 7), a method which allows to interactively explore the data set in an efficient way based on both the histological and the molecular information. In a clinical study, this allows the pathologist to select the mass spectra that are characteristic for a certain disease state, which in a later step can be compared across different patients.

In this chapter, we explain the use of principal component analysis (PCA) and hierarchical clustering (HC) for the efficient interpretation of MALDI imaging data (7, 9). We also outline how these methods can be used to compare specific disease states between patients in the search for biomarkers.

1.1. Molecular Histology

The initial interpretation of a tumor section is always based on histology, with hematoxylin and eosin (H&E) staining being the gold standard. Recognizing the state of a given cell or cell population requires microscopical evaluation at a magnification that allows to see the shape of the nuclei. Unstained tissue sections are not helpful in this regard as they typically do not display sufficient contrast, so histological staining is mandatory to assign tissue features. The result of a MALDI imaging experiment can be interpreted only in conjunction with the histology. Consequently, a major hurdle for MALDI imaging in cancer research is the ability for "multimodal imaging," i.e., the capability to derive different types of image information (H&E stain and MALDI) from the same tissue section.

Therefore, a major breakthrough toward clinical MALDI imaging was the discovery that the MALDI matrix can be washed off after an experiment and histologically stained images can then be obtained from the same section (1, 10, 11). For practical use, the sample has to be prepared on a slide that addresses the requirements for both imaging methods: (1) optical transparency for transmission microscopy and (2) electrical conductivity for MALDI analysis. Conductive indium-tin-oxide (ITO)-coated glass slides fulfill both requirements and have become the de facto standard.

As classical histological images can be directly overlaid with the molecular dimension derived from the same tissue section, this type of experiment can truly be called molecular histology.

Earlier reports demonstrated that some histological stains were compatible with subsequent MALDI imaging analyses (12). Amongst them were methylene blue, cresyl violet but not H&E. Unfortunately, the MALDI-compatible stains do not yield the same information content as the H&E staining. In such pre-staining approaches, unambiguous assignment of MALDI imaging results to tissue features is possible, but to understand the nature of the observed features, an H&E staining of a serial section is still necessary.

The use of consecutive sections for MALDI imaging and the H&E staining has also been reported (3). The section used for MALDI imaging is scanned prior to the matrix application and this image is overlaid with a H&E stain of a consecutive section. Apart from the fact that small features such as carcinoma in situ can be present in one section but absent or significantly shifted in the other section, obtaining consecutive sections of good quality (e.g., not distorted, ripped, or folded) represents a technical challenge, as even slight differences can make the overlay difficult.

In conclusion, the "post-MALDI staining" workflow with MALDI imaging as the first and removal of the matrix, H&E staining, and microscopy as the second step is now the established and preferred approach to MALDI imaging in the clinical context, because it compromises neither the molecular nor the histological dimension of the analysis.

1.2. Interpreting MALDI Images by Principal Component Analysis (PCA)

MALDI imaging data are multidimensional data, with each mass signal defining one molecular dimension. The aim of PCA is to reduce the dimensionality of the molecular information to a small number of relevant dimensions. (For a detailed explanation of PCA, *see* **Section 1.5**) Such PCA results can be turned into image information by displaying the scores of selected principal components for each pixel (i.e., each spectrum) as color density plots. As a result, a rather small number of images (i.e., one per principal component observed) are obtained which combine the relevant information, rather than hundreds of single-mass ion images (*see* **Figs. 22.1a** and **22.2a**).

The main advantage of PCA is that it allows unsupervised feature extraction from MALDI imaging data without any understanding of the tissue. The image based on the first principal component (PC1) will show the main variance in the data set and can be expected to be in good agreement with histology. This is true if the MALDI imaging analysis is performed only within the boundaries of the tissue or if only mass spectra from the tissue region are subjected to the PCA calculation. If the imaging software allows acquisition of spectra from a rectangular region only, the main

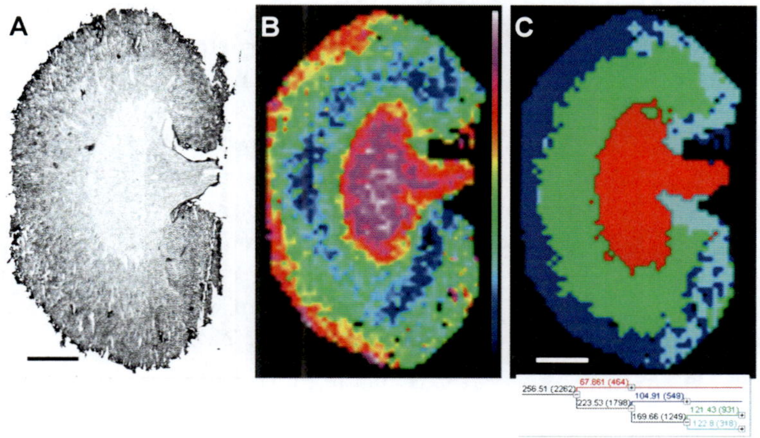

Fig. 22.1. PCA and hierarchical clustering example shown on a kidney sample. (**a**) Optical image of the kidney section prior to matrix application. (**b**) Scores of first principal component. The image is in agreement with the anatomy of the kidney, with the renal pelvis and part of the cortex showing the hot colors. (**c**) Clustering result: Renal cortex, medulla, and pelvis are defined by the highest level clusters. The advantage of this type of analysis is that the dendrogram nodes can be expanded and highlighted until the desired molecular structure is found. Scale bar: 2 mm.

variance in the data comes from "on-tissue" versus "off-tissue" spectra, so PC1 will mainly reflect this difference. In this case the histology of the section will be better represented by higher order principal components.

In a clinical research setting, the person operating the mass spectrometer is typically not able to interpret the results in histological context. Yet the operator of the mass spectrometer may be asked to present initial results of the experiments to the pathologist for a more detailed evaluation. This is where PCA is most helpful. Without understanding of the tissue architecture, one can create a small number of principal component heatmap images that are in good agreement with the histology. Based on these images, it is possible to quickly conclude whether the experiment was successful or if preparation artifacts are present.

Further interpretation of imaging data based on PCA is limited. PCA is not able to classify spectra as similar, and moreover spectra that are similar in one principal component may be very different in another. To determine groups of "similar" spectra, other classification or clustering techniques have to be applied, such as the hierarchical clustering discussed here.

1.3. Interpreting MALDI Images by Hierarchical Clustering (HC)

In hierarchical clustering, mass spectra are grouped by similarity in a dendrogram. The distance of the spectra in the multidimensional space is reflected by the distance and position within the hierarchical tree or dendrogram. Moving from the root toward the branches of the hierarchical tree, mass spectra of increasing similarity are grouped into nodes of lower order. MALDI images can now be reconstructed by selecting all mass spectra (i.e., image pixels) from a given node in the dendrogram and assigning a common color to them. The main structure in a data set can therefore be expected in the top nodes near the root of the dendrogram.

Several features make the hierarchical clustering especially useful for the interpretation of MALDI imaging data:

- The "main" structure of the data set is often found at the top levels. For example, in a tumor data set, top nodes in the dendrogram will most likely separate tumor versus non-tumor (*see* **Fig. 22.2c**).
- A hierarchy of the clusters is maintained, e.g., information on differentiation states of the tumor which are more closely related to each other rather than to non-tumor area in the same tissue is conserved in the result (*see* **Fig. 22.2d**).

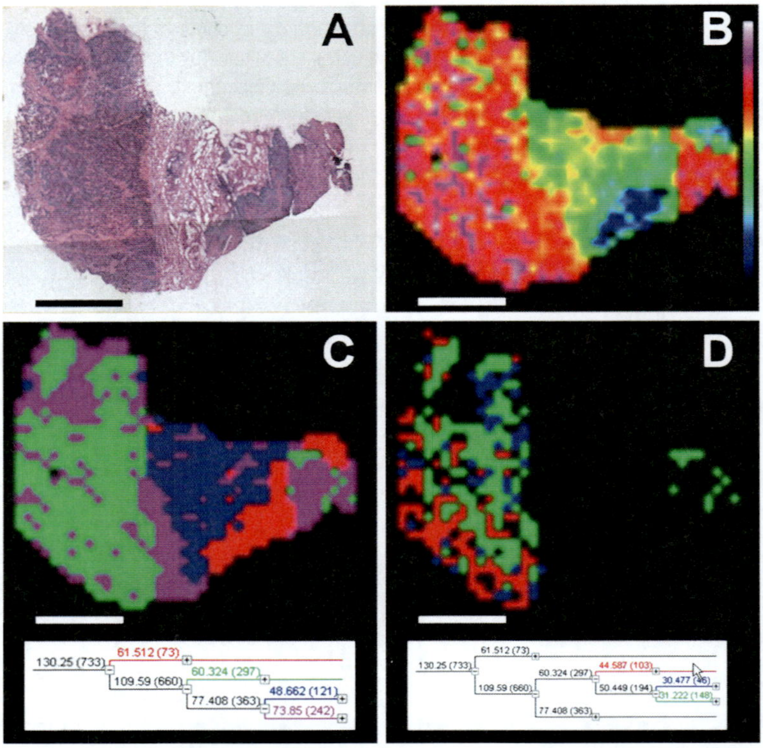

Fig. 22.2. PCA and hierarchical clustering for a gastric cancer section. (**a**) H&E-stained tissue section after MALDI imaging measurement. (**b**) Scores of the first principal component show the hot colors in the tumor area. (**c**) Hierarchical clustering: Top dendrogram nodes differentiate tumor (*green* and magenta) versus non-tumor (*blue*, squamous epithelium in *red*). (**d**) The dendrogram can be expanded down the tumor node to evaluate the molecular differentiation inside the tumor. This can also be directly correlated with the histology. This workflow enables the fast and concise selection of mass spectra representative for specific tissue states. Scale bar: 2 mm.

Interactive exploration of the molecular results is possible in a quick and effective way. For example, if a dendrogram node is found that contains the "tumor" spectra, these can quickly be further differentiated by looking at the match between lower nodes and more detailed histological features.

In summary, MALDI imaging combined with HC has been proven as an efficient way to quickly interpret the molecular information in the histological context.

1.4. Interpretation of Clinical Data

The primary goal of MALDI imaging in clinical research is to put molecular information in the tissue into context with clinical data from numerous patients. Such data include information on survival, outcome, and response to treatment for each patient. Large numbers of analyses have to be performed and evaluated, and the generation of MALDI images is not the endpoint of the analysis, but a tool to assign tissue locations that are specific for a certain tissue (health) state. The comparison of large numbers of samples cannot be done on the image level, but must be performed on selected spectra that are representative for specific states (e.g., invasive tumor, tumor development, or normal epithelium), and MALDI imaging is needed to select these spectra. In these comparisons, PCA and HC can now be used to draw comparisons between patients (*see* **Fig. 22.3**). One can also screen for molecules that are representative for certain health states, which may be biomarker candidates. One of the true advantages of the imaging approach is that once such potential biomarkers are found, it is possible to return to the original images and evaluate the location of the markers in the histological context.

1.5. Understanding PCA and HC

In this section we give a practical, non-mathematical approach to the understanding of PCA and HC analysis. The most crucial point for the understanding of either technique is how mass spectra can be represented as multidimensional coordinate systems.

1.5.1. From Mass Spectra to Multidimensional Coordinate Systems

A mass spectrum consists of a given number of peaks (i.e., mass signals). If the spectrum consists of n peaks, then we can create an n-dimensional coordinate system with each axis representing one peak (i.e., one particular mass). We can now plot the intensity (or area) of each peak onto the corresponding axis. This step translated our mass spectrum into one data point in an n-dimensional space (*see* **Fig. 22.4**). If we repeat this with a large number of mass spectra, they create a "cloud" in that n-dimensional space (*see* **Fig. 22.5a**). The most important part of this concept is that mass spectra that are "similar" are located close together in that n-dimensional space.

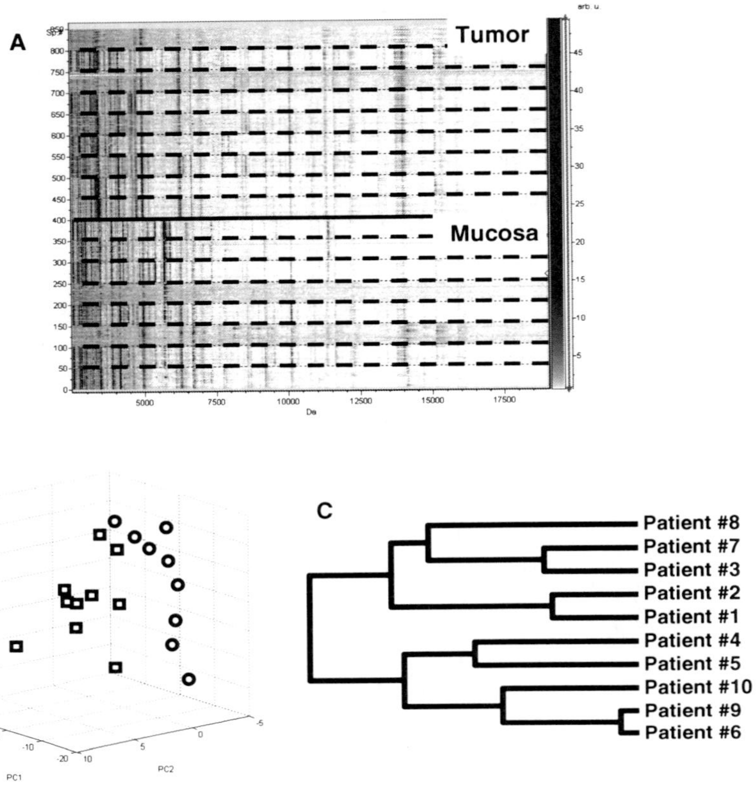

Fig. 22.3. Comparison of imaging data across patients. (**a**) Pseudo-gel view of selected mass spectra from different images/samples. The mass spectra are representative for tumor or tumor-free mucosa. Spectra in between dashed lines are from the same patient, each of these "lanes" represents one patient. Characteristic biomarkers can now be found by visual inspection or by statistical tools such as receiver-operating curves or *p*-values. (**b**) PCA applied to the data set. Each element represents one patient; squares indicate tumor and circle tumor-free mucosa. In this score plot of the first three principal components a separation between tumor and tumor-free mucosa is seen. This indicates that classification based on the MALDI imaging result is possible. (**c**) Hierarchical clustering can be performed, e.g., on tumor spectra from all patients. This allows correlation of patient clusters with clinical meta-information. The data shown here are described in (7).

1.5.2. Principal Component Analysis (PCA)

In mass spectra constituting an imaging data set, there will always be a correlation of peaks, and we can also assume that similar tissue types will generate similar mass spectra. For this reason, the data points in the n-dimensional space (defined by the peaks of the mass spectra) are not randomly distributed but will have an orderly structure. The aim of the PCA is to transform the original coordinate system in a way that better represents this internal structure. This is achieved by shifting and rotating the original coordinate systems (defined by peaks) so that one axis points into the direction of highest variance in the data set (*see* **Fig. 22.5c**). The next axis will be rotated to point in the direction of the second highest variance and so on. The axes of this new

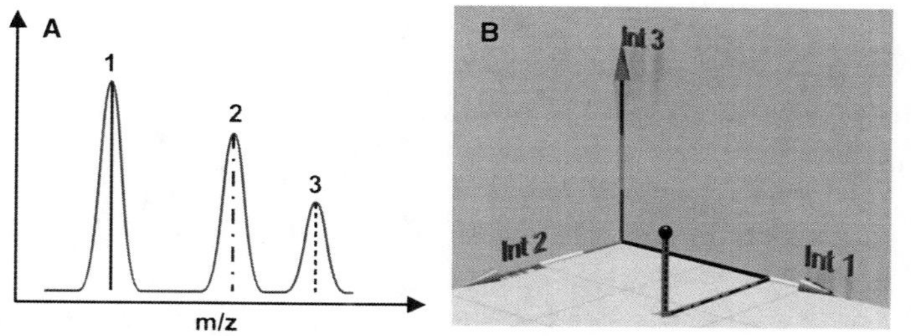

Fig. 22.4. This figure shows the most important concept for interpretation of mass spectra by multivariate statistics. (**a**) Here a mass spectrum with *n* (in this example 3) peaks is symbolized. The mass spectrum can be represented by the intensities of its peaks. (**b**) We can now create an *n*-dimensional coordinate system. Each coordinate represents the intensity of one peak from the original mass spectrum. The original mass spectrum is now represented as a point in an *n*-dimensional space. Note: The same concept applies if there are more than three peaks. The mathematics will stay the same, although visualization in a simple way is no longer possible. If we apply this to a larger number of spectra, then each spectrum will occupy one point in the *n*-dimensional space, and points that are located closely together are spectra that are very similar.

coordinate system are called principal components. The value a spectrum assumes on a principal component, e.g., PC1, is called "score." To reconstruct an image, the score of each spectrum is translated either into a color saturation or into a heatmap representation (*see* **Fig. 22.5e**). Since the principal components are ordered in descending overall variance, PC1 contains the most important information in a data set. While PCA shows similar spectra in similar colors, it does not provide any information on how many groups of similar spectra are present or which spectra belong together based on similarity. For this information a clustering algorithm is much better suited.

The scaling of the variables (in our case peaks) is an additional important concept for PCA. PCA (as well as the HC discussed later) reveals the main variance in the data set. Usually, peaks with a high average intensity across the data set also have a high (absolute) variance when compared to peaks with low intensity. Thus, peak intensity variance must be scaled prior to PCA, otherwise the results will be dominated by the most intensive peaks only.

1.5.3. Hierarchical Clustering (HC)

During the HC analysis clusters (groups of similar spectra) are established based on the distances of the data points in the *n*-dimensional space as discussed above. HC works agglomeratively and builds the dendrogram bottom-up. The algorithm works as follows: The two data points with the smallest distance are identified. These two data points then form one cluster, and this cluster is treated as a new data point for the remainder of the calculation. Now the two data points with the next smallest distance are located and assigned to a cluster (*see* **Fig. 22.6**). This procedure is repeated until all data points have been assigned to

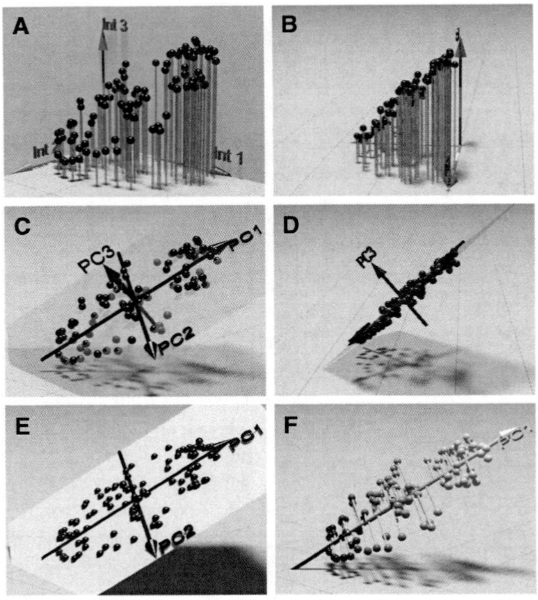

Fig. 22.5. In this figure the concept of PCA and its application to imaging is explained. (**a**) This cloud of points represents a number of mass spectra (*see* **Fig. 22.3**), e.g., from one imaging experiment. Note that there is a clear structure in the data. This is always the case as in reality there are always correlations between peaks. (**b**) The same data set viewed from a different angle. (**c**) The principal components. In the PCA the original coordinate system of the data set is transformed (shifted and rotated) in a way that represents the internal structure of the data better. The axes of this transformed coordinate system are called "principal components" (PC). The principal components are then ordered by descending variance. The first PC points in the direction of highest variance, the second PC points in the direction of second highest variance, and so on. At this stage the PCA has not altered the data. (**d**) Data set after PCA from a different angle. It is now clearly visible that in this example the third principal component contributes only very little information to the explanation of the data set. (**e**) The third PC, which contains only little information, is taken away. By doing so the data set is projected into a lower-dimensional space but the main information is kept. The value that each data point assumes on each principal component is called its "score," and the resulting representation is called a "scores plot." With real-life imaging data sets containing more than a hundred peaks, often the first few PCs already contain 70% or more information of the original data set. (**f**) By selecting one particular principal component and assigning a color saturation to the scores, we can now assign each spectrum a color intensity representing its score for that PC. This can be used to create an image from the principal component analysis results.

clusters. The length of the branches in a dendrogram represents the distance between the two data points (or the two clusters of the lower hierarchy). The result of any HC will always be a full dendrogram containing all data points, and the dendrogram can always be traced down to single spectra. One drawback of this approach is that the algorithm cannot decide how many clusters represent "real" clusters and how many clusters represent just random differences in the data. Mathematical ways to find

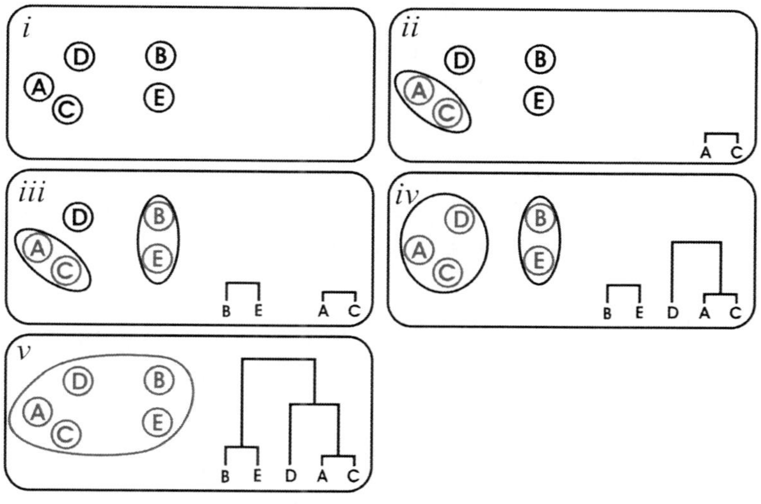

Fig. 22.6. Hierarchical clustering. (i) These five elements shall be subjected to hierarchical clustering. (ii) In a first step, the two elements with the smallest distance are found. Here these are elements A and C. These two elements are then put together into one cluster. The cluster becomes a new element. (iii) Now the elements with the second smallest distance are found. In this case these are elements B and E. These two elements are put together into the second cluster and this cluster also becomes a new element. (iv) Now the elements with the next smallest distance are found, these are cluster AC and the element D. They are grouped together into one cluster. (v) Now only two clusters are left. These are put together into the top-level cluster that contains all elements. Note: The length of the branches in the dendrogram indicates the distance between the two elements in the respective cluster.

the real clusters exist, such as plotting the variance explained by the clusters versus the number of clusters. Such a plot shows an "elbow" at the point where random fluctuation starts ("elbow criterion"). Other methods include repeatedly re-sampling random sub-samples of the original data to locate invariant clusters (i.e., bootstrapping) (13).

For the interpretation of MALDI imaging data this does not matter since here the interpretation can be based on the histological knowledge. The pathologist can simply browse into the dendrogram until the histology is sufficiently explained. Selected nodes in the dendrogram can be further explored if they represent possible additional information that is not found by the histology alone. Then the dendrograms can be annotated with their respective tissue state (such as "invasive tumor," "carcinoma in situ," "lymphocytes,"). Based on these annotations, spectra of a tissue type can now be compared across different patients.

For a more detailed understanding of the HC it has to be noted that there are different options of parameterization. There are multiple ways in which a distance in a multidimensional space can be calculated, with "Euclidean" being the most intuitive one. Other distance metrics include correlation, cosine, Minkowsky, and city-block. Also, there are different options for calculating the distance between two clusters, the so-called linkage. Here the possibilities include the smallest possible distance between two data points in the two clusters (single linkage), the distance between the centers of gravity, the average of all pair-wise distances between two data points (average linkage). The particularly useful "Ward" linkage method tries to minimize the "within cluster" variance. A detailed discussion of these options is beyond the scope of this chapter.

1.5.4. Relationship of PCA and HC

HC and PCA work on any multidimensional data set. Therefore, clustering can be conducted on the original data set as well as on the data set after the PCA transformation. The latter allows to cluster-analyze a data set that is already reduced in the number of dimensions. We found this useful for the clustering calculation with the Euclidean distance.

2. Materials

2.1. Tissue Preparation

1. Conductive indium-tin-oxide (ITO)-coated glass slides (Bruker).
2. Optimum cutting temperature polymer (OCT, Sakura).
3. Two coplin jars with 70% ethanol (HPLC grade). The first jar should contain fresh solvent.
4. One coplin jar with 96% ethanol (HPLC grade).
5. Liquid paper pen, liquid white-out or similar.
6. A scanner suitable to get an optical image of the unstained tissue section (2400 dpi).

2.2. Matrix Preparation

1. Matrix solution: 10 g/l sinapinic acid (puriss. p.a.; Sigma) in 60% acetonitrile (HPLC grade), 0.2% trifluoroacetic acid, 39.8% water (v/v). It is not necessary to use the ultrapure quality of sinapinic acid. The matrix solution should be prepared weekly and stored in the dark.
2. ImagePrep matrix application device (Bruker) for homogeneous matrix coating of tissue samples with fine crystal morphology (~20 μm).

2.3. MALDI Measurements

1. MALDI adapter plate (Bruker) to hold the electrically conductive, optically transparent slides in the MALDI ion source.
2. MALDI-TOF or TOF/TOF mass spectrometer (autoflex III or ultraflex III, Bruker) with ScoutMTP ion source and

high-resolution precision ion source video camera, 200 Hz SmartBeam™ laser with software adjustable 10–100 μm focus diameter. Equipped with FlexImaging 2.1 software or higher for image data acquisition.

2.4. H&E Staining

1. Coplin jar with 70% ethanol.
2. Eosin Y stock solution 1%: 10 g eosin Y (Sigma), 200 ml water, 800 ml 95% ethanol, stir until eosin is dissolved. Store in the dark at room temperature.
3. Eosin Y working solution 0.25%: 250 ml eosin stock solution, 750 ml 80% ethanol, 5 ml glacial acetic acid. Store at room temperature.
4. Quick hardening mounting medium (e.g., Eukitt, Fluka).
5. Hematoxylin solution according to Meyer (Fluka).
6. Coplin jars with deionized water, 70% ethanol, 80% ethanol, 90% ethanol, two jars with absolute ethanol, xylene (histology grade).

2.5. Bioinformatics

1. FlexImaging software (Bruker) version 2.1 or higher.
2. ClinProTools software (Bruker) version 2.2 or higher.

3. Methods

3.1. Tissue Sections

1. Randomize the order of sample preparation (*see* **Note 1**).
2. Put tissue piece onto sample stage, use OCT to glue tissue piece to sample holder. Take care that the OCT does not embed the tissue, in particular it should not touch the cutting plane or blade (*see* **Note 2**).
3. Cool down ITO slides in the cryostat (*see* **Note 3**).
4. Cut tissue section at 8 μm thickness (*see* **Note 4**).
5. Pick up the tissue section with the ITO side of the slide or transfer the tissue section with an artist's brush onto the slide.
6. Use a finger or the back of the hand to warm the slide from underneath, thus thawing and straightening the tissue section. Keep the section warmed until visibly dry. The slide can now be re-frozen to place more sections or taken out to proceed (*see* **Note 5**).
7. Dry the section for 5 min in a desiccator (*see* **Note 6**).
8. Fix the tissue section by subsequently washing it for 1 min in 70% ethanol (fresh), again in 70% ethanol, 96% ethanol (*see* **Note 7**).

9. Dry the section again for 5 min in a desiccator.
10. If the sections have to be shipped or stored for more than 1 week then freeze them at –80°C. It is highly recommended not to freeze the sections and proceed immediately. Sections should always be shipped on dry ice (*see* **Note 8**).
11. With the liquid paper, mark four teach spots roughly rectangular around the tissue section (*see* **Note 9**).
12. Acquire a scan of the unstained tissue section including the teach spots (*see* **Note 10**).

3.2. Matrix Application

1. Coat one slide at a time with the ImagePrep device with the standard method settings according to the manufacturer's instructions (*see* **Notes 11** and **12**).

3.3. MALDI Measurement

1. Place the ITO slide with the matrix-coated tissue section into the MALDI adapter plate.
2. Insert the adapter plate into the ion source of the MALDI mass spectrometer.
3. Import the optical image of the unstained tissue section into FlexImaging.
4. Co-register the liquid paper spots between the optical image and the video image of the mass spectrometer.
5. Define the acquisition raster and region for imaging.
6. Start automated image acquisition (*see* **Note 13**).
7. Submit ITO slides to standard H&E staining and histology.

3.4. H&E Staining and Histology

1. Choose preferred staining protocol or proceed as follows (*see* **Note 14**).
2. Wash off matrix in 70% ethanol (usually 1–2 min) (*see* **Note 15**).
3. Stain the tissue section for 5 min in hematoxylin solution.
4. Remove excess staining solution by a dip-wash in deionized water.
5. Rinse the slide in running tap water for 5 min.
6. Wash slide for 1 min in deionized water.
7. Put slide into eosin working solution until the section is sufficiently stained (*see* **Note 16**).
8. Wash slide in deionized water.
9. Wash slide for 2 min each in 70% ethanol, 80% ethanol, 90% ethanol, 100% ethanol, again 100% ethanol, xylene (*see* **Note 17**).
10. With a glass stir bar put one droplet of mounting medium onto the slide and place coverslip onto tissue.

3.5. PCA and HC on Individual Sections

1. Load all spectra from the data set into the ClinProTools software and calculate a PCA and a hierarchical clustering (*see* **Note 18**).
2. Scan the stained tissue section.
3. Co-register the image of the stained tissue section.
4. Import PCA and clustering results into FlexImaging (*see* **Note 19**).
5. Have a tissue expert interpret the data in the histological context and name the relevant nodes of the dendrogram according to the desired aim of the analysis (*see* **Note 20**).

3.6. Search for Biomarkers by Comparing Multiple Patient Sections

1. Decide on the tissue states to be compared (such as "invasive tumor" versus "tumor-free mucosa," **Fig. 22.3**).
2. In each individual data set, export a list of spectra that are representative of the desired tissue state(s) as determined in **Section 3.5.5**.
3. Create a master list containing the spectra from 2 in the desired tissue classes (*see* **Note 21**).
4. In ClinProTools activate the option "Support spectra grouping." This ensures that spectra from one patient are treated as replicates rather than individual samples (*see* **Note 22**).
5. Now the available statistical tools (*p*-values, ROC curves) can be used to search for tissue state-specific biomarker candidates.
6. Have a tissue expert evaluate the masses of the biomarker candidates in the histological context on the individual tissue sections (*see* **Note 23**).
7. Optionally use PCA or HC at this level to get an idea about biological variance observed in the data set, on whether or not the tissue classes are separated based on the overall variance in the data set, and to correlate patient clusters with clinical meta-information (*see* **Note 24**).

4. Notes

4.1. Tissue Sections

1. Multivariate data analyses are sensitive not only to differences in the investigated groups (such as tumor versus non-tumor in this study) but also to changes in the measurement conditions. Such changes can result from, e.g., aging matrix solution and variations in instrument performance. It is not possible to consider all of these influences

in advance. That observed differences in the results are indeed inherent properties of the samples and not impacts of sample preparation or measurement can only be proven if the samples are randomized as soon as possible in the workflow. For example, it is not recommended to first cut all tumor sections and then all non-tumor sections. Likewise, it is never a good idea to conduct the matrix preparation and measurement for all tumor sections first and for all non-tumor sections thereafter. In these examples, it would not be possible to prove that a difference observed between tumor and control samples did not arise from variations during sample handling or measurement.

2. In a collaboration between a mass spectrometry facility and a pathology the tissue sections are usually prepared in the pathology lab. Although the preparation of tissue sections for MALDI imaging is very similar to the procedure for normal histology, there are small but very important changes that can cause trouble. It is of the utmost importance to discuss and sort out these issues before starting any sample preparation. In our experience, it is also best to discuss and explain the requirements of the sample preparation directly with the person actually doing the cutting. OCT is a polymer solution that is widely used to embed tissue specimens for cryosections. It facilitates cutting and is used to fix the tissue to the sample stage of the cryostat. Because of the detrimental effect of OCT to the MALDI imaging results it is important not to use OCT for embedding the tissue, but is equally important to thoroughly clean all surfaces that have been previously contaminated by OCT. OCT can be used to mount the tissue on the sample stage of the cryostat, as long as it is not cut. If the tissue specimens are already embedded in OCT, as much as possible of it should be trimmed off prior to the cutting.
3. It is important to transfer the tissue sections to the frozen glass slide (or to pick it up with the frozen glass slide) (14). This is in contrast to the widely used routine protocol in histology which simply uses warm glass slides.
4. Thicker tissue sections are easier to prepare, but the histological interpretation becomes more difficult. It is also reported that thinner sections improve mass spectra (15). Thinner sections also stick better to the ITO slides during the H&E staining.
5. Starting to warm the slide next to the tissue section and then slowly moving the finger underneath the section ensures that remaining folds in the section are stretched out.

6. Vacuum pumps and desiccators are not routine equipment in histology labs, but heating plates to dry the sections are. It is quite tempting in the preparation of the sections to stick to standard histology procedures that dry the tissue sections on a heated plate. In our experience, this has always been detrimental for the MALDI imaging results. In many cases, it will be necessary to bring a vacuum pump and desiccator to the histology lab.
7. We recommend to use new coplin jars and to keep them dedicated to the MALDI imaging, since standard equipment in a histology lab is often contaminated by OCT. If one is not satisfied with the quality of the mass spectra, one can consider to increase the washing times up to 5 min.
8. Whenever possible the tissue sections should be prepared fresh, stored in vacuum at room temperature for only a short time, and measured as soon as possible. They should only be re-frozen after the initial preparation, if they have to be shipped or stored for a longer time.
9. Try to make the edges of the teach marks uneven, this facilitates the teaching for the MALDI measurement later. Water-soluble liquid paper should be preferred since it will not fall off during the H&E staining.
10. Office scanners, preferable with transparency option, are sufficient here. Specialized scanners for photographic slides are best suited.

4.2. Matrix Application

11. The matrix should be applied directly before the measurement. Other matrix application devices can also be used; in this case the matrix solution needs to be one optimized for the device used. Sample preparation can also be done by manually spraying the matrix; however, this method will not be reproducible enough to effectively compare different sections.
12. Sinapinic acid is the matrix of choice for protein measurements. For peptides and lipids it is better to use 2,5-dihydroxybenzoic acid (DHB) or apha-cyano-4-hydroxycinnamic acid (HCCA). The latter has the advantage of giving very small crystals, therefore allowing highest lateral resolution.

4.3. MALDI Measurement

13. If multiple sections are available on the sample carrier, a batch acquisition should be considered.

4.4. H&E Staining and Histology

14. The choice of the staining protocol seems to be uncritical, so it is best to use the one routinely used by the pathology lab. The proposed protocol should be considered a

suggestion only. Staining times can be considerably different after MALDI imaging as compared to fresh sections.

15. Occasionally, tissue sections may come off the slide during the staining. In this case they are lost. For this reason it is a good idea to have consecutive backup sections available for the staining. In our experience, if the section detaches from the slide, then this happens during the initial washing step to remove the matrix.
16. Getting a feeling for the right staining intensity requires some practice. In this step the section has to be overstained, since in the subsequent washes in deionized water and 70% ethanol, the eosin is partially washed out.
17. The liquid paper teach marks may come off during the xylene wash. They sometimes move on the surface of the slide when the coverslip is put on and may slip onto the tissue. In doubt, remove the teach mark in the xylene bath with a glass bar. Additional teach marks, etched into the glass with a diamond tip pen, can be used to align the stained image later on. These should be placed before scanning the image as well.

4.5. PCA and HC on Individual Sections

18. Ensure that the peak scaling (by the checkbox "normalize peaks") is switched on. For the clustering, we have the best experience with the settings "Use PCA," "Reduce dimensions to 70% explained variance," "Create full tree," "Distance method: Euclidean," "Linkage method: Ward."
19. The display of PCA results usually benefits from the heatmap representation.
20. The cross-fading tool can help to localize the exact position on the H&E stain. To actually evaluate the histology at that position it is necessary to do microscopy on the stained section.

4.6. Search for Biomarkers by Comparing Multiple Patient Sections

21. For the convenient generation of the master list the tool "SIX" (part of ClinProTools 2.2 SR1) is needed.
22. Not marking spectra from a single patient sample as replicates would result in pseudo-replication, which in turn leads to wrong (and way too small) p-values.
23. This possibility to evaluate the potential biomarker masses directly in the histological context is one of the major strengths of MALDI imaging. This allows immediate evaluation whether the mass in question is indeed a tumor-marker or inflammation marker.
24. On this level each patient is represented by one (or two) data points.

References

1. Walch, A., Rauser, S., Deininger, S.-O., Höfler, H. (2008) MALDI imaging mass spectrometry for direct tissue analysis: a new frontier for molecular histology. *Histochem Cell Biol*, **130**, 421–434.
2. Yanagisawa, K., Shyr, Y., Xu, B. J., Massion, P. P., Larsen, P. H., White, B. C., Roberts, J. R., Edgerton, M., Gonzalez, A., Nadaf, S., Moore, J. H., Caprioli, R. M., Carbone, D. P. (2003) Proteomic patterns of tumour subsets in non-small-cell lung cancer. *Lancet*, **362**, 433–439.
3. Cornett, D. S., Mobley, J. A., Dias, E. C., Andersson, M., Arteaga, C. L., Sanders, M. E., Caprioli, R. M. (2006) A novel histology-directed strategy for MALDI-MS tissue profiling that improves throughput and cellular specificity in human breast cancer. *Mol Cell Proteomics*, **5**, 1975–1983.
4. Aoyagi, S., Kawashima, Y., Kudo, M. (2005) TOF-SIMS imaging technique with information entropy. *Nucl Instrum Methods Phys Res Sect B*, **232**, 146–152.
5. Lockyer, N. P., Vickerman, J. C. (2004) Progress in cellular analysis using ToF-SIMS. *Appl Surf Sci*, **231**, 377–384.
6. Wagner, M. S., Castner, D. G. (2001) Characterization of adsorbed protein films by ToF SIMS with PCA. *Langmuir*, **17**, 4649–4660.
7. Deininger, S.-O., Ebert, M. P., Fütterer, A., Gerhard, M., Röcken, C. (2008) MALDI imaging combined with hierarchical clustering as a new tool for the interpretation of complex human cancers. *J Proteome Res*, **7**, 5230–5236
8. Yao, I., Sugiura, Y., Matsumoto, M., Setou, M. (2008) *In situ* proteomics with imaging mass spectrometry and principal component analysis in the Scrapper-knockout mouse brain. *Proteomics*, **8**, 3692–3701.
9. McCombie, G., Staab, D., Stoeckli, M., Knochenmuss, R. (2005) Spatial and spectral correlations in MALDI mass spectrometry images by clustering and multivariate analysis. *Anal Chem*, **77**, 6118–6124.
10. Crecelius, A. C., Cornett, D. S., Caprioli, R. M., Williams, B., Dawant, B. M., Bodenheimer, B. (2005) Three-dimensional visualization of protein expression in mouse brain structures using imaging mass spectrometry. *J Am Soc Mass Spectrom*, **16**, 1093–1099.
11. Schwamborn, K., Krieg, R. C., Reska, M., Jakse, G., Knuechel, R., Wellmann, A. (2007) Identifying prostate carcinoma by MALDI-imaging. *Int J Mol Med*, **20**, 155–159.
12. Chaurand, P., Schwartz, S. A., Billheimer, D., Xu, B. J., Crecelius, A., Caprioli, R. M. (2004) Integrating histology and imaging mass spectrometry. *Anal Chem*, **76**, 1145–1155.
13. Efron, B., Halloran, E., Holmes, S., (1996) Bootstrap confidence levels for phylogenetic trees. *Proc Natl Acad Sci U S A*, **93**, 13429–13434.
14. Schwartz, S. A., Reyzer, M. L., Caprioli, R. M. (2003) Direct tissue analysis using matrix-assisted laser desorption/ionization mass spectrometry: practical aspects of sample preparation. *J Mass Spectrom*, **38**, 699–708.
15. Sugiura, Y., Shimma S., Setou, M. (2006) Thin sectioning improves the peak intensity and signal-to-noise ratio in direct tissue mass spectrometry. *J Mass Spectrom Soc Jpn*, **54**, 45–48.

Chapter 23

Applications of MALDI-MSI to Pharmaceutical Research

Brendan Prideaux, Dieter Staab, and Markus Stoeckli

Abstract

MALDI-MSI has been demonstrated to be a suitable technique in pharmaceutical research for providing information of the distribution of low molecular weight compounds such as drugs and their metabolites within whole-body tissue sections. Important ADME information can be determined by MALDI-MSI analysis of the distribution of drugs and metabolites in whole-body tissue sections taken from animals killed at a range of time points postdose. In this example we applied MALDI-MSI to the localization of a compound and its primary metabolite in whole-body mouse sections.

Key words: MALDI-MSI, whole-body tissue imaging, drug and metabolite imaging.

1. Introduction

In the drug discovery process essential information can be gleamed from knowing the uptake of a drug to its target as well as its metabolic pathway processes and the sites at which these are occurring within the body following administration. These absorption, distribution, metabolism, and excretion (ADME) data are required to fully understand the efficacy, safety, and thus viability of a compound, and the earlier this is understood during the drug discovery process then increased potential exists for the ability to adapt to potential problems and subsequent time and cost savings.

MALDI-mass spectrometric imaging (MALDI-MSI) has been extensively discussed and reviewed in the literature since its introduction over 10 years ago (1) and has now become an established method of localizing a range of analytes in biological

S.S. Rubakhin, J.V. Sweedler (eds.), *Mass Spectrometry Imaging*, Methods in Molecular Biology 656,
DOI 10.1007/978-1-60761-746-4_23,

tissues. Examples of analytes which have been studied include the imaging of endogenous and exogenous species such as proteins (2), peptides (3, 4), and lipids (5) as well as low molecular weight xenobiotics including (but not exclusive to) drugs (6–10). Studies have been conducted at high resolution (<100 μm) for discrete organs and tissue regions and at lower resolutions to encompass whole bodies of animal models such as mice or rats (6–8, 10). The technology and instrumentation is gaining increasing recognition in industrial pharmaceutical laboratories as an important tool for highly sensitive, high-resolution imaging of compound distribution direct from biological tissue.

Traditional methods of spatially mapping the distribution of compounds and related metabolites in tissue involved a combination approach of whole-body autoradioluminography (WBAL) (11) with metabolite information obtained from LC/MS analysis of tissue extracts. In the WBAL process the compound of interest is labeled with a radioactive isotope and administered to animals which are then killed at a range of time points postdose. Whole-body tissue sections are exposed to a radioactivity detector which allows the distribution of the radioactivity within the tissue section to be visualized. The method enables high-resolution, highly sensitive analysis of the labeled compound and can produce qualitative and quantitative information (QWBAL). For these reasons it has rapidly become the standard method within the pharmaceutical industry for studying tissue distribution and is recommended for use by the FDA and the Organization for Economic Co-operation and Development (OECD) (12). However, there are two major limitations to consider when applying WBAL: the need for radiolabeling of the compound (and the increased cost and time issues that may involve) and the fact that only the radioactive label is measured and thus parent compounds cannot be distinguished from their metabolites. The second is a particular cause of concern when applying WBAL to absorption, distribution, metabolism, and excretion (ADME) studies during the drug development process and is an area where MALDI-MSI can add vital information either as a complementary or as an alternative imaging technology.

MALDI-MSI has previously been used in combination with whole-body autoradioluminography (WBAL) to map the distribution of drugs and metabolites in dosed animals (8). The application of MALDI-MSI methods allows for the simultaneous detection of multiple analytes of interest which can be identified by their molecular weight and specific fragmentation patterns. Subject to instrumental variance and parameter selection (such as raster size and spectra acquisition time and number) whole sections can be imaged rapidly (normally within several hours for a standard 500 μm raster) in comparison to WBAL which may require several days.

2. Materials

1. Embedding media: 2% carboxymethylcellulose (Sigma-Aldrich, Steinheim, Germany) in water, resulting in a semi-liquid gel. Store at 4°C before usage.
2. Double-sided fixation tape (Tesa, Hamburg, Germany).
3. Aluminum foil.
4. MALDI plates: stainless magnetic steel, 0.5 mm thickness. Laser cut to match the MALDI instrument. Edges are polished to enable a flat positioning of the plate.
5. α-Cyano-4-hydroxycinnamic acid (CHCA) (Sigma-Aldrich, Steinheim, Germany) freshly prepared at 10 mg/mL in 50:50 (v/v) ACN:0.1% TFA.
6. Thin-layer chromatography sprayer (TLC sprayer) operated at 0.5 bar pressure using nitrogen gas.
7. All described MS analyses are for the use of a QSTAR Elite mass spectrometer equipped with an oMALDI source utilizing a high repetition 1 kHz Nd:YAG laser (Applied Biosystems/MD Sciex, Foster City, CA, USA). Analyst QS 2.0 and oMALDI Server 5.0 software (Applied Biosystems/MD Sciex, Foster City, CA, USA) are used for image acquisition.
8. BioMap software for data and image processing (freely available from the Web site http://www.maldi-msi.org).

3. Methods

3.1. Small Animal Tissue Sections

1. Animals are dosed with compound and killed at specific time points matching with the kinetics of the compound. A series with eight animals at 0.17, 0.5, 2, 4, 8, 24, 48, and 168 h post is suitable for most compounds. If prior knowledge of compound is available, reduce numbers of animals and adapt the time points.
2. Animals are embedded in carboxycellulose and frozen at –20°C according to the method of Ullberg (13).
3. Sagittal sections of 40 μm thickness are obtained in a cryomicrotome (e.g., CM 3600 PLC, Leica Cryosystems GmbH, D-Nussloch) and lifted off the block using adhesive tape (14).
4. Sections are loosely placed on aluminum foil and transferred to a box holding up to 20 sections. The box is kept in the cryotome during the sectioning process.

5. The container with the section is placed in an airtight plastic bag and transferred to a –80°C freezer.

3.2. Dehydration

1. A copper block having at least the same size as the tape with the tissue section is stored at –80°C.
2. The copper block is transferred to a desiccator and a tape with a section is placed on the block (with the section facing up) and held in place by putting metal weights on the corners of the tape.
3. The desiccator is immediately pumped down using a membrane pump or a rotary pump with a condensation trap until the section is completely dehydrated (approx. 1 h, *see* **Notes 1–3**) (*see* **Fig. 23.1**).

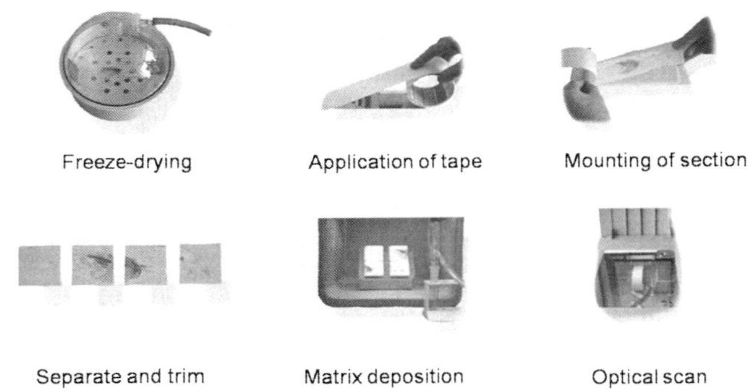

Fig. 23.1. The main stages in sample preparation of whole-body sections for MALDI-MSI. The procedure is explained in detail in **Section 3**, Steps 2–4.

3.3. Mounting

1. If animal does not fit on one single MALDI plate, place a large Post-it(R) (3M) with the adhesive side facing up onto a flat surface. Align the plates on the adhesive strip which will prevent them from shifting during the mounting process.
2. With the plates on a clean flat surface, start attaching the double-sided sticky tape on the short side of the plate and apply strong pressure with a cylindrical object (metal rod or suitable pen). Drag the object firmly across the plate to attach the tape. Ensure no air bubbles become trapped between the tape and the plates.
3. Carefully remove the tissue section from the desiccator and attach to the tape. Start by applying pressure along one edge of the section and progressively press firmly across the entire length of the section to force out any air trapped between the tape and the section. Do not directly apply tools to the tissue surface but use a covering foil to protect it. Ensure that the foil stays in place and lift it up orthogonally to prevent cross-contamination of analytes.

4. Using a scalpel, cut between the plates to separate them and trim any remaining tape or tissue from the outside of each plate to facilitate easy insertion of the plate into the MALDI plate holder.
5. Optically scan each plate using a flatbed scanner. Using a resolution of 762 dpi (30 dots per mm) allows for simple overlay of the optical image with the molecular MALDI-MSI image acquired using a metric raster.

3.4. Matrix Deposition

1. Select suitable matrix based on analysis of compound standards and tissue extracts (an example extraction is explained in **Note 4**). CHCA is frequently a good choice due to small crystal size and high ionization efficiency.
2. Align the plate(s) horizontally in the spray chamber and using a TLC sprayer, operated at 0.5 bar air pressure, apply the CHCA matrix evenly across all plates (holding the TLC sprayer at a distance of 15 cm from the plate and moving at a velocity of ~10 cm/s).
3. Repeat the cycle, allowing the tissue to dry between cycles until 5 mL of matrix solution has been applied to each plate (i.e., for four plates 20 mL of matrix solution is required). A typical coating requires 20 passes in a time period of 20 min (*see* **Note 5**).
4. Use an optical microscope to determine whether the matrix coverage is sufficient and a homogeneous distribution has been achieved. The criterion is a homogeneous coating of matrix over the tissue surface (an example is shown in **Fig. 23.2**). Alternatively, compare the yellow color saturation of a blank area on the plate to a reference plate with a previously applied matrix coating. Apply further matrix if required in 1 mL volumes until a fully homogeneous matrix coverage is achieved.

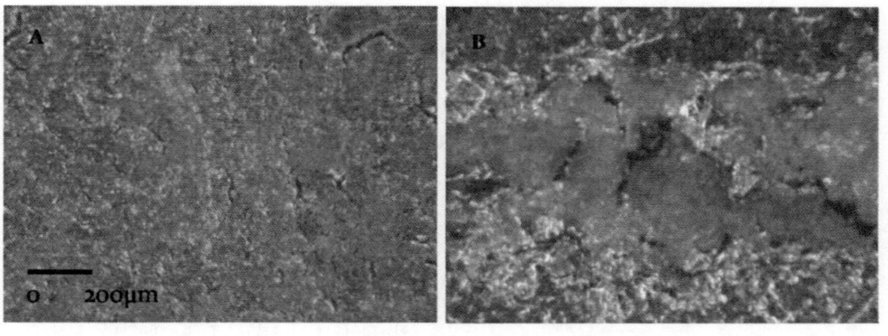

Fig. 23.2. Light microscope image showing (**a**) a homogeneous distribution of CHCA matrix achieved after 20 passes over rat lung tissue by airspray deposition and (**b**) an inhomogeneous coverage of matrix achieved after seven passes of CHCA over a subsequent lung section.

3.5. MALDI-MSI Acquisition

1. Optimize MS and MS/MS imaging parameters for maximum signal intensity of the protonated compound and its fragments using the compound standard and tissue extract samples (following the extraction procedure presented in **Note 4**).
2. Set up an acquisition method to encompass the required number of analyses (full MS scan and MS/MS scans for the compound and any identified metabolites). While full scan does not deliver images with high confidence in identity of analyte, it allows an evaluation of the sample preparation process and the quality of the sample. This scan is combined with MS/MS of parent compound and MS/MS scans of the main metabolites.
3. Select the raster imaging method and define the imaging window to encompass the entire plate. If raster imaging is not available, select discrete pixel mode and define the raster over the plate. In raster mode, select an acquisition speed which results in a spatial resolution of 500 μm along the *x*-axis. As an example, a speed of 1 mm/s acquiring one full scan MS and three MS/MS scans with 125 ms each results in an *x*-raster of 500 μm. Choose a line spacing (*y*-raster) matching with the *x*-raster, in this example 500 μm, to obtain an array of 88 times 88 pixels over each sample plate. Set the laser frequency to 1 kHz and the laser energy to the level which provides the highest quality signal as determined in the optimization step (typically 50% above the intensity required for ablation from a blank metal pate). A frequency of 1 kHz and a dwell time of 125 ms results in spectra with good signal-to-noise ratios (*see* **Note 6**).
4. Convert the resulting data file into the Analyze 7.5 format (in case of QSTAR using the 'WiffToAnalyze_RasterQS' script of the Analyst QS 2.0 software). Using BioMap software, calculate intensity distribution maps by selecting the mass range corresponding to the peak boundaries of the mass of interest and integrating the peak area with baseline subtraction.
5. Normalization of the measured analyte signal against the matrix signal can be performed to compensate for specific tissue suppression (*see* **Note** 7). For the matrix CHCA, selected peaks for normalization are $[M+H]^+$ at *m/z* 190, $[M-H_2O+H]^+$ at *m/z* 172, and the dimer $[2\ M+H]^+$ at *m/z* 379 (**Fig. 23.3**). In this procedure, divide the signal area of the compound peak by the peak area of the matrix measured on the same section using the settings available in the BioMap software. A nominator threshold should be selected as a value above the image noise; typically this is

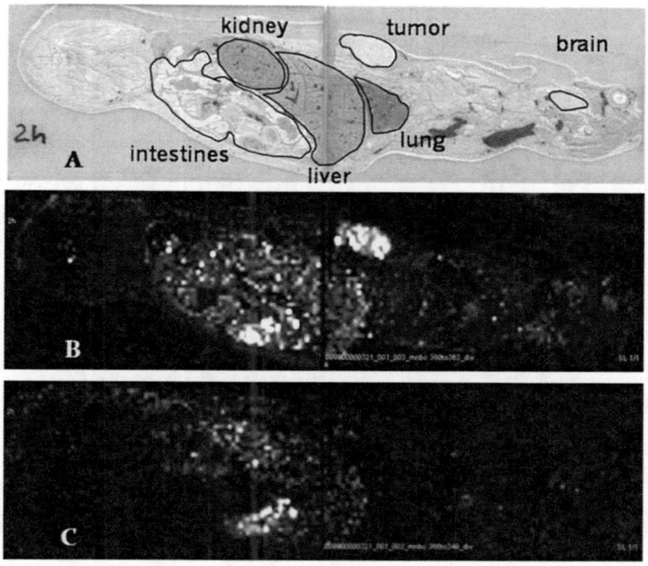

Fig. 23.3. Optical and MS images of whole-body mouse sections 2 h postdosage with a pharmaceutical compound. (**a**) Optical scanned image prior to matrix application. (**b**) and (**c**) Distribution of the compound product ion (*m/z* 363) and the metabolite product ion (*m/z* 349.1), respectively. MS images were normalized with MSI of *m/z* 379.1 (CHCA dimer) as internal standard. All images show the same intensity in gray scale (white = high concentration).

in the region of 30–40 counts. The mass spec image may be overlaid over the scanned optical image using the overlay and co-registration features included in the software enabling regions of interest to be defined and thus region-specific numerical information to be extracted (*see* **Note 8**).

4. Notes

1. Sections must not be dehydrated using the slow protocol suitable for autoradiography.
2. Tissue sections should be processed as soon as possible after the sectioning process. If stored at –80°C in an airtight bag, sections may last for years in a suitable condition for MSI analysis. If sections with prolonged storage are used for MSI measurement, a control experiment is performed and compared to results obtained at an earlier time point to ensure the quality of the section.
3. If a dedicated dehydration chamber is available, use it according to the manufacturer's instructions. It must

facilitate a fast and complete lyophilization, ideally not lasting longer than 1 h.

4. The extraction procedure parameters (such as extraction solvent and volume) will vary depending upon the tissue and compound being investigated. An example of a simple tissue extraction method we use for compounds is spike uncoated tissue sections with 5 μL drops of extraction solvent (50:50 (v/v) ACN:H_2O) and transfer 3 × 1 μL extracts onto a blank metal sample plate using a pipette. Add 1 μL of the matrices for assessment onto each extract, allow to dry, and acquire a spectrum using MALDI-MS. This method allows for direct comparison of three different matrices.
5. It is important that the tissue never becomes fully saturated with the matrix solution which may result in diffusion of the analyte across the tissue and thus loss of spatial integrity in the resulting image. However, the tissue must be sufficiently wet to allow extraction of the analyte from the tissue and co-crystallization with the matrix.
6. Although the described method relates specifically to the QSTAR Elite Mass spectrometer, it can be easily adapted to most commercial instruments optimized for the analysis of low molecular weight compounds. The BioMap software supports all current commercial MALDI-MSI instruments.
7. Normalizing against the matrix will only negate tissue-specific matrix suppression effects. In the case of severe analyte suppression due to competing species, a suitable internal standard may be incorporated into the matrix application step and sprayed homogeneously across the tissue. The distribution of the analyte signal may then be normalized against the distribution of the selected internal standard signal.
8. Regions of interest (ROIs) may be drawn on the optical image and transferred to the MS image enabling distinct tissue areas and organs to be clearly defined. Numerical information relating to analyte distribution and signal abundance can be extracted, plotted graphically, and used for further statistical analysis and the creation of semi-quantitative distribution profiles (8).

References

1. Caprioli, R. M., Farmer, T. B., Gile, J. (1997). Molecular imaging of biological samples: localization of peptides and proteins using MALDI-TOF MS. *Anal Chem,* **69**, 4751–4760
2. Stoeckli, M., Chaurand, P., Hallahan, D. E., Caprioli, R. M. (2001). Imaging mass spectrometry: a new technology for the analysis of protein expression in mammalian tissues. *Nat Med,* 7, 493–496

3. Taban, I. M., Altelaar, A. F., van der Burgt, Y. E., McDonnell, L. A., Heeren, R. M., Fuchser, J., Baykut, G. (2007). Imaging of peptides in the rat brain using MALDI-FTICR mass spectrometry. *J Am Soc Mass Spectrom,* **18**, 145–151
4. Stoeckli, M., Staab, D., Schweitzer, A., Gardiner, J., Seebach, D. (2007). Imaging of a beta-peptide distribution in whole-body mice sections by MALDI mass spectrometry. *J Am Soc Mass Spectrom*, **18**, 1921–1924
5. Touboul, D., Roy, S., Germain, D. P., Chaminade, P., Brunelle, A., Laprévote, O. (2007) MALDI-TOF and cluster-TOF-SIMS imaging of Fabry disease biomarkers. *Int J Mass Spectrom*, **260**, 158–165
6. Rohner, T., Staab, D., Stoeckli, M. (2005). MALDI mass spectrometric imaging of biological tissue sections. *Mech Ageing Dev*, **126**, 177–185
7. Khatib-Shahidi, S., Andersson, M., Herman, J. L., Gillespie, T. A., Caprioli, R. M. (2006). Direct molecular analysis of whole-body animal tissue sections by imaging MALDI mass spectrometry. *Anal Chem*, **78**, 6448–6456
8. Stoeckli, M., Staab, D., Schweizer, A. (2007). Compound and metabolite distribution measured by MALDI mass spectrometric imaging in whole-body tissue sections. *Int J Mass Spectrom*, **260**, 195–202
9. Drexler, D. M., Garrett, T. J., Cantone, J. L., Diters, R. W., Mitroka, J. G., Prieto Conaway, M. C., Adams, S. P., Yost, R. A., Sanders, M. (2007). Utility of imaging mass spectrometry (IMS) by matrix-assisted laser desorption ionization (MALDI) on an ion trap mass spectrometer in the analysis of drugs and metabolites in biological tissues. *J Pharmacol Toxicol Methods*, **55**, 279–288
10. Trim, P. J., Henson, C. M., Avery, J. L., McEwan, A., Snel, M. F., Claude, E., Marshall, P. F., West, A., Princivalle, A. P., Clench, M. R. (2008). Matrix-assisted laser desorption/ionization-ion mobility separation-mass spectrometry imaging of vinblastine in whole body tissue sections. *Anal Chem*, **80**, 8628–8634
11. Johnson, R. F., Pickett, S. C., Barker, D. L. (1990). Autoradiography using storage phosphor technology. *Electrophoresis,* **11**, 355–360
12. Solon, E. G., (2007). Autoradiography: high-resolution molecular imaging in pharmaceutical discovery and development. *Expert Opin Drug Discov* **2**, 503–514
13. Ullberg, S. (1977). The technique of whole-body autoradiography. Cryo-sectioning of large specimens. *Science Tools* (The LKB Instr. Journal). 2–29 (special issue)
14. Schweitzer, A., Fahr, A., Niederberger, W. (1987) A simple method for the quantitation of ^{14}C-whole-body autoradiograms. *Int J Appl Radiot Isotopes*, **38**, 29–333

Chapter 24

Tissue Preparation for the In Situ MALDI MS Imaging of Proteins, Lipids, and Small Molecules at Cellular Resolution

Nathalie Y.R. Agar, Jane-Marie Kowalski, Paul J. Kowalski, John H. Wong, and Jeffrey N. Agar

Abstract

The resolution of MALDI MS imaging is limited by the displacement of analytes during matrix deposition or by laser focal diameter. Here we present three methods that minimize the displacement of analytes during matrix deposition, including a method where image resolution is not limited by the laser focal diameter. The first method, matrix solution fixation, simultaneously fixes tissue while depositing matrix and is optimal for analyzing proteins and for applications requiring a fast preparation time. This method is characterized by compatibility with histology methods and laser focal diameter-limited resolution. The second method, a sensor controlled aerosol, is characterized by aerosol droplet size-limited resolution and is optimal for small molecules, including lipids, peptides, and drug-like molecules. The third method, microinjection with matrix, selectively deposits matrix upon cells of interest, offers cellular resolution and is compatible with most analytes. A flow chart summarizing methods is provided so that users may design a tissue preparation strategy based upon their resources and experimental goals.

Key words: Mass spectrometry, imaging, matrix solution fixation, single cell, optimization.

1. Introduction

The most common solutions used to dissolve MALDI matrix, e.g., 50% H_20, 50% acetonitrile (ACN), 0.1 % trifluoroacetic acid (TFA), also dissolve proteins. When such a matrix solution comes in contact with tissue, diffusion of proteins and lipids occurs and the droplet size defines the lower limit of image resolution (1). Most MALDI imaging sample preparation methods minimize this protein diffusion by decreasing the size of, including altogether

S.S. Rubakhin, J.V. Sweedler (eds.), *Mass Spectrometry Imaging*, Methods in Molecular Biology 656, DOI 10.1007/978-1-60761-746-4_24, eliminating, the matrix solution droplet. More popular methods include the acoustic matrix deposition robots (2), matrix "printers" based upon piezoelectric ink-jet technology (3), nebulization (4) and electrospray (5), and dry deposition methods (6). These methods are listed in order of descending costs and to an extent descending robustness, with costs currently ranging from >$100 k for acoustic deposition to a few dollars for a thin-layer chromatography sprayer, and with methods such as electrospray deposition requiring skilled practitioners. Most of these techniques impose a resolution limit of ~100 μm, which is similar to the resolution limit imposed by the laser focal diameter of commercial mass spectrometers.

A second strategy for minimizing the diffusion of analytes during matrix deposition involves fixing the tissue before (7), or during (8), matrix solution deposition. Such solvent fixation of tissues has been standard practice in histology for over 100 years (9–11). Applying a MALDI matrix dissolved in a solution that denatures and precipitates proteins (8), as is presented here in Method 1 (**Section 3.1**), results in the diffusion of proteins being thermodynamically unfavorable (12). The benefits of combining well-established histology protocols with matrix deposition include minimized disruption of tissue morphology versus common matrix solutions (2) and compatibility with immunohistochemistry methods for validation of mass spectrometry findings. The matrix solution fixation technique presented herein (8) imposes a resolution limit of ~1 μm, making laser focal diameter the determinant of image resolution.

We describe three methods for the preparation of thin tissue sections for in situ MALDI MS imaging of proteins, lipids, and small molecules. Each method can maintain near-cellular resolution during sample preparation. The first method employs a strategy termed "matrix solution fixation," which simultaneously fixes tissue and deposits MALDI matrix. This method enables rapid preparation and high-resolution (~30 μm) MALDI imaging of entire thin tissue sections. The second method uses a sensor controlled vibrational vaporization device for layered deposition of 20–100 μm matrix droplets and limits diffusion of soluble analytes to deposited droplet surface area. The third method employs a modified microinjection setup to selectively deposit matrix upon single cells, offering resolution below the laser spot size (10 μm). Specifically, single neurons are labeled with a mixture of MALDI matrix, sinapinic acid, and rhodamine B (for contrast in light microscopy and for visualization with the MALDI microscope). Fluorescence microscopy is used to guide deposition of matrix and rhodamine B upon single fluorescent cells.

Notably, matrix microinjection overcomes the resolution limit imposed by the MALDI laser focal diameter. Numerous other methods serve a similar purpose including laser and mechanical

Table 24.1
Spatial resolution of mass spectrometry methods described in this book and in this chapter. References to the chapters and the methods are made in parenthesis

Spatial Resolution of Mass Spectrometry Methods

1 μm		10 μm		100 μm		1000 μm
SIMS (Chapters 4,6,11,16)	MALDI MS Microscopy		MALDI MS Imaging* (Methods 1–3)		DESI (Chapters 9,13)	

Matrix Sublimation*	Matrix Solution Fixation* (Method 1)	Microinjection with Matrix* (Method 3)	Nebulization* (Method 2 and Chapter 7)	Accoustic or Microfluidic Application* (Chapter 17)	Manual Application or Dissection* (Chapters 15, 26, 27)
1 μm		10 μm	100 μm		1000 μm

Spatial Resolution of Matrix Deposition Methods

* Image resolution is the greater of the limits imposed by sample preparation and laser spot size (the diffraction-limited spotsize of a nitrogen laser is 170 nm). Microinjection with matrix, MALDI microscopy, tissue stretching (Chapter 27), and laser oversampling (Chapters 9 and 25), can be used to overcome laser diffraction-limited resolution.

dissection of individual cells (13, 14), ordered stretching of tissues (15), laser oversampling (5), improved lasers and optics (16), and ion microscopy (17–19) (**Table 24.1**). These and other "tricks of the trade," including seeding with dry matrix and careful thaw mounting (20), may be considered to augment the methods described herein.

2. Materials

2.1. Matrix Solution Fixation

1. Microm HM525 cryostat (Mikron Instruments Inc., San Marcos, CA, USA).
2. Embedding medium for frozen tissue specimens to ensure optimal cutting temperature (OCT, Sakura Finetek, Torrance, CA, USA).
3. Protein and peptide calibration standards (Bruker Daltonics Billerica, MA, USA).
4. α-Cyano-4-hydroxycinnamic acid (α-CHCA) matrix and sinapinic acid.
5. HPLC grade methanol, water, ethanol, acetonitrile.
6. Trifluoroacetic acid.
7. Indium tin oxide (ITO) coated coverslips with busbars are 1×18 mm with resistivity from 8 to 12 ohms (thickness 1: 0.13–0.17 mm) (SPI Supplies, West Chester, PA, USA).

8. ImmEdge PEN (Vector Laboratories Inc., Burlingame, CA, USA).
9. Custom made coverslip holders. Made from original Bruker's Microflex stainless steel target, with a machined cavity of 25×25 mm and depth of 300 μm to accommodate glass coverslips of dimensions: 22×22 mm, thickness 1 (0.13–0.17 mm). The coverslips are held on the target by minimal conductive adhesive tape. For users of TOF–TOF instruments such as the Bruker Ultraflex, an ITO-coated glass slide holder and ITO-coated glass slides are available as stock items, and therefore a custom made coverslip holder is unnecessary.
10. Biological preparation: transgenic animals with cell-specific fluorescence. Transgenic animal models overexpressing fluorescent proteins in particular subsets of cells are available (The Jackson Laboratories, Bar Harbor, ME, USA). Fluorescent mice are chosen so that the cells of biological interest are labeled, for example, mice overexpressing YFP in motor neurons are bred with ALS transgenic mice, G93A hSOD1, to fluorescently label cells inducing ALS symptoms upon dysfunction (21). This method's animal manipulations are approved by the Brandeis Animal Care and Use Committee and are carried out in the Brandeis University animal care facility in accordance with federal, local, and institutional guidelines.

2.2. Sensor Controlled Aerosol

1. ImagePrep (Bruker Daltonics, Billerica, MA, USA).
2. MALDI-TOF–TOF Ultraflex III mass spectrometer with 200 Hz solid-state laser (Bruker Daltonics, Billerica, MA, USA).
3. Scanning electron microscope Zeiss FESEM Supra55VP (Carl Zeiss SMT Inc., Peabody, MA, USA).
4. Confocal microscope Leica TCS SP2 AOBS (Acousto Optical Beam Splitter) Spectral Confocal Microscope equipped with a 405 nm UV laser (Leica Microsystems, Bannockburn, IL, USA).
5. Vectashield HardSet mounting media with 4′,6-diamidino-2-phenylindole (DAPI) (Vector Laboratories, Burlingame, CA, USA).
6. Biological preparation: multimodal cell tracking using fluorescent dye uptake. Human U87 glioma cells are labeled by fluorescence and mass markers by in vitro uptake of DiI (λ_{ab} 549 and λ_{em} 565 nm; m.w. 933.9 g mol^{-1}) and human neural stem cells HB1.F3 are similarly labeled with DiO (λ_{ab} 484 and λ_{em} 501 nm; m.w. 881.7 g mol^{-1}). Vybrant DiI and DiO solutions are from Invitrogen

(Carlsbad, CA, USA). Aliquots of 120,000 labeled U87 and HB1.F3 cells in 3 μl sterile PBS are surgically injected into the left brain hemisphere of anesthetized nude mice using a stereotactic device on days 1 and 10, respectively. Swiss nude male mice of 4–8 weeks of age (Charles River Laboratories, Wilmington, MA, USA) are anesthetized and killed by intraperitoneal 90 mg/kg ketamine and 10 mg/kg xylazine. Method 2 (**Section 3.2**) animal manipulations are committee approved and carried out in the animal facility at the Brigham and Women's Hospital in accordance with federal, local, and institutional guidelines. Animals are sacrificed on day 14, and the brains readily dissected and flash frozen in liquid nitrogen.

7. Matrices: sinapinic acid and α-cyano-4-hydroxycinnamic acid (α-CHCA).

2.3. Microinjection with Matrix

1. Programmable Micropipette Puller, PMP-102 (MicroData Instrument Inc., Plainfield, NJ, USA).
2. Model PLI-100 Pico-Injector (Harvard Apparatus Inc., Holliston, MA, USA).
3. Microscope with mounted apparatus for microinjection.
4. Borosilicate glass capillaries (GC100-7.5) (Harvard Apparatus Part No. 30-0018, 1.0 mm OD×0.58 mm ID).
5. Micromanipulator with *x*, *y*, *z* control (Standard Manual Mechanical Micromanipulator, Harvard Apparatus Inc., Holliston MA, USA).
6. Biological preparation: same as in Method 1 (**Section 2.1**).
7. Rhodamine B – optional- aids in visualization.

3. Methods

The methods presented herein are histology compatible protocols for use with either proteins (Methods 1 and 3) or smaller molecules, including peptides, lipids, and drugs and their metabolites (Methods 2 and 3). These protocols include matrix solution fixation (**Section 3.1**); a sensor controlled aerosol (**Section 3.2**), and matrix microinjection (**Section 3.3**), a novel method for labeling single cells.

Matrix solution is optimized for proteins, prepares the entire surface of the tissue for imaging, results in tissue deformation and analyte redistribution on the order of ~1 μm, and is among the most rapid of all matrix deposition methods. The sensor controlled aerosol we present is optimized for small

(<1,000 Da) molecules and results in analyte redistribution of less than 100 μm. Matrix microinjection preserves tissue integrity, overcomes laser spot size imposed resolution limits, and has analyte redistribution that is limited to the size of the microinjected matrix droplet (~10 μm). The sensor controlled aerosol and matrix microinjection protocols could conceivably be combined with matrix solution fixation.

Quality control methods for each protocol are also presented. Applications and representative data are presented for the sensor controlled aerosol and matrix microinjection, whereas representative data for matrix solution fixation are reported elsewhere (8). Sensor controlled aerosol is applied to multimodal "cell tracking" by fluorescence and multiple reaction monitoring (MS/MS) of a fluorescent dye incorporated into stem cells and tumor cells surgically implanted into a mouse brain. This protocol enables validation of MSI using confocal microscopy for the tracking of implanted cells, and potentially the effect of a given cell's microenvironment upon its molecular composition. Matrix microinjection is used to acquire a MALDI mass spectrum of a single motor neuron from layer V of a mouse motor cortex.

3.1. Matrix Solution Fixation

1. Prepare fresh matrix fixation solution consisting, by volume, of 8 parts ethanol: 8 parts methanol: 1 part acetonitrile: and 1 part 0.1% TFA in water (*see* **Note 1**). Use this solution to dissolve matrix to a final concentration of 25–30 mg/ml. Sonicate the matrix solution for 30 min and centrifuge for 10 min at 16,000×*g* (*see* **Note 2**). Allow the solution and all other materials, including pipette tips, ITO-coated slides or coverslips, and the tissue block to reach −22°C before using (keep in the cryostat for 30 min prior to use). The temperature is adjusted according to tissue type (*see* **Note 3**).
2. Tissue specimens are mounted in optimal cutting temperature medium (OCT) and sectioned at ~−20°C at 5–16 μm thickness (*see* **Note 3**) and thaw mounted on ITO-coated glass slides, ITO-coated coverslips, or directly upon the metal MALDI target. Thaw mounting should be performed using a cold (−22°C) slide warmed using a finger just enough to affix the sample, but minimizing thawing time to <2 s (*see* **Note 4**).
3. To maintain a constant ratio of matrix solution to tissue volume, an ImmEdge Pen is used to delineate a surrounding physical and hydrophobic barrier at approximately 3 mm away from the tissue's edge and allowed to dry (<10 min at −22°C).
4. For a typical mouse brain section of dimensions 16 μm thickness with a surface of 10× 5 mm, the tissue volume

is ~0.8 μl and 20 μl of fixing solvent (*see* **Note 5**) is applied to the tissue at −22°C and allowed to dry (*see* **Note 6**).

5. A MALDI-TOF mass spectrometer, for example, a Microflex interfaced with FlexImaging Software from Bruker Daltonics (Billerica, MA, USA) is operated in linear mode for m/z greater than 3,000, and in reflectron mode for m/z less than 3,000. The instrument is equipped with a standard nitrogen laser with 337 nm wavelength and a maximum repetition rate of 20 Hz. Standard instrument parameters are used, although delayed extraction times are optimized

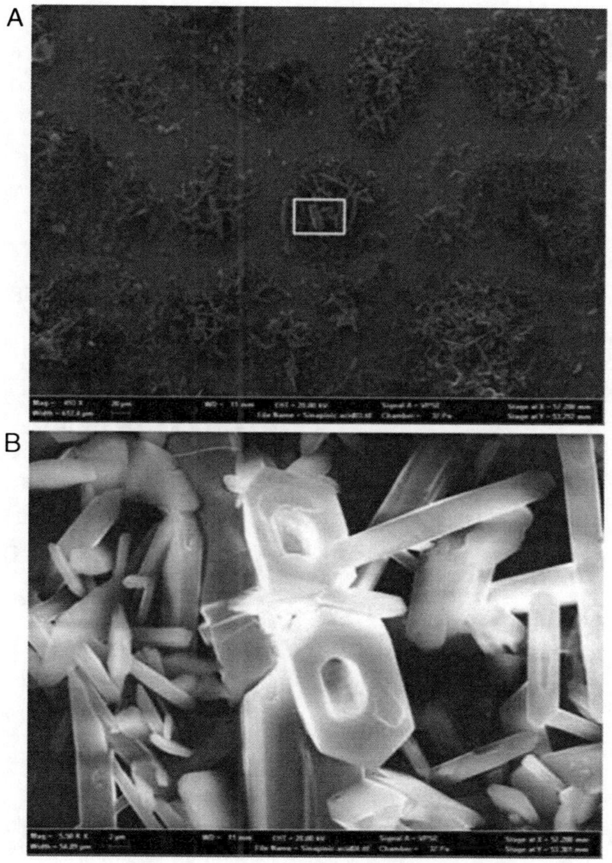

Fig. 24.1. Scanning electron microscopy of sinapinic acid deposition by vibrational vaporization on ITO-coated glass. (**a**) The surface was imaged with a ×493 magnification, providing an image width of 612.4 μm, allowing the visualization of isolated matrix droplets and their surface distribution (scale bar = 20 μm). (**b**) The approximate boxed area in (**a**) was further imaged with a ×5,500 magnification for a full image width of 54.89 μm. This higher magnification view allows the observation of the reported hexagonal crystal lattice of sinapinic acid in its center, as well as for smaller surrounding crystals. Such high magnification also reveals heterogeneity in crystal size and arrangement (scale bar = 2 μm).

(usually within 400–800 ns), depending on the targeted mass range.

6. Quality control (*see* **Note 7**). Due to the use of a single drop of solvent covering the entire tissue's surface, there is the potential for analyte displacement throughout the sample. Users should therefore always analyze an area inside the hydrophobic barrier that was wetted by matrix solution, but that does not contain tissue (i.e., a place in the 3 mm between the tissue and hydrophobic area). If matrix solution fixation was successful, no peak, or a single peak at 4,973 m/z from thymosin β4, is detected. As in standard histology, protocols can be optimized for proteins of interest by adjusting solution composition, temperature, and time.

3.2. Sensor Controlled Aerosol

1. Frozen mouse brains are sectioned at 16 μm and thaw mounted onto ITO-coated glass slides and rinsed with methanol (ethanol may also be used) at −20°C.
2. Matrix solution is 7 mg/ml α-CHCA, 50% ACN, 0.2%TFA. Prepare 5 ml and sonicate for 30 min (*see* **Note 8**).

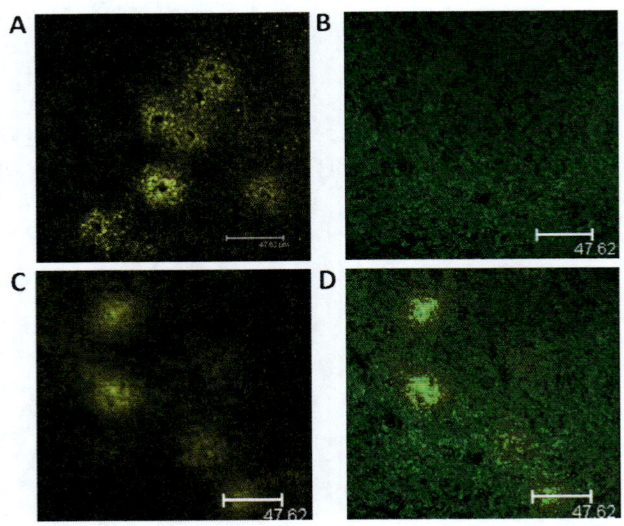

Fig. 24.2. Confocal microscopy for the evaluation of sample preparations. (**a**) A 10 μm mouse tissue section was fixed with −20°C acetone and imaged with a 63× oil objective by confocal microscopy (excitation at 543 nm and detection 570–590 nm). The YFP fluorescence is observed in the cell bodies of genetically labeled neurons. (**b**) A second 10 μm tissue section was prepared by matrix solution fixation with 25 mg/ml sinapinic acid in −20°C acetone (YFP fluorescence is sensitive to alcohols), and the matrix layer imaged by excitation with the 488 nm laser line and detection between 490 and 508 nm through 5.5 μm optical sectioning (10 sections). (**c**) The same specimen was imaged for YFP fluorescence by optical sectioning of the prepared specimen through 9.8 μm (10 sections) and the fluorescent cell bodies observed. The overlay of total matrix and tissue fluorescence is presented in (**d**).

3. The matrix is deposited with the ImagePrep using the manufacturer's default method for approximately 85 thin layers of matrix deposition (1.5 h).
4. Quality control. The quality of matrix deposition on ITO-coated glass is characterized by scanning electron microscopy of the surface. To avoid disturbance of the matrix layer, the SEM analysis is conducted without any additional coating using a variable pressure instrument (*see* **Note 9**, **Fig. 24.1a, b**).
5. For optical images, a sister tissue section as in 3 is fixed in −20°C methanol, covered with a mounting media containing DAPI (fluorescent nuclear stain with λ_{ab} 360 and

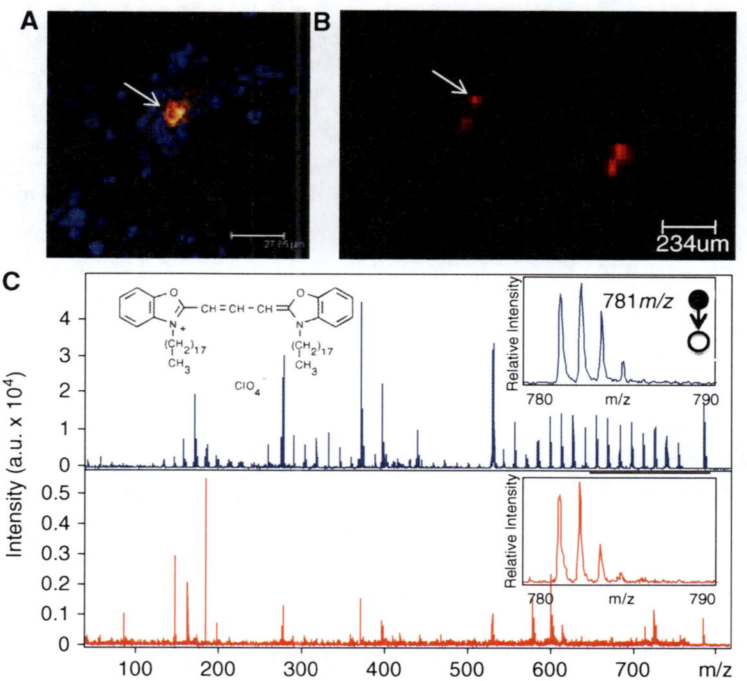

Fig. 24.3. Small molecule spatial distribution and in situ identification by MSI with confocal microscopy validation. (**a**) Confocal microscopy of a site distant from the injection site. The arrow points to a region with overlaid fluorescent signals from both the 500–525 nm and 600–700 nm detection channels and suspected to be migrated cells (scale bar = 27.85 μm. (**b**) MSI rendered for the 781.7 *m/z* of a wider field including both the injection site and the distant region shown in (**a**) (scale bar = 234 μm). (**c**) In situ tandem mass spectrometry is performed from the distant site indicated in both (**a**) and (**b**) for identification of the 781.7 *m/z* signal. The upper spectrum represents the fragmentation pattern of standard DiO and the lower spectrum the fragmentation pattern obtained at the investigated site, for a 40 x 40 μm pixel. The lower spectrum is enlarged on the *y*-axis to show the alignment with the standard fragmentation. Both patterns also agree with typical fragmentation of the aliphatic chains of the molecule shown in the upper left corner of (**c**).

λ_{em} 460 nm; DAPI bound to DNA has a broad emission spectrum >100 nm) and analyzed by confocal microscopy (*see* **Note 10, Fig. 24.2**). Confocal microscopy is performed using a 63× oil objective. The UV laser is used to excite DAPI at 405 nm with detection set between 419 and 456 nm. The 488 nm line of the argon ion laser is used to excite DiO with detection between 500 and 525 nm, and DiI is excited with the helium–neon laser line at 543 nm with detection of emission between 600 and 700 nm. Intensity and contrast of each acquisition channel are adjusted using the "glow over under" option of the Leica Confocal Software (*see* **Fig. 24.3a**).

6. The matrix-coated section is imaged by MALDI mass spectrometry by rastering the laser every 40 μm in reflectron mode, using a mass range of 500–3,100 *m/z* (*see* **Note 11, Fig. 24.2b**).
7. Cells labeled with DiO were localized by a peak at 781.7 ± 0.2 *m/z* with a minimal intensity of 5% and confirmed by MS/MS identification of the 781.7 *m/z* peak from tissue. The fragmentation pattern was compared to that of a DiO standard (*see* **Note 12, Fig. 24.3c**).
8. Class imaging can be used to distinguish cells by building classification models from contralateral regions, with and without injection or fluorescent molecules at λ_{em} 500–520 nm. Using this approach, cells away from the injection site are identified based upon their corresponding molecular signatures (*see* **Note 13**).

3.3. Microinjection with Matrix

1. Pull micropipettes using a modified version of sequence #19 (**Table 24.2**) of the Programmable Micropipette Puller (PMP-102), set pressure 1 (p) = 0.4 and pressure 2 (p) = 1.5, and set the steps as follows (*see* **Note 14**). The internal diameter of the micropipette should be 5–10 μm at the opening and it may be necessary to score the tip of the micropipette to achieve this diameter.
2. Prepare fresh matrix solution consisting, by volume, of 1 part acetonitrile and 1 part 0.1% TFA in water, and use this to dissolve matrix to a final concentration of 25–30 mg/ml. Sonicate the matrix solution for 30 min and centrifuge for 10 min at 16,000×*g* prior to use.
3. Insert the micropipette from the back through the knurled nut, then through the silicon rubber gasket, and then through the metal sleeve. Finally, connect the knurled nut with the gas hose. Set the balance pressure to −0.05 psi or to the lowest setting. Fill the micropipette with matrix using the Model PLI-100 Pico-Injector, by placing the

Table 24.2
Settings for micropipette manufacturing using a modified version of sequence #19 of the Programmable Micropipette Puller (PMP-102)

Step	Operating time (s)	Heat level	Action
T1	L->	H80	PU1:02
T2	09.0	H00	
T3	L->	H80	PU1:02
T4	05.0	H00	
T5	L->	H75	PU1:02
T6	08.0	H00	
T7	L->	H81	PU1:01
T8	0.6	H00	
T9	L->	H00	PU2:05c
T10	00.0	H00	

micropipette tip into the matrix solution and pressing the "fill" button. Allow the matrix solution to fill to half of the observable micropipette (the length of micropipette from tip to knurled nut) (*see* **Notes 15** and **16**).

4. Attach the combined hose and micropipette to a micromanipulator mounted on the microscope stage. Maintain an angle of $\geq 45°$ to prevent the matrix solution from going up the sides of the micropipette (*see* **Note 17**). Place a clean glass slide on the stage. The cover slip with tissue sample is placed onto this glass slide and taped to it to increase stability.

5. Inject cells under medium intensity light, and if fluorescently labeled cells are present, toggle back and forth between fluorescence and white light. Slowly move micropipette near target cell. If needed, increase light intensity in order to visualize the shadow of the micropipette. The shadow may be used as a guideline in order to line up the cell with the tip of the injector. Using the microscope *x*–*y* stage control, move the cell toward the tip of shadow. Now move the micropipette downward and toward the cell (*see* **Note 5**). When the tip of the micropipette touches the cell, a drop of matrix expands outward and encompasses the cell. This process may be repeated after the matrix solution solvent evaporates. Be sure to make a quick tap of the tip of micropipette onto the cell and remove promptly to avoid an overflow of matrix to surrounding cells (*see* **Note 18**).

6. To visualize the injected cell with the MALDI mass spectrometer camera, it is necessary to delineate the injected area by scratching a contour around it with a dry micropipette.
7. Acquire mass spectra from single cell preparation (**Fig. 24.4**).

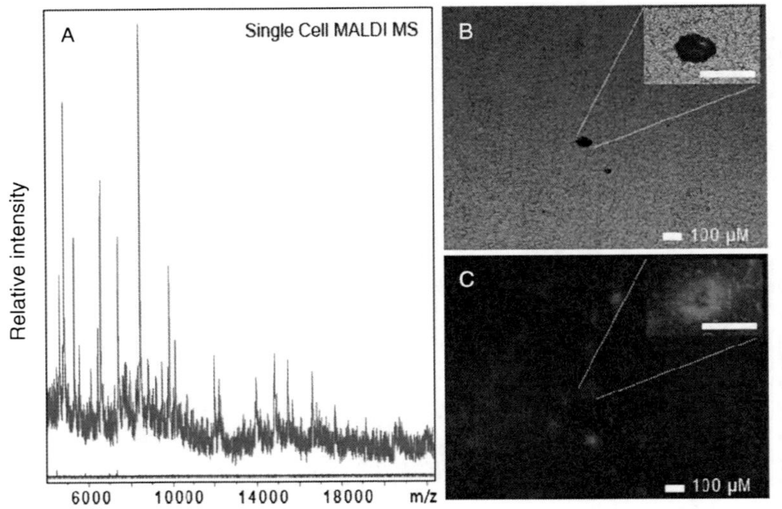

Fig. 24.4. MALDI mass spectrometry of a single neuron. (**a**) Positive ion mode mass spectrum of a single motor neuron from layer V of a mouse motor cortex. (**b**) Using a modified microinjection setup a single neuron was labeled with a mixture of the MALDI matrix, sinapinic acid, and Rhodamine B (for contrast in light microscopy and for visualizing with the MALDI microscope). Light microscopy is used to visualize deposition of the matrix and Rhodamine B. (**c**) The same microscopic field as above visualized using fluorescence microscopy. Microinjection of neurons is guided by neuron fluorescence. Mice are YFPH/WTSOD1 double transgenics with fluorescent motor neurons (described in methods).

4. Notes

1. The following fixative solution was optimized to maximize both quantity and quality of analyte MS signals from mouse brain tissue, while minimizing diffusion of proteins and tissue deformation. This solution was not optimized for lipids or peptides and indeed may solubilize and displace them. Users may substitute solvents or modify the ratio of solvents to optimize MALDI of molecules of interest and preserve tissue integrity. Ethanol, methanol, and acetone are among recognized fixatives, but independently yield limited peak detection, while water, TFA, and acetonitrile contribute to protein displacement, but increase

extraction/detection. An optimized protocol provides a balance between analyte extraction and diffusion to favor matrix/analyte interface formation, while preserving the analyte's spatial distribution.

2. Fresh matrix solution (less than 12 h old) and sonication (22) are critical for reproducible and homogeneous crystallization.
3. A readily available and reliable reference for standard histology methods, including cryosectioning, is the Online Information Center for Immunohistochemistry IHC World (www.ihcworld.com/). While thinner sections offer higher spectral quality, they are also more susceptible to analyte displacement.
4. Excessive thawing times decrease reproducibility (20). The thaw mounting procedure can be circumvented by placing a ~3 μl droplet of matrix fixation solution on a cold target and then placing the tissue on this droplet. OCT suppresses the signal of tissue-derived analytes, so the frozen tissue is held to the sectioning support by a minimal amount of OCT at the base to hold the specimen in place during sectioning (4). Certain types of tissue may require full embedding for sectioning, so the excess OCT deposited on the glass is washed away with subsequent cold ethanol washes.
5. For smaller or larger tissue, use the Online Information Center at www.ihcworld.com to calculate the volume of tissue. Then use a ratio of fixing solvent to tissue of at least 20:1.
6. Drying temperature is a critical parameter. Drying temperature of 0 and 25°C (following 10 min at −22°C) provides higher quality spectra but can result in analyte displacement. If users attempt to dry the matrix solution at a higher temperature (following 10 min fixing at −22°C) care must be taken to keep the samples in dry air since condensation of atmospheric water on the cold surface changes the percentage of water in fixing solvent solution, thereby increasing analyte displacement.
7. The visual appearance of the matrix layer is a reliable indicator of its ionization/desorption efficiency potential. The most efficient matrix layers show a homogenous and shiny appearance whereas layers with a mat finish render lower quality spectral data. The laser spot size is estimated to be between 100 and 150 μm.
8. Sonication of the matrix solution, for any type of deposition, provides increased solution homogeneity and minimizes artifactual nucleation events. It is also recommended

to centrifuge the solution after sonication to remove insoluble particles.

9. The use of high-resolution microscopy allows users to delineate the size of deposited vapor droplets and to resolve crystal arrangement, size, and homogeneity level. With such information, users can begin to correlate analytical characterization of matrix preparations with efficiency of MALDI. Characterization by SEM: The Supra55VP field emission scanning electron microscope (FESEM) allows surface examination down to nanometer scales either in high vacuum or in variable pressure (VP) mode. The instrument uses a low to moderate energy (0.1–30 keV) electron beam to image a sample with resolutions down to 1 nm at 15 keV in high vacuum or 2 nm at 30 keV in VP mode, which enables imaging of the matrix without coating.
10. Confocal microscopy provides molecular imaging with a higher spatial resolution than currently available for MSI, and protocols can be developed to minimize discrepancies in sample preparation for comparing both imaging methods. Confocal microscopy provides users with the ability to optically section a matrix-coated specimen and image a molecular target underneath a layer of matrix to assess the effects of sample preparations (*see* **Fig. 24.2**). Small molecular probes provide the opportunity for the direct detection of fluorescence and mass from the same molecule. The use of DAPI as a nuclear stain minimizes tissue manipulation as it is included in the mounting media and aids to guide the localization of a small population of fluorescent cells at high-resolution imaging of a large tissue specimen.
11. The tissue section was imaged for masses between 500 and 3,100 m/z, and the distribution of mass 781.7 ± 0.2 m/z, which corresponds to the calculated monoisotopic mass of a +1 charge of DiO in solution. The spatial distribution of the signal corresponds to the expected gross anatomical distribution of the stereotactically injected stem cells and to confocal microscopy results.
12. The isolated fluorescence signal indicated with an arrow in **Fig. 24.3a** is questionable as it is distant from the injection site, and tissue typically shows some level of autofluorescence. Even though stem cells can migrate away from the injection site, the presentation of the fluorescent signal is validated by in situ fragmentation of the 781.7 m/z signal with comparison to standard DiO's fragmentation pattern. In this case, the use of MSI offers greater certainty to the visualization of migrating stem cells than confocal microscopy, illustrating the potential of mass spectrometry as an imaging tool.

13. Spectra from a defined region of interest in contralateral regions of the brains, including the fluorescent cells injection site, are saved and used to build classification models in ClinProTools (Bruker Daltonics, Billerica, MA, USA) with genetic algorithm (GA; 96% recognition capability), single neural network (SNN; 94%), and quick classifier (QC; 90%). Four spectra from a third region of the brain showing weak 781.7 *m/z* intensity are classified and recognized as human neural stem cells for 4/4 spectra with both the GA and QC, and 2/4 spectra with the SNN derived model.
14. Try to produce a micropipette with a long taper, which will prevent the tip from breaking too easily. Considerable variation in pipette shape occurs even under identical instrument conditions. Variation depends upon the age of the heating filament, the pressure of the "pressure 1" parameter, and other minor factors (see manual of PMP-102). To maintain pipette shape, alter the heat level of T7 or change the time of T8. A higher heat of T7 will produce a smaller tip, a longer time of T8 will produce a larger tip.
15. If matrix does not move into the micropipette, the tip may be too small. To increase tip size, decrease heat level of T7 or increase operating time on T8.
16. The presence of small volumes of matrix solution in the micropipette helps prevent the matrix from flowing back into the tip.
17. If the micropipette scrapes other cells when being pulled back, either the micropipette tip is too long or the micropipette has been placed at too shallow of an angle. Adjust for a larger angle between micropipette and stage or (while retracting micropipette) move slide toward direction in which micropipette is retracting toward.
18. If droplets of matrix form on the tip, either the balance pressure is too high, the internal diameter of the tip is too large, or air is leaking in from the knurled nut. If decreasing the balance pressure does not remediate the problem, replacing the silicon rubber gasket might (Digitimer, Ltd. Hertfordshire, England Model PLI-SRG-1.0. Silicone Rubber Gasket Replacement for use with PLI-PH1 and PLI-PH1A). Alternatively, use a wide and thin piece of parafilm (Parafilm (M) Laboratory film, Fisher Inc.) to wrap around the area of contact between the micropipette and the knurled nut in order to make it airtight.
19. If after a few injections, matrix does not come out of the micropipette, it is likely that some crystallization has

occurred on the tip of the micropipette. You can either take a piece of wet paper to scrape the sides of the tip without breaking it or slightly increase the balance pressure (0.01–0.07 psi or more) to force the matrix out.

Acknowledgments

This work was made possible by award W81×WH-04-0158 from the Department of Defense to JA; grant 1392 from the Amyotrophic Lateral Sclerosis Society of America to JA; Brain Science Foundation to NYRA; American Brain Tumor Association to NYRA, and the Daniel E. Ponton Fund for the Neurosciences to NYRA. We acknowledge the Brandeis University Mass Spectrometry Resource, and the Brandeis University Animal Care Facility for care of instruments and animals, respectively. This work was performed in part at the Center for Nanoscale Systems (CNS), a member of the National Nanotechnology Infrastructure Network (NNIN), which is supported by the National Science Foundation under NSF award no. ECS-0335765. CNS is part of the Faculty of Arts and Sciences at Harvard University. We also wish to thank Ed Dougherty for maintenance and support of optical microscopy at Brandeis University, Drs. Sacha Nelson and Ken Sugino for transgenic mice, and we also wish to thank Dr. Lata G. Menon for animal surgery and manipulations.

References

1. Gusev, A. I., Vasseur, O. J., Proctor, A., Sharkey, A. G., Hercules, D.M. (1995) Imaging of thin-layer chromatograms using matrix-assisted laser desorption/ionization mass spectrometry. *Anal Chem*, **67**, 4565–4570.
2. Aerni, H. R., Cornett, D. S., Caprioli, R. M. (2006) Automated acoustic matrix deposition for MALDI sample preparation. *Anal Chem*, **78**, 827–834.
3. Sloane, A. J., Duff, J. L., Wilson, N. L., Gandhi, P. S., Hill, C. J., Hopwood, F. G., Smith, P. E., Thomas, M. L., Cole, R. A., Packer, N. H., et al. (2002) High throughput peptide mass fingerprinting and protein macroarray analysis using chemical printing strategies. *Mol Cell Proteomics*, **1**, 490–499.
4. Schwartz, S. A., Reyzer, M. L., Caprioli, R. M. (2003) Direct tissue analysis using matrix-assisted laser desorption/ionization mass spectrometry: practical aspects of sample preparation. *J Mass Spectrom*, **38**, 699–708.
5. Jurchen, J. C., Rubakhin, S. S., Sweedler, J. V. (2005) MALDI-MS imaging of features smaller than the size of the laser beam. *J Am Soc Mass Spectrom*, **16**, 1654–1659.
6. Puolitaival, S. M., Burnum, K. E., Cornett, D. S., Caprioli, R. M. (2008) Solvent-free matrix dry-coating for MALDI imaging of phospholipids. *J Am Soc Mass Spectrom*, **19**, 882–886.
7. Lemaire, R., Wisztorski, M., Desmons, A., Tabet, J. C., Day, R., Salzet, M., Fournier, I. (2006) MALDI-MS direct tissue analysis of proteins: improving signal sensitivity using organic treatments. *Anal Chem*, **78**, 7145–7153.
8. Agar, N. Y., Yang, H. W., Carroll, R. S., Black, P. M., Agar, J. N. (2007) Matrix

solution fixation: histology-compatible tissue preparation for MALDI mass spectrometry imaging. *Anal Chem*, **79**, 7416–7423.
9. Clarke, J. L. (1851) Researches into the structure of the spinal cord. *Phil Trans Roy Soc*, **141**, 601–622.
10. Carnoy, J. B. (1887) La cytodierese de l'oeuf. *Cellule*, **3**, 6.
11. Blum, F. (1893) Der Formaldehyde als Hartungsmittel. *Vorlaufige Mittheilung. Z.w.M.*, **10**, 314–315.
12. Prausnitz, J. M. (2003) Molecular thermodynamics for some applications in biotechnology. *Pure Appl Chem*, **75**, 859–873.
13. Xu, B. J., Li, J., Beauchamp, R. D., Shyr, Y., Li, M., Washington, M. K., Yeatman, T. J., Whitehead, R. H., Coffey, R. J., Caprioli, R. M. (2009) Identification of early intestinal neoplasia protein biomarkers using laser capture microdissection and MALDI MS. *Mol Cell Proteomics*, **8**, 936–945.
14. Neupert, S., Johard, H. A., Nassel, D. R., Predel, R. (2007) Single-cell peptidomics of drosophila melanogaster neurons identified by Gal4-driven fluorescence. *Anal Chem*, **79**, 3690–3694.
15. Monroe, E. B., Jurchen, J. C., Koszczuk, B. A., Losh, J. L., Rubakhin, S. S. Sweedler, J. V. (2006) Massively parallel sample preparation for the MALDI MS analyses of tissues. *Anal Chem*, **78**, 6826–6832.
16. Holle, A., Haase, A., Kayser, M., Hohndorf, J. (2006) Optimizing UV laser focus profiles for improved MALDI performance. *J Mass Spectrom*, **41**, 705–716.
17. Spengler, B., Hubert, M. (2002) Scanning microprobe matrix-assisted laser desorption ionization (SMALDI) mass spectrometry: instrumentation for sub-micrometer resolved LDI and MALDI surface analysis. *J Am Soc Mass Spectrom*, **13**, 735–748.
18. Luxembourg, S. L., Mize, T. H., McDonnell, L. A., Heeren, R. M. (2004) High-spatial resolution mass spectrometric imaging of peptide and protein distributions on a surface. *Anal Chem*, **76**, 5339–5344.
19. Luxembourg, S. L., McDonnell, L. A., Mize, T. H., Heeren, R. M. (2005) Infrared mass spectrometric imaging below the diffraction limit. *J Proteome Res*, **4**, 671–673.
20. Goodwin, R. J., Dungworth, J. C., Cobb, S. R., Pitt, A. R. (2008) Time-dependent evolution of tissue markers by MALDI-MS imaging. *Proteomics*, **8**, 3801–3808.
21. Schaefer, A. M., Sanes, J. R., Lichtman, J. W. (2005) A compensatory subpopulation of motor neurons in a mouse model of amyotrophic lateral sclerosis. *J Comp Neurol*, **490**, 209–219.
22. Lemaire, R., Tabet, J. C., Ducoroy, P., Hendra, J. B., Salzet, M., Fournier, I. (2006) Solid ionic matrixes for direct tissue analysis and MALDI imaging. *Anal Chem*, **78**, 809–819.

Chapter 25

Imaging of Similar Mass Neuropeptides in Neuronal Tissue by Enhanced Resolution MALDI MS with an Ion Trap – Orbitrap^{TM} Hybrid Instrument

Peter D.E.M. Verhaert, Martijn W.H. Pinkse, Kerstin Strupat, and Maria C. Prieto Conaway

Abstract

Several mass spectrometry imaging (MSI) procedures are used to localize physiologically active peptides in neuronal tissue from American cockroach (*Periplaneta americana*) neurosecretory organs. We report how to use this model system to assess, for the first time, the performance of the MALDI LTQ Orbitrap^{TM} XL mass spectrometer to perform MSI of secretory neuropeptides. The method involves the following steps: (1) rapid dissecting of neurosecretory tissue (i.e., insect neurohemal organ) in isotonic sucrose solution; (2) mounting the tissue on a glass slide; (3) controlled spraying of the air-dried tissue with concentrated MALDI matrix solution; (4) loading specimen into the MALDI source of a MS^{n} system equipped with an Orbitrap^{TM} analyzer; (5) setting-up MSI methods by determining tissue areas of interest, spatial resolution, molecular mass range, and molecular mass resolution; (6) acquiring mass spectra; (7) analyzing data using ImageQuest^{TM} MSI software to generate (single or composite) images of the distribution of peptide(s) of interest; (8) confirming the identity of selected peptides by MS^{2} and/or MS^{n} sequencing directly from imaged tissue sample. The results illustrate that high mass accuracy and high mass resolving power of the Orbitrap analyzer are achievable in analyses directly from tissue, such as in MSI experiments. Moreover the mass spectrometric instrumentation evaluated allows for both peptide localization and peptide identification/sequencing directly from tissue.

Key words: Cockroach, mass spectrometry imaging, neuropeptides, Orbitrap detector.

1. Introduction

There is great interest in mass spectrometry imaging (MSI) of secretory peptides, as they are essential molecules for different physiological processes, comprising a major portion of the cellular

S.S. Rubakhin, J.V. Sweedler (eds.), *Mass Spectrometry Imaging*, Methods in Molecular Biology 656,
DOI 10.1007/978-1-60761-746-4_25, © Springer Science+Business Media, LLC 2010

(bio)chemical "language." Whereas the overall fate/tendency of signaling compounds is to eventually end up in the intercellular space or in the general circulation (which makes their detection and visualization at the site of release quite a challenge), it is of great interest to be able to trace the source of these biochemicals. An obvious example is the discovery of potential biomarkers within their site of synthesis, e.g., diseased/malignant tissue, such as a specific tumor.

For obvious methodological and ethical reasons it is unreasonable to obtain fresh unfixed human tumor tissue for the sole purpose of MSI method development. Additionally, tumor biopsies typically come in a wide variety of amorphous shapes, which make proper (immuno)histochemical controls for each sample a must to validate developed MSI protocols. Frequently utilized in such projects are samples from rodent (mouse or rat) brains or pituitaries. However, the sizes and shapes of these tissues require cryotome sectioning, which produces 5–15 μm thick tissue slices amenable for MALDI matrix application and MSI interrogation. Sectioning typically results in significant increase in the duration of sample preparation and the need for freeze and thaw of specimens. To circumvent these drawbacks, we have been focusing on the cockroach retrocerebral neurosecretory system, which has proven to be a truly elegant biological model for evaluating novel investigative approaches in secretory peptide MSI. It is small enough to allow "cryosection free" and relatively quick high spatial resolution tissue analysis, while being large enough to allow facile dissection from the animal.

In evolutionarily ancient insects such as cockroaches, the neuroendocrine glands consist of two pairs of distinctly shaped organs. By their mere shape (**Fig. 25.1**), one easily distinguishes the elongated corpora cardiaca (cc) with a cranial, somewhat thicker glandular part (ccg) and a more caudal storage part (ccs) and the oval/round-shaped corpora allata (ca). The insect cc and ca comprise the animal's major neurosecretory tissue, equivalent with the pituitary or hypophysis gland of higher vertebrates. They are known to contain a great variety of (neuro)secretory peptides, many of which have had their primary sequences elucidated over the past 20 years, with the American cockroach, *Periplaneta americana*, being one of the most extensively studied species (**Table 25.1**).

Typical for arthropod (such as insects, spiders, crustaceans) neurosecretory tissue is that the secretory part of the glands is the outside surface (*see* **Fig. 25.1b**; 1), which, therefore, is readily accessible for MALDI matrix application. Furthermore, the tiny size of the tissue allows for whole mount tissue analysis (i.e., without the need of a cryostat) and for full gland analysis within a relatively short time frame. An additional advantage is that live

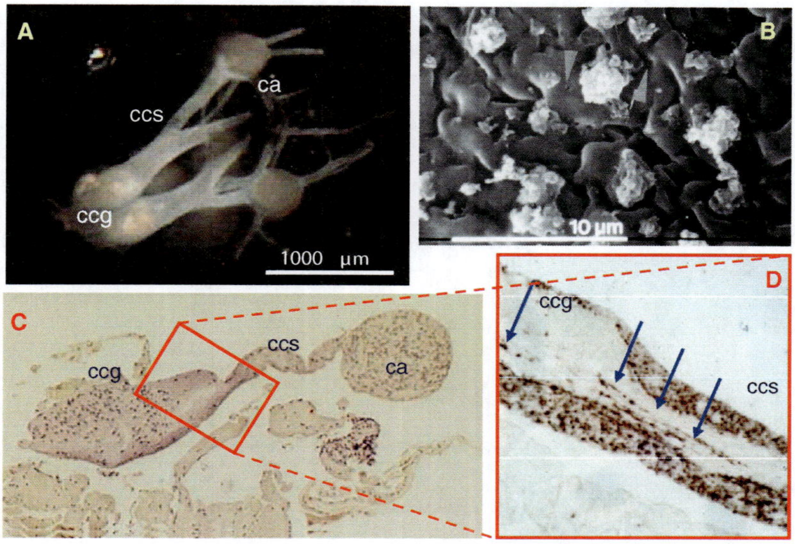

Fig. 25.1. American cockroach retrocerebral complex. (**a**) Tissue complex as dissected out of the animal head; preparation in 250 mM sucrose solution showing the pairwise structure of the glands with cranial glandular lobes of the corpora cardiaca (ccg) linked with the corpora allata (ca) via the storage lobe of the cc (ccs). (**b**) Scanning electron micrograph of rapidly fixed whole mount cc showing (neuro)secretion (white "stuff") taking place at the outer surface (ccs part; picture reproduced from (1) with permission). (**c**, **d**) Different histological sections; (**c**) Light microscopic view (hematoxylin and eosin stain) of cc and ca showing glandular (ccg) and storage lobe (ccs); (**d**) Classic immunohistochemistry stain (monoclonal neuropeptide antibody in immunoperoxidase technique, see (17)), showing neuropeptide containing nerve fibers in the center of the tissue (arrows) which are extensively branching in the gland periphery, particularly in the storage lobe (ccs).

cockroaches are omnipresent throughout the globe. They can be readily obtained virtually everywhere (if not in "the wild," then through purchase at local "pet shops" or "biologicals for education" suppliers).

All these features have made the cockroach neurosecretory system our favorite model for direct tissue (neuro)secretory peptide investigations, for many years (2).

It is obvious that the final quality of MSI data largely, but not solely, depends on the tissue preparation and processing steps. Also, the capabilities and performance of the mass spectrometer used are of importance. Over the past 5 years we have been optimizing a direct tissue (neuro)peptide sample preparation method, in combination with the use of various mass spectrometers, including MALDI time-of-flight (TOF), MALDI-TOF–TOF, MALDI ion traps, MALDI Q TOF, and others (2–5). Here we report the MSI protocol based on a combination of our optimized sample preparation and an ion trap – Orbitrap hybrid mass analyzer (6) sample interrogation, followed by data processing and the creation of molecular images of neuropeptide distributions.

Table 25.1
Known *P. americana* (Pea) cc/ca neuropeptide sequences and accurate monoisotopic (m.i.) masses calculated for their ions as expected (i.e., observed in earlier direct tissue MS analyses (9, 10)). Calculations of theoretical peptide masses are done using protein calculator

Sequence ([a])	[Trivial name]	m.i. Mass [M+H]	(→ Derived masses)
SPPFAPRLamide	[Pea-PK-II]	883.514848	
pQVNFSPNWamide	[Pea-CAH-I]	973.452642	(→ 995.434586 [M+Na], 1011.408524 [M+K])
pQLTFTPNWamide	[Pea-CAH-II]	988.488693	(→ 1010.470637 [M+Na], 1026.444575 [M+K])
LVPFRPRLamide	[Pea-PK-III]	996.646531	
HTAGFIPRLamide	[Pea-PK-I]	1010.589410	
pQDVDHVFLRFamide	[Pea-LMS]	1257.637482	
FDDY(SO3)GHMRFamide	[<Pea-SK]	1266.466653	(→ 1186.509839 [M+H-SO$_3$])
pQSDDY(SO3)GHMRFamide	[Pea-LSK-II]	1317.462296	(→ 1237.505482 [M+H-SO$_3$])
pQTFQYSRGWTNamide	[Corazonin]	1369.628374	
EQFDDY(SO3)GHMRFamide	[Pea-SK]	1523.567824	(→ 1443.611009 [M+H-SO$_3$])
DHLPHDVYSPRLamide	[Pea-PK-IV]	1447.744071	
SESEVPGMWFGPRLamide	[Pea-PKK-VI]	1590.773324	
GGGGSGETSGMWFGPRLamide	[Pea-PK-V]	1651.764550	

Note: [a]pQ = pyroglutamate

2. Materials

1. Adult specimens (both sexes) of American cockroach – *P. americana* L. (Carolina Biologicals, Burlington, NC, USA).
2. Refrigerator or regular ice for cooling cockroaches before dissection.
3. Isotonic sucrose solution: 250 mM sucrose in HPLC grade water.
4. A dissection stereomicroscope with ×30 magnification.

5. Fine microsurgical instruments including insect pins, a scissors, two forceps, a flat piece of cork, and a small dish (e.g., 5.cm diameter) in black glass.
6. Kimwipes®.
7. Conventional microscopy glass slides.
8. Digital photo camera, e.g., Cybershot DSC-H1 (SONY, Tokyo, Japan).
9. Vacuum desiccator.
10. MALDI matrix solution: 100 or 150 mg/ml high purity 2,5-dihydroxybenzoic acid (DHB; LaserBio Labs, Cedex, France and Protea Biosciences, WV, USA) dissolved in 50/50 v/v acetonitrile/water supplemented with 0.1% TFA (*see* **Note 1**).
11. SunCollect MALDI sprayer (SunChrom, Friedrichsdorf, Germany).
12. ScanJet G4050 Flat bed scanner (HP, Palo Alto, CA, USA).
13. Calibration peptide mixture kit – MSCAL4 (Sigma-Aldrich, St. Louis, MO, USA).
14. MALDI LTQ Orbitrap XL (Thermo Fisher Scientific, Bremen, Germany) equipped with 337 nm nitrogen laser (LTB, Lasertechnik Berlin GmbH, Berlin, Germany).
15. ImageQuest 1.0.1 software (Thermo Fisher Scientific, San Jose, CA, USA) for generation of MS images from acquired raw files.
16. Protein Calculator (ThermoFisher Scientific) a software package used for peptide molecular mass calculation using known or predicted peptide sequences including posttranslational modifications and different adducts.

3. Methods

3.1. Tissue Preparation

Cockroaches are "slowed down" by cooling (refrigerator) for 30 min and killed by decapitation. Neurohemal organs (corpora cardiaca [cc] and corpora allata [ca]) are rapidly microdissected under a dissection microscope one at a time. Whereas the initial dissection (cc/ca-complex removal) is done "dry" with the insect head pinned down on a flat piece of cork, the final stages of the tissue preparation (cleaning of the cc/ca from surrounding non-neuronal tissue such as trachea) are performed submerged in isotonic sucrose solution in a small dish in black glass. The wet tissue is then placed on a glass microscope slide at room temperature (RT). On average the time between decapitation of the cockroach and deposition of the tissue onto the glass slide is 5 min. Precise orientation and positioning of the tissue is facilitated in a droplet of the isotonic sucrose solution, after which the liquid (usually not more than a few microliters) is meticulously removed with a Kimwipes® tissue tip (without touching the sample tissue). We find transparent (non-conductive) glass slides to be preferable over metal MALDI sample plates, as the tissue position on the former can be easily marked from the backside using a permanent marker pen. After air-drying the tissue (5–10 min; a vacuum desiccator can be used, but is not necessary if all remaining liquid is properly removed with a Kimwipes tip), MALDI matrix is applied. In our hands, MALDI MSI peptide analysis works well using a concentrated DHB solution. Since covering the tissue with matrix typically renders its outline invisible, macro-lens photographs using the digital camera are taken of the tissue preparation immediately prior to matrix application (*see* **Fig. 25.2**).

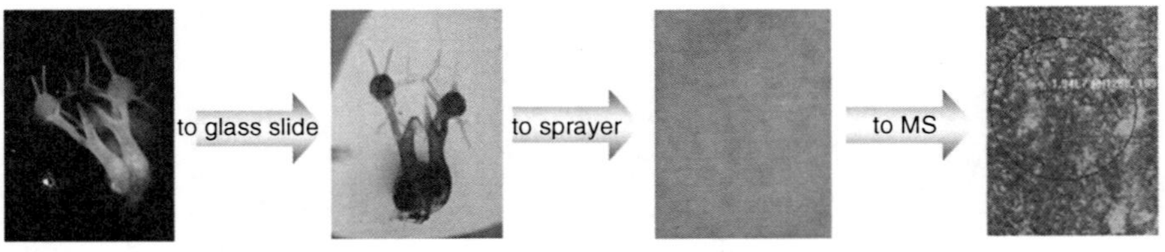

Fig. 25.2. Tissue sample dissected out of animal for investigation with MSI. From left to right: cc/ca microdissected loose from insect brain; cc/ca carefully positioned onto glass slide (position marked on the backside of slide with permanent pen) (*see* **Note 2**).

3.2. MALDI Matrix Application

MALDI matrix solution is immediately applied on top of the dissected tissue after the period of drying, using a "spray" preparation by means of a SunCollect MALDI sprayer (**Fig. 25.3**). This computer automated MALDI sprayer homogeneously sprays a fine mist of matrix through a quartz capillary, consuming minimal amounts of matrix solution (<10 μl per typical insect tissue sample) under control of a straightforward and interactive software user interface. To show the importance and possible differential effect of matrix application with regard to maintaining the original analyte localization, the controlled automated spraying is compared with a dried-droplet preparation. In the latter matrix application, 0.5 μl of DHB matrix is spotted onto the sample. Compared to the "spray" method, the "dried-droplet" preparation results in longer sample drying time, with an enhanced risk for analyte molecule washout (water soluble secretory peptides tend to easily "spill out" of the tissue contours).

The tissue-containing glass slide is mounted on an optimally designed MALDI plate to carry two glass slides at once. The entire plate – its footprint is based on the microtiter plate (MTP) standard – is mounted on a base plate holder (bottom plate)

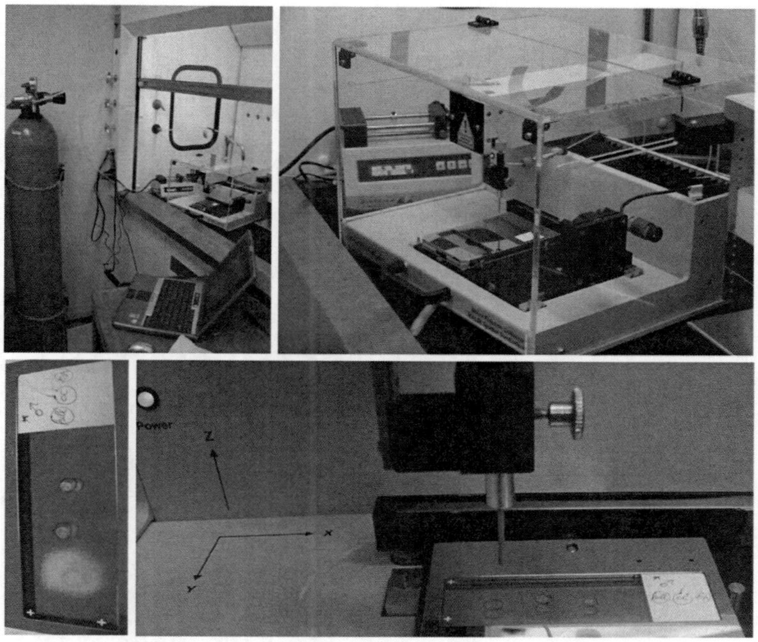

Fig. 25.3. SunCollect MALDI matrix spraying setup. *Clockwise from upper left*: general installation showing nitrogen gas bottle and controlling computer; detail of matrix spraying cabinet; zoom on matrix spraying nozzle; tissue-containing glass slide with sprayed area (*below*) compared to dried-droplet (*middle and upper*) preparations (*see* **Note 3**).

that is capable of carrying several different MALDI plates, both for regular MALDI and for imaging applications (*see* **Note 6**). In MSI applications the sample plate is "photographed" in a photo scanner to obtain a digitized record of the sample prior to plate introduction into the MS instrument. The scanned image (.bmp) is used to quickly localize the tissue of interest by "cropping" around the sample. The tissue to be analyzed can then be photographed in finer detail with the instrument camera or the scanned image stored with the MS data files (**Fig. 25.4a**).

3.3. Mass Spectra Acquisition and MS Image Generation

The mass spectrometric instrumentation used, a MALDI LTQ Orbitrap XL (**Fig. 25.4b**), is described in detail elsewhere (6). It basically consists of a MALDI ion source, a linear ion trap, a curved linear ion trap, a collision cell in elongation to the curved linear ion trap (see further down), and an Orbitrap analyzer which is placed to accept orthogonally extracted ions from the curved linear trap (7). Briefly, molecular ions are produced by MALDI (nitrogen laser at 337 nm, operating at a 60 Hz repetition rate) and are collisionally cooled in a quadrupole held at intermediate (75 m Torr) pressure. Because both detectors, the linear ion trap and the Orbitrap analyzer, are devices working with large ensembles of ion packages, ions are stored in the intermediate pressure quadrupole until a predetermined number of ions (according to

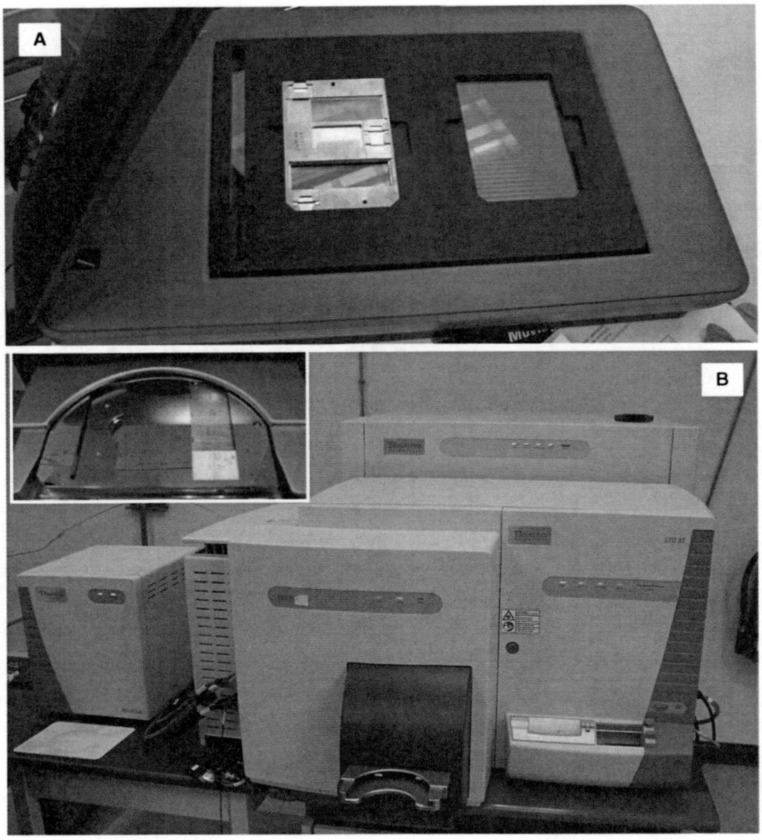

Fig. 25.4. (**a**) Scanning of MALDI sample plate adapter containing glass slide with tissues to be imaged, for rapid localization of tissue sample and digitized record storage. (**b**) MALDI Orbitrap setup. Insert: detail of tissue-containing glass slide in sample plate adaptor (*see* **Note 4**).

an AGC (automatic gain control) pre-scan measurement) for the analytical scan are reached. Thus collected ion packages are sent to the linear ion trap or Orbitrap detector. As reported earlier (6), the instrumentation allows for the acquisition of highly resolution and information-rich spectra with significantly reduced chemical noise in the Orbitrap detection compared to conventional MALDI-TOF instrumentation.

Calibrations of all instrument devices are performed with the MSCAL4 kit. Mass calibration of the Orbitrap detector in particular is done with the peptides therein and two main signals from alpha-cyano 4-hydroxycinnamic acid (HCCA). Mass calibration of the Orbitrap detector is typically done weekly and can be performed from another MALDI sample plate or the calibration sample can be placed next to a tissue imaging sample on the plate. In the herein described experiments, a separate calibration plate was used for mass calibration and the Orbitrap mass calibra-

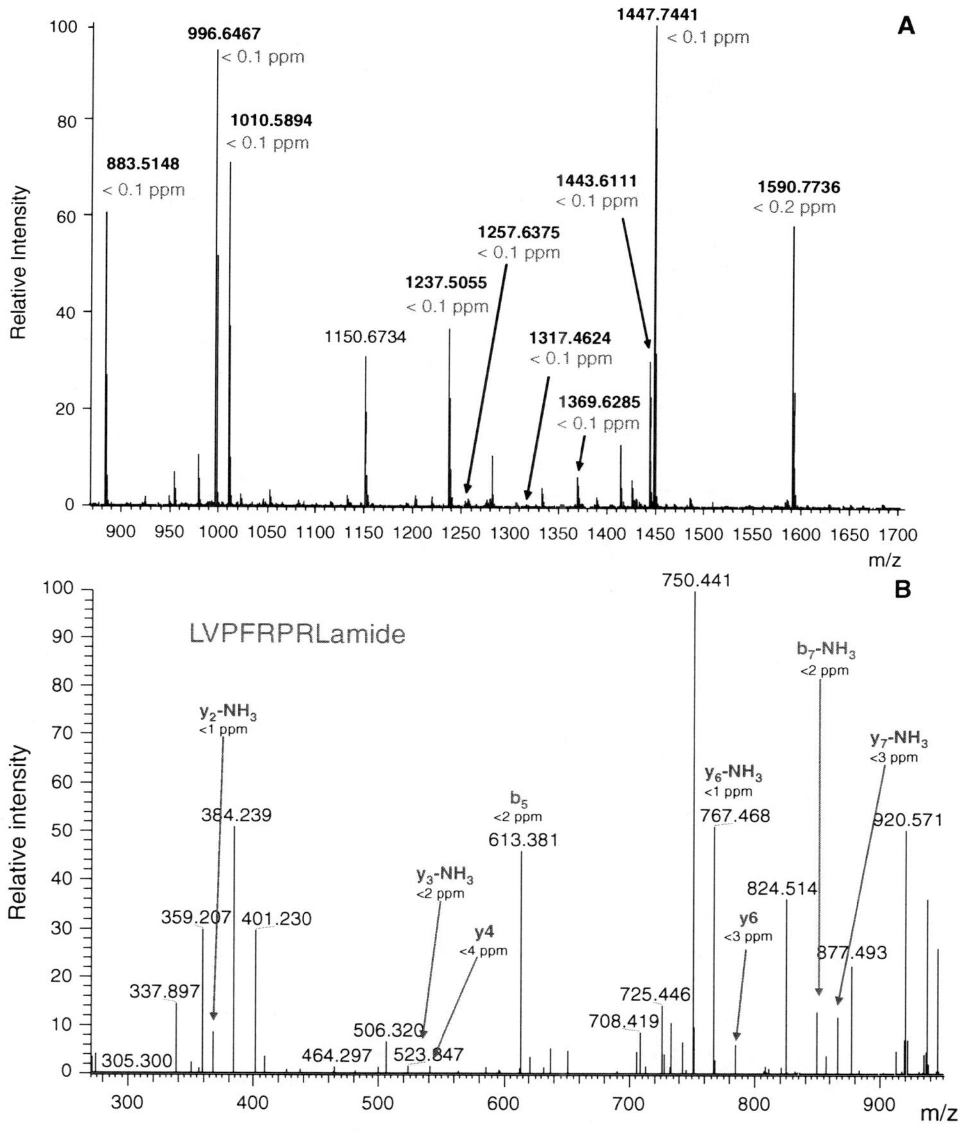

Fig. 25.5. (**a**) Example of direct tissue MS and MS/MS data (American cockroach neurohemal glands (cc and ca-complex) sprayed with the DHB matrix). MS profiling of ca: spectrum averaged over the ca area of the tissue sample (*see* **Note 5**). Molecular mass measurement errors in ppm are calculated (Protein Calculator) and indicated for observed peptides (*see* **Table 25.1** for peptide identifications). (**b**) Orbitrap (FTMS) MS/MS spectrum of *m/z* 996.6467, yielding high accuracy fragment ions confirming peptide sequence [LVPFRPRLamide; Pea-PK-III].

tion was checked every 24 h to ensure sub ppm mass accuracy, as illustrated in **Fig. 25.5**.

Full Fourier transform mass spectrometric (FTMS) data are acquired at various resolving power settings, from 7,500 to 100,000 @ m/z 400 (*see* **7** and **8** for Orbitrap analyzer operation details). FTMS acquisitions are done using a 30 μm raster, meaning the sample plate is moved 30 μm (in *x* and *y* direction) for data acquisition from different locations; the diameter of the laser spot size was measured to be 60–80 μm depending on the laser fluence used. This way a full cc/ca imaging experiment is complete within 200 min.

MSI data are acquired during the 200 min run and recorded as a continuous total ion chromatogram where the intensities of all detected ion signals are summed and plotted versus time. Using the ImageQuest 1.0.1 software, images of analyte distribution are generated by extracting information on individual molecular ion signal intensities and the spatial localization of acquisition points from the data set with a narrow mass tolerance window of 0.005 u (±0.0025 u). In other words, for any user defined *m/z* ratio a color-coded, spatially (*x*–*y*) resolved extracted ion chromatogram can be created from the accurate (highly resolved) FTMS *m/z* data. This allows us to determine the spatial distribution of peptides of very similar molecular masses. Peptide MSI data are displayed as single or combined images, with different peptide ions presented in different colors. Colors are selected arbitrarily to display the best contrast among the peptides (**Figs. 25.7** and **25.8**) and the extracted signal is smoothed (spline algorithm). The images are aligned with the actual photographs of the tissues at the time of deposition onto the glass slide prior to MALDI matrix application.

3.4. MS/MS Data Acquisition from Imaged Tissue

FTMS/MS data are acquired at a resolving power of 60,000 (@ *m/z* 400). Both collision-induced dissociation (CID) and higher energy collisional dissociation (HCD) are performed for structural identification of analyte molecules. For the latter, the collision cell placed in elongation to the curved linear ion trap is used (*see* **6**). Precursor ion isolation is achieved in the linear ion trap for both kinds of fragmentation techniques. As for the molecular (MS mode) ions, also the fragment (MS/MS mode) ions are analyzed in the Orbitrap detector in these experiments.

Indeed, as shown before (4), neuropeptide tandem MS (sequence) data can be obtained directly from tissue prepared for imaging. By MS/MS potential uncertainties about the identity of the detected peptide can be elucidated. For the peptides imaged here (well-known American cockroach system, *see* **Table 25.1**) in all cases the highly accurate Orbitrap measurement provides sufficient evidence for the peptide identity. Yet to obtain further validation, extra MS/MS sequence information for a peptide is achieved easily (previously identified peptides do not require more than a few typical "signature" sequence ions or MS/MS peaks) for unambiguous peptide identification. This was successfully done following the imaging experiment for selected peptides (example in **Fig. 25.5b**). In principle, an imaging experiment in which both MS and MS/MS data are collected *in the same "run"* can be conducted provided that sufficient signal is available from a sample position (analyte abundance and matrix deposition dependent). The visualization software can then extract data from all the available scan filters. However, the main goal of this work was to obtain images of peptide localization and sequence confirmation and, therefore, this alternate experiment will be conducted in future work.

3.5. Selected Results of the MS Imaging Protocol Application

This chapter is focused on the methodology, and we here show only a selected set of results to illustrate the type of data that can be achieved with the described MSI procedure. The complete data set will be described in more detail elsewhere. Due to space limitations, we restrict ourselves here to MS data, but we want to emphasize that we also used the system described here to produce high quality MS^n data to unequivocally confirm the identities of selected peptide ions imaged.

3.5.1. MALDI Matrix Application: Dried Droplet Versus Spray

It is clear that matrix application by the dried-droplet method is not the preferred method for (neuro)peptide imaging

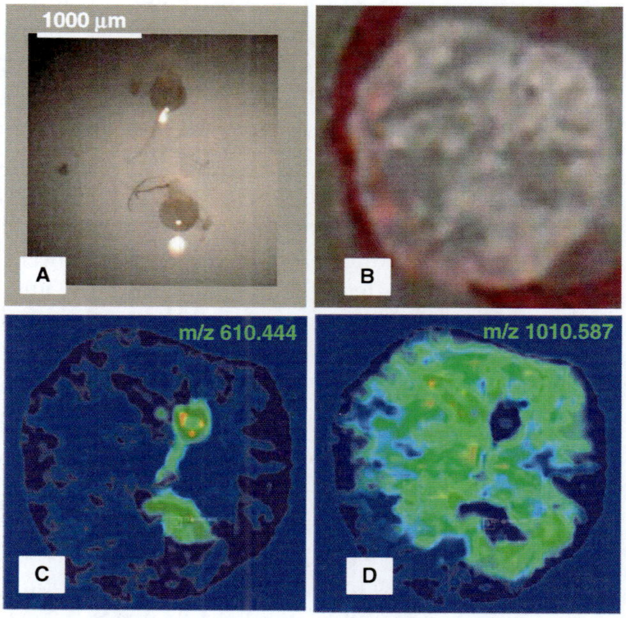

Fig. 25.6. Cockroach ca prepared with dried-droplet matrix deposition; (**a**) before matrix addition, (**b**) after matrix addition. (**c**, **d**) MALDI LTQ Orbitrap images: *green color* represents selected lipid ion at *m/z* 610.444 (**c**), and Pea-PK-I peptide MH+ ion at *m/z* 1,010.587 (**d**). *Dark blue color* shows intense DHB matrix crystals from the overlaid photograph (optical image). In contrast to the 610.444 ion, the PK-I peptide is clearly diffused out of the tissue into the matrix droplet. The MS image is indicated using the "*Rainbow" color* scheme, where relative abundance is coded as: red>yellow>green>blue = zero intensity (*see* **Notes 6** and **7**).

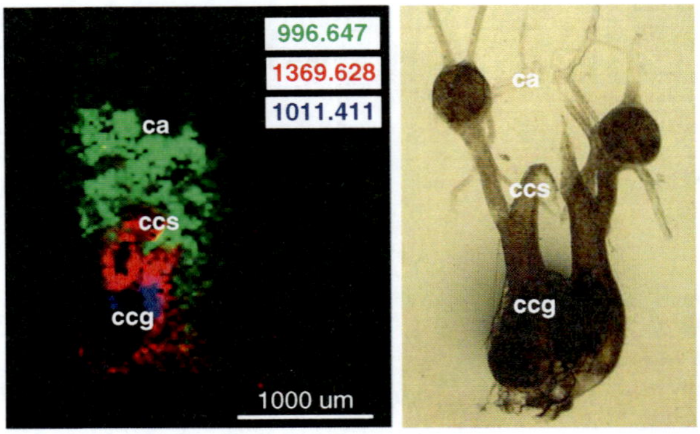

Fig. 25.7. Example of combined ImageQuest MSI image of three different peptide ion species (*see* **Table 25.1**), representing Pea-PK III (MH+ 996.647), corazonin (MH+ 1,369.628), and Pea-CAH (MK+ 1,011.411). Peptides are associated with three distinct distributions in the American cockroach retrocerebral complex, corresponding to ccg, ccs, and ca (*see* **Note 7**). Picture on *right* shows the tissue photographed prior to matrix deposition.

(**Fig. 25.6**). As exemplified for Pea-PK-I, the peptide is clearly washed out of the tissue of origin, whereas in the sprayed sample, the peptide is still located on the tissue (*see* **Figs. 25.7** and **25.8**).

This does not mean that dried-droplet preparations cannot be imaged. Indeed, clearly dependent on the chemical structure, certain compounds do remain in the tissue (e.g., the (prenol) lipid ion at 610.444). For specific analyte molecules such as these latter compounds, the dried-droplet method can thus be used as a specific sample preparation method to wash away unwanted compounds. In this respect, it is interesting to note that the corpora cardiaca are the main synthesis sites of the insect juvenile hormone (JH), also a prenol lipid of the class of terpenoids. JH as such is not secreted in the adult stage of the insect imaged here, but the presence of other terpenoids in this tissue is expected.

3.5.2. MSI Comparative Investigation of cc and ca

Using ImageQuest software, images are extracted from the data sets of the sprayed cc/ca samples. This is done for all peptides listed in **Table 25.1**. Mass accuracy obtained from these "off-tissue peptide MSI experiments" is well within the instrument specifications (better than 3 ppm), as exemplified in **Fig. 25.5**. Multiple peptide ion images can be combined in a single image using different colors (*see* **Figs. 25.7** and **25.8**).

Our results show that different peptide species yield different images, whereas related peptide ions, such as peptide truncations, loss of sulfate, cation adducts, are associated with virtually identical images. In this respect the peptides (*see* **Table 25.1**) imaged can be divided into three classes. Pea-PK peptides are predominantly associated with the ca. Pea-CAH I and Pea-CAH II appear

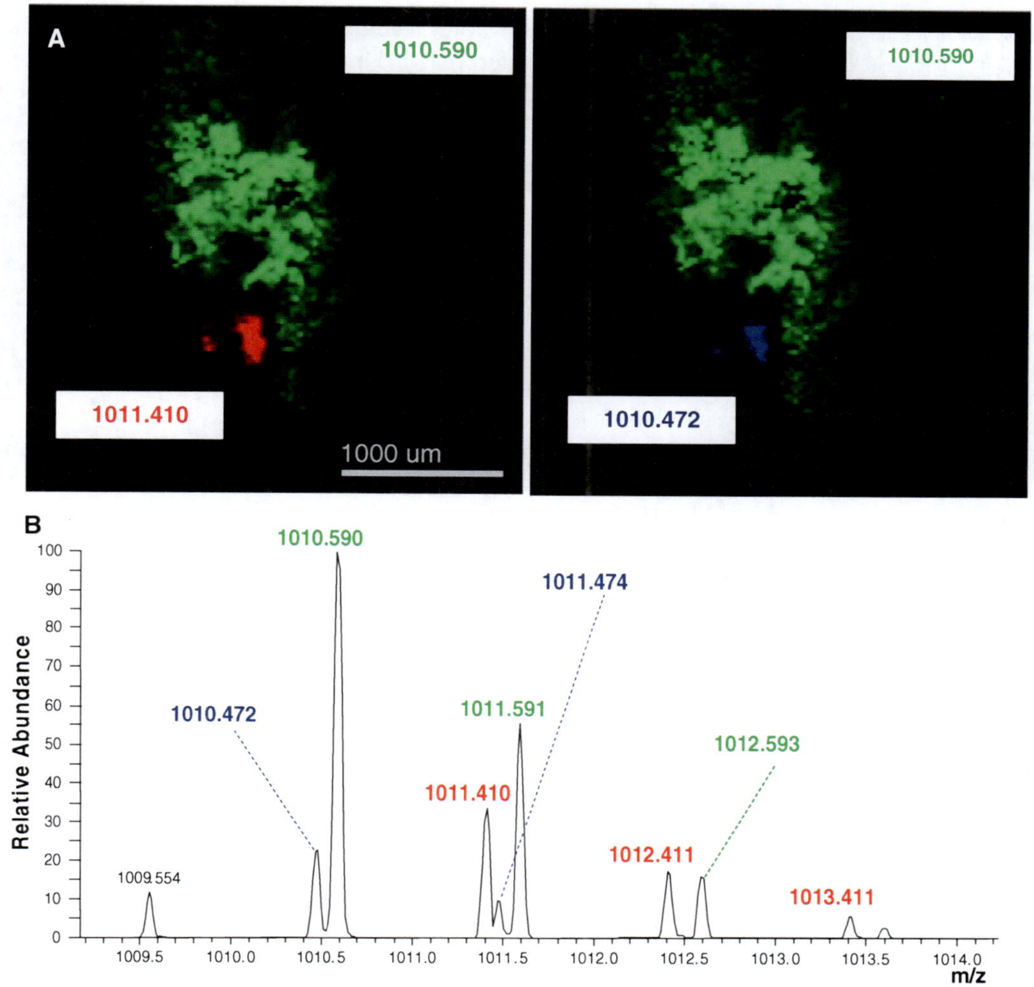

Fig. 25.8. Differential images of the distribution of peptides with similar molecular masses (peptide ion images extracted at *m/z* ±0.0025 Da). Note that these images demonstrate the distributions of single isotopes; i.e., the 1,011.410 molecular species is clearly resolved from the second isotopes of 1,010.472 and 1,010.590 (respectively 1,011.472 and 1,011.590) (*see* **Note 7**). The higher relative intensity of the 1,011.410 signal vs. the 1,010.472 signal is in good agreement with previously detected by different methods variety in abundances of these peptides (Pea-CAH-I is more than twice as abundant as Pea-CAH-II, **12**) (see also **Note 9** on relative quantification).

to be limited to the ccg area, whereas the others are mainly ccs associated.

As observed earlier (9, 10), most peptides show up as protonated species (see example in **Fig. 25.5**), whereas the peptides Pea-CAH-I and II predominantly yield Na^+- and K^+-adducts. Images of the respective *m/z* values of these two Pea-CAH-related quasi-molecular ion species confirm that they originate from identical sites in the tissue. Their images nicely reflect previous immunohistochemical work confirming the ccg (glandular lobe of the cc) as the site of synthesis (11, 12).

3.5.3. Creating Molecular Ion Images of Distributions of Neuropeptides with Similar Molecular Masses

It shall be clear that biological tissues are biochemically complex samples, and different MSI methods are therefore needed to investigate this complexity. One way to approach this is using sample preparation steps selective for certain analyte molecules over others, by which the complexity of the sample is effectively reduced. Different matrices can be employed to preferentially ionize analytes of interest, whereas different washing steps can be included to remove unwanted analytes. Moreover, recently yet another dimension can be incorporated in the MS analysis, such as ion mobility (which can separate analytes with close (even identical) molecular weight, on the basis of their difference in collisional cross sections (13)). If, however, analytes with similar physicochemical characteristics occur in the tissue with very similar molecular weights, perhaps the only way to achieve an "unmixed" analysis is by high-resolving MS.

The high-resolving power of the MSI system described here and the capability to extract m/z ratios with narrow tolerance windows allows us to "see in color-coded manner" that peptides with similar m/z ratios can be localized to different areas of the tissue. This is exemplified for three different peptides with monoisotopic (singly charged) m/z ratios of 1,010.590 (green), 1,010.472 (blue), 1,011.410 (red) in **Fig. 25.8** (*see* **Note 8**).

3.6. Discussion

It is clear that mass spectrometry imaging has great potential as a comprehensive analysis technique for endogenous peptides, particularly with a hardware configuration as evaluated in this chapter, where MALDI produced ions are analyzed in an ion trap – Orbitrap hybrid instrumentation. A single experiment can provide high mass accuracy data in combination with the spatial distribution of peptides in the tissue sample. With MS/MS experiments the molecular identity (accurate mass measurements complemented with MS^n sequence data) can be confirmed from a single or few scans only and from a few laser shots in total.

The data obtained allow us to conclude that different peptides are differently localized within the cc and ca, even with multiple precursor masses being closely spaced within a single mass unit. Mass accuracy, as calculated from over 10 peptides of known sequences, is well within or better than the 3 ppm instrument specification, accuracies rarely associated with MSI experiments, particularly for peptides.

The 30 μm rastering of the tissue (so-called "oversampling" technique with a <~100 μm diameter laser spot size), nicely allows for clear differential localization of peptide families, including ones exclusively localized in the ca, and others which are predominant in the cc. Within the cc, even a distinction can be made between peptide distributions over the glandular and the storage lobes. The only way this type of peptide localization data could be obtained in the past was by tedious individual dissection of the minute subparts of the little glands followed by peptide profiling approaches, or by immunohistochemistry, which obviously implied the production of immunospecific (non-cross-reacting) peptide antibodies for each individual peptide. Particularly in the case of the insect peptide families with numerous members sharing essential amino acid sequences, this is seldom possible. Now the data are obtained in a single MSI experiment (200 min run), where every peptide can be localized and identified individually.

The dried-droplet experiment clearly demonstrates that this sample preparation makes peptides migrate out of the tissue, rendering the procedure unsuitable for peptide localizations. However, a set of lipid ions remain in the tissue (as shown for the typical 610.444 MH+ ion), nicely demarcating the tissue on the MS target plate/slide.

The ImageQuest software is capable of combining multiple peptide data in one image.

Together with the delivery of highly accurate data, a strong point of the Orbitrap analyzer is the availability of all MS^n options (both CID and HCD) for peptide sequence confirmation (and for short peptides, there is the potential for de novo sequencing).

With more improvements regarding spatial resolution, MSI may well out-compete immunohistochemistry for neuropeptide localizations in the not too far distant future...

4. Notes

1. We tried using less concentrated DHB matrix (75 mg/ml as opposed to 150 mg/ml) in the sprayed application, but poorer quality images resulted. The spraying technique deposits less quantity of DHB crystals per area and is a much drier matrix application than the dried-droplet method. Good spectra (abundant ions, good signal-to-noise) were obtained using 100 mg/ml of DHB with the dried-droplet technique. In our experience, the need for high matrix concentration only appears necessary for peptide measurements in freshly dissected tissue samples.
2. It is advantageous to take a photograph with the instrument camera before applying matrix as it makes for a better overlaying of optical and MS images later on. An offline digital camera can also be used, as long as the image can be imported into the mass spectrometer and register properly with the *XY* plate coordinates. We did not have such a setup. Because of the nature of the analytes under study (rapid degradation after dissection) it was opted to apply matrix as soon as possible.
3. The spraying of MALDI matrix should always be done inside a chemical hood. Laboratory coats, gloves, and safety goggles should be used when working with MALDI matrices.

4. Placing a black background is necessary to provide contrast between the tissue and the glass when scanning glass slides. Care should be taken to ensure that the template holding the sample plate is positioned against the top corner, as indicated by an arrow in the scanner.
5. Using the "Averaging" tool in ImageQuest, one can average spectra within a selected area of the tissue.
6. Overlaying of optical image (photograph) over MS images is done in ImageQuest with the "Combo" tool. The contribution of one or the other can be easily manipulated.
7. The selected mass range can be normalized by the total ion current (TIC) or not. MS images in **Fig. 25.6c**, **d** were normalized by the TIC. MS images in **Figs. 25.7** and **25.8** were optimized using ImageQuest histogram view. This histogram view displays intensity in the X axis vs. number of pixels in the Y axis and is used as a guideline to tune the display intensity range. It helps to optimize the brightness and contrast of the 2D image (minimizing zero data points for the extracted m/z) and maximize the view of fine structure within the tissue. When comparing images this histogram tuning is kept constant (as in the images in **Fig. 25.8a**).
8. The MS spectrum generated in an Orbitrap analyzer is extremely rich in information. As one zooms into the baseline, a lot of "buried" peaks become apparent. The isotopes belonging to a molecular entity can be easily matched and followed, even when mixed with similar molecular masses (as in **Fig. 25.8b**). Even though ImageQuest offers a quick way of extracting data, generating MS images from spectra with this much information could benefit from software automation.
9. Relative MALDI quantitation has been shown for lipids (14) and small molecules in the MALDI linear ion trap. Good precision in MALDI quantitation requires normalizing for an internal standard. MS^2 and MS^3 improve precision by providing specificity (15). Including a deuterated internal standard in the isolation window for MS^2 has also been shown to improve precision (16).

Acknowledgments

This work is partly funded by the Netherlands Genomic Initiative.

References

1. Verhaert, P. D., De Loof, A. J. (1996) Scanning electron microscopy of an active neurohaemal area, the cockroach (*Periplaneta americana*) corpora cardiaca: looking at neurosecretion from an unprecedented viewpoint. *The Peptidergic Neuron*, in (Krisch, B., Mentlein R., eds.), **Birkhäuser Verlag**, Basel, 73–79
2. Verhaert, P. D. E. M., Pinkse, M. W. H., Prieto-Conaway, M., Kellmann, M. (2008) A short history of insect (neuro)peptidomics – a personal story of the birth and youth of an excellent model for studying peptidome biology. *Peptidomics Methods and Applications*, in (Soloviev, M., Shaw, C., Andren, P., eds.), John Wiley & Sons, New York, NY, 25–54
3. McDonell, L. A., Piersma, S. R., Altelaar, A. F. M., Mize, T. H., Luxembourg, S. L., Verhaert, P. D., van Minnen, J., Heeren, R. M. (2005). Subcellular imaging mass spectrometry of brain tissue. *J. Mass Spectrom*, **40**, 160–168
4. Verhaert, P. D., Prieto Conaway, M. C., Pekar, T. M., Miller, K. (2007) Neuropeptide imaging on an LTQ with vMALDI source: the complete "all-in-one" peptidome analysis. *Int J Mass Spec*, **260**, 177–184
5. Prieto Conaway, M., Verhaert, P., Kovtoun, V., Bui, H., Izgarian, N., Moehring, T., Strupat, K. (2007) Elucidation of closely-spaced neuropeptides directly off tissue by enhanced resolution MALDI ionization with Orbitrap detection. *Proceedings of the 55th ASMS Conference on Mass Spectrometry and Allied Topics*, Indianapolis, IN, June 3–7, 2007
6. Strupat, K., Kovtoun, V., Bui, H., Viner, R., Stafford, G., Horninga, S. (2009) MALDI produced ions inspected with a linear ion trap-orbitrap hybrid mass analyzer. *J Am Soc Mass Spec*, **20**, 1451–1463.
7. Makarov, A., Denisov, E., Kholomeev, A., Balschun, W., Lange, O., Strupat, K., Horning, S. (2006) Performance evaluation of a hybrid linear ion trap/Orbitrap mass spectrometer. *Anal Chem*, **78**, 2113–2120
8. Scigelova, M., Makarov, A. (2006) Orbitrap mass analyzer – overview and applications in proteomics. *Proteomics*, **6**(**S2**), 16–21
9. Verhaert, P., Uttenweiler-Joseph, S., de Vries, M., Loboda, A., Ens, W., Standing, K. G. (2001) Matrix-assisted laser desorption/ionization quadrupole time-of-flight mass spectrometry: an elegant tool for peptidomics. *Proteomics*, **1**, 118–131
10. Predel, R., Gaede, G. (2005) Peptidomics of neurohemal organs from species of the cockroach family Blattidae: how do neuropeptides of closely related species differ? *Peptides*, **26**, 3–9
11. Eckert, M., Gabriel, J., Birkenbeil, H., Greiner, G., Rapus, J., Gäde, G. (1996) A comparative immunocytochemical study using an antiserum against a synthetic analogue of the corpora cardiaca peptide Pea-CAH-I (MI, neurohormone D) of *Periplaneta americana. Cell Tissue Res*, **284**, 401–413
12. Scarborough, R. M., Jamieson, G. C., Kalish, F., Kramer, S. J., McEnroe, G. A., Miller, C. A., Schooley, D. A. (1984) Isolation and primary structure of two peptides with cardioacceleratory and hyperglycemic activity from the corpora cardiaca of *Periplaneta americana. Proc Natl Acad Sci U S A*, **81**, 5575–5579
13. McLean, J. A., Fenn, L. S., Enders, J. R. (2009) Structurally selective imaging mass spectrometry by imaging ion mobility-mass spectrometry. *Mass Spectrometry Imaging*, Springer, New York, NY.
14. Landgraf, R. R., Garrett, T. J., Stacpoole, P. W., Yost, R. A. (2008) Quantitation of lipids in nerve tissue using intermediate-pressure MALDI/linear ion trap mass spectrometry. *Proceedings of the 56th ASMS Conference on Mass Spectrometry and Allied Topics*, Denver, CO, June 1–5, 2008
15. Manier, L., Caprioli, R. M. (2009) Rapid screening of opiate and benzodiazepine drugs in biological fluids using electrophoretic fractionation and MALDI linear ion trap mass spectrometry. *Proceedings of the 59th ASMS Conference on Mass Spectrometry and Allied Topics*, Philadelphia, PA, May 31–June 4, 2009
16. Richard, F. Reich, R. F., Cudzilo, K., Yost, R. A. (2008) Quantitative imaging of cocaine and its metabolites in postmortem brain tissue by intermediate-pressure MALDI/linear ion trap tandem mass spectrometry. *Proceedings of the 56th ASMS Conference on Mass Spectrometry and Allied Topics*, Denver, CO, June 1–5, 2008
17. Verhaert, P., De Loof, A. (1986) Substances resembling peptides of the vertebrate gonadotropin system occur in the central nervous system of *Periplaneta americana* L. Immunocytological and some biological evidence. *Insect Biochem*, **16**, 191–197

Chapter 26

Mass Spectrometric Imaging of Neuropeptides in Decapod Crustacean Neuronal Tissues

Ruibing Chen, Stephanie S. Cape, Robert M. Sturm, and Lingjun Li

Abstract

The emerging technology mass spectrometric imaging (MSI) provides an attractive opportunity to detect and probe the molecular content of tissues in an anatomical context. This powerful methodology has been applied extensively to the localization of proteins, peptides, pharmaceuticals, metabolites, lipids, and other biological and chemical compounds in tissues. Herein, we present a method developed specifically for mapping neuropeptides in crustacean neuronal tissues. Both cryostat tissue sectioning and whole-mount tissue blotting techniques are highlighted. Careful sample preparation is essential for obtaining sufficient analyte/matrix mixing while retaining the spatial localization of the neuropeptides. Several matrix application apparatus and techniques are described and compared. Furthermore, three-dimensional (3D) imaging has been developed to provide detailed information about the distribution of neuropeptides within 3D structure of a crustacean brain.

Key words: Imaging mass spectrometry, mass spectrometric imaging, MALDI-TOF/TOF, neuropeptide, crustacean.

1. Introduction

The application of mass spectrometry (MS) to the field of neuroscience has enabled the discovery and characterization of many neuropeptides and neurohormones (1–6). This is usually achieved by tandem MS analysis of nerve tissue extracts. Alternatively, direct tissue methods, in which the tissue is coated with matrix and probed via matrix-assisted laser desorption/ionization (MALDI)-MS analysis, enable the sensitive

S.S. Rubakhin, J.V. Sweedler (eds.), *Mass Spectrometry Imaging*, Methods in Molecular Biology 656,
DOI 10.1007/978-1-60761-746-4_26, © Springer Science+Business Media, LLC 2010

detection of neuropeptides in single organs (7, 8) and even single cells (9, 10). Because direct tissue analysis does not necessitate disruption of tissue morphology, this technique can be extended for MALDI-mass spectrometric imaging (MSI) of neuropeptides (11–13). MSI reduces the time-consuming steps of sample extraction, purification, and separation, while maintaining the topographical information about molecular distribution (14, 15). Of the current techniques, MALDI-MSI enables the study of a broad mass range of molecular species from small molecules to proteins, including neuropeptides. This technique creates distribution maps of selected compounds in a similar manner to the more traditional biochemical techniques such as chemical staining, immunocytochemistry, and radiochemistry (16–18). However, MSI does not require a priori knowledge of the target analytes, and the inherent nature of mass spectrometry to detect with high chemical specificity enables the simultaneous differential localization of numerous peptides even when significant sequence homology exists. Furthermore, MS/MS fragmentation can be performed for unambiguous identification or even de novo sequencing.

The crustacean stomatogastric nervous system (STNS) consists of several linked, motor-pattern-generating networks that produce coordinated motor output. Numerous anatomical, electrophysiological, immunocytochemical, and mass spectral studies have been performed on components of this system (2, 19, 20). Therefore, the STNS is an excellent test-bed for MSI method development, as well as a useful model nervous system from which general neurobiological concepts may be derived. The challenges of MSI of neuropeptides from crustacean nerve tissue come from low abundance of analytes, high sample complexity, and the likelihood of diffusion of some neuropeptides (21). Therefore, proper sample handling from animal dissection through analysis is essential for obtaining useful and relevant results. Herein we present a simple scheme to map the neuropeptides in two important nerve organs in a crustacean. As outlined in **Fig. 26.1**, after dissection, the neuronal organs are laid on the MALDI plate. Depending on the property of the specific organ, it can be arranged on target directly, such as with the pericardial organ, or for larger organs, like brain, it should be thinly sectioned prior to being transferred to the plate. The tissue is then desiccated and matrix is applied by airbrush and allowed to dry. An array of mass spectra is then collected from multiple locations throughout the tissue according to a predefined Cartesian grid. This array of mass spectra is processed into a cohesive image where each pixel contains the information from the corresponding spectrum. **Figure 26.2** shows MALDI images of neuropeptides located in the pericardial organ and brain of *Cancer borealis*. Additionally, a simple method

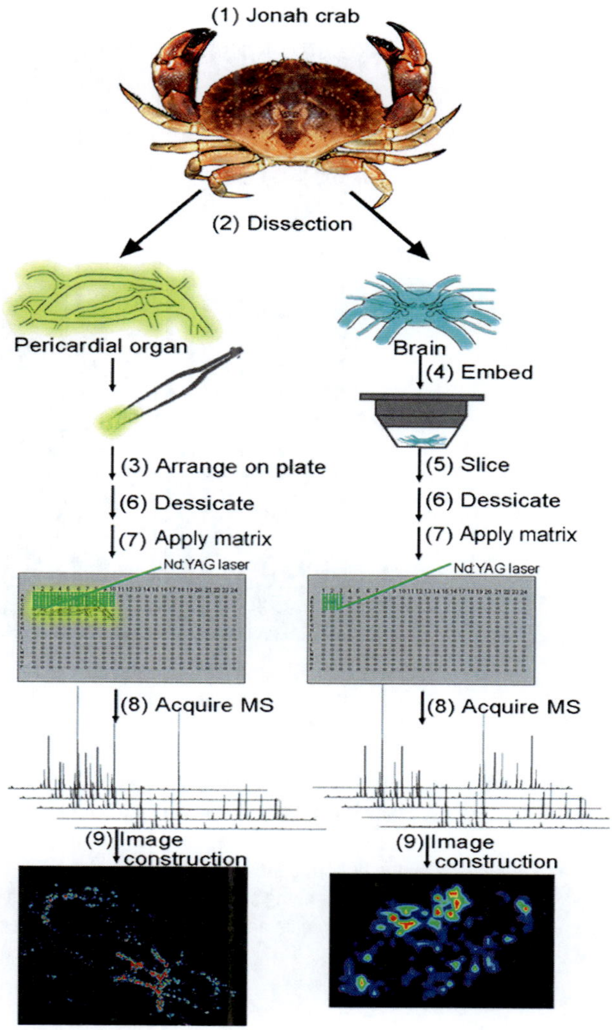

Fig. 26.1. A schematic outline of MALDI imaging of pericardial organ (PO) and brain in *C. borealis*. After dissection, the neuronal organs are laid out on the MALDI plate. Depending on the property of the specific organ, it can be arranged directly on the plate, such as for the PO, or thinly sectioned and then transferred to the plate, which is necessary for the brain. Once the tissue is on the plate, it is desiccated prior to matrix application. Mass spectra are then collected according to a predefined Cartesian grid and processed into a cohesive image where each pixel contains the information from the corresponding spectrum.

to generate three-dimensional images of crustacean brain is also presented here. 3D MSI is achieved by acquiring images of serial sections from a single specimen to provide a depth dimension to the data set (22, 23). The in-depth profiling provides more comprehensive spatial information, allowing 3D representation of multiple analytes of interest in the target organs.

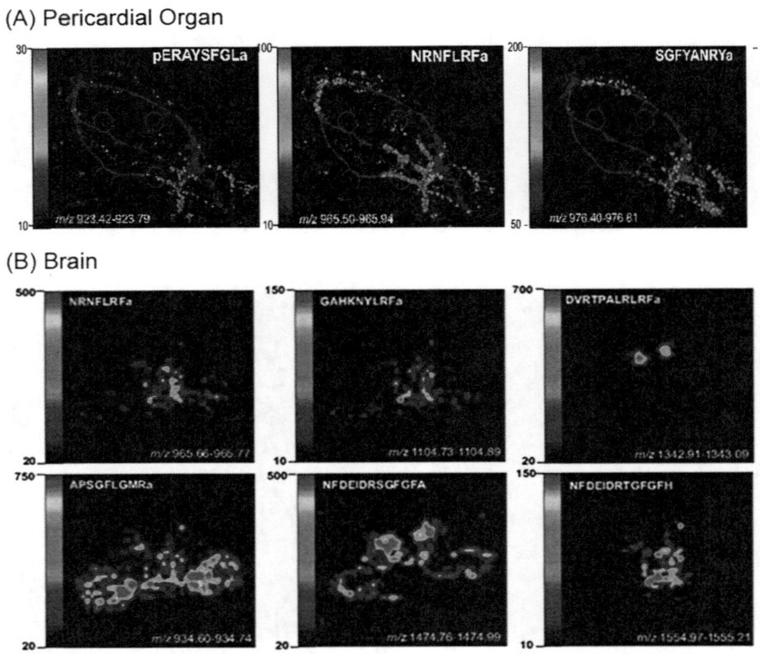

Fig. 26.2. Differential localization of neuropeptides in two neural organs from *C. borealis*. (**a**) MALDI images of three different neuropeptide families in pericardial organ, including A-type allatostatin, Arg-Phe-amide (RFamide), Arg-Tyr-amide (RYamide) peptides from left to right. Peptides from different families show differential distribution patterns in the pericardial organ (PO). (**b**) MALDI images of three different neuropeptide families in the brain, including three RFamide isoforms on top three panels, crustacean tachykinin-related peptide (CabTRP 1a), and two orcokinin peptides from left to right on the bottom panels. MALDI-MS images of select peptides demonstrate colocalization of members of the same neuropeptide family as well as exceptions to this trend. The amino acid sequence and mass to charge ratio of each peptide is labeled in each image.

2. Materials

2.1. Reagents/Equipment

1. Physiological saline: 440 mM NaCl, 11 mM KCl, 26 mM $MgCl_2$, 13 mM $CaCl_2$, 11 mM Trizma base, 5 mM maleic acid; adjust pH to ~7.45, store at 4°C.
2. Methanol: purge and trap grade (Fisher Scientific, Fair Lawn, NJ, USA).
3. Ammonia (Roundy's Inc., Milwaukee, WI, USA).
4. 2,5-Dihydroxybenzoic acid (DHB) matrix (ICN Biomedicals, Aurora, OH, USA).
5. α-Cyano-4-hydroxycinnamic acid (CHCA) (10 mg/ml in 60% acetonitrile/0.1% formic acid, v/v) (Sigma-Aldrich, St. Louis, MO, USA).

6. Deionized water (Millipore, Bedford, MA, USA).
7. Gelatin (100 mg/ml in water).
8. 0.22 μm syringe filter (Millipore, Bedford, MA, USA).
9. Cryostat (Leica, Wetzlar, Germany).
10. Airbrush (Paasche Airbrush Company, Chicago, IL, USA) coupled with 75 ml steel container.
11. Epson Stylus Photo R280 printer and Epson print CD software (Epson America, Inc., Long Beach, CA, USA).
12. Epson R280 refillable ink cartridges with auto reset chip kit (www.inkproducts.com).

2.2. Instrumentation

A model 4800 MALDI-TOF/TOF analyzer (Applied Biosystems, Framingham, MA, USA) equipped with a 200 Hz, 355 nm Nd:YAG laser (spot diameter of 75 μm) was used for imaging. Other types of MALDI mass spectrometers may also be used. Acquisitions were performed in positive ion reflectron mode. Instrument parameters were set using the 4000 Series Explorer Software (Applied Biosystems). The tissue region to be imaged and the raster step size were controlled using the 4800 Imaging application (Novartis, Basel, Switzerland) available through the MALDI-MSI web site (www.maldi-msi.org). To generate images, spectra were collected at 100 μm intervals in both the x and y dimensions across the surface of the sample. Each mass spectrum was generated by averaging 250 laser shots over the mass range m/z 800–2,000. Individual spectra were acquired using 1.0 ns binning to yield 27,812 data points per spectrum. Mass spectra were externally calibrated using peptide standards applied directly to the stainless steel MALDI target.

3. Methods

3.1. Tissue Handling

Careful sample preparation is crucial for obtaining biologically relevant mass spectral images. Variables, including embedding medium, tissue slice thickness, and fixing reagents, greatly affect the final quality of the acquired image. Depending on the properties of specific organs, the sample can be arranged on the MALDI target directly after dissection or snap-frozen and cryostat sliced. Commonly used optimal cutting temperature (OCT) medium severely interferes with MALDI detection of neuropeptides. Alternatively gelatin provides stability for the tissue during sectioning while giving manageable interference. Typically, tissue thickness of 10–20 μm is appropriate for imaging applications. Based on our experiences, the thinner the tissue, the better the sensitivity achieved, but slices thinner than 12 μm usually result in poor tissue integrity.

3.1.1. Tissues Which Require Cryostat Slicing (i.e., Brain and Thoracic Ganglion)

1. Immediately following dissection, rinse the tissue briefly in water and gently blot dry.
2. Submerge the organ in a cryostat cup of gelatin (100 mg/ml aqueous) (*see* **Notes 1** and **2**).
3. Use the forceps to orient the organ as desired and to untangle any nerves that have become entwined (*see* **Note 3**).
4. Freeze the gelatin by placing the cup in a dry ice/ethanol bath and store in –80°C freezer until use (*see* **Note 4**).
5. Prior to cryostat slicing allow tissue embedded in gelatin to warm in the cryostat box to the appropriate temperature (*see* **Note 5**).
6. Cryostat section into slices with 4–20 μm thickness as appropriate.
7. Thaw mount each slice onto the MALDI plate (*see* **Notes 6** and 7).
8. For 3D imaging, obtain multiple slices that are evenly distributed through the *z*-axis of the organ of interest (**Fig. 26.3**) (*see* **Note 8**).
9. Thaw mount each slice individually on the MALDI-TOF/TOF plate.
10. Place the MALDI plate in a desiccator for at least 30 min before applying matrix.

3.1.2. Tissues Which Do Not Require Cryostat Slicing (i.e., Pericardial Organ)

1. Dissect the tissue free of the animal and rinse it by dipping in small amounts of water quickly.
2. Arrange the tissue to lay flat in a small pool of water on the MALDI plate. Then carefully remove the residual water with a piece of Kimwipe paper (*see* **Note 9**).
3. Place the MALDI plate in a desiccator for at least 30 min before applying matrix.

3.2. Matrix Application

Choosing a proper matrix is essential for successful MALDI imaging experiments. We compared two commonly used organic matrices, including DHB (150 mg/ml in 50% methanol, v/v) and CHCA (10 mg/ml in 60% acetonitrile/0.1% formic acid, v/v). DHB provides better sensitivity and also better coverage of the tissue surface. Careful control of the humidity level during matrix application is crucial for achieving good detection sensitivity without causing significant sample diffusion. Two matrix application methods are presented: airbrush and inkjet printer. Airbrushing the matrix on the sample has been a widely used matrix application technique for MALDI imaging, which is relatively easy and fast. Recently, our laboratory also set up an automated

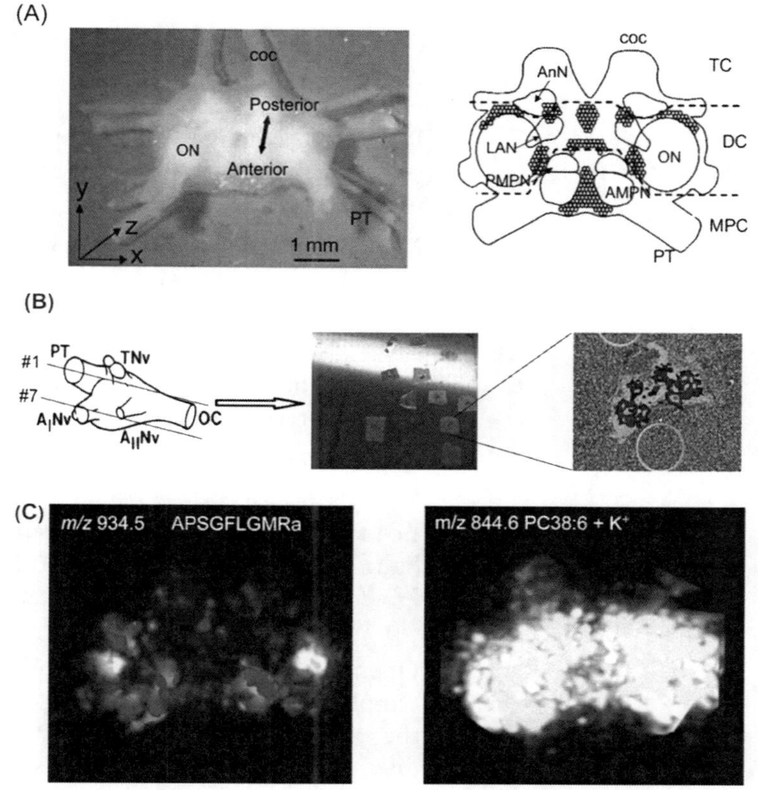

Fig. 26.3. Three-dimensional imaging of a *C. borealis* brain. (**a**) Micrograph of an isolated supraesophageal ganglion (brain) and representation of the ventral surface of the isolated *C. borealis* brain with labeled neuropil regions showing distinctions between the fused ganglia. The most anterior portion of the brain, the protocerebrum, is a distributed structure with the protocerebral tract (PT) linking the lateral protocerebrum (not shown) with the median protocerebrum (MPC), which contains two paired neuropils, the anterior (AMPN) and the posterior medial protocerebral neuropil (PMPN). The tritocerebrum (TC) is the most posterior region and contains the antenna I neuropil (AnN). Located between these fused ganglia is the deutocerebrum (DC), which includes the olfactory lobe (ON) and the lateral II antenna neuropil (LAN). (**b**) A schematic diagram of 3D imaging experiments showing serial sectioning along the *z*-axis and the brain slices in gelatin arranged on MALDI plate. (**c**) 3D distribution of neuropeptides CabTRP 1a (*m/z* 934.5) and a lipid phosphatidylcholine (PC 38:6) in the *C. borealis* brain (adapted from ref. (26)).

matrix application technique by the use of an inkjet printer, which has been previously demonstrated to be inexpensive, more reproducible than airbrushing, and also enhances the signal intensity compared to the conventional airbrush matrix application (24).

3.2.1. Airbrush Application of MALDI Matrix

1. Thoroughly clean the airbrush every time before matrix application using 70% ethanol (v/v).
2. Fill the container with about 70 ml DHB matrix solution (150 mg/ml in 50% methanol, v/v) and place the airbrush about 35 cm from the MALDI plate (*see* **Note 10**).

3. Apply five coats of matrix on the surface of the tissue. Spray duration for each coat was 30 s with 1 min dry time between each cycle.

3.2.2. Inkjet Printer Application of MALDI Matrix

1. Purchase a commercial inkjet printer that has a CD/DVD printer tray. In this case an Epson Stylus Photo R280 printer was used.
2. Modify the CD/DVD printer tray and compact disk to accommodate your MALDI sample plate using a rotary Dremel tool (or equivalent) (*see* **Note 11**).
3. Make up 40 ml of DHB matrix (100 mg/ml, 60% methanol, v/v) and filter the solution by passing it through a 0.22 μm syringe filter.
4. Replace the Epson ink cartridges with refillable inkjet cartridges.
5. Load and charge the cyan, magenta, yellow, light cyan, and light magenta refillable ink cartridges with deionized water. Load and charge the black ink cartridge with filtered matrix (*see* **Note 12**).
6. Open the Epson Print CD software and create a new print template for the modified compact disk. Outline and fill in the location of the MALDI plate on the template compact disk displayed on-screen with the color black. Save the template for later use.
7. Print the CD template just created to an unmodified compact disk to confirm that the MALDI matrix solution deposition pattern meets the experimenter's needs and to confirm that the printhead is not clogged (*see* **Note 13**).
8. Proceed to print MALDI matrix to the tissue using the modified CD/DVD printer tray and modified compact disk. A range from 12 to 18 printer cycles is often required to obtain tissue coverage of matrix similar to airbrush application. This will take approximately 20–30 min. Proceed to the next printer cycle only after MALDI matrix has crystallized on the tissue surface.

3.3. Image Acquisition

1. Before putting the plate in the mass spectrometer, carefully record where the tissue is located. The sample wells may be used as a guide.
2. Set up an image acquisition file using the 4000 Series Explorer Software (Applied Biosystems) and check multiple regions of the tissue to ensure that sufficient neuropeptide signals can be detected (*see* **Note 14**).
3. Open the image acquisition software 4800 Imaging application (Novartis, Basel, Switzerland) available through the MALDI-MSI web site (www.maldi-msi.org). Specify the

step size for both the *x* and *y* dimensions in the appropriate boxes and choose a directory in which to save your data (*see* **Note 15**).

4. Depending on the resolution required, 50–100 μm is usually used as acquisition step size.
5. Specify the area you would like to image by pressing the appropriate bars to set the upper, lower, left, and right borders.
6. Begin acquisition. The software will calculate the estimated time the experiment will take and display this in the lower right corner.
7. For 3D imaging, perform acquisition as described above for each tissue section.

3.4. Image Processing

3.4.1. 2D Image Generation

1. Open TissueView™ software 1.0 (Applied Biosystems) (*see* **Note 16**).
2. Under File go to "open MSI" and import your data file.
3. You may change the intensity indicator by clicking on it and choosing the intensity scale you desire.
4. Under Analysis, open the mass spectrum, then you can click on each pixel of the image to view the mass spectrum from each *x*–*y* coordinate. You may use the middle mouse key to highlight a peak of interest to view its location in the tissue.
5. Images can be saved by using the edit copy function and pasting the image in Microsoft PowerPoint or other similar programs.

3.4.2. 3D Image Generation

1. Open Image J software (http://rsbweb.nih.gov/ij/, NIH) (*see* **Note 17**).
2. Under File go to "open" and import a consecutive series of 2D images of a specific peptide of interest on each layer of the target organ generated by TissueView (*see* **Note 18**).
3. Align the orientation and position of each image and make slight adjustments according to the anatomy of the brain using "Translate" and "Rotate" functions under "Image."
4. Under "Image" function "Stacks," open "Images to stack" to combine 2D images from different sections into one stack.
5. Go to "Stacks" again and view the stack in three dimensions by clicking on "3D projection." Parameters such as axis of rotation and slice spacing as well as several others can be adjusted by changing values in the 3D projection window. Two examples of 3D imaging of neuropeptides and lipids in *C. borealis* brain are shown in **Fig. 26.3**.

4. Notes

1. Gelatin produces less interference for mass spectral analysis than optimal cutting temperature (OCT) media that contains a high concentration of polyethylene glycol (PEG). Keep the gelatin warm before submerging the tissue, otherwise it will become sticky.
2. Push the organ down into the cup so that it is near the bottom and does not require as much trimming to reach, but ensure the tissue is still surrounded on all sides by gelatin.
3. It is generally a good idea to mark the cryostat cup and sketch the orientation of the organ in your notebook for future reference. Start slicing the organ from the bottom side of the cup. Remember to also note which portion of the organ will be sliced first. For example, if the dorsal side is down toward the bottom of the cup, the first slice will be from the dorsal side of the brain.
4. You may use larger forceps to hold the cup upright. Be careful to not get ethanol in the cryostat cup or the gelatin will dissolve in the ethanol and not freeze completely, thus complicating slicing.
5. For crab brain slicing, setting the cryostat box temperature to −25°C instead of the commonly used −20°C encourages better slicing.
6. For some tissue it is beneficial to sandwich the tissue between two layers of matrix. To do this, use an artist's airbrush to spray a thin coating of matrix on the MALDI plate before thaw mounting the tissue slice on the plate.
7. To thaw mount slices, one can either place a room temperature plate near a cold slice and allow the slice to melt onto the plate or use fine paint brushes to place a cold slice on a cold plate and warm both the plate and the slice together by removing them from the box into the room temperature air. The latter method has been shown to result in less peptide loss (25).
8. Brains from crustaceans are generally smaller than from mammals such as mice (**Fig. 26.3a**). Usually about seven sections with a distance of 100–150 μm between each individual slice are enough for covering most regions of the brain.
9. Try to touch the water with the edge of Kimwipe and remove the water carefully without letting the tissue come into contact with the Kimwipe.

10. Use the airbrush in a hood or other enclosed environment to avoid inhaling aerosolized matrix solution. Spray parameters will need to be adjusted based on air currents and other factors. In the fume hood, try holding the airbrush perpendicular to the MALDI plate at a distance of 35 cm and adjust the flow rate of matrix so that most of the matrix solvent evaporates before reaching the plate. However, a certain amount of solvent is crucial for extracting peptides from tissue samples to yield detectable signals.
11. The Epson Print CD will not print if the CD/DVD printer tray disk recess is not completely full. Therefore it is important to only remove excess material from the compact disk holder that the MALDI plate will occupy.
12. The refillable ink cartridges come with directions on how to properly load and charge them.
13. If the printhead clogs it is likely due to crystallized DHB. To unclog, remove the DHB from the black ink cartridge and replace with deionized water. Next, run cleaning cycles followed with nozzle checks using the printer utility function. If the printhead remains clogged, place a small volume of 50/50 ammonia/deionized water (or only ammonia if the clog is difficult to remove) in the printhead resting seat. Turn off the printer and let it sit overnight. The clog should be resolved the following day. Once unclogged, remove excess ammonia from the printhead resting seat and run a few cleaning and nozzle check cycles before reloading the black ink cartridge with MALDI matrix.
14. To build an image acquisition file, experiment with different laser powers and number of shots to get optimal data. Set the laser position to fixed.
15. The image processing software requires that the computer on which the data are viewed has more than twice as much RAM as the data file size. It may be necessary to decrease the image resolution or image smaller areas to work within this limit.
16. BioMap software (Novartis, Basel, Switzerland) available through the MALDI-MSI web site (www.maldi-msi.org) can also be used for the same purpose. The functionalities of these two programs are very similar.
17. Several other 3D software packages may also be useful for 3D construction of MALDI imaging data, such as the LSM Viewer (Zeiss, Germany), Metamorph (Molecular Devices, Sunnyvale, CA, USA), and Amira (Mercury Computer Systems, Chelmsford, MA, USA).
18. The images must be in gray scale to be processed by Image J.

19. The forum on the web site (www.maldi-msi.org) contains helpful videos which should be reviewed for further information.

Acknowledgments

The authors wish to thank the University of Wisconsin-Biotechnology Mass Spectrometry Facility and Drs. Amy Harms and Michael Sussman for access to the MALDI-TOF/TOF instrument as well as Dr. Jeffrey Johnson of University of Wisconsin-Madison Pharmaceutical Sciences Division for use of a cryostat. We also thank Dr. Junhua Wang from the Li laboratory for assistance with taking optical pictures of *C. borealis* brain. This work was supported by a National Science Foundation CAREER Award (CHE-0449991) and the National Institutes of Health through grant 1R01DK071801.

References

1. Fu, Q., Kutz, K. K., Schmidt, J. J., Hsu, Y. W., Messinger, D. I., Cain, S. D., de la Iglesia, H. O., Christie, A. E., Li, L. (2005) Hormone complement of the *Cancer productus* sinus gland and pericardial organ: an anatomical and mass spectrometric investigation. *J Comp Neurol*, **493**, 607–626.
2. Fu, Q., Goy, M. F., Li, L. (2005) Identification of neuropeptides from the decapod crustacean sinus glands using nanoscale liquid chromatography tandem mass spectrometry. *Biochem Biophys Res Commun*, **337**, 765–778.
3. Li, L., Pulver, S. R., Kelley, W. P., Thirumalai, V., Sweedler, J. V., Marder, E. (2002) Orcokinin peptides in developing and adult crustacean stomatogastric nervous systems and pericardial organs. *J Comp Neurol*, **444**, 227–244.
4. Bulau, P., Meisen, I., Schmitz, T., Keller, R., Peter-Katalinic, J. (2004) Identification of neuropeptides from the sinus gland of the crayfish *Orconectes limosus* using nanoscale on-line liquid chromatography tandem mass spectrometry. *Mol Cell Proteomics*, **3**, 558–564.
5. Fu, Q., Tang, L. S., Marder, E., Li, L. (2007) Mass spectrometric characterization and physiological actions of VPNDWAHFRGSWamide, a novel B type allatostatin in the crab, *Cancer borealis*. *J Neurochem*, **101**, 1099–1107.
6. Ma, M., Chen, R., Sousa, G. L., Bors, E. K., Kwiatkowski, M. A., Goiney, C. C., Goy, M. F., Christie, A. E., Li, L. (2008) Mass spectral characterization of peptide transmitters/hormones in the nervous system and neuroendocrine organs of the American lobster *Homarus americanus*. *Gen Comp Endocrinol*, **156**, 395–409.
7. Kutz, K. K., Schmidt, J. J., Li, L. (2004) *In situ* tissue analysis of neuropeptides by MALDI FTMS in-cell accumulation. *Anal Chem*, **76**, 5630–5640.
8. Stemmler, E. A., Cashman, C. R., Messinger, D. I., Gardner, N. P., Dickinson, P. S., Christie, A. E. (2007) High-mass-resolution direct-tissue MALDI-FTMS reveals broad conservation of three neuropeptides (APSGFLGMRamide, GYRKPPFNGSIFamide and pQDLDHVFLRFamide) across members of seven decapod crustaean infraorders. *Peptides*, **28**, 2104–2115.
9. Rubakhin, S. S., Churchill, J. D., Greenough, W. T., Sweedler, J. V. (2006) Profiling signaling peptides in single mammalian cells using mass spectrometry. *Anal Chem*, **78**, 7267–7272.
10. Neupert, S., Predel, R. (2005) Mass spectrometric analysis of single identified neurons

of an insect. *Biochem Biophys Res Commun*, **327**, 640–645.

11. Monroe, E., Annangudi, S. P., Hatcher, N. G., Gustein, H. B., Rubakhin, S. S., Sweedler, J. V. (2008) SIMS and MALDI MS imaging of the spinal cord. *Proteomics*, **8**, 3746–3754.
12. Boonen, K., Landuyt, B., Baggerman, G., Husson, S.J., Huybrechts, J., Schoofs, L. (2008) Peptidomics: the integrated approach of MS, hyphenated techniques and bioinformatics for neuropeptide analysis. *J Sep Sci*, **31**, 427–445.
13. Altelaar, A. F. M., Klinkert, I., Jalink, K., deLange, R. P. J., Adan, R. A. H., Heeren, R. M. A., Piersma, S. R. (2006) Gold-enhanced biomolecular surface imaging of cells and tissue by SIMS and MALDI mass spectrometry. *Anal Chem*, **78**, 734–742.
14. Cornett, D. S., Reyzer, M. L., Chaurand, P., Caprioli, R. M. (2007) MALDI imaging mass spectrometry: molecular snapshots of biochemical systems. *Nat Methods*, **4**, 828–833.
15. Chaurand, P., Norris, J. L., Cornett, D. S., Mobley, J. A., Caprioli, R. M. (2006) New Developments in profiling and imaging of proteins from tissue sections by MALDI mass spectrometry. *J Proteome Res*, **5**, 2889–2900.
16. Gallus, L., Bottaro, M., Ferrando, S., Girosi, L., Ramoino, P., Tagliafierro, G. (2006) Distribution of FMRFamide-like immunoreactivity in the alimentary tract and hindgut ganglia of the barnacle *Balanus amphitrite* (Cirripedia, Crustacea). *Microsc Res Tech*, **69**, 636–641.
17. Sousa, G. L., Lenz, P. H., Hartline, D. K., Christie, A. E. (2008) Distribution of pigment dispersing hormone- and tachykinin-related peptides in the central nervous system of the copepod crustacean *Calanus finmarchicus*. *Gen Comp Endocrinol*, **156**, 454–459.
18. Polanska, M., Yasuda, A., Harzsch, S. (2007) Immunolocalisation of crustacean-SIFamide in the median brain and eyestalk neuropils of the marbled crayfish. *Cell Tissue Res*, **330**, 331–344.
19. Marder, E., Bucher, D. (2007) Understanding circuit dynamics using the stomatogastric nervous system of lobsters and crabs. *Annu Rev Physiol*, **69**, 291–316.
20. Skiebe, P., Dietel, C., Schmidt, M. (1999) Immunocytochemical localization of FLRFamide-, proctolin-, and CCAP-like peptides in the stomatogastric nervous system and neurohemal structures of the crayfish, *Cherax destructor*. *J Comp Neurol*, **414**, 511–532.
21. DeKeyser, S. S., Kutz-Naber, K. K., Schmidt, J. J., Barrett-Wilt, G. A., Li, L. (2007) Imaging mass spectrometry of neuropeptides in decapod crustacean neuronal tissues. *J Proteome Res*, **6**, 1782–1791.
22. Crecelius, A. C., Cornett, D. S., Caprioli, R. M., Williams, B., Dawant, B. M., Bodenheimer, B. (2005) Three-dimensional visualization of protein expression in mouse brain structures using imaging mass spectrometry. *J Am Soc Mass Spectrom*, **16**, 1093–1099.
23. Andersson, M., Groseclose, M. R., Deutch, A. Y., Caprioli, R. M. (2008) Imaging mass spectrometry of proteins and peptides: 3D volume reconstruction. *Nat Methods*, **5**, 101–108.
24. Baluya, D. L., Garrett, T. J., Yost, R. A. (2007) Automated MALDI matrix deposition method with inkjet printing for imaging mass spectrometry. *Anal Chem*, **79**, 6862–6867.
25. Schwartz, S. A., Reyzer, M. L., Caprioli, R. M. (2003) Direct tissue analysis using matrix-assisted laser desorption/ionization mass spectrometry: practical aspects of sample preparation. *J Mass Spectrom*, **38**, 699–708.
26. Chen, R., Hui, L., Sturm, R. M., Li, L. (2009) Three dimensional mapping of neuropeptides and lipids in crustacean brain by mass spectral imaging. *J Am Soc Mass Spectrom*, **20**, 1068–1077.

Chapter 27

Mass Spectrometry Imaging Using the Stretched Sample Approach

Tyler A. Zimmerman, Stanislav S. Rubakhin, and Jonathan V. Sweedler

Abstract

Matrix-assisted laser desorption/ionization (MALDI) mass spectrometry imaging (MSI) can determine tissue localization for a variety of analytes with high sensitivity, chemical specificity, and spatial resolution. MS image quality typically depends on the MALDI matrix application method used, particularly when the matrix solution or powder is applied directly to the tissue surface. Improper matrix application results in spatial redistribution of analytes and reduced MS signal quality. Here we present a stretched sample imaging protocol that removes the dependence of MS image quality on the matrix application process and improves analyte extraction and sample desalting. First, the tissue sample is placed on a monolayer of solid support beads that are embedded in a hydrophobic membrane. Stretching the membrane fragments the tissue into thousands of nearly single-cell sized islands, with the pieces physically isolated from each other by the membrane. This spatial isolation prevents analyte transfer between beads, allowing for longer exposure of the tissue fragments to the MALDI matrix, thereby improving detectability of small analyte quantities without sacrificing spatial resolution. When using this method to reconstruct chemical images, complications result from non-uniform stretching of the supporting membrane. Addressing this concern, several computational tools enable automated data acquisition at individual bead locations and allow reconstruction of ion images corresponding to the original spatial conformation of the tissue section. Using mouse pituitary, we demonstrate the utility of this stretched imaging technique for characterizing peptide distributions in heterogeneous tissues at nearly single-cell resolution.

Key words: Mass spectrometry imaging, matrix-assisted laser desorption/ionization, nervous tissue, pituitary, mouse, stretched sample, image reconstruction, automated data acquisition.

S.S. Rubakhin, J.V. Sweedler (eds.), *Mass Spectrometry Imaging*, Methods in Molecular Biology 656,
DOI 10.1007/978-1-60761-746-4_27,

1. Introduction

Both invertebrate and mammalian nervous systems exhibit high levels of biochemical and morphological heterogeneity. Neighboring neurons often possess different sets of intercellular signaling peptides, with several prohormones encoding multiple peptides that are expressed differently among individual neurons. Investigation of the mammalian nervous system, where neurons number in the billions, presents a significant challenge when using classical labeling approaches to examine one or more cell-to-cell signaling molecules at a time. In contrast to other bioimaging techniques, mass spectrometry imaging (MSI) can uncover the distribution of a variety of analytes within tissues while simultaneously determining their chemical identities, without the need for specific labeling or immunostaining (1–5). MSI has broad applications in academic, clinical, and industrial research, having had significant impact on cancer studies (6–9), the search for new pharmaceuticals (10), and investigations of the nervous system (11). A variety of MSI approaches targeting different types of analytes have been developed over the years. MALDI MSI has become one of most successful technologies for investigation of peptide and protein distributions in fixed and freshly prepared tissues. The analyte desorption and ionization processes occurring during exposure of the MALDI matrix/analyte layer to UV or IR laser light allows detection of intact/unfragmented analytes. Not only is the entire sequence of ion desorption, formation, separation, and detection fast, the laser beam can be focused to sub-micrometer diameters. However, because the amount of proteins and peptides present decrease concomitantly with the size of the area being probed, the smallest laser diameters are not commonly used. Obviously, larger laser spot sizes allow desorption of increased amounts of analyte. Typical chemical images are generated with 25–50 μm spatial resolution.

In MALDI-MSI, a liquid or powder matrix is deposited on top of the sample, incorporating the analyte into the matrix crystals. When illuminated with the laser, the matrix and analyte are vaporized and ionized. Although longer exposure of the sample to matrix facilitates extraction of the analyte of interest from the tissue, it can also delocalize the analyte. Shorter extraction times ameliorate this problem, but result in poorer signal. This issue is particularly problematic when investigating small hydrophilic substances that diffuse during matrix application. Recent advances in MALDI matrix solution application approaches have helped to create uniform MALDI matrix layers; these include spray coating

(12), electrospray deposition (13), and automated acoustic deposition (14). Using these techniques, the imaging of fine analyte spatial distributions has been achieved. However, each type of sample and even class of analytes requires individualized optimization of the MALDI matrix exposure duration and drying time.

The stretched sample protocol resolves these issues by eliminating redistribution of analytes during the matrix application stage. A tissue slice is placed on top of a monolayer comprised of ~40 μm diameter glass beads, which has been partially embedded into a layer of Parafilm M (15). As the Parafilm M layer is manually stretched to a ~16-fold increase in area, the beads separate from each other and the tissue, which adheres to the beads, is fragmented into thousands of islands. Because each bead contains only one or a few cells, chemically and spatially separated by areas of hydrophobic membrane, the sample can be exposed to the MALDI matrix solution for a longer period of time without sacrificing spatial resolution. The spatial isolation of tissue fragments allows rare signals from small cellular clusters, single cells, and even subcellular regions to be better detected and spatially distinguished. Furthermore, multiple MALDI matrix wetting/recrystallization cycles can be accomplished via temperature-dependent condensation of solvents onto the stretched sample, which allows for increased incorporation of the analyte into the matrix, contributing to further signal enhancement.

Effective MS imaging of stretched samples demands new methods of data acquisition, along with image reconstruction protocols, to register the spectral data with the corresponding conformation of the tissue before stretching. Although classical MSI experiments collect data in a regular raster pattern over the sample (16), the small tissue/cell islands in the stretched sample occur in irregular spatial patterns on the Parafilm M membrane. Incorporating a step to identify bead positions from optical images of the sample via light thresholding allows automated MS data acquisition from the individual bead positions. Image reconstruction is done in silico with a free transform process that mimics the actual stretching process (17). During stretching, the beads tear sizable craters partially through the layer of Parafilm M, visible in an optical image of the stretched sample. Image reconstruction is performed by aligning an image of the initial positions of the embedded beads with an image of the tissue sample after stretching. The spectral data taken from the stretched sample are assigned to the nearest corresponding initial bead positions to reconstruct an ion image of the tissue in its original conformation. This novel stretched imaging method shows increased potential for identifying rare signals from heterogeneous tissue samples (18).

2. Materials

2.1. Preparation of Parafilm M Substrate

1. Parafilm M (Pechiney, Neenah, WI, USA).
2. Glass slides, 25×75×1.1 mm (Delta Technologies, Stillwater, MN, USA).
3. At least 100 mg of ~40 μm diameter clear glass beads (Mo-Sci Corp., Rolla, MO, USA). Blue beads are optional and are used as markers to aid image reconstruction. Beads of other types and sizes including liquid chromatography solid phase materials can be also used.
4. A heated aluminum block.

2.2. Tissue Preparation and Sample Stretching

1. Four-month-old C57BL/6 mice obtained from an in-house colony bred by the Greenough group, University of Illinois at Urbana-Champaign, were used. Animals of similar strains can be purchased for research purposes from the Jackson (http://www.jax.org) or Harlan laboratories (http://www.harlan.com). A variety of tissues from different animals can be investigated using the protocol presented here.
2. SPECTRA-SONIC (or similar) solution, pH 7 (Spectrum Surgical Instruments Corp., Stow, OH, USA) for surgical instrument clean up.
3. Modified Gey's balanced salt solution (mGBSS): 1.5 mM $CaCl_2$, 4.9 mM KCl, 0.2 mM KH_2PO_4, 11 mM $MgCl_2$, 0.3 mM $MgSO_4$, 138 mM NaCl, 27.7 mM $NaHCO_3$, 0.8 mM Na_2HPO_4, 25 mM HEPES, and 10 mM glucose, pH 7.2 adjusted with NaOH.
4. Dissection tools including forceps, scissors (available online from Fine Science Tools http://www.finescience.com or World Precision Instruments – http://www.wpiinc.com), and a properly sharpened guillotine.
5. Cryostat capable of keeping the specimen temperature at –15 to –20°C and of cutting 10 μm sections, e.g., Microm HM550 (Thermo Scientific, Waltham, MA, USA).
6. Indium tin oxide (ITO)-coated glass slides, 25×75×1.1 mm (Delta Technologies, Stillwater, MN, USA).
7. A piece of firm paper.
8. Liquid nitrogen and dry ice.
9. Vials for specimen storage.
10. Protective lab coat, gloves, and goggles.

2.3. MALDI Matrix Application

1. MALDI matrix solution: 300 mg of 2,5-dihydroxybenzoic acid (Sigma-Aldrich, St. Louis, MO, USA) in 10 ml of 75:25 acetone:water.
2. Artist's spray brush (Badger, Franklin Park, IL, USA).
3. In-house built condensation chamber consisting of a Peltier device (Melcor, Trenton, NJ, USA) connected to a cooling basin of water, and a thermocouple connected to a CN77000 temperature controller (Omega, Stamford, CT, USA).
4. Acetone.

2.4. Mass Spectrometry and Automated Imaging

1. Optical stereomicroscope.
2. Inverted transmission light microscope with 2.5–10× magnification and equipped with a digital camera (e.g., AxioCam MRc camera controlled by the AxioVision digital image processing software package, Carl Zeiss, Bernreid, Germany; AxioVision LE is free and a sample version of the full package is available at http://www.zeiss.com/).
3. Ultraflex II MALDI-TOF mass spectrometer (Bruker Daltonics, Billerica, MA, USA) with a solid-state UV laser.
4. MTP slide adapter (Bruker Daltonics) for insertion of slides into the MS instrument.
5. ImageJ, version 1.38 (National Institutes of Health, http://rsb.info.nih.gov/ij/).
6. Java SDK, version 1.6.0 (Sun Microsystems, http://java.sun.com).
7. FlexControl 3.0 (Bruker Daltonics).
8. Bead geometry application (free at http://neuroproteomics.scs.illinois.edu/imaging.html).

2.5. Data Conversion

1. Software tool: CompassXport (Bruker Daltonics, free at http://www.brukerdaltonics.com; for more information see: http://www.ionsource.com/functional_reviews/CompassXport/CompassXport.htm).
2. Software package: MATLAB R2006a, version 7.2, and the Bioinformatics Toolbox 3.0 (The MathWorks, Natick, MA, USA).
3. Batch conversion MATLAB code (free at http://neuroproteomics.scs.illinois.edu/imaging.html).

2.6. Image Reconstruction

1. Photoshop CS, version 8.0 (Adobe Systems).
2. Java-based code to create image of dots at the initial bead positions (free at http://neuroproteomics.scs.illinois.edu/imaging.html).

3. MSIReconstructor application (free at http://neuroproteomics.scs.illinois.edu/imaging.html).

3. Methods

3.1. Preparation of Parafilm M Substrate

1. Parafilm M is cut into a square measuring approximately 5×5 cm and placed on top of a glass slide (*see* **Note 1**). In this step, the slide is used as a clean solid support and so does not require a conductive ITO-coated slide.
2. Approximately 100 mg of beads are transferred to the Parafilm M surface. Another glass slide is placed on top of the beads and vertical pressure is manually applied to partially embed the beads into the Parafilm M layer.
3. Application of a nitrogen stream to the substrate removes loose beads, ensuring an even monolayer is attached to the Parafilm M surface.
4. Placing the substrate between two glass slides and heating it on top of an aluminum block at ~60°C for 10–15 s with downward manual pressure allows the beads to become more strongly and uniformly attached to the Parafilm M. Care must be taken to ensure that peripheral parts of the Parafilm M section, which might touch the metal block, do not melt onto it. A small separate piece of Parafilm M can be used to test if the temperature is such that the Parafilm M might be melted by the metal block.
5. An optical image of the initial bead/Parafilm M substrate is taken in transmission mode (*see* **Note 2**).

3.2. Tissue Preparation and Sample Stretching

1. Experimental animals are selected and euthanized by decapitation. Importantly, work performed on animals should comply with local and federal rules and regulations for the humane care and treatment of animals.
2. Surgical/dissection instruments are cleaned and sterilized by ultrasonic treatment in SPECTRA-SONIC (or similar) solution for 5–10 min, followed by autoclaving according to the manufacturer's manual.
3. Vials and paper are prepared; protective lab coat, gloves, and goggles are worn.
4. The animal is decapitated using a sharp guillotine.
5. The cranium is exposed by pushing the skin in a rostral direction using a piece of firm paper.
6. The cranial bones in the frontal plane are cut using long, thin scissors.
7. The brain is carefully lifted and discarded after removal of the previously cut dorsal part of the cranium. The pituitary

typically remains in the scull, held in place by connective tissue.

8. The connective tissue surrounding the pituitary should be removed first; the pituitary is quickly removed using fine forceps.
9. The pituitary is briefly washed in ice-cold mGBSS. The tissue is quickly frozen in liquid nitrogen and stored in a vial over dry ice for transport to the cryostat environment.
10. The pituitary is placed on a cooled (to −20°C) sample stage inside of the cryostat, without addition of embedding solution (*see* **Note 3**).
11. Tissue sections are made (10 μm thick).
12. Within the cryostat, the room temperature bead substrate is positioned onto the tissue section and briefly pressed using an index finger or an artist's brush handle. This ensures transfer of the tissue from the cryostat surface onto the substrate surface. Using a magic marker, the orientation and perimeter of the tissue section within the bead substrate are marked on the back of the Parafilm M substrate.
13. An ITO-coated glass slide is mounted onto a tall, thin vertical support (the slide box cover works well) with tape, conductive side facing upwards (**Fig. 27.1a**). A digital multi-

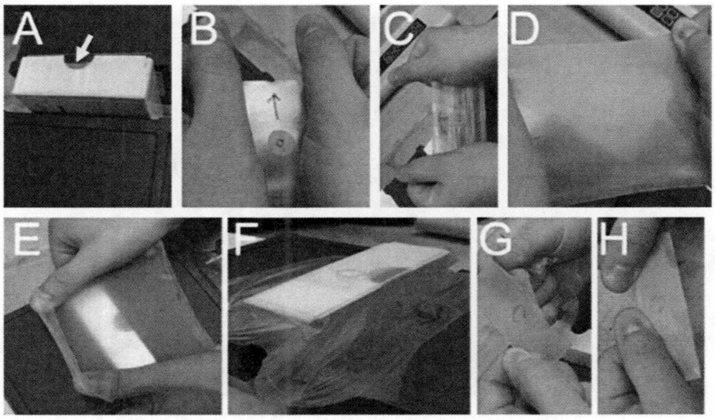

Fig. 27.1. Manual stretching of the sample. (**a**) An apparatus holds the sample, so that both hands can be used during stretching; here this device consists of the glass slide (*arrow*) taped to the thin side of the slide box cover, laterally stabilized by heavy objects. (**b**) The bead substrate, with the location and orientation of the tissue marked, is manually stretched along one axis, (**c**) rotated 90° and stretched, and (**d**) rotated and stretched again. (**e**) The stretched membrane is placed so that the marked sample area is on the glass slide. Use of thumbs provides a last bit of stretching before the sample is applied to the surface. (**f**) The arrow marking the orientation of the sample is clearly visible after stretching and can be marked again on the area of the glass slide if necessary. (**g**) The excess Parafilm M is removed from the edges of the glass slide. (**h**) A finished slide is ready for MALDI matrix coating.

meter can be used to determine which side is conductive. The glass slide box cover is the appropriate shape to support the slide as the stretched substrate is pushed onto the slide. This vertical support enables stretching without having to also hold the glass slide and can be stabilized by placing it between two holders such as large books, as illustrated in **Fig. 27.1a**.

14. The sample is stretched by hand and attached to the ITO-coated slide (**Fig. 27.1b–e**), and the excess Parafilm M is manually torn off of the sides of the slide (*see* **Note 4**). To ease the subsequent process of image reconstruction, the sample should be stretched with the maximum directional uniformity possible. The magic marker label along the perimeter (described above) helps when visually inspecting the sample to ensure it retains gross shape after the stretching process (*see* **Fig. 27.1f**).

3.3. MALDI Matrix Application

1. MALDI matrix is applied using the artist's spray brush at an ~25 cm distance from the sample (*see* **Note 5**). The spray brush is washed with pure acetone after use.
2. Water is condensed onto the sample with the condensation chamber at 14°C for 60 s and the sample evaporated at 28°C for 90 s. This cycle is repeated three times for increased analyte extraction, followed by returning the sample to room temperature (*see* **Note 6**).
3. The specimen is loaded into the mass spectrometer. Mass spectral profiling (15) is used to assess the quality of peptide signal received from the specimen before MS imaging of the sample.

3.4. Mass Spectrometry and Automated Imaging

1. The glass slide with the stretched sample is loaded into the mass spectrometer. Although a calibration bar is typically used (18), we determined that regularly spaced laser-melted holes in the Parafilm M serve as more accurate spatial calibration markers. Provided that the sample is stretched to sufficient thinness, the mass spectrometer's UV laser beam is used to melt several ~100 μm diameter holes through the Parafilm M surface at several of the ordered positions found in the "MTP Slide Adapter II" geometry file included in the Bruker FlexControl software. The location of these points should be chosen so that they span the area of the tissue sample; depending on the size of the sample, three to four points are sufficient.
2. The specimen is unloaded from the mass spectrometer and a transmission mode optical image is taken with a digital camera coupled to an optical microscope. If several optical

images are needed to cover the entire area of the sample, the Photomerge function in Photoshop can be used to stitch multiple images together.

3. ImageJ, along with the color threshold plugin, is used to automatically report the pixel coordinates of the beads (*see* **Note 7**). The computational steps for bead identification by thresholding are summarized in **Fig. 27.2a**. The Analyze–>Set Scale function is used to specify the units of the coordinates as pixels. The results of the threshold are viewed by

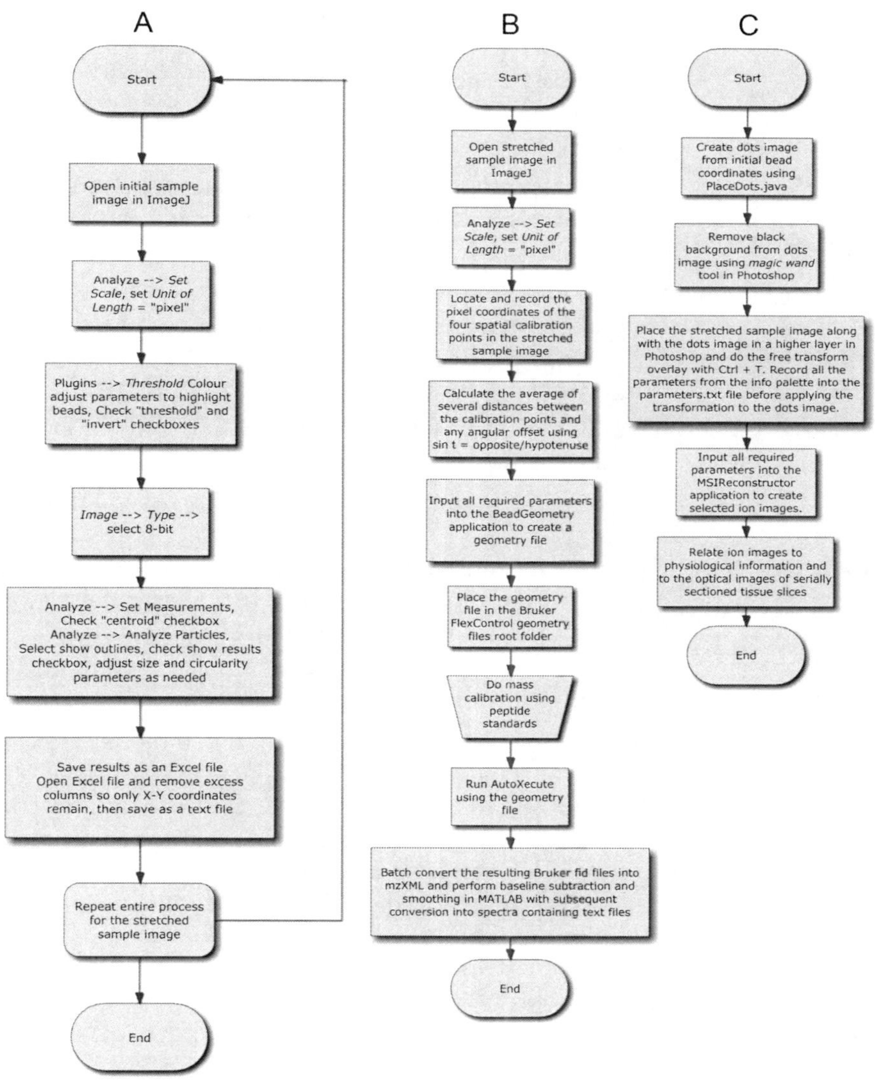

Fig. 27.2. Flow chart for the computational steps in the stretched imaging method, including the process of (**a**) bead identification by light thresholding, (**b**) geometry file creation, and (**c**) image reconstruction. The custom software routines can be found online at http://neuroproteomics.scs.illinois.edu/imaging.html.

selecting Analyze–>Analyze Particles–>Show Outlines. The circularity and size parameters can be adjusted and the process repeated until the outputted outlines image appears not to be highlighting non-bead regions and irregular shapes.

4. In ImageJ, the pixel coordinates of the center of the melted calibration regions are recorded and the distances between them are calculated using the Cartesian distance formula (*see* **Note 8**). If several equivalent distances can be calculated between the various calibration points, the variations in these distances are averaged (*see* **Note 9**).
5. The coordinates of the calibration points and stretched sample bead positions are entered into an in-house written Java-based application available on the Web (*see* **Note 10**). The steps for geometry file creation are summarized in **Fig. 27.2b**.
6. The resulting geometry file is placed in the FlexControl software's geometry files root folder and can be easily found and automatically loaded by the software.
7. The sample is loaded into the mass spectrometer and mass calibration is performed using peptide standards.
8. An AutoXecute sequence is created using the geometry file, specifying an appropriate maximum laser intensity, a value of 100 laser shots per spot, and a 50 Hz repetition rate before starting the MS imaging run.
9. The region of interest is imaged using MALDI-TOF MS.

3.5. Data Conversion

1. The data must be converted from the Bruker ftd file format to the more general mzXML format. The CompassXport software is run along with the – multiName tag at an MS-DOS prompt to create a file called new.mzXML within each spectrum directory.
2. MATLAB is used along with the bioinformatics toolbox and the batch conversion wrapper code available online (*see* **Note 10)** to convert mzXML files into spectra-containing text files. During this step, the data may be processed by baseline subtraction and smoothing with the Bioinformatics Toolbox functions in MATLAB to eliminate noise and create more uniform ion images.

3.6. Image Reconstruction

1. The coordinates of the initial bead positions are found in the same manner as for the stretched sample image, as outlined above in **Section 3.4**, Step 3 (*see* **Note 11**).
2. A simple code (*see* **Note 10**) is compiled in Java and used to create a separate image that places small dots at the initial positions. These dots are easier to see and aid the free

transform process. The computational steps for image reconstruction are summarized in **Fig. 27.2c**.

3. The small dots image and the stretched sample optical images are opened into Photoshop. The free transform command (Ctrl + T) is used to report the centroid coordinates of each of the two images from the options bar. A duplicate background layer is created, and the black background is removed from the small dots image to create a transparent image using the Magic Wand tool in Photoshop. A new blank Photoshop image file is created, large enough (in the range of 5,000×5,000 pixels) to hold both images when placed side-by-side in separate layers, allowing adequate work space to manipulate the images when aligning on top of each other. The small dots image must be placed in a layer above the stretched image layer. The new centroid positions of each image in the blank image file are recorded from the options bar. The free transform command is used to translate, rotate, and resize the small dots image until it is

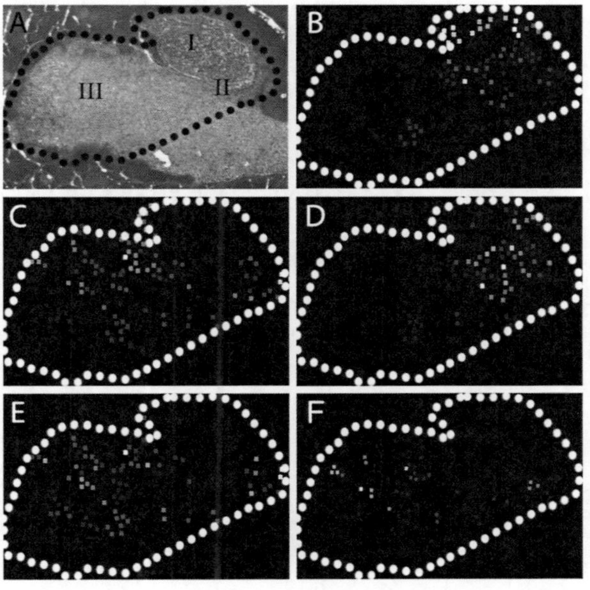

Fig. 27.3. Reconstructed MALDI-MSI ion images from a 10 μm section of mouse pituitary prepared with the stretched sample method. (**a**) Optical photomicrograph of the pituitary section showing (I) the posterior lobe, (II) a darker band corresponding to the intermediate lobe, and (III) the anterior lobe. Only the outlined region tissue was imaged. Reconstructed ion images correspond to the outlined tissue area showing: (**b**) oxytocin, 1,007 *m/z*; (**c**) di-Ac-α-MSH, 1,707 *m/z*; (**d**) vasopressin, 1,083 *m/z*; (**e**) POMC J-peptide, 1,883 *m/z*; and (**f**) Arg-CLIP [1–22], 2,505 *m/z*. The intermediate lobe is a small band that is highlighted by signals from the di-acetylated-α-MSH and J-peptide ion images.

appropriately aligned on top of the stretched sample image. The small dots should each align within one of the bead-torn regions of the stretched sample image.

4. Before applying the transformation to the transformed images in Photoshop, the final width, height, and rotation angle are recorded into a text file from the Info palette, along with the final centroid position of each image.
5. The text files of the initial positions and the image reconstruction parameters, recorded both before and after the free transformation, are inputted into another in-house Java program (*see* **Note 10**) to create reconstructed ion images at select *m/z* ratios of interest as seen in **Fig. 27.3** (*see* **Note 12**). The example shown in **Fig. 27.3** is with a thin tissue section from mouse pituitary.

4. Notes

1. Optionally, Parafilm M may be soaked for 1 h in either acetic acid (100%) or ammonium hydroxide solution (28.8%) to soften the film (19). After drying, this treatment allows the film to be stretched by a greater degree into an approximately sevenfold increase in each dimension. Use of the more elastic film results in formation of small, concentrated droplets of solution upon matrix application with less spatial spreading. In addition, soaking can reduce polymer signals resulting from the Parafilm M.
2. If only performing mass spectral profiling without imaging on stretched samples, as described in (15), this step and the steps related to image reconstruction and geometry file creation can be omitted.
3. Most embedding media interferes with obtaining good signals in mass spectrometry investigations. One exception to this is embedding of tissues in low melting point agarose gel blocks or gelatin. It was found that a block of solidified saturated agarose solution that is freeze mounted onto the dissection stage, followed by sectioning through the top layers of the gel, creates a flat surface for better freeze mounting and orienting of the tissue.
4. For mass spectrometry imaging, it is important to prevent the tissue from completely drying as this will cause bead clumping and may reduce incorporation of analyte into the MALDI matrix crystals. Excess sample drying is prevented by immediately applying MALDI matrix after stretching while the tissue is still partially wet. Thus, if more than one

section is to be taken from the tissue, these sections are sectioned with the microtome after immediately applying MALDI matrix to the preceding stretched section.

5. A larger sprayer-to-sample distance helps in not over-wetting the sample, as larger (~0.5 mm) droplets can cause spreading, even in a stretched sample. A light microscope can be used to monitor the drying process, so that the sample is completely dry before the next spray application. The light microscope also helps to visually monitor the amount of matrix applied. Generally, several spraying-drying cycles over 10–20 min is sufficient. Alternatively, matrix application can be done by capillary deposition to control the size of the matrix spots and prevent the spatial redistribution of analytes (19).
6. The solvent condensation/MALDI matrix recrystallization procedure has shown the ability to improve mass spectra quality by reducing the number and intensity of inorganic salt ion adducts typical for traditional MS imaging sample preparations. This reduction in potassium and sodium salt adducts creates less complex mass spectra (20).
7. The success rate of the bead position identification depends on the quality of the optical images. Transmission-mode images are easier to threshold for bead positions as they appear brighter than the background Parafilm M.
8. As the FlexControl "MTP Slide Adapter II" geometry file uses a fractional distance coordinate system where the distance between each point in the regular array is separated by exactly 0.086957 units, a new geometry file can be created in this coordinate system to acquire data at the bead positions.
9. Variations in the distances between calibration points that are larger than bead diameter signify inaccuracies in the optical image of the stretched sample such that the resulting geometry file may not accurately represent the bead positions. Inaccuracies sometimes occur because of errors in stitching of images by Photoshop and can be prevented by taking well-focused images with sufficient spatial overlap.
10. All in-house written Java software is available online, along with an example dataset with step-by-step instructions, at http://neuroproteomics.scs.illinois.edu/imaging.html.
11. Alternatively, image reconstruction can be done before geometry file creation so that some time is saved in the rare event that image reconstruction is not successful, upon which the sample is discarded. Any difficulties with image reconstruction using the free transform approach arise from

highly non-uniform stretching of the Parafilm M that can be prevented by visually adjusting for the shape of the marked perimeter of the sample while stretching. Overall, image reconstruction is fairly reproducible, as a set of six samples resulted in a classification rate of 84.1% for bead position matching between the stretched and the initial samples, with the remaining portion being only near-neighbor mismatches (18).

12. The success of image reconstruction can be verified using an in-house written code (http://neuroproteomics.scs.illinois.edu/imaging.html) that plots an image of the calculated transformed initial positions. This image can be checked against the transformed image in Photoshop to verify for any calculation or positional errors.

Acknowledgments

We thank Georgina M. Aldridge, University of Illinois at Urbana-Champaign, for providing the animals. The project described was supported by Award No. P30 DA018310 and Award No. 5RO1DA017940 from the National Institute On Drug Abuse and Award No. 5RO1DE018866 from the National Institute of Dental and Craniofacial Research (NIDCR) and the Office of Director (OD), National Institutes of Health (NIH). The content is solely the responsibility of the authors and does not necessarily represent the official views of the NIDA, NIDCR, or NIH.

References

1. Seeley, E. H., Caprioli, R. M. (2008) Molecular imaging of proteins in tissues by mass spectrometry. *Proc Natl Acad Sci U S A*, **105**, 18126–18131.
2. McDonnell, L. A., Heeren, R. M. (2007) Imaging mass spectrometry. *Mass Spectrom Rev*, **26**, 606–643.
3. Pacholski, M. L., Winograd, N. (1999) Imaging with mass spectrometry. *Chem Rev*, **99**, 2977–3006.
4. Rubakhin, S. S., Jurchen, J. C., Monroe, E. B., Sweedler, J. V. (2005) Imaging mass spectrometry: fundamentals and applications to drug discovery. *Drug Discov Today*, **10**, 823–837.
5. Becker, J. S. (2007) *Inorganic Mass Spectrometry: Principles and Applications*, John Wiley & Sons, Hoboken, NJ.
6. Lemaire, R., Menguellet, S. A., Stauber, J., Marchaudon, V., Lucot, J.-P., Collinet, P., Farine, M.-O., Vinatier, D., Day, R., Ducoroy, P., et al. (2007) Specific MALDI imaging and profiling for biomarker hunting and validation: fragment of the 11S proteasome activator complex, reg alpha fragment, is a new potential ovary cancer biomarker. *J Proteome Res*, **6**, 4127–4134.
7. Chandra, S., Tjarks, W., Lorey, D. R., Barth, R. F. (2008) Quantitative subcellular imaging of boron compounds in individual mitotic and interphase human glioblastoma cells with imaging secondary ion mass spectrometry (SIMS). *J Microsc*, **229**, 92–103.
8. Chaurand, P., Norris, J. L., Cornett, D. S., Mobley, J. A., Caprioli, R. M. (2006) New developments in profiling and imaging of

proteins from tissue sections by MALDI mass spectrometry. *J Proteome Res*, **5**, 2889–2900.

9. Chaurand, P., Rahman, M. A., Hunt, T., Mobley, J. A., Gu, G., Latham, J. C., Caprioli, R. M., Kasper, S. (2008) Monitoring mouse prostate development by profiling and imaging mass spectrometry. *Mol Cell Proteomics*, 7, 411–423.
10. Hsieh, Y., Chen, J., Korfmacher, W. A. (2007) Mapping pharmaceuticals in tissues using MALDI imaging mass spectrometry. *Pharmacol Toxicol Methods*, **55**, 193–200.
11. Rubakhin, S. S., Hatcher, N. G., Monroe, E. B., Heien, M. L, Sweedler, J. V. (2007) Mass spectrometric imaging of the nervous system. *Curr Pharm Design*, **13**, 3325–3334.
12. Chaurand, P., Norris, J. L., Cornett, D. S., Mobley, J. A., Caprioli, R. M. (2006) New developments in profiling and imaging of proteins from tissue sections by MALDI mass spectrometry. *J Proteome Res*, **5**, 2889–2900.
13. Kruse, R., Sweedler, J. V. (2003) Spatial profiling invertebrate ganglia using MALDI MS. *Am Soc Mass Spectrom*, **14**, 752–759.
14. Aerni, H. R., Cornett, D. S., Caprioli, R. M. (2006) Automated acoustic matrix deposition for MALDI sample preparation. *Anal Chem*, **78**, 827–834.
15. Monroe, E. B., Jurchen, J. C., Koszczuk, B. A., Losh, J. L., Rubakhin, S. S., Sweedler, J. V. (2006) Massively parallel sample preparation for the MALDI MS analyses of tissues. *Anal Chem*, **78**, 6826–6832.
16. Clerens, S., Ceuppens, R., Arckens, L. (2006) Createtarget and analyze this!: new software assisting imaging mass spectrometry on Bruker Reflex IV and Ultraflex II instruments. *Rapid Commun Mass Spectrom*, **20**, 3061–3066.
17. Decker, J. D. (2004) Image editing. *Am J Orthod Dentofacial Orthop*, **125**, 215–219.
18. Zimmerman, T. A., Monroe, E. B., Sweedler, J. V. (2008) Adapting the stretched sample method from tissue profiling to imaging. *Proteomics*, **8**, 3809–3815.
19. Wang, J., Chen, R., Ma, M., Li, L. (2008) MALDI MS sample preparation by using paraffin wax film: systematic study and application for peptide analysis. *Anal Chem*, **80**, 491–500.
20. Monroe, E. B., Koszczuk, B. A., Losh, J. L., Jurchen, J. C., Sweedler, J. V. (2007) Measuring salty samples without adducts with MALDI MS. *Int J Mass Spectrom*, **260**, 237–242.

SUBJECT INDEX

S.S. Rubakhin, J.V. Sweedler (eds.), *Mass Spectrometry Imaging*, Methods in Molecular Biology 656, DOI 10.1007/978-1-60761-746-4, © Springer Science+Business Media, LLC 2010

E

F

G

H

I

L

M

N

O

P

Q

R

S

T

V

W

Z